ENCYCLOPEDIA OF
ANIMAL BEHAVIOR

ENCYCLOPEDIA OF
ANIMAL BEHAVIOR

EDITORS-IN-CHIEF

PROFESSOR MICHAEL D. BREED
University of Colorado, Boulder, CO, USA

PROFESSOR JANICE MOORE
Colorado State University
Fort Collins, CO, USA

AMSTERDAM • BOSTON • HEIDELBERG • LONDON • NEW YORK • OXFORD
PARIS • SAN DIEGO • SAN FRANCISCO • SINGAPORE • SYDNEY • TOKYO
Academic Press is an imprint of Elsevier

ACADEMIC
PRESS

In Memoriam

PREFACE

Ancient drawings on the walls of caves speak for the ageless intrigue that animal behavior holds for human beings. In those days, the fascination was certainly motivated in part by survival; our ancestors were both predators and prey. There is some evidence that early humans also found animal behavior to be intrinsically interesting; the myths and stories that come down to us from prehistory contain elements of what animals do in the world and what they mean to people. These are the oldest statements of human relationship with the natural world and the living things that inhabit it.

Our ancestors would not recognize the far-flung universe of the modern science of animal behavior. Only 14 decades (approximately) have passed since Darwin first published *The Expression of Emotions in Man and Animals* – generally acknowledged as the starting point for the scientific study of animal behavior – and behavioral biologists now ask questions about topics ranging from the relationship of immunological phenomena and behavioral disorders in dogs, rats, and people to the integration of animal behavior and conservation. Experts in animal behavior provide commentary on the mating displays of rare primates, on television, where entire channels are devoted to the sensory worlds of insects and the ability of octopus to disappear in plain sight. In short, human fascination with animal behavior has produced a field that is rich beyond imagination, and frustratingly beyond the full embrace of any one person.

The almost hopelessly dispersed primary literature of animal behavior reflects the reticulated evolution of the field, which comes to us from field studies; from laboratory experiments; from our understanding of nerves, muscles, and hormones; and from our grasp of social interactions and ecology. It is difficult to think of a major area of biological inquiry that has not been touched by a behavioral tendril or two. A temptation exists to surrender to this fragmentation – allowing our intellectual landscape to reflect increasingly small and disjunct patches of thought and discovery.

Such surrender is, of course, distasteful to any scholar, but there is a more penetrating reason that makes it unacceptable: Anthropogenic change is occurring at a higher rate than ever before, and if we are to preserve our own habitat – the world that the ancients felt compelled to explain in their stories about animals – we must not fail in our attempts to understand its inhabitants. Those residents sustain our own habitat, and their requirements are varied, going far beyond calories and oxygen. They migrate and forage, choose mates, and defend territories, and all this behavior is influenced by hormones, external physical stimuli, trophic and social interactions, and eons of fitness outcomes. A fully integrated knowledge of animal behavior will be indispensable as scientists analyze changing populations, communities, ecosystems, and landscapes. Indeed, it will be indispensable for anyone who seeks to be an honest custodian of nature.

This encyclopedia offers over 300 authoritative and accessible synopses of topics ranging from dolphin signature whistles to game theory. As library reference material, the encyclopedia serves a public that is increasingly challenged to be aware of scientific advances. It is designed as a first stop for the curious advanced undergraduate or graduate student, as well as for the researcher desiring to learn about developments in fields related to his or her own study or to enter a new phase of inquiry.

In compiling this work, we contacted internationally known scientists in the broad array of fields that inform animal behavior. These accomplished men and women are the section editors, and they, in turn, invited some of the best scholars and rising stars in their subject areas to write for the encyclopedia. Thus, every contribution has been reviewed by experts. In short, the articles approach the best that our field has to offer, written by people whose passion for animal behavior is equaled only by their expertise.

Creating the list of sections was as daunting as it was enjoyable. Of course, we included traditional, major areas like foraging, predator–prey interactions, mate choice, and social behavior, along with endocrinology, methods, and neural processes, to name a few. You will see those and more as you survey these volumes. We also included areas that have recently captured the attention of an increasing number of behavioral biologists; these include infectious disease, cognition, conservation, and animal welfare. Looking to the future, we invited contributors from robotics and applied areas. We realized that we could not do the study of animal behavior justice without some exploration of the model systems – the landmark studies – that have molded and continue to guide the development of the field.

In general, you will not find human behavioral studies in this collection, although some articles are tangentially related to such work. That exclusion was a difficult decision, but was motivated not so much by some parochial commitment to a human/non-human divide as by the fact that the non-human literature itself is rich beyond description. Limiting the collection to non-human studies in no way removed the danger of intellectual gluttony.

We remember two eminent behavioral biologists in the dedication of this work – Professors Christopher J. Barnard and Ross H. Crozier. Both of these men played important roles in the creation of this work, and neither lived to see it come to fruition. Professor Barnard (1952–2007) was the first editor-in-chief of the encyclopedia and developed the initial overview of topics, but had to step back from the process because of the illness that eventually took his life. Professor Crozier (1943–2009) was the section editor of Genetics until his untimely death in late 2009. Their immense and varied contributions enriched our knowledge of animal behavior and are cataloged in numerous locations. Those contributions are remarkable in their scope and influence, but they are nonetheless dwarfed by the legions of students, friends, and family members who feel fortunate to have known these scientists and who will carry their legacy forward.

We are grateful to Dr. Andrew Richford, formerly the Senior Acquisitions Editor, Life Sciences Books, Academic Press, who guided us through the formative part of this project. His expertise in and enthusiasm for animal behavior provided significant momentum, not to mention some good fun. Simon Wood, Major Reference Works Development Editor, was indispensable to the project. He answered an amazing variety of questions from contributors and editors, kept the project organized and moving forward, and did all this without losing his fine sense of humor. Nicky Carter, Project Manager, guided us through the completion of the project, providing a pleasantly seamless interface between the scientific scribblers and other publishing professionals. We thank Kristi Gomez and Will Smaldon, also of Academic Press, for their roles in bringing the project to completion. Finally, working with the section editors (see pp. ix) was a real treat; their expertise and devotion to animal behavior is reflected in every page of this work. We particularly thank James Ha, Joan Herbers, James Serpell, and David Stephens for attending an organizational meeting to set the stage for the development of the project.

We are pleased to see this culmination of effort on the part of hundreds of authors and co-authors. Each article is the distillation of expert understanding, acquired over many years. We are excited to be part of such a remarkable collaboration, one that opens so many doors to the future of animal behavior for undergraduates and professionals alike.

Michael Breed, Boulder, CO
Janice Moore, Fort Collins, CO
August 2010

SECTION EDITORS

Bonnie Beaver

Bonnie is internationally recognized for her work in the normal and abnormal behaviors of animals. She has given over 250 scientific presentations to veterinary and veterinary student audiences on subjects of animal behavior, animal welfare, and the human–animal bond, as well as discussed many areas of veterinary medicine for the public media. In addition, she has authored over 150 scientific articles and has nine published books, including *The Veterinarian's Encyclopedia of Animal Behavior* (Blackwell Press), *Feline Behavior: A Guide for Veterinarians* (Saunders), and the newly released second edition of *Canine Behavior: Insights and Answers* (Saunders).

Bonnie is a member of numerous local, state, and national professional organizations and has served as president or chair of several organizations, including the American Veterinary Society for Animal Behavior, the American College of Veterinary Behaviorists, Phi Zeta, and the Texas Veterinary Medical Association. She is board certified by the American College of Veterinary Behaviorists and currently serves as its Executive Director. In addition, Bonnie is the President of the Organizing Committee for the American College of Animal Welfare.

Bonnie is a past president of the American Veterinary Medical Association and has served as Chair of the AVMA Executive Board. She has also served on several AVMA committees, including the Animal Welfare Committee, Council on Education, Committee on the Human–Animal Bond, and American Board on Veterinary Specialties.

In addition, she chaired the AVMA's Canine Aggression and Human–Canine Interactions Task Force, and the Panel on Euthanasia.

Professionally, Bonnie has been honored by being elected as a Distinguished Practitioner of the National Academies of Practice, named as the recipient of the 1996 AVMA Animal Welfare Award, awarded the 2001 Friskies PetCare Award in Animal Behavior, and received the 2001 Leo K. Bustad Companion Animal Veterinarian of the Year Award. She has been recognized for outstanding professional achievement in more than 150 editions of over 50 publications, including *Who's Who in America*, *The World Who's Who of Women*, *Who's Who in the World*, and *American Men and Women of Science*.

Michael Breed

After receiving his PhD from the University of Kansas in 1977, Michael came to Colorado to work as a faculty member at the University of Colorado, Boulder, where he has been ever since. He is currently a Professor in the Department of Ecology and Evolutionary Biology, and teaches courses in general biology, animal behavior, insect biology, and tropical biology. Michael's research program focuses on the behavior and ecology of social insects, and he has worked on ants, bees, and wasps. He has studied the nestmate recognition, the genetics of colony defense, the behavior of defensive bees, and communication, during colony defense. He was the Executive Editor of *Animal Behaviour* from 2006 to 2009.

Jae Chun Choe

After receiving his PhD from Harvard University in 1990, Jae became a Junior Fellow at the Michigan Society of Fellows. He then returned to his home country, Korea, to work in the School of Biological Sciences at Seoul National University. In 2006, he moved to Ewha Womans University to take the post of university chair professor and the director of its natural history museum. He served as the president of the Ecological Society of Korea and is currently serving as the co-president of the Climate Change Center. Since his return to Korea, he has been conducting a long-term ecological research of magpies while continuing to study insects. Quite recently, he began a field study of Javan Gibbons in the Gunuung Halimun-Salak National Park of Indonesia.

Nicola Clayton

Nicola is Professor of Comparative Cognition in the Department of Experimental Psychology at the University of Cambridge, and a Fellow of Clare College. She received her undergraduate degree in Zoology at the University of Oxford and her doctorate in animal behavior at St. Andrews University. In 1995, she moved to the University of California Davis where she gained her first Chair in Animal Behaviour in 2000. She moved to Cambridge and was appointed a personal Chair in 2005. She has 185 publications to her credit.

Nicola studies the development and evolution of intelligence. For example, she addresses the question of whether animals can plan for the future and what they remember about the past, as well as when these abilities develop in children. She is also interested in social and physical intelligence, such as whether animals can differentiate between what they know and what other individuals know. Nicola's work deals mainly with the members of the crow family (e.g., rooks and jays), and comparisons between crows, nonhuman apes, and young children.

Jeff Galef

After receiving his Ph.D. from the University of Pennsylvania in 1968, Jeff moved as an Assistant Professor to McMaster University in Hamilton, Ontario where, for 38 years, his research focused on understanding social influences on the feeding behavior of Norway rats and the mate choices of Japanese quail. Empirical work in his laboratory on social learning in animals has resulted in the publication of more than 100 scientific articles, (www.sociallearning.info) and his scholarly pursuits have produced three co-edited volumes (*Social Learning: Psychological and Biological Perspectives* (with TR Zentall), *Social Learning and Imitation: the Roots of Culture* (with CM Heyes), and *The Question of Animal Culture* (with KN Laland)) as well as a special issue of the journal *Learning & Behavior* (2004, 32(1) (with CM Heyes)). He was honored with the Lifetime Contribution Award of the Social Learning Group, St. Andrews University, Scotland, in 2005, and in 2009, was elected a Fellow of the Royal Society of Canada.

Sidney Gauthreaux

Sidney received his PhD in 1968 and did a post-doctorate at the Institute of Ecology at the University of Georgia in the following 2 years. He joined the zoology faculty at Clemson University in 1970 and retired as Centennial Professor of Biological Sciences in 2006. In 1959, he began working with weather surveillance radar at National Weather Service installations in an effort to detect, quantify, and monitor migrating birds in the atmosphere.

His research has focused on radar studies of bird migration across the Gulf of Mexico and over much of the United States in spring and fall. Since 1992, modern Doppler weather radar has 'revolutionized' the study of bird migration, and he has used it to monitor the flight behavior of birds in the surveillance areas of approximately 150 weather radar stations throughout the United States and explore the interrelationships of bird movements at different spatial scales in relation to geography, topography, habitat, weather, and climatic factors. Recent work with high-resolution surveillance radar (modified marine radar) and thermal imaging and vertically pointing radar (TI-VPR) has greatly enhanced his capability to work at small spatial scales and explore the behavior of migrating birds within 12 km of the radar.

Sidney was President of the Animal Behavior Society from 1987 to 1988 and was elected a Fellow in the American Association for the Advancement of Science in 1988. In October 2006, he received the William Brewster Memorial Award of the American Ornithologists' Union, and in April 2009, the Margaret Morse Nice Medal of the Wilson Ornithological Society.

Deborah M. Gordon

After receiving her PhD from Duke University in 1984, Deborah joined the Harvard Society of Fellows. She did her postdoctoral research at Oxford and at the Centre for Population Biology at Silwood Park, University of London. She came to Stanford in 1991 and is currently a Professor in the Department of Biology. She teaches courses in ecology and behavioral ecology. Deborah's research program focuses on the organization and ecology of ant colonies, and how colonies, without central control, use interaction networks to regulate colony behavior. Her projects include a long-term study of a population of harvester ant colonies in Arizona, studies of the invasive Argentine ant in northern California, and ant–plant mutualisms in Central America.

Patricia Adair Gowaty

Patricia is a Distinguished Professor of Ecology and Evolutionary Biology – UCLA and a Distinguished Research Professor *Emerita* of Ecology at the University of Georgia. After receiving her PhD in 1980, she supported herself with funding from NSF and NIH, until her first tenure track job as an Associate Professor of Zoology in 1993 at the University of Georgia. She studied social behavior, demography, and ecology of eastern bluebirds in the field for 30 years. She pioneered studies of extra-pair paternity in socially monogamous species. She studied fitness outcomes of reproduction under experimentally imposed social constraints in flies, mice, ducks, and cockroaches. Her theoretical work includes papers on the evolution of social systems, forced copulation, compensation, and sex role evolution. Currently, she is completing studies in the genetic mating system of eastern bluebirds, experiments on the fitness variation of males and females in the three species of *Drosophila*, and a book on reproductive decisions under ecological and social constraints. She was President of the Animal Behavior Society in 2001. She is a Fellow of the American Association for the Advancement of Science, the American Ornithologists' Union, the International Ornithologists' Union, and the Animal Behavior Society.

James Ha

James has a 1989 Ph.D. in Zoology/Animal Behavior from Colorado State University and has been on the faculty of the University of Washington since 1992. He is actively involved in research on the social behavior of Old World monkeys and their management in captivity, Pacific Northwest killer whales, local and Pacific island crows, and domestic dogs. He is also certified as an Applied Animal Behaviorist by the Animal Behavior Society and has his own private practice in dealing with companion animal behavior problems in the Puget Sound area.

Joan M. Herbers

Joan is a Professor of Evolution, Ecology, and Organismal Biology at The Ohio State University in Columbus Ohio. She has studied social evolution in ants for many years, with contributions to queen-worker conflict, sex ratio theory, and coevolution. She is currently serving as the Secretary-General of the International Union for the Study of Social Insects and also as the President of the Association for Women in Science.

Jeffrey Lucas

Jeffrey received a Ph.D. from the University of Florida in 1983, studying under Dr. H. Jane Brockmann. He then took a postdoc position in Dr. John Kreb's lab at Oxford University. After teaching at the College of William & Mary and Redlands University, he came to Purdue University in 1987, where he is currently a professor of Biological Sciences. Jeffrey teaches courses in ecology, animal behavior, sensory ecology, and animal communication. His research program focuses on the chick-a-dee call of chickadees and a comparison of auditory physiology in a variety of birds. He has worked on seed dispersal, antlions, and fish, and has published dynamic programming models of a number of systems. He is a past Executive Editor of *Animal Behaviour* and is a fellow of the Animal Behavior Society.

Constantino Macías Garcia

Constantino has been interested in animal behavior ever since he joined Hugh Drummond's laboratory to study the feeding habits of snakes for his BSc and MSc. His main research has been on sexual selection and the evolution of ornaments, which he has studied mainly in Goodeid fish. He was careless not to follow the early forays of his PhD supervisor, Bill Sutherland, into the hybrid field of behavior and conservation. But time, as well as the increasingly grim reality of Mexican fauna, has led him to investigate the links between behavior and conservation in fish, frogs, and birds.

Justin Marshall

Justin's interest in biology and the sea came from his parents, both marine biologists and keen communicators of the ocean realm. He was then fortunate to begin learning about sensory biology in aquatic life during his undergraduate degree in Zoology at The University of St Andrews. The Gatty Marine Laboratory and its then director, Mike Laverack, introduced him to the diversity of marine life and the challenges of different sensory environments under water. Enjoying the cold clear waters of Scotland, he also began to take interest in tropical biodiversity and traveled to Australia and The Great Barrier Reef toward the end of his undergraduate degree. Currently, he is the President of The Australian Coral Reef Society and lives in Australia working at The University of Queensland. He holds a position of Professor at The Queensland Brain Institute and is an Australian Research Council Professorial Research Fellow. Before moving into the superb sensory environment of Jack Pettigrew's Vision Touch and Hearing Research Centre, he did his D.Phil and spent his initial postdoctoral years at The University of Sussex in the UK., Mike Land and The University of Maryland's

Tom Cronin were his mentors during these years and Justin developed an enthusiasm for the amazing world of invertebrate vision only because of them. His work now focuses on the visual ecology of a variety of animals, mostly aquatic, and has branched out to include fish, reptiles, and birds. Animal behavior and questions, such as 'why are animals colorful?', form a large section of his current research.

Janice Moore

As an undergraduate, Janice was inspired by parasitologist Clark P. Read to think about the ecology and evolution of parasites in new ways. She was especially excited to learn that parasites affected animal behavior, another favorite subject area. Most biologists outside the world of parasitology were not interested in parasites; they were relegated to a nether world between the biology of free-living organisms and medicine. After peregrination through more than one graduate program, she completed her PhD studying parasites and behavior at the University of New Mexico. Janice did postdoctoral work on parasite community ecology with Dan Simberloff at Florida State University, and then accepted a faculty position at Colorado State University, where she has remained since 1983. She is currently a Professor in the Department of Biology where she teaches courses in invertebrate zoology, animal behavior, and the history of medicine. She studies a variety of aspects of parasite ecology and host behavior ranging from behavioral fever and transmission behavior to the ecology of introduced parasite species.

Daniel Papaj

After receiving his PhD from the Duke University in 1984, Daniel engaged in postdoctoral research at the University of Massachusetts at Amherst and at Wageningen University in The Netherlands. He joined the faculty of the Department of Ecology and Evolutionary Biology at the University of Arizona in 1991, where he has been ever since. His research focuses on the reproductive dynamics of insects, with special attention to the role of learning by the insect in its interactions with plants. Daniel's focal organisms have included butterflies, tephritid fruit flies, parasitic wasps, and more recently, bumble bees. Recent projects in the lab include the costs of learning in butterflies, the dynamics of social information use in bumble bees, the thermal ecology of host preference in butterflies, ovarian dynamics in fruit flies, multimodal floral signaling, and bumblebee learning. He teaches courses in animal behavior, behavioral ecology, and introductory biology.

Ted Stankowich

Ted grew up in suburban Southern California where opportunities to observe macrofauna in nature were few, but still found ways to observe and enjoy the animals that he could find in his own backyard. While his initial interests in biology were in biochemistry and genetics, after taking introductory courses at Cornell University, he quickly realized that these disciplines were not his calling. He developed interests in ecology and evolution after taking introductory courses and working in George Lauder's functional morphology lab for a summer at the University of California, Irvine, but he took an abiding interest in animal behavior after taking a course as a junior at Cornell and joined Paul Sherman's naked mole-rat lab, where he completed an honors thesis on parental pup-shoving behavior. Ted entered the Animal Behavior graduate program at the University of California, Davis to work with Richard Coss. He spent three field seasons working on predator recognition, flight decisions, and antipredator behavior, in Columbian black-tailed deer, and completed his dissertation in 2006. Ted served as the Darwin Postdoctoral Fellow at the University of Massachusetts, Amherst from 2006 to 2008, investigating escape behavior in jumping spiders. Since completing his tenure, he has continued to work as a postdoc and teach at UMass.

David W. Stephens

David received his PhD from Oxford University in 1982. Currently, David is a Professor at the University of Minnesota in the Twin Cities. His research takes a theoretical and experimental approach to behavior ecology. His research focuses on the connections between evolution and animal cognition, especially the evolutionary forces that have shaped animal learning and decision-making. His work makes connections with many disciplines within the behavioral sciences, and he has presented his work to groups of psychologists, economists, anthropologists, mathematicians, and neuroscientists. He is the author, with John Krebs, of the well-cited book *Foraging Theory*, and the editor (with Joel Brown and Ronald Ydenberg) of *Foraging: behavior and ecology*. He served as an editor of *Animal Behaviour* from 2006 to 2009.

John C. Wingfield

John's undergraduate degree was in Zoology (special honors program) from the University of Sheffield and he did his Ph.D. in Comparative Endocrinology and Zoology from the University College of North Wales, UK. Although John is trained as a comparative endocrinologist, he has always interacted with behavioral ecologists and has strived to integrate ecology and physiology down to cellular and molecular levels. The overarching question is how animals cope with a changing environment – basic biology of how environmental signals are perceived, transduced into endocrine secretions that then regulate morphological, physiological, and behavioral responses. The diversity of mechanisms is becoming more and more apparent and how these evolved is another intriguing question. He was an Assistant Professor at the Rockefeller University in New York and then spent over 20 years as a Professor at the University of Washington. Currently, he is a Professor and Chair in Physiology at the University of California at Davis.

Harold Zakon

Harold received a B.S. degree from Marlboro College in Vermont. He worked as a research technician at Harvard Medical School for 2 years and realized his love for doing research. He earned a Ph.D. from the Neurobiology & Behavior program at Cornell University, working with Robert Capranica, studying the regeneration of the frog auditory nerve. He did postdoctoral work at the University of California, San Diego with Theodore Bullock and Walter Heiligenberg. There, he began working on weakly electric fish. He established his laboratory at the University of Texas in Austin, Texas where he has been studying communication in electric fish, and the regulation and evolution of ion channels in electric fish and other organisms. He was the first chairman of the then newly established Section of Neurobiology at UT. He has been Chairman for Gordon Research Conference on Neuroethology and organizer for the International Congress in Neuroethology. His hobbies include playing guitar and piano. He, his wife Lynne (mandolin), and son Alex (banjo), have a band called Red State Bluegrass. Their goal is to perform on Austin City Limits one day.

CONTRIBUTORS

J. S. Adelman
Princeton University, Princeton, NJ, USA

E. Adkins-Regan
Cornell University, Ithaca, NY, USA

J. F. Aggio
Neuroscience Institute and Department of Biology, Atlanta, GA, USA

M. Ah-King
University of California, Los Angeles, CA, USA

I. Ahnesjö
Uppsala University, Uppsala, Sweden

J. Alcock
Arizona State University, Tempe, AZ, USA

L. Angeloni
Colorado State University, Fort Collins, CO, USA

B. R. Anholt
University of Victoria, Victoria, BC, Canada; Bamfield Marine Sciences Centre, Bamfield, BC, Canada

C. J. L. Atkinson
University of Queensland, St Lucia, QLD, Australia

F. Aureli
Liverpool John Moores University, Liverpool, UK

A. Avarguès-Weber
CNRS, Université de Toulouse, Toulouse, France; Centre de Recherches sur la Cognition Animale, Toulouse, France

K. L. Ayres
University of Washington, Seattle, WA, USA

J. Bakker
University of Liège, Liège, Belgium

G. F. Ball
Johns Hopkins University, Baltimore, MD, USA

J. Balthazart
University of Liège, Liège, Belgium

L. Barrett
University of Lethbridge, Lethbridge, AB, Canada

A. H. Bass
Cornell University, Ithaca, NY, USA

D. K. Bassett
University of Auckland, Auckland, New Zealand

M. Bateson
Newcastle University, Newcastle upon Tyne, UK

G. Beauchamp
University of Montréal, St. Hyacinthe, QC, Canada

B. V. Beaver
Texas A&M University, College Station, TX, USA

P. A. Bednekoff
Eastern Michigan University, Ypsilanti, MI, USA

M. Beekman
University of Sydney, Sydney, NSW, Australia

J. A. Bender
Case Western Reserve University, Cleveland, OHIO, USA

G. E. Bentley
University of California, Berkeley, CA, USA

A. Berchtold
University of Lausanne, Lausanne, Switzerland

I. S. Bernstein
University of Georgia, Athens, GA, USA

S. Bevins
Colorado State University, Fort Collins, USA

D. T. Blumstein
University of California, Los Angeles, CA, USA

C. R. B. Boake
University of Tennessee, Knoxville, TN, USA

R. A. Boakes
University of Sydney, Sydney, NSW, Australia

W. J. Boeing
New Mexico State University, Las Cruces, NM, USA

N. J. Boogert
McGill University, Montréal, QC, Canada

T. Boswell
Newcastle University, Newcastle upon Tyne, UK

A. Bouskila
Ben-Gurion University of the Negev, Beer Sheva, Israel

R. M. Bowden
Illinois State University, Normal, IL, USA

E. M. Brannon
Duke University, Durham, NC, USA

M. D. Breed
University of Colorado, Boulder, CO, USA

M. R. Bregman
University of California, San Diego, CA, USA

J. Brodeur
Université de Montréal, Montréal, QC, Canada

E. D. Brodie, III
University of Virginia, Charlottesville, VA, USA

A. Brodin
Lund University, Lund, Sweden

D. M. Broom
University of Cambridge, Cambridge, UK

J. L. Brown
University at Albany, Albany, NY, USA

J. S. Brown
University of Illinois at Chicago, Chicago, IL, USA

M. J. F. Brown
Royal Holloway University of London, Egham, UK

H. Brumm
Max Planck Institute for Ornithology, Seewiesen,
Germany

R. Buffenstein
University of Texas Health Science Center at San Antonio,
San Antonio, TX, USA

J. D. Buntin
University of Wisconsin-Milwaukee, Milwaukee, WI, USA

J. Burger
Rutgers University, Piscataway, NJ, USA

G. M. Burghardt
University of Tennessee, Knoxville, TN, USA

R. W. Burkhardt, Jr.
University of Illinois at Urbana-Champaign, Urbana,
IL, USA

N. T. Burley
University of California, Irvine, CA, USA

S. S. Burmeister
University of North Carolina, Chapel Hill, NC, USA

D. S. Busch
Northwest Fisheries Science Center, National Marine
Fisheries Service, Seattle, WA, USA

R. W. Byrne
University of St. Andrews, St. Andrews, Fife, Scotland, UK

R. M. Calisi
University of California, Berkeley, CA, USA

J. Call
Max Planck Institute for Evolutionary Anthropology,
Leipzig, Germany

U. Candolin
University of Helsinki, Helsinki, Finland

J. F. Cantlon
Rochester University, Rochester, NC, USA

C. E. Carr
University of Maryland, College Park, MD, USA

C. S. Carter
University of Illinois at Chicago, Chicago, IL, USA

F. Cézilly
Université de Bourgogne, Dijon, France

E. S. Chang
University of California-Davis, Bodega Bay, CA, USA

J. W. Chapman
Rothamsted Research, Harpenden, Hertfordshire, UK

J. C. Choe
Ewha Womans University, Seoul, Korea

J. A. Clarke
University of Northern Colorado, Greeley, CO, USA

N. S. Clayton
University of Cambridge, Cambridge, UK

B. Clucas
University of Washington, Seattle, WA, USA;
Humboldt University, Berlin, Germany

R. B. Cocroft
University of Missouri, Columbia, MO, USA

J. H. Cohen
Eckerd College, St. Petersburg, FL, USA

S. P. Collin
University of Western Australia, Crawley, WA, Australia

L. Conradt
University of Sussex, Brighton, UK

W. E. Cooper, Jr.
Indiana University Purdue University Fort Wayne, Fort
Wayne, IN, USA

R. G. Coss
University of California, Davis, CA, USA

J. T. Costa
Western Carolina University, Cullowhee, NC, USA;
Highlands Biological Station, Highland NC, USA

I. D. Couzin
Princeton University, Princeton, NJ, USA

N. J. Cowan
Johns Hopkins University, Baltimore, MD, USA

R. M. Cox
Dartmouth College, Hanover, NH, USA

J. Crast
University of Georgia, Athens, GA, USA

S. Creel
Montana State University, Bozeman, MT, USA

W. Cresswell
University of St. Andrews, St. Andrews, Scotland, UK

D. Crews
University of Texas, Austin, TX, USA

K. R. Crooks
Colorado State University, Fort Collins, CO, USA

J. D. Crystal
University of Georgia, Athens, GA, USA

S. R. X. Dall
University of Exeter, Cornwall, UK

D. Daniels
University at Buffalo, State University of New York, Buffalo, NY, USA

J. M. Davis
Vassar College, Poughkeepsie, NY, USA

K. Dean
University of Maryland, College Park, MD, USA

J. Deen
University of Minnesota, St. Paul, MN, USA

R. J. Denver
University of Michigan, Ann Arbor, MI, USA

C. D. Derby
Neuroscience Institute and Department of Biology, Atlanta, GA, USA

M. E. Deutschlander
Hobart and William Smith Colleges, Geneva, NY, USA

F. B. M. de Waal
Emory University, Atlanta, GA, USA

D. A. Dewsbury
University of Florida, Gainesville, FL, USA

A. Dickinson
University of Cambridge, Cambridge, UK

J. L. Dickinson
Cornell University, Ithaca, NY, USA

A. G. Dolezal
Arizona State University, Tempe, AZ, USA

B. Doligez
Université de Lyon, Villeurbanne, France

R. H. Douglas
City University, London, UK

K. B. Døving
University of Oslo, Oslo, Norway

V. A. Drake
University of New South Wales at the Australian Defence Force Academy, Canberra, ACT, Australia

L. C. Drickamer
Northern Arizona University, Flagstaff, AZ, USA

H. Drummond
Universidad Nacional Autónoma de México, México

J. P. Drury
University of California, Los Angeles, CA, USA

J. E. Duffy
Virginia Institute of Marine Science, Gloucester Point, VA, USA

R. Dukas
McMaster University, Hamilton, ON, Canada

F. C. Dyer
Michigan State University, East Lansing, MI, USA

W. G. Eberhard
Smithsonian Tropical Research Institute; Universidad de Costa Rica, Ciudad Universitaria, Costa Rica

N. J. Emery
Queen Mary University of London, London, UK; University of Cambridge, Cambridge, UK

C. S. Evans
Macquarie University, Sydney, NSW, Australia

S. E. Fahrbach
Wake Forest University, Winston-Salem, NC, USA

E. Fernández-Juricic
Purdue University, West Lafayette, IN, USA

J. R. Fetcho
Cornell University, Ithaca, NY, USA

J. H. Fewell
Arizona State University, Tempe, AZ, USA

G. Fleissner
Goethe-University Frankfurt, Frankfurt, Germany

G. Fleissner
Goethe-University Frankfurt, Frankfurt, Germany

T. H. Fleming
University of Miami, Coral Gables, FL, USA

A. Florsheim
Veterinary Behavior Solutions, Dallas, TX, USA

E. S. Fortune
Johns Hopkins University, Baltimore, MD, USA

R. B. Forward, Jr.
Duke University Marine Laboratory, Beaufort, NC, USA

S. A. Foster
Clark University, Worcester, MA, USA

D. M. Fragaszy
University of Georgia, Athens, GA, USA

O. N. Fraser
University of Vienna, Vienna, Austria

P. J. Fraser
University of Aberdeen, Aberdeen, Scotland, UK

T. M. Freeberg
University of Tennessee, Knoxville, TN, USA

K. A. French
University of California, San Diego, La Jolla, CA, USA

A. Frid
Vancouver Aquarium, Vancouver, BC, Canada

C. B. Frith
Private Independent Ornithologist, Malanda, QLD, Australia

D. J. Funk
Vanderbilt University, Nashville, TN, USA

L. Fusani
University of Ferrara, Ferrara, Italy

C. R. Gabor
Texas State University-San Marcos, San Marcos, TX, USA

R. Gadagkar
Indian Institute Science, Bangalore, India

B. G. Galef
McMaster University, Hamilton, ON, Canada

C. M. Garcia
Instituto de Ecología, UNAM, México

S. A. Gauthreaux, Jr.
Clemson University, Clemson, SC, USA

F. Geiser
University of New England, Armidale, NSW, Australia

T. Q. Gentner
University of California, San Diego, CA, USA

H. C. Gerhardt
University of Missouri, Columbia, MO, USA

M. D. Ginzel
Purdue University, West Lafayette, IN, USA

L.-A. Giraldeau
Université du Québec à Montréal, Montréal, QC, Canada

M. Giurfa
CNRS, Université de Toulouse, Toulouse, France; Centre de Recherches sur la Cognition Animale, Toulouse, France

J.-G. J. Godin
Carleton University, Ottawa, ON, Canada

J. Godwin
North Carolina State University, Raleigh, NC, USA

E. Goodale
Field Ornithology Group of Sri Lanka, University of Colombo, Colombo, Sri Lanka

M. A. D. Goodisman
Georgia Institute of Technology, Atlanta, GA, USA

C. J. Goodnight
University of Vermont, Burlington, Vermont, USA

P. A. Gowaty
University of California, Los Angeles, CA, USA; Smithsonian Tropical Research Institute, USA

W. Goymann
Max Planck Institute for Ornithology, Seewiesen, Germany

P. Graham
University of Sussex, Brighton, UK

T. Grandin
Colorado State University, Fort Collins, CO, USA

M. D. Greenfield
Université François Rabelais de Tours, Tours, France

G. F. Grether
University of California, Los Angeles, CA, USA

A. S. Griffin
University of Newcastle, Callaghan, NSW, Australia

M. Griggio
Konrad Lorenz Institute for Ethology, Vienna, Austria

T. G. G. Groothuis
University of Groningen, Groningen, Netherlands

R. Grosberg
University of California, Davis, CA, USA

C. M. Grozinger
Pennsylvania State University, University Park, PA, USA

R. D. Grubbs
Florida State University Coastal and Marine Laboratory, St. Teresa, FL, USA; George Mason University, Fairfax, VA, USA

R. R. Ha
University of Washington, Seattle, WA, USA

J. P. Hailman
University of Wisconsin, Jupiter, FL, USA

I. M. Hamilton
Ohio State University, Columbus, OH, USA

R. R. Hampton
Emory University, Atlanta, GA, USA

I. C. W. Hardy
University of Nottingham, Loughborough, Leicestershire, UK

B. L. Hart
University of California, Davis, CA, USA

L. I. Haug
Texas Veterinary Behavior Services, Sugar Land, TX, USA

M. Hauser
Harvard University, Cambridge, MA, USA

L. S. Hayward
University of Washington, Seattle, WA, USA

S. D. Healy
University of St. Andrews, St. Andrews, Fife, Scotland, UK

E. A. Hebets
University of Nebraska, Lincoln, NE, USA

M. R. Heithaus
Florida International University, Miami, FL, USA

H. Helanterä
University of Sussex, Brighton, UK; University of Helsinki, Helsinki, Finland

J. M. Hemmi
Australian National University, Canberra, ACT, Australia

L. M. Henry
University of Oxford, Oxford, UK

J. M. Herbers
Ohio State University, Columbus, OH, USA

M. R. Heupel
James Cook University, Townsville, QLD, Australia

H. Hoi
Konrad Lorenz Institute for Ethology, Vienna, Austria

K. E. Holekamp
Michigan State University, East Lansing, MI, USA

R. A. Holland
Max Planck Institute for Ornithology, Radolfzell, Germany

A. G. Horn
Dalhousie University, Halifax, NS, Canada

L. Huber
University of Vienna, Vienna, Austria

M. A. Huffman
Kyoto University, Inuyama, Aichi Prefecture, Japan

H. Hurd
Keele University, Staffordshire, UK

P. L. Hurd
University of Alberta, Edmonton, AB, Canada

A. Jacobs
University of California, Riverside, CA, USA

V. M. Janik
University of St. Andrews, St. Andrews, Fife, Scotland, UK

K. Jensen
Queen Mary University of London, London, UK

C. Jozet-Alves
University of Caen Basse-Normandie, Caen, France

J. Kaminski
Max Planck Institute for Evolutionary Anthropology, Leipzig, Germany

L. Kapás
Washington State University, Spokane, WA, USA

A. S. Kauffman
University of California, San Diego, La Jolla, CA, USA

J. L. Kelley
University of Western Australia, Crawley, WA, Australia

A. J. King
Zoological Society of London, London, UK; University of Cambridge, Cambridge, UK

S. L. Klein
Johns Hopkins Bloomberg School of Public Health, Baltimore, MD, USA

M. J. Klowden
University of Idaho, Moscow, ID, USA

J. Komdeur
University of Groningen, Groningen, Netherlands

M. Konishi
California Institute of Technology, Pasadena, CA, USA

J. Korb
University of Osnabrueck, Osnabrück, Germany

I. Krams
University of Daugavpils, Daugavpils, Latvia

R. T. Kraus
Florida State University Coastal and Marine Laboratory, St. Teresa, FL, USA; George Mason University, Fairfax, VA, USA

W. B. Kristan, Jr.
University of California, San Diego, La Jolla, CA, USA

J. M. Krueger
Washington State University, Spokane, WA, USA

C. W. Kuhar
Cleveland Metroparks Zoo, Cleveland, OH, USA

C. P. Kyriacou
University of Leicester, Leicester, UK

F. Ladich
University of Vienna, Vienna, Austria

K. N. Laland
University of St Andrews, St Andrews, Fife, Scotland, UK

P. H. L. Lamberton
Imperial College Faculty of Medicine, London, UK

A. V. Latchininsky
University of Wyoming, Laramie, WY, USA

L. Lefebvre
McGill University, Montréal, QC, Canada

J. E. Leonard
Hiwassee College, Madisonville, TN, USA

M. L. Leonard
Dalhousie University, Halifax, NS, Canada

G. R. Lewin
Max-Delbrück Center for Molecular Medicine, Berlin, Germany

F. Libersat
Institut de Neurobiologie de la Méditerranée, Parc Scientifique de Luminy, Marseille, France

A. E. Liebert
Framingham State College, Framingham, MA, USA

C. H. Lin
University of British Columbia, Vancouver, BC, Canada

J. A. Linares
Texas A&M University, Gonzales, TX, USA

J. Lind
Stockholm University, Stockholm, Sweden

T. A. Linksvayer
University of Copenhagen, Copenhagen, Denmark

C. List
London School of Economics, London, UK

N. Lo
Australian Museum, Sydney, NSW, Australia; University of Sydney, Sydney, NSW, Australia

C. M. F. Lohmann
University of North Carolina, Chapel Hill, NC, USA

K. J. Lohmann
University of North Carolina, Chapel Hill, NC, USA

Y. Lubin
Ben-Gurion University of the Negev, Beer Sheva, Israel

J. Lucas
Purdue University, West Lafayette, IN, USA

S. K. Lynn
Boston College, Chestnut Hill, MA, USA

K. E. Mabry
New Mexico State University, Las Cruces, NM, USA

D. Maestripieri
University of Chicago, Chicago, IL, USA

D. L. Maney
Emory University, Atlanta, GA, USA

T. G. Manno
Auburn University, Auburn, AL, USA

S. W. Margulis
Canisius College, Buffalo, NY, USA

L. Marino
Emory University, Atlanta, GA, USA

T. A. Markow
University of California at San Diego, La Jolla, CA, USA

C. A. Marler
University of Wisconsin, Madison, WI, USA

P. P. Marra
Smithsonian Migratory Bird Center, National Zoological Park, Washington, DC, USA

L. B. Martin
University of South Florida, Tampa, FL, USA

M. Martin
North Carolina State University, Raleigh, NC, USA

J. A. Mather
University of Lethbridge, Lethbridge, AB, Canada

K. Matsuura
Okayama University, Okayama, Japan

T. Matsuzawa
Kyoto University, Kyoto, Japan

K. McAuliffe
Harvard University, Cambridge, MA, USA

E. A. McGraw
University of Queensland, Brisbane, QLD, Australia

N. L. McGuire
University of California, Berkeley, CA, USA

N. J. Mehdiabadi
Smithsonian Institution, Washington, DC, USA

R. Menzel
Freie Universität Berlin, Berlin, Germany

J. C. Mitani
University of Michigan, Ann Arbor, MI, USA

J. C. Montgomery
University of Auckland, Auckland, New Zealand

J. Moore
Colorado State University, Fort Collins, CO, USA

J. Morand-Ferron
Université du Québec à Montréal, Montréal, QC, Canada

J. Moreno
Museo Nacional de Ciencias Naturales, Madrid, Spain

K. Morgan
University of St. Andrews, St. Andrews, Fife, Scotland, UK

R. Muheim
Lund University, Lund, Sweden

C. A. Nalepa
North Carolina State University, Raleigh, NC, USA

D. Naug
Colorado State University, Fort Collins, CO, USA

D. A. Nelson
Ohio State University, Columbus, OH, USA

R. J. Nelson
Ohio State University, Columbus, OH, USA

I. Newton
Centre for Ecology & Hydrology, Wallingford, UK

K. Nishimura
Hokkaido University, Hakodate, Japan

J. E. Niven
University of Cambridge, Cambridge, UK; Smithsonian Tropical Research Institute, Panamá, República de Panamá

P. Nonacs
University of California, Los Angeles, CA, USA

A. J. Norton
Imperial College Faculty of Medicine, London, UK

B. P. Oldroyd
University of Sydney, Sydney, NSW, Australia

T. J. Ord
University of New South Wales, Sydney, NSW, Australia

M. A. Ottinger
University of Maryland, College Park, MD, USA

D. H. Owings
University of California, Davis, CA, USA

J. M. Packard
Texas A&M University, College Station, TX, USA

A. Pai
Spelman College, Atlanta, GA, USA

T. J. Park
University of Illinois at Chicago, Chicago, IL, USA

L. A. Parr
Yerkes National Primate Research Center, Atlanta, GA, USA

Y. M. Parsons
La Trobe University, Bundoora, VIC, Australia

G. L. Patricelli
University of California, Davis, CA, USA

M. M. Patten
Museum of Comparative Zoology, Cambridge, MA, USA

A. Payne
Tufts University, Medford, MA, USA

I. M. Pepperberg
Harvard University, Cambridge, MA, USA

M.-J. Perrot-Minnot
Université de Bourgogne, Dijon, France

S. Perry
University of California-Los Angeles, Los Angeles, CA, USA

K. M. Pickett
University of Vermont, Burlington, VT, USA

N. Pinter-Wollman
Stanford University, Stanford, CA, USA

D. Plachetzki
University of California, Davis, CA, USA

G. S. Pollack
McGill University, Montréal, QC, Canada

G. D. Pollak
University of Texas at Austin, Austin, TX, USA

R. Poulin
University of Otago, Dunedin, New Zealand

S. C. Pratt
Arizona State University, Tempe, AZ, USA

V. V. Pravosudov
University of Nevada, Reno, NV, USA

G. H. Pyke
Australian Museum, Sydney, NSW, Australia; Macquarie University, North Ryde, NSW, Australia

D. C. Queller
Rice University, Houston, TX, USA

M. Ramenofsky
University of California, Davis, CA, USA

C. H. Rankin
University of British Columbia, Vancouver, BC, Canada

F. L. W. Ratnieks
University of Sussex, Brighton, UK

D. Raubenheimer
Massey University, Auckland, New Zealand

S. M. Reader
Utrecht University, Utrecht, Netherlands

H. K. Reeve
Cornell University, New York, NY, USA

J. Reinhard
University of Queensland, Brisbane, QLD, Australia

L. Rendell
University of St Andrews, St Andrews, Fife, Scotland, UK

A. N. Rice
Cornell University, Ithaca, NY, USA

J. M. L. Richardson
University of Victoria, Victoria, BC, Canada

H. Richner
University of Bern, Bern, Switzerland

T. Rigaud
Université de Bourgogne, Dijon, France

R. E. Ritzmann
Case Western Reserve University, Cleveland, OHIO, USA

A. J. Riveros
University of Arizona, Tucson, AZ, USA

D. Robert
University of Bristol, Bristol, UK

G. E. Robinson
University of Illinois at Urbana-Champaign, Urbana, IL, USA

I. Rodriguez-Prieto
Museo Nacional de Ciencias Naturales, Madrid, Spain

B. D. Roitberg
Simon Fraser University, Burnaby, BC, Canada

L. M. Romero
Tufts University, Medford, MA, USA

T. J. Roper
University of Sussex, Brighton, UK

G. G. Rosenthal
Texas A&M University, College Station, TX, USA

C. Rowe
Newcastle University, Newcastle upon Tyne, UK

L. Ruggiero
Barnard College and Columbia University, New York, NY, USA

G. D. Ruxton
University of Glasgow, Glasgow, Scotland, UK

M. J. Ryan
University of Texas, Austin, TX, USA

R. Safran
University of Colorado, Boulder, CO, USA

W. Saltzman
University of California, Riverside, CA, USA

R. M. Sapolsky
Stanford University, Stanford, CA, USA

L. S. Sayigh
Woods Hole Oceanographic Institution, Woods Hole, MA, USA

A. Schmitz
University of Bonn, Bonn, Germany

H. Schmitz
University of Bonn, Bonn, Germany

J. Schulkin
Georgetown University, Washington, DC, USA; National Institute of Mental Health, Bethesda, MD, USA

H. Schwabl
Washington State University, Pullman, WA, USA

A. M. Seed
Max Planck Institute for Evolutionary Anthropology, Leipzig, Germany

M. R. Servedio
University of North Carolina, Chapel Hill, NC, USA

J. C. Shaw
University of California, Santa Barbara, CA, USA

S.-F. Shen
Cornell University, New York, NY, USA

B. L. Sherman
North Carolina State University, Raleigh, NC, USA

T. N. Sherratt
Carleton University, Ottawa, ON, Canada

D. M. Shuker
University of St. Andrews, St. Andrews, Fife, Scotland, UK

R. Silver
Barnard College and Columbia University, New York, NY, USA

B. Silverin
University of Göteborg, Göteborg, Sweden

A. M. Simmons
Brown University, Providence, RI, USA

S. J. Simpson
University of Sydney, Sydney, NSW, Australia

U. Sinsch
University Koblenz-Landau, Koblenz, Germany

H. Slabbekoorn
Leiden University, Leiden, Netherlands

P. J. B. Slater
University of St. Andrews, St. Andrews, Fife, Scotland, UK

C. N. Slobodchikoff
Northern Arizona University, Flagstaff, AZ, USA

A. R. Smith
Smithsonian Tropical Research Institute, Balboa, Ancon, Panamá

G. T. Smith
Indiana University, Bloomington, IN, USA

J. E. Smith
Michigan State University, East Lansing, MI, USA

B. Smuts
University of Michigan, Ann Arbor, MI, USA

E. C. Snell-Rood
Indiana University, Bloomington, IN, USA

C. T. Snowdon
University of Wisconsin, Madison, WI, USA

R. B. Srygley
USDA-Agricultural Research Service, Sidney, MT, USA

T. Stankowich
University of Massachusetts, Amherst, MA, USA

P. T. Starks
Tufts University, Medford, MA, USA

C. A. Stern
Cornell University, Ithaca, NY, USA

J. R. Stevens
Max Planck Institute for Human Development, Berlin, Germany

P. K. Stoddard
Florida International University, Miami, FL, USA

J. E. Strassmann
Rice University, Houston, TX, USA

C. E. Studds
Smithsonian Migratory Bird Center, National Zoological Park, Washington, DC, USA

L. Sullivan-Beckers
University of Nebraska, Lincoln, NE, USA

R. A. Suthers
Indiana University, Bloomington, IN, USA

J. P. Swaddle
College of William and Mary, Williamsburg, VA, USA

R. Swaisgood
San Diego Zoo's Institute for Conservation Research, Escondido, CA, USA

É. Szentirmai
Washington State University, Spokane, WA, USA

M. Taborsky
University of Bern, Hinterkappelen, Switzerland

Z. Tang-Martínez
University of Missouri-St. Louis, St. Louis, MO, USA

E. Tauber
University of Leicester, Leicester, UK

D. W. Thieltges
University of Otago, Dunedin, New Zealand

F. Thomas
Génétique et Evolution des Maladies Infectieuses, Montpellier, France; Université de Montréal, Montréal, QC, Canada

C. V. Tillberg
Linfield College, McMinnville, OR, USA

M. Tomasello
Max Planck Institute for Evolutionary Anthropology, Leipzig, Germany

A. L. Toth
Pennsylvania State University, University Park, PA, USA

B. C. Trainor
University of California, Davis, CA, USA

J. Traniello
Boston University, Boston, MA, USA

K. Tsuji
University of the Ryukyus, Okinawa, Japan

G. W. Uetz
University of Cincinnati, Cincinnati, OH, USA

M. Valentine
University of Vermont, Burlington, VT, USA

A. Valero
Instituto de Ecología, UNAM, México

J. L. Van Houten
University of Vermont, Burlington, VT, USA

M. A. van Noordwijk
University of Zurich, Zurich, Switzerland

C. P. van Schaik
University of Zurich, Zurich, Switzerland

S. H. Vessey
Bowling Green State University, Bowling Green, OH, USA

G. von der Emde
University of Bonn, Bonn, Germany

H. G. Wallraff
Max Planck Institute for Ornithology, Seewiesen, Germany

R. R. Warner
University of California, Santa Barbara, CA, USA

E. Warrant
University of Lund, Lund, Sweden

R. Watt
University of Edinburgh, Edinburgh, Scotland, UK

J. P. Webster
Imperial College Faculty of Medicine, London, UK

M. Webster
Cornell Lab of Ornithology, Ithaca, NY, USA

N. Wedell
University of Exeter, Penryn, UK

E. V. Wehncke
Biodiversity Research Center of the Californias, San Diego, CA, USA

M. J. West-Eberhard
Smithsonian Tropical Research Institute, Costa Rica

G. Westhoff
Tierpark Hagenbeck gGmbH, Hamburg, Germany

C. J. Whelan
Illinois Natural History Survey, University of Illinois at Chicago, Chicago, IL, USA

A. Whiten
University of St. Andrews, St. Andrews, Fife, Scotland, UK

A. Wilkinson
University of Virginia, Charlottesville, VA, USA

D. M. Wilkinson
Liverpool John Moores University, Liverpool, UK

S. P. Windsor
University of Auckland, Auckland, New Zealand

J. C. Wingfield
University of California, Davis, CA, USA

K. E. Wynne-Edwards
University of Calgary, Calgary, AB, Canada

D. D. Yager
University of Maryland, College Park,
MD, USA

R. Yamada
University of Queensland, Brisbane, QLD,
Australia

J. Yano
University of Vermont, Burlington, VT, USA

K. Yasukawa
Beloit College, Beloit, WI, USA

J. Zeil
Australian National University, Canberra, ACT,
Australia

T. R. Zentall
University of Kentucky, Lexington, KY, USA

E. Zou
Nicholls State University, Thibodaux, LA, USA

M. Zuk
University of California, Riverside, CA, USA

GUIDE TO USE OF THE ENCYCLOPEDIA

Structure of the Encyclopedia

The material in the Encyclopedia is arranged as a series of articles in alphabetical order.

There are four features to help you easily find the topic you're interested in: an alphabetical contents list, a subject classification index, cross-references and a full subject index.

1. Alphabetical Contents List

The alphabetical contents list, which appears at the front of each volume, lists the entries in the order that they appear in the Encyclopedia. It includes both the volume number and the page number of each entry.

2. Subject Classification Index

This index appears at the start of each volume and groups entries under subject headings that reflect the broad themes of Animal Behavior. This index is useful for making quick connections between entries and locating the relevant entry for a topic that is covered in more than one article.

3. Cross-references

All of the entries in the Encyclopedia have been extensively cross-referenced. The cross-references which appear at the end of an entry, serve three different functions:

i. To indicate if a topic is discussed in greater detail elsewhere
ii. To draw the readers attention to parallel discussions in other entries
iii. To indicate material that broadens the discussion

Example

The following list of cross-references appears at the end of the entry Landmark Studies: Honeybees

See also: Communication: Social Recognition; Invertebrate Social Behavior: Ant, Bee and Wasp Social Evolution; Invertebrate Social Behavior: Caste Determination in Arthropods; Invertebrate Social Behavior: Collective Intelligence; Invertebrate Social Behavior: Dance Language; Invertebrate Social Behavior: Developmental Plasticity; Invertebrate Social Behavior: Division of Labor; Invertebrate Social Behavior: Queen-Queen Conflict in Eusocial Insect Colonies; Invertebrate Social Behavior: Queen-Worker Conflicts Over Colony Sex Ratio.

4. Index

The index includes page numbers for quick reference to the information you're looking for. The index entries differentiate between references to a whole entry, a part of an entry, and a table or figure.

5. Contributors

At the start of each volume there is list of the authors who contributed to the Encyclopedia.

SUBJECT CLASSIFICATION

Anti-Predator Behavior

Section Editor: *Ted Stankowich*

Antipredator Benefits from Heterospecifics
Co-Evolution of Predators and Prey
Conservation and Anti-Predator Behavior
Defensive Avoidance
Defensive Chemicals
Defensive Coloration
Defensive Morphology
Ecology of Fear
Economic Escape
Empirical Studies of Predator and Prey Behavior
Games Played by Predators and Prey
Group Living
Life Histories and Predation Risk
Predator Avoidance: Mechanisms
Parasitoids
Predator's Perspective on Predator–Prey Interactions
Risk Allocation in Anti-Predator Behavior
Risk-Taking in Self-Defense
Trade-Offs in Anti-Predator Behavior
Vigilance and Models of Behavior

Applications

Section Editor: *Michael D. Breed* and *Janice Moore*

Conservation and Animal Behavior
Robot Behavior
Training of Animals

Arthropod Social Behavior

Section Editor: *Jae Chun Choe*

Ant, Bee and Wasp Social Evolution
Caste Determination in Arthropods
Collective Intelligence
Colony Founding in Social Insects
Crustacean Social Evolution
Dance Language

Developmental Plasticity
Division of Labor
Kin Selection and Relatedness
Parasites and Insects: Aspects of Social Behavior
Queen–Queen Conflict in Eusocial Insect Colonies
Queen–Worker Conflicts Over Colony Sex Ratio
Recognition Systems in the Social Insects
Reproductive Skew
Sex and Social Evolution
Social Evolution in 'Other' Insects and Arachnids
Spiders: Social Evolution
Subsociality and the Evolution of Eusociality
Termites: Social Evolution
Worker–Worker Conflict and Worker Policing

Behavioral Endocrinology

Section Editor: *John C. Wingfield*

Aggression and Territoriality
Aquatic Invertebrate Endocrine Disruption
Behavioral Endocrinology of Migration
Circadian and Circannual Rhythms and Hormones
Communication and Hormones
Conservation Behavior and Endocrinology
Experimental Approaches to Hormones and Behavior: Invertebrates
Female Sexual Behavior and Hormones in Non-Mammalian Vertebrates
Field Techniques in Hormones and Behavior
Fight or Flight Responses
Food Intake: Behavioral Endocrinology
Hibernation, Daily Torpor and Estivation in Mammals and Birds: Behavioral Aspects
Hormones and Behavior: Basic Concepts
Immune Systems and Sickness Behavior
Invertebrate Hormones and Behavior
Male Sexual Behavior and Hormones in Non-Mammalian Vertebrates
Mammalian Female Sexual Behavior and Hormones
Maternal Effects on Behavior
Memory, Learning, Hormones and Behavior

Cognition

Section Editor: *Nicola Clayton*

Communication

Section Editor: *Jeffery Lucas*

Conservation

Section Editor: *Constantino Macías Garcia*

Decision Making by Individuals

Section Editor: *David W. Stephens*

Evolution

Section Editor: *Joan M. Herbers*

Evolution: Fundamentals
Isolating Mechanisms and Speciation
Levels of Selection
Microevolution and Macroevolution in Behavior
Nervous System: Evolution in Relation to Behavior
Phylogenetic Inference and the Evolution of Behavior
Reproductive Success
Specialization

Foraging

Section Editor: *David W. Stephens*

Caching
Digestion and Foraging
Foraging Modes
Habitat Selection
Hormones and Breeding Strategies, Sex Reversal, Brood
    Parasites, Parthenogenesis
Hunger and Satiety
Internal Energy Storage
Kleptoparasitism and Cannibalism
Optimal Foraging and Plant-Pollinator Co-Evolution
Optimal Foraging Theory: Introduction
Patch Exploitation
Wintering Strategies: Moult and Behavior

Genetics

Section Editor: *Ross H. Crozier*

Caste in Social Insects: Genetic Influences Over Caste
    Determination
Dictyostelium, the Social Amoeba
Drosophila Behavior Genetics
Genes and Genomic Searches
Kin Recognition and Genetics
Marine Invertebrates: Genetics of Colony Recognition
Nasonia Wasp Behavior Genetics
Orthopteran Behavioral Genetics
Parmecium Behavioral Genetics
Social Insects: Behavioral Genetics
Unicolonial Ants: Loss of Colony Identity

History

Section Editor: *Michael D. Breed*

Animal Behavior: Antiquity to the Sixteenth Century
Animal Behavior: The Seventeenth to the Twentieth
    Centuries
Behavioral Ecology and Sociobiology
Comparative Animal Behavior – 1920–1973
Ethology in Europe

Future of Animal Behavior: Predicting Trends
Integration of Proximate and Ultimate Causes
Neurobiology, Endocrinology and Behavior
Psychology of Animals

Infectious Disease and Behavior

Section Editor: *Janice Moore*

Avoidance of Parasites
Beyond Fever: Comparative Perspectives on Sickness
    Behavior
Conservation, Behavior, Parasites and Invasive Species
Ectoparasite Behavior
Evolution of Parasite-Induced Behavioral Alterations
Intermediate Host Behavior
Parasite-Induced Behavioral Change: Mechanisms
Parasite-Modified Vector Behavior
Parasites and Sexual Selection
Propagule Behavior and Parasite Transmission
Reproductive Behavior and Parasites: Vertebrates
Reproductive Behavior and Parasites: Invertebrates
Self-Medication: Passive Prevention and Active Treatment
Social Behavior and Parasites

Landmark Studies

Section Editor: *Michael D. Breed*

Alex: A Study in Avian Cognition
Aplysia
Barn Swallows: Sexual and Social Behavior
Betta Splendens
Boobies
Bowerbirds
Chimpanzees
Cockroaches
Domestic Dogs
Hamilton, William Donald
Herring Gulls
Honeybees
Locusts
Lorenz, Konrad
Norway Rats
Octopus
Pheidole: Sociobiology of a Highly Diverse Genus
Pigeons
Rhesus Macaques
Sharks
Spotted Hyenas
Swordtails and Platyfishes
Threespine Stickleback
Tinbergen, Niko
Tribolium

CONTENTS

VOLUME 1

A

B

C

D

E

F

VOLUME 2

G

H

I

K

L

M

VOLUME 3

W

Z

Queen–Queen Conflict in Eusocial Insect Colonies

M. A. D. Goodisman, Georgia Institute of Technology, Atlanta, GA, USA

Introduction

A great deal can be learned about the ecology and evolution of social behavior by studying the origin, nature, and resolution of conflict among queens within insect societies. Queen conflict fundamentally concerns the struggle for reproductive success among society members. Thus, conflict among queens is a pervasive feature of eusocial insect colonies.

Eusocial insects dominate ecological communities and rank among the most successful of animal taxa. The success of eusocial insects stems primarily from their cooperative and helping behaviors. Members of eusocial insect societies work together to efficiently rear young, defend the colony, and forage for food. The efficiency of eusocial insect colonies arises, in part, from the fact that colony members belong to different castes, which undertake distinct tasks.

The queen and male castes are generally responsible for reproduction and dispersal. In contrast, the worker and soldier castes are subfertile. The formation of societies composed of distinct reproductive and subfertile castes represents a major evolutionary transition in biological organization. Such transitions, from independent replication to replication as part of a larger group, occur only rarely in evolution. In the case of eusocial insects, independent organisms, the workers and soldiers, gave up the ability to reproduce separately and instead came to rely on others, the queens and males, to successfully pass on their genes to future generations.

Eusocial insect colonies have justifiably been called 'superorganisms' because of the tremendous cooperative and integrative nature of their societies. However, eusocial insect colonies are also subject to internal conflict. Conflicts arise because members of insect societies are not clonal and thus their genetic interests differ. Under such conditions, selection acts to promote selfish behaviors. However, such behaviors are often kept in check by mechanisms that balance selection on individuals and selection at the colony level, leading to the formation of

relatively stable groups. Consequently, eusocial insect colonies represent a dynamic equilibrium that arises between cooperation among individuals and potential conflict among them that could destroy the society.

This article concerns conflict that occurs among queens within eusocial insect colonies. Here, queens are defined as the primary reproductive females within insect societies. Queens may be contrasted with workers and soldiers, which are often morphologically distinct from queens and, by definition, have lower reproductive potential than queens (**Figure 1**).

It is important to note that, although many invertebrates show varying levels of social behavior, this article will be restricted to the traditional eusocial insects, which fall into two insect orders, the Hymenoptera (ants, bees, and wasps) and the Isoptera (termites). Moreover, the major focus of discussion will be on the eusocial Hymenoptera, because far more is known about eusociality in this group than in termites. Finally, this article deliberately avoids discussion of other types of conflict, including queen–worker conflict and conflict between the sexes, which are reviewed elsewhere.

Importance of Understanding Queen–Queen Conflict

Queen conflict is an outcome of cooperative breeding among queens. Cooperative breeding is a type of social structure whereby a group of individuals assists in caring for offspring that are not exclusively their own. Cooperative breeding represents a special difficulty for evolutionary theory because it results in individuals helping raise the offspring of others. Rearing the offspring of other individuals would seem to decrease the direct fitness of the individual providing the help and increase the direct fitness of potential competitors in the population. Both of these outcomes should lead to a decrease in the frequency of helping behavior in the population.

1

Figure 1   Red imported fire ant, *Solenopsis invicta*, queen and workers. The queen is morphologically and behaviorally distinct from the workers and specializes in reproduction and dispersal. In contrast, the workers are sterile and specialize in brood rearing, foraging, and colony defense.

However, despite the potential negative effects on individuals, cooperative breeding can still evolve under appropriate ecological conditions. In general, ecological constraints impeding solitary reproduction select for individuals to work together to reproduce. As will be discussed later, this constraint is central to explaining the origin of multiple-queen societies in eusocial insects. In addition, cooperative breeders also receive direct or indirect fitness benefits from membership in the group.

Direct benefits are gains in the helpers' own reproduction. Such gains may arise if large groups are more secure or stable and therefore actually provide greater fitness benefits to an individual than solitary reproduction. Alternatively, helpers may assist dominant individuals in order to remain on a territory. Remaining on an established territory may be directly beneficial to the helper in the long term if the helper ultimately gets to reproduce. Finally, cooperative breeding may act as a signal that helpers are of high quality, which may attract breeding attempts that directly benefit the helper.

In addition to direct fitness benefits, cooperative breeders may receive indirect fitness benefits. Indirect fitness benefits result from increased reproductive success for relatives of the helping individual. Cooperative behaviors that lead to fitness changes for relatives evolve through the action of kin selection. Although kin selection is common, the most spectacular example of kin-selected traits can be found in advanced eusocial insect workers. Workers of these species are unable to mate and generally do not reproduce, but they still receive indirect fitness benefits by helping related individuals to produce offspring.

Consequently, understanding potential conflicts among queens in eusocial insect colonies also provides insight into the nature of kin selection and kin groups. Indeed, as will be seen below, kin structure is a key factor affecting conflict within colonies. However, the characteristics of conflict also depend fundamentally on the ecological circumstances surrounding the interactions.

Types of Queen–Queen Conflict

Conflict necessarily originates from selection acting to promote reproductive success among individuals. Here, however, the types of conflict have been divided into categories based upon when they occur in the eusocial insect life cycle. It is important to note that different types of conflict should not be viewed as mutually exclusive. Indeed, one type of conflict often leads to another. Nevertheless, the categories represent convenient ways to consider how conflict occurs among queens within complex societies.

Conflict over Joining a Group

One of the most important areas of conflict among queens is centered on whether a queen may join an existing colony. The potential joining together of eusocial insect queens represents a major change to the social structure of a typical insect society. Indeed, the formation of multiple-queen groups represents a second type of cooperative breeding that can arise within insect societies, the first instance being the origin of the sterile worker caste that defines eusocial insects.

Many types of queen conflict within colonies begin with the acceptance of multiple queens. For example, queens within polygyne (multiple-queen) colonies potentially compete for reproductive dominance. In addition, workers within polygyne colonies are typically less related to their colonymates than workers in monogyne (single-queen) colonies. Thus, having multiple queens within a colony also affects worker fitness.

How can multiple queens come to coexist within colonies? The first mechanism, known as primary polygyny, occurs when multiple queens join together to initiate a new colony. The second method, known as secondary polygyny, occurs when existing queens attempt to enter and join an already established queenright colony. Primary and secondary polygyny are treated separately here because the origins and resolution of the two situations typically differ greatly.

Primary polygyny

Queens of several eusocial insect taxa form foundress associations when initiating new colonies. The formation of multiqueen associations is known as pleometrosis, to be contrasted with haplometrosis, which occurs when queens found colonies independently. Primary polygyny resulting from pleometrotic associations is unusual from the perspective of typical cooperative breeding associations because of the way that groups of queens form and how reproductive conflicts among them are ultimately resolved.

Pleometrotic associations typically occur when unrelated queens join together and attempt to initiate a new colony. Aggregations of queens may form without queens actively searching for other queens if, for example, new queens are attracted to similar nesting habitats. However, selection may favor formation of foundress associations if there are strong constraints on independent colony founding and if groups of queens succeed at founding colonies better than solitary queens. In this case, kinship need not play a role in the formation of associations. Instead, mutualism drives pleometrosis because a queen's direct fitness is higher on average when she joins a group than if she attempts to found a colony independently. In addition, selection may not drive queens that already belong to a pleometrotic group to keep other queens out, provided that the group has not reached a size at which benefits from membership begin to diminish.

Pleometrotic associations occur in many ant species, including members of the subfamilies Myrmicinae, Dolichoderinae, and Formicinae. Ants that form pleometrotic associations are typically territorial. Workers from newly founded colonies frequently destroy or raid colonies in their immediate vicinity. Thus, it is beneficial for new colonies to produce as many workers and as quickly as possible. Colonies founded by multiple queens typically produce more brood than those founded by single queens, thereby providing a clear numerical advantage in competitive interactions with adjacent colonies.

Surprisingly, pleometrotic associations in ants almost always end with monogyny. Queens that peacefully coexisted during the initiation of pleometrotic groups often become highly aggressive toward each other as the colony matures. The cohabiting queens may even enter into mortal combat that only ends when a single queen is left.

Queens may also compete indirectly during the seemingly peaceful stage when they are producing the first brood. For example, evidence suggests that *Solenopsis invicta* fire ant queens, which use their own internal body reserves to produce eggs, vary in weight loss during reproduction. Such variance in weight loss may be associated with ultimate success within pleometrotic associations, as heavy queens are more likely to survive fights than light queens. Queens in pleometrotic associations of ants, including species such as *S. invicta*, *Myrmecocystus mimicus*, and *Formica podzolica*, also sometimes eat eggs produced by rivals, thereby increasing their food intake and potentially forestalling loss of mass. Thus, queens may be preparing for fighting even while they coexist without direct aggression. The workers may or may not take part in the culling of queens. However, in *S. invicta*, workers do not treat their mother differently from other unrelated queens within the colony.

Although primary polygyny is best studied in ants, it also occurs in other eusocial insects. For instance, some eusocial wasps in the genus *Polistes* form multiple foundress associations. These associations frequently lead to a single queen dominating reproduction within a colony. In addition, some termites, including species in the genera *Macrotermes*, *Microcerotermes*, and *Nasutitermes*, form pleometrotic associations. These multiple-queen colonies may grow more rapidly than those founded by single sets of reproductives, thereby providing an advantage to joining the group. Beyond this, little is known about pleometrotic associations in termites.

A final, related example of competition among queens is found in honeybees (*Apis mellifera*). Honeybee colonies are almost always headed by a single queen. New colonies arise through colony fission when the original mother queen leaves with part of the colony to find a new nesting site. After the mother queen departs, the remaining colony rears several of the original queens' daughters as potential replacement queens. These potential replacement queens kill their potential rival sisters while they are still developing. Alternatively, surviving replacement queens engage in lethal duals so that only a single queen ultimately remains in the old nest. As Darwin noted in *The Origin of Species*, "It may be difficult, but we ought to admire the savage instinctive hatred of the queen-bee, which urges her instantly to destroy the young queens her daughters [sic, sisters] as soon as born, or to perish herself in the combat; for undoubtedly this is for the good of the community."

Secondary polygyny

Many mature eusocial insect colonies contain multiple-functional queens. These colonies typically come to be polygyne through secondary adoption of queens as the colony ages. Thus, the processes leading to secondary polygyny differ from those that give rise to primary polygyny. Moreover, in contrast to primary polygyny, secondary polygyny often leads to permanently polygynous nests.

The primary factor promoting secondary polygyny is the low probability of new queens founding a colony independently. A variety of ecological circumstances can conspire to make colony foundation difficult for new queens. High risk of predation, lack of nesting sites, high levels of competition, or high risk of colony loss – all should select for queens to attempt to enter established colonies. Consequently, from the standpoint of a new queen, joining a mature colony should be a relatively attractive option under many circumstances.

However, the decision to allow a new queen to join an established colony is not so simple from the standpoint of the queens and workers that already belong to the focal colony. Polygyny involves shared reproduction among queens and tends to decrease per-capita reproductive output, which should lead to queen conflict. In addition, the presence of multiple queens generally decreases the relatedness of workers to the brood they rear, which should lead to conflict among families within colonies. So why would a mature colony accept new queens?

Evolution favors colonies that accept new queens if there is a strong likelihood that a colony will lose its queen or if queen lifespan is short relative to colony lifespan. Under these circumstances, the colony may get potential long-term reproductive benefits by accepting new queens as reproductives. Moreover, established colonies may be selected to accept additional queens produced by that colony if the probability of these daughter queens surviving colony foundation on their own is very low. Finally, recruitment of multiple queens may be selected for genetic reasons. For example, workers in genetically diverse colonies may undertake a greater range of tasks or may better resist parasites than workers in less diverse colonies. Indeed, genetic diversity mediated by queen mate number benefits honeybee (*A. mellifera*), bumblebee (*Bombus terrestris*), and harvester ant (*Pogonomyrmex occidentalis*) colonies.

Regardless, under almost all of these circumstances, colonies should only accept their own relatives as new queens. By doing so, workers within the nest still obtain indirect, kin-selected benefits from rearing brood of the new queens. In contrast, acceptance of unrelated queens into nests would generally be detrimental to members of the established colony as this leads to helping behavior directed at nonrelatives and produces no indirect fitness benefits for workers. Unrelated new queens can thus be viewed as social parasites within the colony.

Empirical evidence shows that cohabitating queens in most eusocial insect colonies are related, as predicted. However, there are exceptions. Some ants form supercolonies that contain large numbers of unrelated queens. For example, invasive, polygyne *S. invicta* fire ants sometimes form large unicolonial populations where individuals move freely among nests. Introduced Argentine ants, *Linepithema humile*, represent an even more extreme example, as their supercolonies can span hundreds of kilometers. Conflict among queens and workers under these circumstances is expected and should disrupt the social structure within colonies. Thus, the evolutionary persistence of such colonies represents somewhat of a puzzle.

The formation of polygyne colonies frequently leads to a great number of changes in the life history and behavior of colony members. For example, queens in polygyne species tend to have shorter lifespans than queens in related monogyne species. Polygyny also entails queens having varied dispersal and mating strategies that include adoption into established nests and sometimes mating within the nest. Polygyne queens are also usually smaller and less fecund than their monogyne counterparts.

Queen number also affects worker morphology and behavior. For example, workers in polygyne species tend to be smaller and display lower levels of polymorphism than workers in monogyne species. In addition, new polygyne colonies sometimes form through the budding off of existing colonies, in contrast to monogyne colonies, which are almost always established through independent founding events.

Nevertheless, polygyny is taxonomically widespread. Many ants, such as some species in the genera *Formica*, *Leptothorax*, *Linepithema*, *Myrmica*, and *Solenopsis*, are polygyne for part of their life cycle or show variation in queen number among colonies. Many eusocial wasps, particularly in the subfamily Polistinae, also form polygyne associations. For instance, multiple queens often jointly initiate colonies in some *Ropalidia*, *Polistes*, *Parapolybia*, or *Belonogaster* wasp species. Eusocial wasp colonies may also become polygyne through adoption of daughter queens as they mature. For example, *Vespula* wasp colonies are typically initiated and headed by only a single queen. But multiple, daughter queens may be recruited as reproductives in some species if the colony persists for more than a single season (**Figure 2**). In addition, some species of *Ropalidia* and *Mischocyttarus* wasps display serial polygyny, whereby only a single queen exists within a colony at any one time, but is frequently replaced by relatives from within the nest. A few species of Halictid bees also form semipermanent, polygyne associations, although these tend to be rather limited in size and scope. Finally, some termites form colonies containing multiple-functional queens, kings (male termite primary reproductives), or other secondary reproductives. The secondary reproductives, which are a common feature of many termite species, frequently arise

Figure 2 Inside the nest of a yellowjacket (*Vespula maculifrons*) social wasp. *Vespula* colonies are generally headed by a single reproductive queen (lower right). However, the colonies of some species can adopt the queen's daughters (center) as reproductives under unusual circumstances such as when a colony survives for more than a single season.

through inbreeding events and allow colonies to persist long after the primary reproductives have died.

Conflict over Becoming a Queen

A major source of conflict within eusocial insect societies centers on whether a given individual will develop into a queen or worker. Workers, by definition, are subfertile. In contrast, queens are reproductively capable. Thus, theory predicts that females should develop into queens if possible. So why do individuals develop into workers?

A likely proximate reason why most developing females develop into workers is that individuals typically cannot control their own development in most eusocial insect species. Rather, developing individuals are almost always reared directly by workers that control the caste fate of offspring and rear brood according to the interests of the colony. Consequently, the castes of individuals produced in colonies typically result from resolution of conflicts between existing queens and workers, rather than the developing individuals.

Nevertheless, in a few eusocial insect species, developing individuals can control their own fate. For example, female *Melipona* bees decide their own caste. This unusual situation arises because *Melipona* larvae are not progressively provisioned with food by workers as they develop. Instead, the larvae receive all of the food necessary for growth at the beginning of their developmental phase. Thus, larvae are capable of choosing how much food to ingest and determining their final body size, which is related to whether they become queens or workers. This situation leads to an overproduction of queens, as individuals vie to become reproductives within the population. Excess queens, however, are killed by resident workers. Caste may also be self-determined in a few other eusocial insects including some termites. In these cases, colonies also tend to greatly overproduce reproductives because individuals attempt to selfishly develop into queens and kings.

Conflict over Reproduction

When multiple queens coexist within a single colony, conflict necessarily arises over how reproduction is shared. Indeed, queens should be under strong selection to dominate reproduction in polygyne colonies. Variation in reproductive success is quantified as reproductive skew. High-skew societies have one or a few queens that dominate reproduction at the expense of other queens. In contrast, queens in low-skew societies share reproduction more equally.

Theoretical analyses dissecting the factors that affect skew have been very helpful in understanding queen conflict. Skew theory integrates genetic, ecological, and social factors to reveal what features of social living affect reproductive success. Moreover, skew models predict conditions under which queens should peacefully coexist, partition reproductive success, fight with one another, or disperse from the colony to attempt to found colonies independently. Results of modeling suggest that important factors affecting skew include (1) the expected reproductive success of a queen that chooses to reproduce alone, (2) a colony's productivity if a new queen joins an existing colony, (3) the relatedness between interacting queens, and (4) the probability that a new queen will win a fight with an already existing queen.

Data from several species of eusocial wasps, eusocial bees, and ants suggest that skew is associated with ecological or genetic factors, as expected under the theoretical models. For example, skew among *Polistes fuscatus* wasp queens is directly related to the relatedness among queens and the overall colony productivity. High skew in *Halictus ligatus* bee associations is correlated with a low probability of successful nest founding, as expected if females that join an existing colony must accept low reproductive success if they have little hope of founding a colony on their own. Similar results have been observed in *Leptothorax* ants, in which skew increases with greater ecological constraints on independent colony founding.

In addition, aggressive interactions among queens arising from reproductive partitioning are observed in some polygyne insect societies. For example, antagonistic behaviors among *Polistes* wasp queens often lead to dominance hierarchies, whereby one queen physically dominates her rivals and ultimately obtains greater reproductive success. Alternatively, aggression can lead to eviction of queens, ovarian regression, or even death. As predicted under some models, aggression in eusocial wasps is sometimes associated with factors such as relatedness among queens and ecological constraints. Interestingly, more complex polygyne societies, like those found in some ants, show reproductive skew but no obvious signs of aggression. In these cases, conflict is apparently resolved indirectly through pheromones.

Queen conflict may also occur through variation in the fate of brood. Eggs produced by queens may be eaten by rivals or workers, or otherwise show differential viability leading to variation in queen reproductive success. Competition among queens may even take place during larval development. In particular, queens may vary in the proportion of the different castes they produce. Queens that exclusively produce workers ultimately obtain low fitness, because workers are, by definition, subfertile. However, queens whose brood is reared as new queens or males achieve higher fitness. Investigations in several polygyne ant taxa have discovered that queens may indeed vary in their contributions to queens, workers, and males within colonies.

A final, very unusual, means of indirect queen conflict is found in the ant *Cardiocondyla*. Queens in polygyne colonies produce sons very early in the colony life cycle. These males kill competitor males within the colony and subsequently mate with existing virgin females. Thus, conflict among queens in this species occurs through the production of precocious, aggressive sons and leads to greater reproductive success of some queens than others.

Conflict over Nepotism

The presence of multiple queens within colonies greatly complicates the dynamics of insect societies. In particular, polygyne colonies contain multiple different families. Thus, polygyny may lead to conflict among subfamilies within colonies. Such conflict can potentially be resolved if individuals help or cooperate solely with members of their own subfamilies. Workers produced by particular queens may attempt to direct their helping behavior to other offspring of that queen rather than to individuals that are less related. Such nepotism can be viewed as an indirect mechanism of queen conflict, which gets played out through the queens' offspring.

Surprisingly, there is very little evidence of nepotism within eusocial insect colonies. The lack of nepotism may be a byproduct of workers' inabilities to differentiate close relatives. This failure to discriminate based on relatedness may actually be selected at the colony level if nepotistic acts ultimately prove detrimental to colony performance. In addition, recognition errors could be costly leading to failure to correctly help relatives.

Conclusion

Members of eusocial insect colonies normally display remarkable helping and cooperative behaviors. However, the selective pressures associated with possible gains in reproductive success that arise through conflict always exist. Indeed, the possibility that potential conflict could erupt into actual conflict represents a pervasive problem within insect societies. Conflict among queens is one of the most important and obvious types of conflict within insect colonies. Queen conflict plays a central role in shaping the life cycle of eusocial insect queens and the dynamics within eusocial insect societies. Ultimately, social order within insect colonies partly rests on a delectate balance between selection operating at the level of individual queens, which promotes selfish behaviors that could break down the social order, and selection operating at the level of the colony, which maintains cooperative and helping behaviors.

See also: Ant, Bee and Wasp Social Evolution; Colony Founding in Social Insects; Cooperation and Sociality; Division of Labor; Kin Selection and Relatedness; Levels of Selection; Queen–Worker Conflicts Over Colony Sex Ratio; Reproductive Skew; Social Insects: Behavioral Genetics; Social Selection, Sexual Selection, and Sexual Conflict; Unicolonial Ants: Loss of Colony Identity.

Further Reading

Beekman M and Ratnieks FLW (2003) Power over reproduction in social Hymenoptera. *Philosophical Transactions of the Royal Society of London* 358: 1741–1753.

Bernasconi G and Strassmann JE (1999) Cooperation among unrelated individuals: The ant foundress case. *Trends in Ecology & Evolution* 14: 477–482.

Bourke AFG (2005) Genetics, relatedness and social behaviour in insect societies. In: Fellowes MDE, Holloway GJ, and Rolff J (eds.) *Insect Evolutionary Ecology*, pp. 1–30. Cambridge, MA: CABI Publishing.

Bourke AFG and Franks NR (1995) *Social Evolution in Ants*. Princeton, NJ: Princeton University Press.

Crozier RH and Pamilo P (1996) *Evolution of Social Insect Colonies: Sex Allocation and Kin Selection*. Oxford: Oxford University Press.

Hölldobler B and Wilson EO (1990) *The Ants*. Cambridge, MA: The Belknap Press of Harvard University Press.

Johnstone RA (2000) Models of reproductive skew: A review and synthesis. *Ethology* 106: 5–26.

Keller L (1993) *Queen Number and Sociality in Insects*. Oxford: Oxford University Press.

Keller L (1999) *Levels of Selection in Evolution*. Princeton, NJ: Princeton University Press.

Queller DC and Strassmann JE (1998) Kin selection and social insects. *Bioscience* 48: 165–175.

Queller DC and Strassmann JE (2002) The many selves of social insects. *Science* 296: 311–313.

Ratnieks FLW, Foster KR, and Wenseleers T (2006) Conflict resolution in insect societies. *Annual Review of Entomology* 51: 581–608.

Reeve HK and Keller L (2001) Tests of reproductive-skew models in social insects. *Annual Review of Entomology* 46: 347–385.

Ross KG and Matthews RW (1991) *The Social Biology of Wasps*. Ithaca, NY: Comstock Publishing Associates.

Queen–Worker Conflicts Over Colony Sex Ratio

N. J. Mehdiabadi, Smithsonian Institution, Washington, DC, USA

Introduction

Studies of sex-ratio conflict in hymenopteran insect societies (ants, bees, and wasps) highlight the importance of understanding how colony members resolve conflicts to maintain cooperation. The typical and simplest colony type consists of the sole reproductive (i.e., the queen) and her nonreproductive daughters (i.e., the workers) that perform various tasks for colony maintenance and survival. One of these tasks includes brood care of their immature brothers and sisters, some of which become reproductives of future colonies. Understanding why workers forgo their own reproduction to help their mother queen produce more young is a major question in evolutionary biology and animal behavior.

Theory

In their seminal paper in 1976, Trivers and Hare predicted that extreme altruism by workers is more likely to evolve in Hymenoptera than in most other animals because of their unusual sex determination system. Hymenoptera, unlike diploid organisms, are haplodiploid, meaning that females develop from fertilized eggs and are thus diploid, whereas males develop from unfertilized eggs and are haploid. This sex determination system renders workers more closely related to their sisters ($r = 0.75$) than they would be to their own daughters ($r = 0.50$), assuming the simplest case of a hymenopteran society with a singly mated queen and her workers (i.e., a monogynous colony). To maximize genetic representation in future generations, selection should favor workers helping their mother queen produce more young (particularly their sisters) rather than reproducing themselves.

Despite this extreme level of altruism exhibited by workers, the haplodiploid system of sex determination also sets up the potential for conflict between the queen and workers over sex ratios (**Figure 1**), workers being more closely related to their sisters than to their brothers (on average, three times more related in monogynous colonies) and the queen equally related to her sons and daughters. Expanding on Fisher's sex-ratio theory and Hamilton's kin selection theory, Trivers and Hare therefore predicted that workers should bias sex ratios toward female reproductives and the queen should counter this female bias. Fisher's basic sex-ratio argument predicted for diploid species that sex ratios should stabilize at 1:1

(female:male) for the following reasons: each offspring derives from the pairing of a female and a male, and each sex, thus, produces overall the same total number of offspring; any deviations from an even sex ratio are unstable, because negative frequency-dependent selection gives the rarer sex a reproductive advantage over the more common sex, ultimately leading to equal sex ratios at the population level. However, hymenopterans are haplodiploid and different sex ratio predictions arise because of asymmetries in relatedness of colony members to one another. Trivers and Hare combined their knowledge of Fisher's sex ratio theory with Hamilton's kin selection theory (inclusive fitness theory), which states that altruists can gain indirect fitness benefits by helping their relatives. Thus, in colonies headed by a singly mated queen, workers should be selected to rear three reproductive females for every reproductive male (assuming an equal cost in the production of male and female reproductives), given that workers share three times as many genes with their sisters as with their brothers. The equilibrium is 3:1 (female:male) because genetic value balances reproductive value: the genetic value of a sister is three times as valuable as that of a brother, whereas the reproductive value of a brother is three times the reproductive value of a sister (i.e., an average male is expected to have three times the offspring compared to the average female).

To test their theory, Trivers and Hare examined empirical data from 20 species of monogynous ants and found that population-level sex ratios were indeed female-biased with an average value of about 3:1 (female:male). With this evidence, the authors then concluded that queen–worker conflict over the sex ratio had been demonstrated and that such conflict was resolved in favor of workers in these ant species.

This work stimulated numerous empirical and theoretical studies. In 1986, Nonacs later expanded the dataset of Trivers and Hare and confirmed the predicted 3:1 (female:male) bias. However, one paper in particular by Alexander and Sherman a year after Trivers and Hare's study claimed that another potential factor, other than genetic relatedness, could also explain the observed sex ratio patterns, namely local mate competition where brothers compete for matings with future queens. Such competition for mates for hymenopteran males reduces the relative value of son reproductives compared to daughter reproductives, and thus, sex-ratio interests of queens and workers become more closely aligned and female-biased sex ratios are considered optimal for both

Figure 1 Queen and workers of fungus-growing ant species *Acromyrmex coronatus*. Photo by Jeremy Harrison.

parties. Under local mate competition, brothers compete more with each other than sisters compete with each other. It can be looked at as a kin-selected process where mothers try to minimize the negative effects that offspring have on each other. Because sons compete with each other and daughters do not compete with each other (or compete less), mothers minimize negative effects between relatives (i.e., offspring) by biasing the sex ratio. Local resource enhancement, another hypothesis proposed to explain sex-ratio bias in Hymenoptera, predicts the opposite: under local resource enhancement, mothers maximize the positive effects that offspring have on each other; daughters cooperate synergistically to enhance their joint reproduction, increasing the relative value of daughters.

Distinguishing between which of these different hypotheses modulate sex ratios in various Hymenoptera still continues in the literature. However, most theories of sex-ratio conflict focus on how relatedness asymmetry affects sex allocation. The degree of relatedness asymmetry varies across the Hymenoptera, depending on the social structure of the colony. For example, the presence of more than one reproductive queen in a colony (i.e., polgyny) and/or a queen that has mated with multiple males (i.e., polyandry) reduces the asymmetry in relatedness of workers to the average female versus the average male produced by a colony and more closely aligns sex-ratio interests of queens and workers. However, it is important to note the following: In a polygynous colony, not all males produced by a colony are true brothers to all workers; some males are unrelated to some workers (assuming that the queens are unrelated). More importantly, in a polygynous colony with unrelated queens, each worker should still bias the sex ratio to 3:1 (female:male); by doing so, the worker will bias the sex ratio of the reproductives to which a worker is related (and the incidental effect on the sex ratio of the unrelated sexuals is irrelevant). Thus, polygyny matters only if the queens are related; then the relatedness

asymmetry changes with the number of related queens. This highlights the importance of discounting all potential confounding factors that could explain sex ratio patterns before claiming that queen–worker conflict over the sex ratio exists for a given species.

Effects of Environmental Factors on Sex Ratios

Social factors, such as genetic relatedness, have played an important role in explaining sex allocation patterns in Hymenoptera. However, recent studies are elucidating the importance of testing the effects that various ecological factors may have on sex ratios, for example, food availability, nest density, temperature, and colony size. In bumblebees, which are usually monandrous but sometimes show a small degree of polyandry, colonies with greater nectar and pollen are more likely to produce female reproductives than colonies with less nectar and pollen. A similar pattern on sex ratios has been found in the ant *Formica podzolica*; colonies with supplemented food tended to produce female-biased sex ratios and colonies without supplemented food tended to produce male-biased sex ratios. This effect is consistent with the hypothesis that colonies should invest more in the cheaper sex (i.e., males, which are generally smaller than females in Hymenoptera) when resources are limited. However, this Trivers–Willard resource constraint hypothesis can actually be argued as important only for species that make one (or a few) reproductives per reproductive bout (e.g., red deer). In bumblebees, dozens of males and females are produced at the same time, and the Trivers–Willard effect seems irrelevant; if food-constrained, why not just make a few less sexuals but keep the sex ratio constant? However, making fewer offspring is not an option for red deer when producing only one offspring per reproductive bout; there the only option is to make either a male or a female, depending on what resources the mother can afford in a given year and which sex will convert the investment into the greatest number of grand-offspring. In addition to food, temperature can also influence sex ratios. In a population study over multiple years of a primitively social halictine bee, *Halictus rubicundus*, warmer temperatures correlated with increasing male-biased sex ratios. However, in the Argentine ant, *Linepithema humile*, cooler temperatures resulted in more male-biased sex ratios than warmer temperatures. Thus, studies that test the independent and interactive effects of social and ecological factors on sex ratios are needed.

Mechanisms of Sex-Ratio Bias

One important way that queen–worker conflict can be empirically tested is to determine the mechanism(s) by which the sex-ratio bias is achieved by the queen or workers. Sex ratios can be manipulated by the queen at the egg-laying stage (i.e., the proportion of unfertilized vs. fertilized reproductive-destined eggs laid by the queen) also known as the primary sex ratio, or presumably by workers at later developmental stages also known as the secondary sex ratio. Thus, a comparison between primary and secondary sex ratios can provide a powerful test of whether queen–worker conflict exists over the sex ratio given that both the queen and workers have power to bias sex ratios at different developmental stages of the reproductive brood (**Figure 2**).

Because workers care for the brood, they have the ability to control the sex ratios of queen-laid eggs (**Figure 3**). Several mechanisms of worker control have been empirically demonstrated in various Hymenoptera. One such example includes the selective elimination of male reproductives.

Figure 2 Queen and workers of fungus-growing ant species *Acromyrmex coronatus* shown with brood. Photo by Jeremy Harrison.

Figure 3 Workers of *Cyphomyrmex* spp. embedded in their fungal garden with winged adult reproductives. Photo by Jeremy Harrison.

This has been shown in a number of different ant species, and furthermore, in the wood ant *Formica exsecta*, workers appear to do this facultatively; male brood were eliminated by workers in colonies with a singly mated queen, but not by workers in colonies with a doubly mated queen. However, a recent study in honeybees, which have multiply mated queens, found facultative male brood elimination even though theory predicts that queen–worker conflict should be minimal or absent. Workers were more likely to cannibalize males if older male brood was more abundant. This behavior was attributed to environmental factors rather than to queen–worker conflict – male elimination was beneficial to both the queen and workers because of increased colony efficiency. Thus, this work on honeybees emphasizes the importance of teasing apart and testing various factors that could possibly affect sex ratios and points out the value of manipulative experiments for understanding cooperation and conflict in hymenopteran societies.

Other mechanisms of worker control include preferential rearing and feeding of female reproductives, influencing female caste fate to force worker-destined brood to become reproductive females, imprisoning males to limit their access to food, and distancing themselves and brood from the queen (to prevent sex-ratio manipulations by the queen). Most of these sex-ratio biasing mechanisms employed by workers imply that workers can distinguish between male and female brood in order to manipulate sex ratios in their favor. However, virtually nothing is known on the chemicals or cues that workers might use to discriminate between male and female brood; this could be a fruitful area for future research.

The queen also has power to control sex ratios. However, this occurs at the earlier developmental stages of the reproductive brood (i.e., the egg stage). Studies of hymenopteran populations characterized by split sex ratios have recently demonstrated mechanisms of queen control over sex ratios. Split sex ratios occur in populations with two types of colonies, one type specializing in the production of female reproductive broods (i.e., female-specialist colonies) and other colonies that specialize in the production of only male reproductive broods (i.e., male-specialist colonies). Under a standard split-sex ratio scenario, one type of colony makes 100% sexuals of one kind (this is called the biasing class of colonies); the other type (the so-called balancing class) produces both males and females, depending on their best interests in adjusting the reproductive value of their offspring. This type of specialization in the production of one sex or the other provides a powerful test of queen–worker conflict over the sex ratio. However, it is important to note that split sex ratios can arise because of resource differences as explained earlier (small colonies across ants tend to make males, while large colonies tend to have a higher proportion of females).

The bumblebee *Bombus terrestris* exhibits split sex ratios and is a model for investigating queen–worker conflict. The society consists of a singly mated queen and workers are capable of laying haploid male eggs. Queens appear to have some control over sex allocation by forcing workers to rear her males in male-specialist colonies. Queens achieve this control by laying male eggs during the early phase of colony development when workers are presumably unable to distinguish the sex of brood. Later, it is too costly for the workers to replace her males with their own males.

Similar to *B. terrestris*, the monogynous form of the notorious invasive pest *Solenopsis invicta* is also characterized by split sex ratios. Passera and colleagues performed a clever laboratory transplant experiment in which they took queens from male-specialist colonies and put them with workers from female-specialist colonies and vice versa. After some time, they recorded the type of reproductive brood produced by these experimental colonies and found that male-specialist colonies with female-specialist queens produced female-biased sex ratios, and female-specialist colonies with male-specialist queens produced male-biased sex ratios. This finding demonstrates that queens, not workers, control sex ratios, and that queens manipulate sex ratios by limiting the number of female eggs laid, forcing workers to rear males in male-specialist colonies.

In another ant species characterized by split sex ratios, *Pheidole desertorum*, queens can influence sex ratios by controlling their oviposition. Queens from male-specialist colonies affect female caste determination by producing worker-destined eggs instead of female reproductive-destined eggs. Nevertheless, other experiments have shown that workers of *S. invicta* and *P. desertorum* exert some control over sex ratios by preferentially rearing female reproductive brood over male brood. These findings demonstrate queen–worker conflict over the sex ratio in these two ant species.

A recent study showed that the queen has the potential to influence female caste fate in *Pogonomyrmex* harvester ants. Once again, by performing cross-fostering experiments, Schwander and colleagues revealed that maternal, rather than nutritional, factors influence female caste determination: female reproductives were produced when the queen was exposed to cold temperatures and when the queen was at least 2 years old. These results emphasize the need for a reevaluation of the traditional view that queens have little influence over female caste fate, which can have important consequences in queen–worker conflict over the sex ratio.

Other Intracolony Conflicts

Queen–worker conflict over the sex ratio is just one of several examples of potential conflicts among colony members. Other major types of conflict that can occur include conflict among workers or multiple queens over queen rearing, conflict among workers or between workers and the queen over the production of males, conflict among totipotent individuals over reproduction, and conflict over life-history decisions.

Conflict over Queen Rearing

Conflict over queen rearing among workers or multiple queens can occur if a queen has mated with multiple males or if there are multiple reproductive queens in a colony. Such conflict has been studied in a variety of Hymenoptera, including bees, wasps, and ants. However, most studies have failed to find strong evidence for nepotism. Lack of recognition or potential recognition errors in discriminating full sisters from half sisters may explain why evidence for conflict over queen rearing is weak or absent in Hymenoptera.

Conflict over Male Production

Another form of conflict among colony members is the potential for conflict over the production of males. In some hymenopterans, workers can lay unfertilized male eggs and thus compete with the queen (or even with other workers) over male production. This type of conflict has been frequently observed in various Hymenoptera, including bees, wasps, and ants. Chemical cues and ritualized behaviors are mechanisms that have been shown to signal conflict over male production in ants and bees. Such conflict can arise because of haplodiploidy, where, for example in monogynous colonies, workers are more related to their own sons ($r = 0.50$) than they would be to their brothers ($r = 0.25$), yet the queen is more related to her own sons ($r = 0.50$) than she would be to her grandsons ($r = 0.25$). Relatedness asymmetries change depending on the social structure of the colony. For example, if a queen has mated with multiple males, then workers are on average more related to the queen's sons ($r = 0.25$) than to other worker-produced sons ($r < 0.25$), which selects for workers preventing one another from laying male eggs (i.e., worker policing). In line with this theory, worker policing of male production has been found to occur in polyandrous species, such as honeybees, but not in monandrous species (e.g., stingless bees). However, recent work by Wharton, Dyer, and Getty claims that other factors, including colony efficiency, may also limit worker reproduction. That is, worker policing may be selected for if colony growth and efficiency is reduced because of workers spending their energy on reproducing rather than on performing tasks for colony maintenance and survival.

Conflict Among Totipotent Females over Reproduction

Unlike typical hymenopteran societies, in some bees, wasps, and queenless ant species, females in a colony are not specialized for worker or queen roles. These females have little morphological differences, are able to mate, and thus produce male and female offspring. This is referred to as totipotency, and conflict among totipotent individuals over reproduction can occur. Examples of such conflict are similar to those that can occur among queens in a polygynous colony. For instance, totipotent females can compete over reproduction, resulting in aggressive interactions over who becomes the 'breeder.' Cases of totipotency are particularly useful for understanding the evolution of morphological castes and caste conflict.

Conflict over Life-History Allocation

Life-history allocation refers to a colony's investment of resources into growth (i.e., production of workers) versus reproduction (i.e., production of sexuals). This type of queen–worker conflict is argued to occur in hymenopteran societies with complex social structures because queen and worker interests are more aligned in monogynous, monandrous colonies. For example, for societies where queen replacement can occur (e.g., facultative polygynous colonies), queens and workers are predicted to be in conflict over life-history allocation with the existing queen favoring investment into workers as long as possible so that newly produced reproductives cannot possibly replace her; workers, on the other hand, favor the opposite scenario.

Integrating Intracolony Conflicts

As already noted, queen–worker conflict over the sex ratio is just one of several forms of conflict that can occur in hymenopteran societies, yet each are interrelated with one another. All potential conflicts are predicted to arise because of workers and queens having different selective interests that result in an increase in their own fitness, the ultimate measure of evolutionary success. The magnitude of this difference presumably affects the level of conflict observed between colony members. For example (as noted), based on the relatedness hypothesis, queen–worker conflict over the sex ratio is more likely to be observed in monogynous colonies with a singly mated queen than in monogynous colonies with a multiply mated queen because of the increased relatedness asymmetry of workers to their sisters versus their brothers in monogynous, monandrous societies versus monogynous, polyandrous societies. This in turn, results in queens and workers

having very different sex-ratio interests, and in this circumstance, theory predicts conflict over sex ratio of reproductives produced by the colony.

Conclusion and Future Directions

Over three decades of research on queen–worker conflict over the sex ratio have enlightened our understanding of conflict and its resolution in hymenopteran societies. This work has also provided some powerful tests of Hamilton's kin-selection theory. However, so much more is yet to be discovered.

Areas of future research include integrating sex-ratio conflict with other queen–worker conflicts for a better understanding of conflict resolution in hymenopteran societies. The integration of multiple queen–worker conflicts allows for a more realistic approach toward understanding conflict evolution and might reveal new predictions on how conflict resolution is achieved. Some recent theoretical models have already taken such an approach.

Future research will undoubtedly need to address further the mechanisms by which queens and workers achieve control over sex ratios – more specifically, what chemical, physiological, and genetic cues underlie these mechanisms of control and conflict. Advances in chemical ecology are likely to reveal novel mechanisms of how queens and/or workers achieve control, which presumably have some sort of chemical basis given that this is the primary mode of communication for social insects. For example, what exactly are the recognition cues that queens and/or workers use to distinguish between male and female brood, and thus, allows them to manipulate colony sex ratios in their favor? Also, endocrine analyses are increasingly being performed to investigate how hormone levels (e.g., juvenile hormone) influence queen–worker conflict. Finally, as the field of genomics increases at a rapid pace, molecular tools will become more available to uncover the genes behind such conflict. Once the genes, chemicals, and hormones are discovered, then these genetic, chemical, and physiological studies in combination with traditional animal behavior experiments are likely to provide much promise in improving our understanding of how conflicts of interest get resolved.

See also: Ant, Bee and Wasp Social Evolution; Kin Selection and Relatedness; Queen–Queen Conflict in Eusocial Insect Colonies; Worker–Worker Conflict and Worker Policing.

Further Reading

Alexander RD and Sherman PW (1977) Local mate competition and parental investment in social insects. *Science* 196: 494–500.
Boomsma JJ (1989) Sex-investment ratios in ants: Has female bias been systematically overestimated? *American Naturalist* 133: 517–532.

or logic in thinking out a problem' or 'Endowed with the capacity to reason.' Both these definitions refer to reasoning, which in turn is defined as the ability to think, or to draw conclusions from known facts. These layman's definitions of rationality correspond quite closely with the way philosophers and cognitive psychologists use the term. Interestingly, they are also similar to the definition adopted by Charles Darwin who wrote in his notes, "Rational actions ... are actions which are required to meet circumstances of comparatively rare occurrence in the life-history of the species, and which therefore can only be performed by an intentional effort of adaptation ... rational actions ... serve to meet novel exigencies which may never before have occurred even in the life-history of the individual." Darwin went on to argue that rational action, "Implies the conscious knowledge of the relation between means employed and ends attained" (Darwin cited in Romanes (1882)). In defining rational actions, Darwin contrasts them with what he refers to as 'reflexes' and 'instinctive actions.'

Thus, in describing the process of choice as rational, Darwin and others are implying the use of cognitive mechanisms that we might describe as 'clever' or 'intelligent.' By this, we mean mechanisms that represent information about the state of the world and the goals of the animal, and use this information in a flexible way to solve novel problems effectively. Darwin's definition also implies that conscious intention has to be present for rationality. However, most modern biologists are not happy with the notion of ascribing conscious intentions a causal role in the generation of behavior. Consciousness is a private experience, and consequently we can never objectively observe or measure it in animals. Therefore, most modern research in animal cognition distinguishes between the study of information processing in animals and the study of consciousness. We can ask how animals acquire, represent, and use information in the generation of behavior without asking whether or not this happens via some conscious process. Thus, for the purposes of this study, we will define a *choice process* as rational if the resulting behavior displays evidence of flexible, goal-directed information processing based on representations of the state of the world.

Rationality of Choice Outcomes

The second use of the term 'rationality' focuses on the alternatives an animal actually chooses, as opposed to the processes responsible for choice. This use of rationality therefore refers to directly observable behavior rather than unobservable cognitive processes. An individual's behavior is defined as rational if it is compatible with the individual maximizing a currency of some type, resulting in internally consistent decisions. This definition of rationality has its roots in microeconomic theory and

has only relatively recently been explicitly considered in the context of animal behavior. I will therefore start by describing what rationality means in economics before exploring how we can apply the concept in biology.

Economic rationality

The theory of individual decision making developed in microeconomics starts by considering the problem of choosing from among a set of mutually exclusive alternatives (similar to the egg-choice problem with which I opened this study). Economic models of choice assume that when making such choices human consumers maximize a quantity called 'utility.' One can think of utility as a measure of the relative satisfaction an individual derives from a specific resource. However, it is important to realize that utility cannot be measured independent of what people actually choose. Rational choice is simply defined as choice behavior that is compatible with the maximization of utility. If an individual maximizes utility, or indeed any other currency, their choice behavior will be internally consistent in various ways that are considered to be hallmarks of rational choice. These hallmarks include the properties of transitivity, independence from irrelevant alternatives, and regularity. I will briefly describe each of these properties in the following paragraphs.

Transitivity is a property that applies specifically to binary choices. Preferences are transitive between the three options A, B, and C if A is preferred to B, B is preferred to C, and A is preferred to C. For example, if binary choices reveal that I prefer a cherry to a pear, and a pear to an apple, then if my choices are transitive I should prefer a cherry to an apple (see **Figure 1(a)** for an example). If I showed the opposite preference and preferred the apple to the cherry, this would constitute a violation of transitivity (see **Figure 1(b)** and **1(c)**).

Independence from irrelevant alternatives is a property that applies when a choice set is expanded. It implies that the preference between two options should be independent of the presence of additional inferior alternatives. For example, if A is preferred to B in the binary choice of A versus B, then the introduction of option C should not affect the preference for A over B (see **Figure 2(a)** for an example). If the relative preference for A over B is altered by the addition of C, this is referred to as a 'violation of the constant ratio rule' (**Figure 2(b)**). If the absolute preference for either A or B increases when C is added to the choice set, this is referred to as a 'violation of regularity' (**Figure 2(c)**).

Thus, we can summarize the economists' definition of rationality as follows. In economics, rationality describes the internal consistency in an individual's choices that results if they are maximizing a currency known as 'utility.' Economists consider transitivity and regularity to be fundamental features of rational choice. Given that one cannot measure utility directly, assessing transitivity and regularity

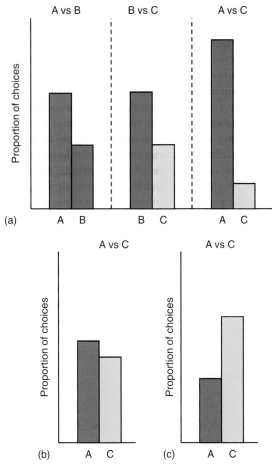

Figure 1 (a) An example of transitive choice: A is preferred to B, B to C, and A to C; (b) an example of a violation of strong stochastic transitivity: the preference for A over C is less than the preference for A over B or B over C; (c) an example of a violation of weak stochastic transitivity: C is preferred to A.

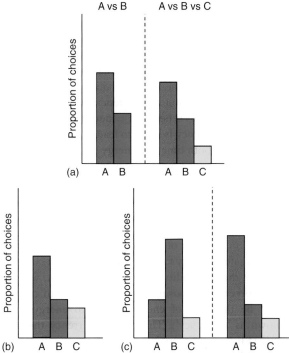

Figure 2 (a) An example of independence from irrelevant alternatives. When the inferior option, C, is added to the binary choice of A and B, the relative preference for A over B remains unchanged; (b) an example of a violation of independence from irrelevant alternatives: the addition of C leaves the proportion of choices for A unchanged but reduces the proportion of choices for B, leading to a violation of the constant ratio rule; (c) two examples of violations of regularity: the addition of option C either increases the proportion of choices for B (first panel) or A (second panel).

will often be the only way to test whether human consumers maximize utility.

Biological rationality and optimal foraging theory

Given the strong superficial similarity between the kinds of choices faced by humans and animals, it is perhaps not surprising that research on animal choice has drawn heavily on the theories developed to model the behavior of human consumers in microeconomics. However, a major difference between biological and economic models of choice is that they assume different currencies of maximization. In animal behavior, we start with the basic assumption that an animal's behavioral repertoire is ultimately the product of evolution by natural selection. Natural selection favors genetic variants with the highest inclusive fitness; thus, the behavior of an animal observed in the context in which it has evolved should ultimately maximize its inclusive fitness. By analogy with the economists' definition earlier, we can think of an individual that behaves in a way that maximizes its inclusive fitness as biologically rational. Since we assume

that all animals should be ultimately biologically rational, this is in some sense a trivial definition. Ultimately, however this does not matter, because behavioral ecologists take biological rationality as their starting point for more detailed analyses of behavior; the basic assumption of ultimate biological rationality is not under test.

Unlike utility, biologists can, in principle, measure inclusive fitness. However, inclusive fitness is unlikely to be an appropriate currency for computing the costs and benefits of alternative decisions in many circumstances. For example, assume that we want to understand the moment-to-moment flower choices of a foraging hummingbird. The inclusive fitness consequences of the bird choosing a specific flower type are hard for us to measure, because in order to estimate these it would be necessary to record the lifetime reproductive success of birds that fed on this flower type compared with birds that fed on another flower type. Similarly, the hummingbird cannot use its inclusive fitness as the currency it is maximizing when it is making foraging decisions, because the consequences of its choices are not immediately translated into detectable changes in lower in the scale of psychological evolution and development" (Morgan, 1903, p. 59). Following this rule, there is little evidence for process rationality in animals that we could not explain by some simpler mechanism.

Evidence for Outcome Rationality in Animals

Outcome rationality is simpler to test than process rationality because tests rely on direct observations of what animals choose. As explained earlier, the assumption of biological rationality underpins the whole of behavioral ecology and is generally not directly tested. Instead, behavioral ecologists have focused on testing specific hypotheses about the proximate currencies animals maximize, and this approach has been extremely successful in showing how animal behavior is evolutionarily rational. However, a small number of more recent studies have set out to test whether animals are rational in the economists' sense.

Tests of economic rationality in animals have been inspired by examples of human irrationality. Experiments on human decision making have shown that we tend to make irrational choices when alternative options differ in more than one attribute (as in the egg example with which I started this article). When faced with decisions of this type, a rational decision maker should combine all the attributes into a single currency and choose the alternative that yields the highest value. For example, a hummingbird might choose from a set of flowers that differ in nectar volume, nectar concentration, and handling time. Under an optimal foraging account, we can summarize all these attributes in the single currency of net rate of energy intake. Using this currency, the hummingbird could compare the flowers and make a choice that maximizes net rate of energy intake. However, when humans face complex, multidimensional decisions, they often show violations of transitivity and tend to be influenced by the presence of irrelevant alternatives. For example, an experiment found that purchases of large cans of a high-quality, high-price brand of baked beans increased, and purchases of large cans of a low-quality low price brand decreased, when smaller, relatively more expensive cans of the same high-quality brand are added to the choice set. The small-but-expensive option is an irrelevant alternative, being more expensive and of no better quality than one of the other options, making this result a clear violation of regularity.

One explanation for this irrationality is that rather than combining the attributes into a single currency, we instead resort to simple heuristics for decision making. For example, we might simply choose the option that ranks highest on the greatest number of attributes, ignoring the absolute values of the various attributes. Such heuristics have the benefit of being fast and easy to compute, but they sometimes result in economically irrational choices. Therefore, experiments designed to look for economic rationality in animals have specifically focused on situations in which animals face choices between options that simultaneously differ in multiple attributes of interest.

Tests for transitivity of choice

Transitivity (not to be confused with transitive inference) is a property of a series of binary choices made between pairs of simultaneously presented mutually exclusive alternatives. Thus, tests of transitivity typically present animals with pairs of choices and study which option the animal prefers. In the first experiment explicitly designed to test economic rationality in animals, Sharoni Shafir presented foraging honeybees (*Apis melifera*) with a series of binary choices between pairs of artificial flowers varying in two attributes both known to affect bees preference: the corolla length and the nectar volume. He found some individual bees that preferred flower A to B, B to C, C to D, but also D to A. Preferring D to A violates what is known as 'weak stochastic transitivity.' Bees that violated weak stochastic transitivity also violated strong stochastic transitivity, meaning that the strength of preference between two flowers adjacent on the scale of utility (e.g., A and B) was larger than that between two more widely separated flowers (e.g., A and C). Similar results have also been found in foraging gray jays (*Perisoreus Canadensis*).

Tests for independence from irrelevant alternatives and regularity

Independence from irrelevant alternatives and regularity are properties of choice that emerge when we increase the number of alternatives in the choice set. Tests of regularity ask how adding a third alternative to the choice set (a ternary choice) affects preference between two options (a binary choice). My colleagues and I tested the preferences of foraging rufous hummingbirds presented with artificial flowers that offered different volumes and concentrations of nectar. In the binary treatment, birds chose between a high concentration flower (20 μl of 40% sucrose) and a high volume flower (40 μl of 20% sucrose), whereas in two ternary treatments we added a third flower type that was worse than either the high concentration or the high volume flower (10 μl of 30% sucrose and 30 μl of 10% sucrose, respectively) (see **Figure 4(a)**). The additional flowers should be irrelevant to the birds' preference between the high concentration (C) and high volume flowers (V) because they are clearly worse than one of these flowers on both dimensions. However, we found that the third flower type affected both the relative and the absolute preferences for the two options compared in the binary treatment (**Figure 4(b)**). Thus, the birds' preferences violated both independence from irrelevant alternatives and regularity. Interestingly, our results support the idea that the birds use a simple heuristic that ranks concentration and volume dimensions independently,

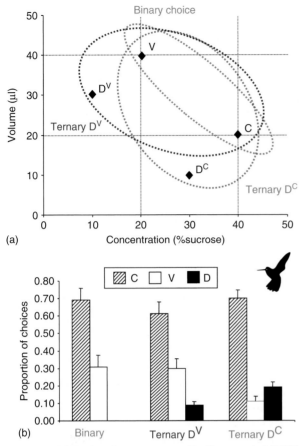

(a)

(b)

Figure 4 (a) The four flower types used by Bateson et al. (2003) in their study of hummingbird foraging decisions; (b) results from the experiment: relative preference for V increases in the treatment with D^V, whereas relative and absolute preference for C increases in the treatment with D^C. Redrawn from Bateson M, Healy SD, and Hurly TA (2003) Context-dependent foraging decisions in rufous hummingbirds. *Proceedings of the Royal Society B* 270: 1271–1276.

because in both ternary choices, preferences shifted toward the flower with the highest relative ranks on both dimensions. Studies using foraging gray jays, honeybees, and starlings, as well as in female green swordtails (*Xiphophorus helleri*) and fiddler crabs (*Uca mjoebergi*) choosing their mates have reported similar results.

What Does It Mean if Behavior Is Irrational?

The experiments described earlier show that animals are sometimes irrational in the economists' sense. These results imply that animals do not necessarily assign absolute values to alternatives options, but instead the values assigned can depend on the specific set of alternatives available at the time of choice. However, it is important to understand that this in no way threatens our view as behavioral ecologists that animals are ultimately biologically rational. Context dependency could occur for a

number of different reasons. Earlier, I suggested that simple heuristics – such as preferring the option with the highest rank on all dimensions – could explain human and non-human irrationality. Although these heuristics might sometimes lead an animal to prefer a poor alternative (e.g., one that yields a lower rate of energy intake), we assume that natural selection has favored these heuristics because on balance they benefit the animal. Benefits could occur either via increased speed of decision making or a reduced requirement for computational resources in the brain. Other explanations for context dependency have also been suggested. In most studies, for example, the animal makes a sequence of choices, so the options chosen early in the sequence could change the animals state (e.g., reduce its hunger) and thus change the nature of optimal decisions later in the sequence. Thus, it might be possible to accommodate some apparently irrational behavior within a conventional optimal foraging framework.

In summary, although animal behavior can sometimes appear economically irrational, when we consider it in its full ecological context, the biological rationality should become apparent. The value of studying economic irrationality in animal decision making lies in what these studies can tell us about the proximate mechanisms underlying animal choices.

See also: Kin Selection and Relatedness; Niko Tinbergen; Optimal Foraging Theory: Introduction.

Further Reading

Bateson M (2004) Mechanisms of decision-making and the interpretation of choice tests. *Animal Welfare* 13: S115–S120.

Bateson M, Healy SD, and Hurly TA (2003) Context-dependent foraging decisions in rufous hummingbirds. *Proceedings of the Royal Society B* 270: 1271–1276.

Hurley S and Nudds M (2006) *Rational Animals.* Oxford: Oxford University Press.

Kacelnik A (1984) Central place foraging in starlings (*Sturnus vulgaris*). I. Patch residence time. *Journal of Animal Ecology* 53: 283–299.

McGonigle BO and Chalmers M (1992) Monkeys are rational! *The Quarterly Journal of Experimental Psychology* 45B: 189–228.

Morgan CL (1903) *An Introduction to Comparative Psychology,* 2nd edn. London: W. Scott.

Romanes GJ (1882) *Animal Intelligence.* London: Kegan Paul Trench & Co.

Schuck-Paim C, Pompilio L, and Kacelnik A (2004) State-dependent decisions cause apparent violations of rationality in animal choice. *Public Library of Science Biology* 2: e402.

Shafir S (1994) Intrasitivity of preferences in honey bees: Support for 'comparative' evaluation of foraging options. *Animal Behaviour* 48: 55–67.

Stephens DW (2008) Decision ecology: Foraging and the ecology of animal decision making. *Cognitive, Affective, & Behavioral Neuroscience* 8: 475–484.

Stephens DW and Krebs JR (1986) *Foraging Theory.* Princeton: Princeton University Press.

Vasconcelos M (2008) Transitive inference in non-human animals: An empirical and theoretical analysis. *Behavioural Processes* 78: 313–334.

Wynne CDL (2004) *Do Animals Think?* Princeton, NJ/Oxford: Princeton University Press.

Recognition Systems in the Social Insects

A. Payne and P. T. Starks, Tufts University, Medford, MA, USA

Introduction: A Wide-Angle View of Recognition Systems

Readers of scientific journals are no strangers to recognition: a recent keyword search by the authors turned up over 100 000 papers related to that subject. Of these, only a fraction were from the animal behavior literature; the balance came from journals of molecular biology, immunology, cellular medicine, or cognitive psychology. There were papers on human facial recognition and on the immune detection of viral proteins, papers on nepotism in ground squirrels, and on the specificity of restriction enzymes – papers, in other words, from almost every branch of biology. The only question was what, if anything, they had in common.

During the last 15 years, some researchers have begun to converge upon an answer. They argue that these studies are united both by a common theme and by the need for a common framework, and that the time has come for a unified approach to the study of recognition. They point out that all recognition research, no matter what the model system, is about explaining how evaluators identify and discriminate among entities. To these scientists, it makes little difference whether that evaluator is a red-tailed hawk or a human lymphocyte; both bird and blood cell have evolved to recognize significant entities accurately and efficiently. In this expanded view – what we call the 'wide-angle' approach to recognition – all biological recognition systems are variations on a theme, a theme best understood through a shared vocabulary and a common theoretical framework.

And yet, despite the promise of new insights and new collaborations, this wide-angle perspective remains rare. Even after well-reasoned arguments published by Blaustein and Porter in 1996 and Sherman and colleagues in 1997, many still think of kin recognition not as a limited case, but rather as the *only* case, of animal recognition. While this may be due to poor coverage in the textbooks – recognition is often given little more than a short mention in a chapter on kin selection – it also owes something to long-standing inconsistencies in the technical language, inconsistencies that persist despite efforts at standardization.

In this review, we explore how students of behavioral ecology might benefit by adopting the wide-angle approach. We offer a quick review of current terms and make the case for recognition as a ubiquitous biological process. We then illustrate the approach by bringing it to bear upon the study of social insects, and by showing how four seemingly unrelated behaviors – selecting nest sites, choosing mates, recognizing relations, and detecting parasites – all depend on the ability to recognize and discriminate.

Describing Recognition: The Need for a Common Vocabulary

Despite its widespread use in the literature, 'recognition' remains a vague term. Almost all of us have some sense of what it means to 'recognize an opportunity' or to 'recognize a face in a crowd,' but few, if pressed, could provide a rigorous definition. What do we really mean when we say we recognize an old friend on the train? For that matter, what do we mean when we say that a honeybee recognizes olfactory cues, or that a restriction enzyme recognizes a nucleotide sequence? If we expect to get a handle on recognition systems, we must first get a handle on the terms we use to describe them.

To that end, we propose the following definition: recognition occurs whenever an evaluating entity, regardless of its level of biological organization, identifies another entity with reference to a previously existing template. In short, the ability to recognize is the ability to identify encountered entities. While no one would argue that an enzyme identifies its substrate through the same mechanisms that we use to identify a car in a parking lot, the overall structures and the ultimate outcomes of those processes are the same. A recognition system does not require cognitive processing to achieve these goals: Neither gated channels nor restriction enzymes nor tRNA molecules require ghosts in the machines – or in this case, in the alpha helices – to meet the basic requirements for recognition.

Any attempt at synthesis must also designate those features that all recognition systems share. Perhaps the most obvious of these is the condition of having two or more participants, one that does the recognizing and at least one other that gets recognized. Following Liebert and Starks, we call these the 'evaluator' and the 'cue bearer,' respectively. These terms are particularly helpful as they replace a confusing set of words ('signaler' vs. 'receiver'; 'recipient' vs. 'actor') that grew out of the exigencies of communication and kin recognition work.

The act of recognizing also means that evaluators must possess criteria against which to judge cue bearers. Taken together, these criteria form the 'template,' a term we use regardless of whether those criteria are learned or fixed. Learned templates may crystallize early on or they may require constant updates throughout the evaluator's lifetime, but either way they only form after exposure to example stimuli, the so-called referents. For example, one of the first things that newly emerged *Polistes* paper wasps are exposed to is the odor of the natal nest. This unique hydrocarbon signature becomes the referent by which the wasps form nestmate recognition templates, and future interactions between adults depend on how well newly encountered wasps and those templates match (see section 'Recognizing kin').

Though we usually think of recognition as a single behavior, it is often helpful to break it down into three essential components: expression, perception, and action. The expression component refers to the production or acquisition of identity cues by the cue bearer, the perception component to the detection and interpretation of those cues, and the action component to all those behaviors elicited by recognition on the part of the evaluator. Expression thus falls to the cue bearer (which need not be a biological entity), while the other two are exclusive to the evaluator. Each of these components is explored in detail elsewhere by Starks.

There is one more subtle distinction in the terms used to describe recognition systems, namely the difference between 'discrimination' and 'recognition.' While authors have sometimes used them interchangeably, the former properly refers to observable behavioral changes, while the latter refers only to an invisible process occurring inside the evaluator. In other words, only one, discrimination, is amenable to traditional behavioral analysis. Some organisms almost certainly recognize objects or other individuals without outwardly changing their behavior, but short of brain-imaging assays or EEGs, we have no way of knowing for sure. (These techniques are, however, becoming more common. See the discussion of pheromone detection in section 'Identifying Mates.')

In the next sections, we show how these terms apply to some behaviors of the social insects and how the wide-angle view helps us make sense of those behaviors.

Evaluating Nest Sites

This is a common enough scene in the summertime forests of North America: a swarm of honeybees, *Apis mellifera*, is dangling from the high branches of a tree, a compact mass made up of thousands of homeless workers. Hours before, this swarm split from its parent hive, leaving the old nest in the hands of a young queen and striking out for a new place to call home. Now, it hangs in a moment of indecision, exploring its surroundings, weighing its options, and figuring out where to begin the task of building a new colony.

This decision is not to be taken lightly. Many hives do not survive their first year, and much of their success or failure depends on finding a quality nest site. But how do the bees even begin to decide? The forest and surrounding fields are full of potential homes: every tree cavity, every old stump, every farmer's empty bee box is a possibility. To complicate matters, most workers are inexperienced at choosing real estate; hardly any of them are more than a few weeks old, and for many the swarm marks the first real journey into the outside world. What's more, this corporation of workers must arrive at a decision without the help of central management because the queen, buried deep within a writhing mass of bees, is in no position to affect the process.

So how to begin? First, the scout bees, the workers charged with finding the new nest site, must know what to look for and how to rank the cue-bearing sites they encounter. In other words, it must possess a nest template. Seeley and Morse first investigated this template in the late 1970s by evaluating the nest sites of feral bees in the forests around Ithaca, New York. They found that bees tended to prefer tree cavities that fell within a circumscribed range of volumes, usually between thirty and sixty liters, and that they tended to choose cavities with small exterior openings over those with larger entrances. The data did not imply any preference for one tree species over another and suggested only a small preference for living versus dead trees.

Following their forest observations, Seeley and Morse designed a series of experiments to see what other cues might influence nest site preference. The researchers built a series of side-by-side next boxes, each modified according to a single variable, and presented them as choice experiments to bee swarms. Scout bees would encounter both cue-bearing boxes at the same time, evaluate the cues with reference to the nest site template, and then convince the swarm to settle in one of the two boxes. When one type of box was chosen significantly more often than another, the researchers interpreted this as evidence of a preference.

The results showed that scout bees possess far more complicated templates than one might expect given their small brains. As it turns out, the scout bees were looking not only at cavity volume and entrance diameter, but also at nest height (higher boxes were preferable to lower ones), distance from the previous nest (somewhere between 400 and 1000 m was ideal) and nest entrance direction (south facing was better than north facing). Data for nest entrances supported the previous observation that bees preferred small over large holes, but the bees showed no preference when it came to the shape of the entrance. Meanwhile, the position of that entrance, either near the bottom or the top of the nest box, did seem to matter, while the bees showed no preference when it came to cavity shape, dryness, or draftiness. Through careful observation and well-designed experiments, Seeley and Morse were beginning to shed light on the nature of the honeybee nest template.

Of course, there is more to the perception component of a recognition system than just the template; on an even more fundamental level, evaluators must first be able to perceive identity cues. In another project, Seeley investigated the proximate mechanisms behind the scout bees' ability to perceive differences in nest site cavity volume. Clearly, the bees were able to accurately and consistently discriminate between cavity sizes, but the means by which they did so were far from clear. Seeley had previously observed that the bees spent a great deal of time walking and flying around the interior of the nest cavity; through a series of elegant experiments, he was able to demonstrate that this movement was actually a method for measuring and calculating interior volumes. When the bees were deprived of visual information by means of a light baffle over the cavity entrance, they were still able to choose those nests closest to the ideal volume. However, when Seeley placed rotating cylinders inside the nest boxes and created a sort of treadmill for incoming bees, he was able to confuse their spatial assessments. Regardless of whether the bees were forced to do more or less walking than the cavity size required, their perceptions of volume correlated to the distances they walked, and not to the real size of the cavity. Thus, Seeley showed that the bees measured not with their eyes, as one might expect, but rather through a complex calculus of distances walked and angles turned.

Recent studies by Seeley and colleagues have begun to shed light on the action component of nest site recognition, the decision-making process that leads a swarm to move into a site. Removing colonies to a mostly treeless island off the coast of Maine, the researchers offered scout bees their choice of several nest boxes, only one of which was a high quality site. Previous studies had demonstrated that individual scouts visit only a single site, evaluate it thoroughly, and then return to the hive to share their assessment with the other workers. The scout bees then advertise their site through a series of waggle dances containing information about both the location and the quality of the cue bearer. Stronger dances lead more bees to investigate the sites, until eventually the swarm becomes a miniature political convention, each faction dancing in support of its nest site choice. When a quorum is reached, the scouts begin to make faint piping noises, and the swarm takes off in the direction of their new home. (Pratt covers this process in more detail and with a specific focus on collective decision making in another chapter in this volume.)

The problem of finding a nest site is not, of course, limited to bees. Starks investigated some of the same questions using European paper wasps, *Polistes dominulus,* and found that they too recognize quality nest sites. Given a choice between long, medium, and short nest boxes, the wasps preferred to initiate nests in the medium ones. While several foundresses started nests in the small boxes, almost none chose the large ones, perhaps because more exposed sites raise the risk of predation by birds or brood parasites. Unlike honeybees, however, these wasps had no preference for higher versus lower nest sites and settled equally often into boxes at all elevations in the enclosure. Thus, as expected, recognition templates vary between species and almost certainly reflect the unique selection pressures faced by each.

As an interesting side note, Starks found that female wasps emerging from hibernation preferentially stopped to perch on fragments of the nests they were raised on, even when those fragments had been moved from their original sites. Since these nest materials were in new locations and had been cut up into smaller pieces, the wasps must have used chemical cues to discriminate between them. It is well known that wasps acquire hydrocarbon signatures from their nests and use those signatures as referents when creating nestmate recognition templates; Starks suggests that returning to the nest may be a way of reconnecting with the previous season's sisters or of updating a learned template by new exposure to the original referent. Either way, returning to the natal nest may be a proximate mechanism by which wasps facilitate cooperation between kin, a behavior that is facilitated in turn by the ability to recognize.

Despite advances such as these, we still have much to learn about the expression, perception, and action components of nest site recognition systems. While we now know that honeybees possess a nest cavity template that includes ideal measurements of volume, entrance size, and height, we know practically nothing about the development of that template. Somehow, by the time scout bees are searching the landscape for appropriate nest cavities, they have developed a Platonic vision of their ideal home; how they and other insects are able to form such templates, sometimes in the absence of clear referents, is an exciting area for future research.

Identifying Mates

Among the social insects, sex tends to be a limited affair. For the workers who make up the bulk of any given colony, life is a chaste, and often brief, exercise in altruism. That much is to be expected. But even among reproductives – the winged queens and frenetic males that fill the skies in early spring and autumn – the mating season is often little more than a brief interlude in a longer life cycle.

Of course, we must not mistake brief for boring; more often than not, these mating periods are every bit as extravagant as they are short. Reproductive female paper wasps fly out to meet some of their male counterparts on high, well-lit structures where the latter congregate en masse. Honeybees meet each other high above the ground where males hone in on and mate with polyandrous queens midflight. Winged reproductive ants rendezvous in swirling, carousing clouds, several hundred strong, and naturalists lucky enough – or unlucky enough – to wander into them do not soon forget the experience. In the high stakes game of finding a mate under such chaotic conditions, recognition is a critical component of fitness.

This is especially true for males. With the exception of one genus of ant (*Cardiocondyla*, see later), all social Hymenoptera species seem to have sperm-limited males with life expectancies that are significantly shorter than their queens'. Once the males' sperm supplies are exhausted, so too is their biological relevance, and selection seems to favor individuals who bow out gracefully (though "gracefully" may be too charitable a word for a species like the honeybee: its drones are famous for explosively rupturing immediately after they mate.) Honeybee males not only mate just once before they die; they also have just one *opportunity* to mate, a single high-altitude flight in which they must compete with hundreds of other males to find and inseminate a new queen. It makes sense for them to maximize that opportunity and to locate receptive females as quickly as they can.

To this end, selection has favored honeybee males with large eyes and large antennae to serve as well-honed, queen-detection devices. Recent studies by Wanner and colleagues investigated the perception component of queen recognition and showed that drones possess sex-specific odorant receptors that respond to a chemical (9-oxo-2-decenoic acid (9-ODA)) found in the queen retinue pheromone. Meanwhile, using calcium-imaging techniques, Sandoz demonstrated that brain regions found only in the males respond specifically to this and to two other components of the queen pheromone. These long-distance chemical signals, along with a highly developed detection apparatus, help drones perceive flying cue bearers long before they can detect them visually. Their sterile sisters, meanwhile, recognize and respond to some of the same chemicals, but with an entirely different behavior: in their case, recognition of 9-ODA leads them to gather around the queen and to suppress the development of their ovaries. The same expression component, the production and emission of 9-ODA, leads to two very different action component outcomes in the honeybee.

While males have a lot invested in finding females, they do sometimes misinterpret cues. An interesting example comes from *Cardiocondyla obscurior*, an ant species that exhibits strong dimorphism within the male population of each nest. Some males are wingless, or ergatoid, individuals who stay within their natal nests, mate with their relatives, and produce sperm throughout their lives.

Others are winged, sperm-limited dispersers who mate with the same females as the ergatoids, but who also leave the nest to mate. Interestingly, while ergatoid males show extreme aggression toward each other and often kill new males before they emerge, they are highly accepting of the winged males. In fact, they not only tolerate them, but also frequently mount them in attempts at copulation. Cremer and colleagues have recently shown that the wingless males escape the lethal aggression of their nestmates by mimicking the chemical signatures of virgin queens. By covering themselves in mimicked cues, these ants are able to fool the ergatoid recognition system, thus escaping aggression while still competing for mates.

While hymenopteran males must do everything they can to maximize their reproduction, most of them do not, at least, have to live with the consequences. Not so their newly inseminated queens, who can live years or even decades longer than their mates. While some species, notably the honeybee and some bumblebees, are able to mate multiply with different males, many hymenoptera appear to be monandrous and to store their sperm for life. This leads to the curious instance, rare among animals, of sperm outliving the males that created them. It should also lead to a great deal of pickiness on the part of the females, and to the development of highly precise mechanism for recognizing quality males. Unfortunately, the high-altitude nature of the nuptial flights means that we know far less than we would like about female mate choice in these species. Baer provided tantalizing details in a recent review of bumblebee male sexual selection – apparently females in laboratory settings show a great deal of choosiness among potential mates and occasionally sting undesirable males to death – but much of the necessary research remains to be done.

For termites, lifetime mating and the continuous production of sperm create a different set of selection pressures, but the need to recognize and evaluate potential mates remains the same. In a recent study of the dampwood termite, *Zootermopsis nevadensis*, Shellman-Reeve demonstrated that they were able to recognize and avoid close relatives during laboratory mating trials. Although termites often replace dead mates with offspring or other close relatives, they appear to avoid such inbreeding during initial colony formation.

As a result of their haplodiploid reproductive system, some hymenopteran species suffer disastrous consequences after inbreeding. When these females mate with close relatives, they increase their chances of producing diploid offspring that are homozygous at the complementary sex-determination locus; since heterozygosity at the *CSD* (complementary sex determination locus) is required to develop as a female, these individuals grow up to become genetically abnormal males. If they do not die early or get identified and removed from the nest (as in honeybees), they may mate with females, produce sterile triploid offspring, and thus increase the colony's genetic load. We know that several of these species demonstrate some form of kin recognition, and so we expect some of them to recognize and avoid close relatives when choosing mates. Intriguingly, this kin discrimination may relax under extreme circumstances, such as those found in the genetic bottlenecks associated with invasions. A study by Keller and Fournier tested whether or not nonnative French populations of Argentine ant *Linepithema humile* avoided inbreeding with siblings. They concluded that these individuals did not recognize their siblings in mate choice situations, but suggested that this might not be true of the populations in the native range.

Of course, inbreeding avoidance is just one way that animals benefit from the ability to recognize their kin. In the next section, we look at some more ways that the social insects rely on these abilities to survive, reproduce, and maintain their fitness advantages.

Recognizing Kin

Few evolutionary quandaries have caused as much puzzlement, or as much grief, as the question of how altruism persists in a Darwinian world. Consider, for instance, how long it took for an elegant and, more importantly, a mathematically rigorous solution to emerge. In the early 1960s, over a hundred years after Darwin hinted at it in the *Origin*, William Hamilton published his vision of inclusive fitness, a gene-centered view of natural selection that explains why some animals sacrifice so much for their relatives. The concept is simple: Hamilton argued that an altruism promoting gene could spread through a population so long as it caused organisms to preferentially direct care toward their relatives. If an organism sacrifices its own fitness, and if that sacrifice means that more copies of the sacrifice-inducing gene wind up in the next generation via the reproduction of close relatives, then the altruistic gene has an evolutionarily successful strategy. Inclusive fitness is the lens that clarifies, and allows us to see for the first time, the true nature of altruism.

Hamilton also suggested that if an animal could recognize its relatives and then discriminate between them and other individuals, then that ability to recognize would play a central role in the evolution of altruism. In the decades since the debut of inclusive fitness, behavioral ecologists have confirmed that kin recognition exists in multiple species and that it seems to be adaptive in ways predicted by theory. In fact, the recognition system framework described in this chapter largely grew out of that work and has blossomed most fully within it; while the examples that follow are covered by the authors elsewhere in this volume, it is worthwhile to briefly revisit them here in the context of social insect recognition.

Many insects are, it turns out, surprisingly smelly creatures – at least to other individuals of the same species. Most of the scent cues studied so far belong to a class of molecules found embedded in the waxy outer surfaces of their exoskeletons and known to biochemists as cuticular hydrocarbons (CHCs). These molecules have proven to be quite diverse, and the normal variability found between species and even between individuals is sufficient to provide high-resolution signals of identity. As such, CHCs are well suited to act as the cues by which insects recognize differences between kin and nonkin.

As predicted, researchers have indeed uncovered a central role for CHCs in the recognition systems of social insects. A classic example comes from the work of Gamboa and his colleagues on recognition in the primitively eusocial *Polistes* paper wasps. Females of the temperate species emerge in the spring and found nests dominated by a single reproductive queen; often this queen also receives a great deal of help from nonreproductive assistants who do everything from collect food to defend and enlarge the nest. By Hamiltonian logic, this kind of altruistic behavior makes little sense unless it is directed toward close relatives and leads to an increase in the helper's inclusive fitness, even as she reduces her personal fitness. Paper wasps, like most other animal altruists, would do well to tell relatives apart from other reproductive females.

This discrimination between kin and nonkin, or more accurately between nestmates and non-nestmates, is based on recognizing CHC cues specific to wasps from the same natal nest group. While there is some evidence to suggest that genetics may play a role in these chemical signatures, by far the most important influence appears to come from the papery nest itself. *Polistes* wasps are much more likely to accept individuals that eclosed on their own nest than they are to associate with foreign wasps, even if those foreign individuals are more genetically related to themselves. Experimental manipulation of larval origin supports these findings and shows that wasps preferentially associate with nestmates even when overall relatedness is low. The nest carries the cues which, when acquired by a female, provide a passport to interacting with nestmates found off the original nest. While this kin recognition system is clearly susceptible to error, nestmate status is probably linked to kinship status often enough in nature that nest origin serves as a decent proxy.

There are at least two proximate mechanisms at work in *Polistes* nestmate recognition: first, the acquisition of cues by newly eclosed wasps (an expression component mechanism), and second, the learning of cues and the formation of a nestmate template by evaluator wasps (a perception component mechanism). For the expression component, each wasp appears to acquire and then bear the specific odor cues of the nest material on which it was raised; this means that while odor cues are homogenized across nestmates, no meaningful intracolonial distinctions can be made on the basis of CHC dissimilarity. Experiments bear this out and demonstrate that wasps appear to follow an 'all-or-none' rule when encountering previously unmet individuals: when other wasps smell like nestmates, they are accepted as kin, but when they do not, they are treated as intruders and often met with aggression.

When it comes to the perception component, wasps seem to develop their nestmate templates in much the same way that they develop their own chemical cue profiles, namely through exposure to the natal nest. The nest odor itself appears to be a mixture of environmental- and wasp-based odors, and taken together these cues form the referent on which the nestmate recognition template is based. In this way, an inanimate referent (the nest) ends up having a profound effect on the way a wasp evaluates and discriminates between animate cue bearers (other wasps).

Of course, this particular model of cue expression and template formation is not applicable across all social insects: that polyphyletic taxon is simply too diverse to allow for generalizations. Ants, for instance, seem to have evolved an array of different expression mechanisms that result in nest- or colony-specific CHC profiles. In some groups, for example in species of the genus *Camponotus*, the queens appear to be the sole source of colony-specific odors; in others, for example, *Cataglyphis iberica* and *C. niger*, the queen has little influence on the cues, and the colony's odor is derived instead from a 'gestalt' mixture of individual worker-produced chemicals. For the vast majority of species, the mechanisms remain unknown; while research continues, the only conclusion we can draw so far is that there are no easily generalizable conclusions.

It appears that insects need not rely solely on chemical cues to recognize differences between conspecific individuals. To return to *Polistes* wasps, Tibbetts studied the mechanisms by which females identify individuals and maintain dominance hierarchies. After observing that individual brown paper wasps, *P. fuscatus*, varied greatly in individual facial markings, she decided to modify these markings with paint and then observe how other wasps responded. All focal individuals received paint treatments on their faces, though those treatments did not affect the previously existing patterns on control wasps. As it turned out, individuals with altered facial patterns received more aggression from their nestmates than the controls did. None received aggression of the sort that would be used against non-nestmates – presumably because they still bore chemical cues that tied them to that specific nest – but they did lose their places in the dominance hierarchy and had to fight to regain positions of power. After a short span characterized by aggressive interactions, the wasps sorted themselves back into the previous dominance order, apparently familiarizing themselves with the new facial patterns. It seems that the wasps have flexible and updatable visual templates for who is in charge in the colony; when the disconnect between facial cues and other indicators of status, such as aggression and fighting ability, becomes too great, the wasps learn to look for a different set of cues.

While the existence of visual templates raises intriguing questions for future research, chemical recognition still seems to be the sine qua non of insect kin recognition. It also plays a vital role in another set of recognition systems, systems that have evolved to protect those resources that social insects work so hard to accumulate.

Avoiding Parasites

Social insect colonies are centers of wealth in the insect world; with hidden food stores, nutritious larvae, and armies of committed workers, they are cities ripe for pillaging. The committed workforce is particularly appealing, and often essential, to a certain type of insect opportunist, the so-called social parasite. These organisms steal workers and occasionally the entire nests of other species, and use those resources to benefit their own offspring. But such coups require a certain delicacy. Most social insects have highly developed mechanisms by which to recognize and discriminate against other species, and the successful parasite must somehow find a way to subvert these systems. The means with by which they do so are the focus of this section.

While social parasites exist in many different groups of insects, nowhere have they evolved to such an extraordinary degree as in the ants. In some of these species, the so-called dulotic or slave-making ants, workers routinely leave their own nests to make raids on the pupae of neighboring species. The captured pupae are returned to the slave-makers' nests, where they emerge as adults and, using their new masters as referents for template formation, quickly accept their faux nestmates as kin. Thus, an action component behavior suitable under normal conditions becomes terrifically maladaptive after a recognition error.

In one genus of dulotic ants (*Polyergus*), the slave makers are so highly specialized to their raiding tasks that they have lost most of the abilities required to take care of themselves and have thus become obligatory social parasites. *Polyergus* species often raid the nests of closely related *Formica* species, returning the pupae to eclose inside their own colonies; occasionally, however, they have been observed to enter queen-right host colonies, kill the host queen, and usurp her position among the workers. Recent work by Tsuneoka investigated the nature of these usurpations by the Japanese pirate ant, *Polyergus samurai*, against its host, *Formica japonica*.

Tsuneoka introduced *P. samurai* queens into *F. japonica* colonies raised in the laboratory and observed the parasite queens' behaviors under three experimental conditions: in queen-right, queenless, and workerless host colonies. Under none of these circumstances did the slave-making queen physically attack her host workers; instead, in both the queen-right and queenless colonies, she was able to stave off *F. japonica* attacks by raising her gaster and presumably emitting behavior modifying pheromones (which immobilized and sometimes killed host workers). Within a few hours, the host workers gradually ceased to attack the parasite and instead accepted her presence.

The relatively low level of aggression directed toward host workers was not seen in interactions between the queens in worker-filled colonies. When *P. samurai* females entered queen-right nests, they attacked the *F. japonica* queen immediately upon contact, grasping and biting the host with their mandibles. During these attacks, the parasite queens also directed their gasters toward the host queens, and spent some time after the attack licking their victims and grooming their own bodies; shortly after, the workers gradually accepted the new queen. In queenless colonies, meanwhile, *P. samurai* queens appeared to have more difficulty usurping colonies. Finally, in workerless colonies, most introduced queens simply ignored the host queens, presumably avoiding conflicts in which there were no workers to be gained.

All this adds up to a picture of *P. samurai* as a highly skilled chemical cue mimic, a not uncommon type among the social parasites. Indeed, some social insects seem to use similar methods against their own species: the Cape honeybee, *Apis mellifera capensis*, has become a major pest for South African beekeepers who rely on colonies of its host, *A. m. scutella*. The workers of *A. m. capensis* are unusual among honeybee subspecies in their ability to lay diploid female eggs via asexual reproduction; most honeybee workers are only able to lay haploid male eggs and even then only successfully in colonies without queens. In queen-right colonies, workers usually police each other's egg-laying and remove eggs that they have not lain themselves. Thus, not only have Cape honeybees managed to evolve a novel method of asexual reproduction, but they have also developed a way to avoid detection by African honeybee workers intent on policing. Presumably they do so by means of chemical mimicry of queen-laid eggs.

Finally, the mixing and the acceptance of heterospecifics are not limited to parasitic relationships. Mixed colonies of ants often live together in fungus gardens, symbiotic associations between ants and certain species of epiphytes; Orivel and colleagues investigated one such

relationship between two species of ant, *Crematogaster limata parabiotica* and *Odontomachus mayi*. These two species coexist peacefully inside their ant gardens, sharing food resources and odor trails, but keeping their broods separate. Remarkably, the relationship between these species goes beyond tolerance of each other to the exclusion of any individual not associated with the ant garden. Indeed, both species will readily attack members of their own species if these individuals come from outside the shared nests. Equally remarkable is that each of these species manages to maintain distinct hydrocarbon profiles throughout their association. This implies that these ants are able to build recognition templates using the individuals around them as nonexclusive referents and, in this respect, they share some aspects of their behavior with those ants that fall prey to dulotic parasites.

Conclusions

This article provides only a limited tour of the many recognition behaviors performed by social insects; it does not address the identification of foreign debris within the nest, the recognition of specific floral odors, or the discovery and removal of diploid male larvae, to name but a few. Nevertheless, each of these − and scores of other behaviors − are best understood as acts of recognition.

The recognition systems framework first developed to study kin recognition is still the best way to approach these topics. Future research will no doubt reveal even more applications, and will hopefully explore the mechanisms behind template formation, cue expression, and behavioral modification in even more detail.

See also: Collective Intelligence.

Further Reading

Hölldobler B and Carlin NF (1991) The role of the queen in ant nestmate recognition: Reply to Crosland. *Animal Behaviour* 41: 525–527.

Seeley TD and Morse RA (1978) Nest site selection by the honey bee, *Apis mellifera*. *Insectes Sociaux* 25: 323–337.

Sherman PW, Reeve HK, and Pfennig DW (1997) Recognition systems. In: Krebs JR and Davies NB (eds.) *Behavioural Ecology: An Evolutionary Approach*. Oxford: Blackwell Science.

Starks PT (2004) Recognition systems: From components to conservation. *Annales Zoologici Fennici* 41: 689–690.

Wilson EO (1971) *The Insect Societies*. Cambridge, MA: Harvard University Press.

Referential Signaling

C. S. Evans, Macquarie University, Sydney, NSW, Australia
J. A. Clarke, University of Northern Colorado, Greeley, CO, USA

Introduction

Darwin argued strongly that language had evolved from precursors in the natural signals of animals. This quintessentially human trait formed part of his case for continuity which remains controversial today. Critics typically take the Cartesian position that language is unique and that, unlike language, animal signals are simply a read-out of emotional state. Thus, traditional models of animal communication suggest that signals principally encode motivational information. This theoretical position implies that variation in the sender's internal state will be reflected by continuous gradation in the physical properties of the signal produced.

In contrast, some animal calls have the unusual property of seeming to denote environmental events. These 'referential signals' are produced in response to specific stimuli (e.g., approach of a particular predator, discovery of food) and are sufficient to evoke from companions a full suite of appropriate responses (e.g., adaptive escape behavior, food search).

Given the cognitive sophistication implied by such systems, it was logical for initial research to concentrate on non-human primates, beginning with Struhsaker's studies of vervet monkeys. Reports accumulating over the last three decades have revealed that referential signaling may be relatively widespread. For example, it is also present in other cercopithecines, tufted capuchins, lemurs, suricates, and several species of birds, including fowl, ravens, and ptarmigan. In each of these systems, call playbacks are sufficient to evoke adaptive responses, implying that receiver behavior is mediated by a predictive relationship between signal structure and a particular type of salient event. Referential signals hence have properties consistent with the idea that they encode relatively specific information about the external environment and which are plainly incompatible with models relying on variation in motivational state, at least if this is conceived of in terms of general arousal.

Recognizing Referential Signals

The discovery of referential signals in the natural behavior of a wide range of social animals invites comparative and developmental studies. An essential prerequisite for such a program is the development of agreed criteria for recognizing the properties of functional reference. Current consensus is that this should involve consideration both of the caller's behavior and of the effects of the signal on companions. Studies of signal production and perception thus assume equal importance.

The principal considerations with regard to production are that referential signals should be structurally discrete and that they should have a degree of stimulus specificity. Eliciting stimuli should belong to a coherent category, although the absolute size of this category, and hence the degree of referential specificity, can vary considerably. The key point is thus not the absolute level of specificity, but rather the relationship between event class and signal type. We would not expect the same class of referential signal to be produced in response to stimuli that are clearly drawn from qualitatively different categories.

The importance of this distinction is illustrated by Owings and Morton's work on California ground squirrels. These rodents have a series of alarm calls forming a continuum from broad-band 'chatters' to tonal 'whistles,' evoked by terrestrial and aerial predators, respectively. However, there are exceptions to this pattern of usage suggesting that this call system does not describe predator class, but rather encodes differences in response urgency perceived by the caller. Squirrels being closely pursued by terrestrial predators sometimes produce whistles, and chatters are given to distant raptors. Similar patterns of call usage have been described in sciurid rodents such as marmots which have call systems designed to elicit a more or less rapid response of fleeing to a burrow refuge, which is their sole response tactic.

Referential signals must also meet a perception criterion. They should be sufficient to permit receivers to select appropriate responses, in the absence of the eliciting stimulus and of other normally available cues. Macedonia and Evans termed this property 'context independence.' The most common technique for assessing perception of putative referential signals is playback experiments in which recorded sounds are presented to conspecific receivers. This approach, by design, strips away the contextual cues that might normally be provided by the nonvocal behavior of the sender.

The full diversity of animal signals is not captured by the simple dichotomous classification scheme adopted in some early papers, in which calls were considered to be either affective or referential. Current practice instead models the properties of animal signals as points falling along a continuum, with signals that principally reflect the motivational state of the sender at one end and denotative labels at the other. Categorizing a signal as referential is equivalent to postulating a threshold value on such an underlying continuum and then demonstrating empirically that the properties of the signal are such that it is exceeded.

The strategy for exploring the properties of a system of putative referential signals involves mapping systematically the relationship between eliciting events (**Figure 1(a)**) and signal morphology (**Figure 1(c)**), and then assessing the effects of variation in signal type (**Figure 1(c)**) on receiver response (**Figure 1(e)**). In addition to these three observable levels (stimulus characteristics, signal structure, and receiver response), two hypothetical levels are included that are necessarily hidden and have properties that can only be inferred. We assume for heuristic purposes that visual stimuli do not evoke call production directly, but rather that they are first recognized as members of a category (**Figure 1(b)**), which is then linked to a particular call type. Also included is an analogous level in which categorization of call type by receivers occurs (**Figure 1(d)**). Evidence exists for such processing in birds and mammals.

Information Content

Struhsaker's pioneering field studies of vervet monkeys established that they have acoustically distinct alarm calls corresponding to their three principal classes of predator: eagles, leopards, and snakes. Cheney and Seyfarth followed up this work with playback experiments, demonstrating convincingly that calls are sufficient to evoke responses appropriate to the type of predator that had originally elicited the sound. Macedonia's studies using playbacks to ring-tailed lemurs provide similar evidence of predator-class-specific alarm calls.

Studies of alarm calling and food calling in birds also reveal a high degree of communicative complexity and specificity. Work by Evans and Marler demonstrates that

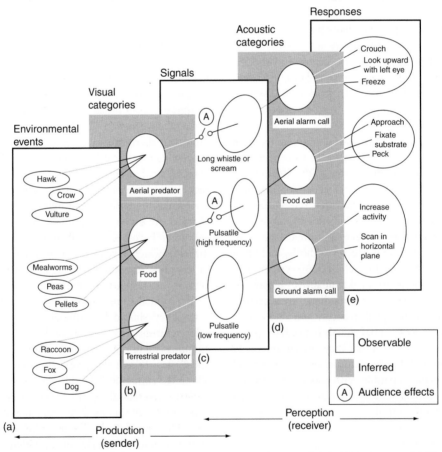

Figure 1 Functional reference in vocal signals. Examples summarize findings obtained from experimental analyses of the production and perception of fowl calls. Note the strategy of mapping relationships between eliciting events and signal structure, and between signal structure and receiver response. The system has both observable and inferred properties (see text for details). Audience effects are unique to call type. Aerial alarm calling is potentiated by the presence of any conspecific. There is no audience effect for ground alarm calls, which are probably in part predator directed. Males that have found something edible produce food calls at a low rate when they are alone, but they call much more when in the presence of a hen, particularly if she is an unfamiliar. In contrast, food calling is completely suppressed in the presence of a rival male.

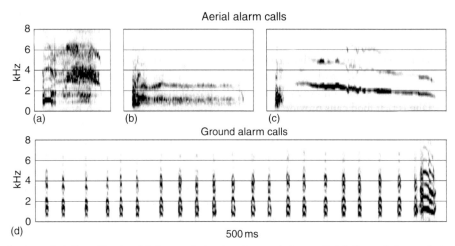

Figure 2 Acoustic characteristics of fowl aerial alarm calls (a–c) and ground alarm calls (d). The calls in panels (a–c) were produced by three different males in response to a computer-generated 'hawk' stimulus. Note the individual variation in call structure. Panel (d) is part of a long bout of ground alarm calling.

fowl (*Gallus gallus*) have an alarm call system broadly similar to that of ring-tailed lemurs. They produce two structurally distinct signals, one to terrestrial predators (e.g., canids) and the other to aerial predators (e.g., raptors) (**Figure 2**). These birds tolerate handling and behave naturally in the laboratory, so it is possible to create a virtual environment in which the stimulus attributes necessary for call production can be systematically explored. Computer-generated animations of raptor silhouettes produce the full gamut of responses to an aerial predator. This technique allowed characterization of key parameters of stimuli that elicit alarm calls. The size and the speed of a simulated predator play an important role, and there is also a degree of sensitivity to stimulus shape and spatial location.

Differences in ecology have been implicated as one determinant of signal specificity. While chicken aerial alarm calling depends upon simple characteristics, producing a system in which eliciting stimuli are quite broadly defined, the alarm calls of some avian species are as highly specific as those of vervet monkeys. For example, Walter's studies of lapwings demonstrate that both South American and African species make subtle discriminations between raptors that are visually very similar. Lapwings inhabit open terrain, which affords them the opportunity to examine approaching raptors for some time before producing an alarm. In contrast, red junglefowl, which are the ancestral species for domesticated chickens, live in forest and dense brush where visibility is limited and the time available for responding to potential avian predators will consequently be brief. This ecological constraint may have selected for a simple, general rule for predator recognition, which, although it entails some loss in accuracy, facilitates rapid response. Other reports describing the antipredator behavior of forest-dwelling birds also suggest a fairly high frequency of false alarms.

It is possible that the range of events evoking alarm calls is the product of an evolutionary trade-off between two types of error: calling when the approaching bird is not dangerous (called Type I error) and failing to respond to an approaching predator (called Type II error). Birds living in open habitats may have been subject to selection for low Type I error rates, as the cost of the frequent false alarms that would otherwise be produced in an environment where aerial objects are visible much of the time would be prohibitive. This process would give rise to relatively specific alarm calls. Species inhabiting habitats where visibility is restricted and response times must be short may have been selected to reduce Type II error rates and their alarm calls may be less specific as a consequence. The logic of this argument is closely analogous to that of classical signal detection theory.

What about signal receivers – are these sounds sufficient for selection of appropriate escape responses? As Macedonia and Evans have argued, referential signals should have the property of context independence, that is, they should be sufficient to permit receivers to select appropriate responses in the absence of the eliciting stimulus or other normally available cues. When hens, isolated in a cage with a small area of cover, were played terrestrial and aerial alarm calls, they responded to terrestrial alarm calls by drawing themselves up into an erect alert posture and scanning back and forth in the horizontal plane as though trying to detect a threat approaching on the ground. When they heard aerial alarm calls, they ran to cover, crouched, and looked upward, precisely as though they were trying to detect an object moving above. Each of these responses is appropriate to the type of predator that originally elicited the call, indicating that chicken alarm calls, like those of vervets and lemurs, are referential signals.

Predators are not the only environmental events that chickens respond to with distinctive calls. When a male

If a field site spans the boundary between two UTM zones, a researcher may choose to use latitude/longitude coordinates instead, or to convert locations from one UTM zone so that they are compatible with those collected in the other zone. Conversion between the UTM and the longitude/latitude systems is possible but not straightforward.

Finally, researchers may create their own coordinate systems for use in small field sites or lab studies. This approach is usually taken when the spatial resolution of a GPS is large relative to the spatial activity of the study animal. For example, if the movements of a 1-cm long insect are being tracked, the 5-m error associated with the locations from most handheld GPS units will be too large. The main disadvantage to this approach is its inconsistency across studies. Furthermore, for maximum accuracy, one should use proper distance measuring tools in the field, which may not always be possible.

How Do We Study Behavior Remotely?

There are many methods for gathering behavioral data remotely. All methods reviewed here involve affixing a tag to an animal and tracking the signals emitted from the tag, or downloading data stored on a tag. There are several types of tags, categorized by the wavelengths they transmit. A tag may contain more than one type of tracking mechanism. For example, a GPS recording tag can also be equipped with a VHF transmitter. The suitability of each tag type will depend on the study species, habitat, and the study aims. Here, we present several tagging methods, explain how they operate, and provide some advantages and disadvantages of each tag type. We will summarize this section with general considerations relevant to all tags.

Radio Tracking (VHF)

Radio tracking is probably the most widely used method for acquiring behavioral data remotely. Radio tags constantly transmit a radio signal at a set frequency in the very high frequency (VHF) range (142–230 MHz). Each tag transmits a unique radio frequency (e.g., 150.020, 148.800 MHz) used for distinguishing between different tagged individuals; these signals are detected using a receiver (**Figure 1**). Both tag and receiver are equipped with antennas, the size of which will determine the distance from which the tag can be detected.

Receivers can tune into different frequencies either manually or automatically (using a programmable receiver). When a tagged animal is within the range of the receiver, a beeping tone is heard and its strength is used to determine the direction of the animal. The antenna connected to the receiver is used for scanning the landscape by carefully

Figure 1 R1000 Communications Specialists handheld receiver connected to a three-element Yagi antenna. Photo by Karen Mabry.

'sweeping' it from side to side. The direction in which the tone is strongest is recorded as the bearing to the animal. Researchers can determine the location of a tagged animal using the following techniques.

Ground tracking

One simple method for locating a tagged animal is homing in to its signal. This is achieved by following the strength of the tag's signal until the animal is seen. This method is not always feasible because the animal may hear the observer approaching and move away, or the terrain may be impassable. In addition, this tracking method is likely to disturb and alter the tagged animal's behavior. Thus, researchers may choose to obtain the location of a tagged animal indirectly, using triangulation.

Triangulation requires at least two directional bearings toward a tag from known locations. The estimated location of the tagged animal is based on the intersection of the bearings (**Figure 2(a)**). Bearings are obtained using a handheld antenna, vehicle-mounted antenna, or antennas on fixed towers. The simplest way to obtain a bearing is to rotate the antenna 360° and record the direction in which the signal is the strongest, using a compass. When using a handheld compass, it is important to account for magnetic declination – the difference between true north and magnetic north, which varies across the globe. Just two intersecting bearings are needed to generate an estimated location, but at least three bearings are needed to estimate the associated location error (**Figure 2(a)**). However, more bearings are not necessarily better; rather than resulting in a more accurate location, too many bearings may create confusion (**Figure 2(b)**). Location accuracy will depend on the intersection angle of the bearings, the number of bearings, and the time interval between them. Note that animals may move while obtaining bearings. Thus, the size and speed of the study animal should be carefully

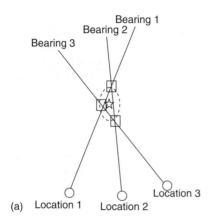

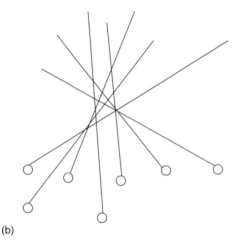

(a)

(b)

Figure 2 Triangulation to VHF tags. (a) Two bearings might be enough for obtaining an estimate of the animal's location, but notice that different bearing pairs may produce different locations (squares). Therefore, at least three bearings are needed for estimating a more accurate location and its measurement error (star – location and oval – error). (b) Too many bearings might not always be helpful – removing some of the bearings in this figure will not affect the estimated location of the animal.

considered when deciding on the number of bearings to use and the spatial and temporal distance between them. LOCATE III is a useful computer program for estimating an animal's location using triangulation.

Aerial tracking

When animals move large distances, are in inaccessible terrain, or many animals need to be tracked at once, aerial tracking may be used. Similar to homing in from the ground, an aircraft can be used to home in on a tagged animal. Light aircrafts used for wildlife telemetry are usually mounted with two antennas, one on each wing (**Figure 3**). A switch box connected to the two antennas is used for activating them. Initially, both antennas are activated. Once a signal is detected, the listener can alternate between the antennas to determine the signal's direction, depending on its strength. When the signal is very strong from

Figure 3 Light aircraft with antennas connected to each wing for wildlife tracking. Photo by Roy Wollman.

both sides, the aircraft can begin circling to visually search for the tagged animal.

A major advantage to radio tracking is its common use as a tracking method. There is a great deal of information about radio tracking in the scientific literature. It is possible that radio tracking has become so widely used because of its simplicity. VHF tag and receiver architectures are simple and therefore relatively inexpensive compared with other tags discussed later. In addition, the tag's simplicity allows for small tags that can be attached to small animals such as mice, frogs, and even large beetles. Despite its widespread use, radio tracking does have some disadvantages. The tracking equipment itself is relatively inexpensive, but tracking in the field may become very costly in terms of researcher time and fuel costs, especially if aerial tracking is needed. In addition, if triangulation is used, locations might be inaccurate. Terrain and vegetation cover can greatly influence signal accuracy because VHF signals may bounce off from hills, disappear in valleys, or be absorbed by heavy forest cover. Finally, accessibility should be carefully considered when planning ground tracking – extensive road or trail systems and high topographical features for bearing stations can be useful.

Ultrasonic Telemetry

In aquatic environments, radio frequencies (VHF) can be greatly distorted and are hard to detect. Therefore, tags for tracking aquatic animals utilize ultrasonic frequencies (30–80 kHz, 75 kHz being the most popular). These frequencies can be used only in fresh water environments.

Tracking ultrasonic tags is similar to radio (VHF) tracking, with the primary difference being that tracking is conducted by boat, or by using fixed underwater receiver stations, because the tagged animal is underwater. Water-submerged antennas are often used to avoid signal distortion arising from the transition between water and air. In addition, homing in on the tagged animal, as described for radio tracking, can be used to position a boat or an aircraft directly above the tagged animal. The advantages and disadvantages of ultrasonic telemetry are similar to the ones mentioned for radio tracking.

Satellite Tracking

Some animals, such as migrants, cannot be feasibly tracked using radio telemetry because of their large-scale movements. Platform Transmitter Terminals (PTTs) are tags used for satellite-based tracking. The 'Argos system' is used for detecting the locations of PTTs and for transmitting these locations directly to researchers, with no need for tracking the animals in the field. PTTs transmit a signal at a specific ultra high frequency (UHF) (401.650 MHz). Currently, five polar-orbiting satellites are equipped with receivers for this particular frequency. Based on how often a tag transmits a signal and the speed of the passing satellite, the tagged animal's location is calculated using the Doppler effect. Animals are distinguished from one another based on identification numbers that are specific to the tag's transmission electronics. In addition to calculating location, the satellite can also download data recorded by other devices attached to the tag such as a GPS, pressure sensor, and others discussed later. The satellites transmit the data to a receiving station on the ground which in turn sends the information to a processing center. The processing center consolidates the data and prepares them for presentation to the end user, the researcher. Data can be viewed on the internet, received by email, text message to a cell phone, and other communication channels.

PTTs are often deployed on marine and migrating animals (**Figure 4**). When aquatic animals are deep in the ocean, the satellites cannot receive the UHF signal. Therefore, PTTs for aquatic animals are often equipped with a device that activates them only when the animal surfaces, thus preserving battery power. This may reduce the number of locations per day that can be obtained for these animals.

A great advantage of satellite tracking is that animals can be tracked under all weather conditions, and at all times of day. Furthermore, once the tag is attached, there is no need to follow the animals in the field, thus greatly reducing the expenses of field work. This point is particularly important for long-ranging animals, such as migrants that can travel from pole to pole and for which VHF tracking is not practical. However, the location accuracy varies (from 250 m to 1.5 km error) depending on the number of fixes the satellite obtained when passing over the tagged animal. In addition, the number of locations obtained per animal depends on its position around the world. At the poles, a tag can be detected up to 14 times a day, but this number declines at lower latitudes. Another disadvantage to satellite tracking is its price. Because PTTs use a very specific frequency, manufacturers are under stringent technical constraints. As a result, PTT tags are larger and more expensive than VHF tags. Still, PTT tag sizes are getting smaller and the cost of tags and the associated data processing fees should be carefully weighed against the expenses of tracking a VHF tag.

GPS

All the earlier-mentioned tags can be equipped with a global positioning system (GPS) unit. GPS units automatically record the animal's location at fixed, predetermined, time intervals and store the data on-board the unit. The frequency at which locations are obtained will determine the battery life of the unit and therefore the tracking duration. To preserve battery life, GPS units must have a clear view of the sky, to enable communication with the GPS satellites (**Figure 5(a)** and **5(b)**). Heavy vegetation cover (e.g., thick forests) and aquatic environments may distort or disable the use of a GPS unit.

There are several ways to retrieve the data stored onboard the GPS unit.

Manual retrieval

One straightforward method for data retrieval is to remove the tag from the animal at the end of the study and directly connect to the GPS unit (using hardware and software provided with the tag). If recapture is impossible or unwanted, tags can be equipped with a drop-off device (**Figure 5(c)**). The drop-off device is scheduled to detach a tag at a predetermined time, and the researcher is left to find it. Having a VHF unit on such tags is useful for successfully locating them. After detaching from aquatic animals, tags float to the surface and can either be retrieved by the researcher or the data can be remotely downloaded through a satellite link (see below).

Remote retrieval

An alternative method to retrieve data from a GPS unit is linking to it remotely. Remote download permits multiple

Figure 4   A satellite-tagged Franciscana dolphin. Chicago Zoological Society's Sarasota Dolphin Research Program. Courtesy of Randall S. Well.

Figure 5 African elephants collared with VHF/GPS collars with a drop-off unit. (a) A collared female elephant and her calf. The GPS unit is on top, between her ears, and the VHF unit is hanging below the elephant's chin. The drop-off unit is behind the elephant's ear and cannot be seen. (b) The GPS unit can be seen between the elephant's ears, facing the sky to allow easy connection with the GPS satellites. (c) An anesthetized elephant immediately after collar deployment. The VHF unit can be seen at the bottom with black cloth covering the VHF unit's antenna. The silver box is the drop-off unit and the GPS unit cannot be seen. Photos by Noa Pinter-Wollman.

data downloads throughout the study period, allowing adaptive management decisions and troubleshooting of data acquisition rate. Furthermore, remote downloading eliminates the need for recapturing an animal or depending on unreliable drop-off mechanisms. It also eliminates the challenge of finding a dropped tag in dense vegetation. Three ways to remotely link and download GPS data are:

UHF: Short-range receivers which rely on UHF remotely connect to the GPS unit and download the data, requiring a stable connection for several minutes.

Satellite: A satellite link can be used for downloading data from GPS units mounted on PTT tags, through the 'Argos system.' The data are relayed to the researcher in a similar manner to that described earlier in the satellite-tracking section. For downloading data using a satellite link from aquatic animals, the tag must be at the water surface to allow a stable connection.

GSM: Recent advances in the cellular phone technology allow remote downloading using the Global System for Mobile communications (GSM) network. The location data are then received as a Short Message Service (SMS) to a cell phone. This remote download method is available only in field sites with GSM coverage.

The major advantage to using a GPS unit is the excellent location accuracy (5–55 m error) at high temporal resolution. In addition, GSM and satellite retrieval methods allow obtaining real-time location of tagged animals, which is especially useful for making adaptive management decisions.

Some disadvantages to using GPS units include the tradeoff between data temporal resolution and battery lifetime, distorted or missing locations in dense habitats, and the GPS unit size, which currently constrains their use to animals heavier than 2 kg. Some potential problems with manual data downloading include inability to recapture the animal, failure of the drop-off mechanism, inaccessibility to the drop-off location, or inability to find the tag. For these reasons, GPS units should always be deployed along with VHF/PPT/Ultrasonic units to allow tag retrieval. These other tracking devices can also be used for obtaining spatial data at a lower temporal resolution, as backup for GPS failures.

Additional Data Collection Devices

In addition to a GPS unit, other sensors can also be attached to tags for collecting information about the environment and the animal's physiology. The data are often stored on-board these devices and can be retrieved using the techniques mentioned in the GPS section. Sensors may record water salinity, pressure, air and body temperature, heart rate, and activity. Marine animals are often equipped with pressure sensors to provide information on the animal's swimming depth, which is an additional

Use Committee (IACUC) at US universities) should be consulted to ensure that all research is consistent with the applicable laws, and that appropriate capture and anesthesia protocols are followed.

How large can a tag be? The rule of thumb used by most researchers allows for tags that weigh less than 5% of a terrestrial mammal's body weight and less than 2% of body weight for birds and bats. Aquatic animals can carry slightly heavier weights, but the tag's effect on their hydrodynamics is usually the confounding factor. Battery weight usually contributes the most to tag weight, and is positively correlated with the tag's life span and thus with the study duration. GPS and PTT fixes require considerable battery power, leading to a negative correlation between the number of GPS or PTT fixes and the tag's life expectancy. The development of solar-powered PTT and GPS units may increase tag life span. It is also important to match the battery life with the longevity of the attachment method. It makes little sense to invest in a large tag that can last for 2 years if the animal can remove it after 1 week.

Finally, once the study is completed, the tags should be removed to ensure the animal's well-being. Tags are often deployed for short periods of time, but may adversely affect the animal if left on for too long. Drop-off mechanisms and pop-off units are one way to achieve tag removal, but animals may need to be recaptured and even anesthetized for removing their tag. It may not always be possible to remove the tag at the end of the study, but every effort should be made to do so. As a practical consideration, many tags can be refurbished and reused, thus reducing the cost of tracking in future studies.

See also: Amphibia: Orientation and Migration; Bat Migration; Bats: Orientation, Navigation and Homing; Bird Migration; Fish Migration; Habitat Selection; Insect Migration; Insect Navigation; Irruptive Migration; Magnetic Compasses in Insects; Magnetic Orientation in Migratory Songbirds; Maps and Compasses; Migratory Connectivity; Pigeon Homing as a Model Case of Goal-Oriented Navigation; Sea Turtles: Navigation and Orientation; Spatial Orientation and Time: Methods; Vertical Migration of Aquatic Animals.

Further Reading

Cooke SJ, Hinch SG, Wikelski M, et al. (2004) Biotelemetry: A mechanistic approach to ecology. *Trends in Ecology and Evolution* 19: 334–343.
Kenward RE (2001) *A Manual for Wildlife Radio Tagging*. London: Academic Press.
Millspaugh JJ and Marzluff JM (2001) *Radio Tracking and Animal Populations*. New York: Academic Press.
Priede IG and Swift SM (1992) *Wildlife Telemetry, Remote Monitoring and Tracking of Animals*. Chichester: Ellis Horwood Limited.
White GC and Garrott RA (1990) *Analysis of Wildlife Radio-Tracking Data*. London: Academic Press.

Relevant Websites

http://www.locateiii.com/ – Radio-telemetry Triangulation Program, Locate III.
http://www.argos-system.org/ – The Argos system.
http://las.pfeg.noaa.gov/TOPP_recent/index.html – Tracking of Pacific Pelagics – near real time animal tracks.
http://www.biotrack.co.uk/.
http://www.sirtrack.com/.
http://www.telonics.com/.
http://www.biomark.com/.
http://www.holohil.com/.

Reproductive Behavior and Parasites: Invertebrates

J. C. Shaw, University of California, Santa Barbara, CA, USA

Introduction

Every parasite confronts the same dilemma: how to extract ample host resources for growth and reproduction while allowing the host enough resources to survive – or face certain death as well. Specific variations of this quandary depend on the species of parasite and its particular life stage, but all parasites need time to grow or mature inside intermediate and/or final hosts. Parasites should maintain an optimal balance between minimizing host damage and maximizing nutrient availability during this time. The host reproductive system provides an ideal resource for such a strategy.

Reproduction constitutes the primary objective of all living organisms. However, reproduction is energetically costly, and animals in suboptimal health often forego breeding in order to maintain overall somatic health. For example, birds that are highly parasitized by feather mites or endoparasitic worms will skip the winter migration to breeding grounds. Although these individuals sacrifice a year's reproductive gain, the reallocated energy goes toward somatic repair, possibly extending the animal's overall reproductive output via increased longevity. Similar situations occur when other animals are infected with parasites. Parasitic infection can result in a decrease or cessation of host reproduction, and the subsequently available energy is shifted toward other physiological demands, such as increased immune function or nutritional requirements. Furthermore, reproductive organs and structures are not essential for an organism's survival, and the tissues provide an energy-rich source of nutrients that can be diverted toward host growth or energy storage, or – as is the focus of this study – parasite growth and reproduction.

It remains debated whether alterations of host reproduction ultimately prove beneficial for parasites or hosts. Manipulations of host reproductive behavior and reproduction represent a large portion of parasite-induced host modification, especially for invertebrate hosts. That a wide variety of parasitic taxa exploit host reproduction suggests that it may be a successful strategy for maximizing parasite fitness. However, there are situations where hosts might benefit from the resultant decrease in reproduction. A host might limit reproductive activity upon infection to minimize the subsequent nutritional stress. Reductions in reproductive effort could be advantageous for hosts if such measures somehow increase host longevity, and if the temporary decrease in reproduction results in greater overall reproductive output over the extended lifespan.

This generally applies to systems where hosts can potentially resume reproduction after overcoming or surviving a parasitic infection.

There appear to be more host–parasite examples demonstrating that parasites benefit – largely in the form of greater fecundity – from altered host reproduction. For example, bacteria of the genus *Wolbachia* induce sex reversals in their intermediate arthropod hosts (aquatic and terrestrial), so that males become functional females. *Wolbachia* is transmitted vertically through eggs, and its drastic modification of males ensures greater transmission to new hosts.

Sometimes, the beneficiary of the altered host behavior is unclear. In one unusual example, the physical presence of ectoparasites actually prevents the host from reproducing. Desert flies (*Drosophila nigrospiracula*) are host to ectoparasitic mites (*Macrocheles subbadius*) that attach to the underside of the fly's abdomen (**Figure 1**). Infection loads of several mites create a physical obstruction large enough to prevent the male fly from mating successfully. Heavily infected males still attempt to mate by mounting females, but the male's efforts to engage copulatory structures are hindered by the mites situated under his abdomen. It remains unclear how or if the mites benefit from this impediment of host reproduction. The mites are not transmitted sexually, so increased mating attempts by the male would not facilitate new infections. Furthermore, the transient nature of ectoparasitic mites means that infected males may not always be prevented from mating, suggesting that this altered reproduction may simply be a side effect of the infection.

In contrast to ectoparasites, a large proportion of parasites that affect host reproductive activity are endoparasites that are long-lived inside their hosts. Many of these parasites manipulate host reproduction using methods that include energetic drain and/or mechanical and neurochemical disruption. There are three general types of parasite-induced reproductive changes: altered reproductive behavior, parasitic castration, and changes in host reproductive effort. Specific host–parasite examples are discussed in the following sections.

Changes in Host Reproductive Behaviors

Altered host reproductive behavior falls into two broad categories: changes in behavioral displays and behavioral castration. Many of the altered behavioral displays exhibited by hosts are not completely novel but rather existing behaviors displayed in unconventional circumstances. Helluy and Holmes noted that gammarids (*Gammarus lacustris*) infected with the acanthocephalan *Polymorphus paradoxus* display a peculiar clinging behavior. In the wild, uninfected gammarids normally reside down near the sand or mud bottom, where they spend much of their time in burrows. Infected gammarids, however, become positively phototactic and hover near the water's surface, where they often cling to floating vegetation. Here, they are more likely to be ingested by grazing ducks, which serve as final hosts for the acanthocephalan.

The clinging posture displayed by infected gammarids is actually a male reproductive behavior that is displayed out of context. During breeding, an uninfected male will cling to an ovigerous female for several days, waiting for the opportunity to fertilize her eggs. Incredibly, the acanthocephalan parasite is able to induce its host to display this male reproductive behavior, regardless of host gender. Thus, the physiological state required to produce clinging behavior is present in both male and female gammarids, and *P. parodoxus* somehow reproduces the required condition(s) in females and in males out of context.

There is strong evidence that alterations of the neurotransmitter serotonin underlie the clinging behavior and positive phototactism of infected gammarids. Helluy and Holmes were able to elicit clinging behavior in uninfected gammarids after directly injecting serotonin into the body cavity. Additionally, serotonergic neurons of infected gammarids have increased serotonin varicosities, which may serve as storage locations for the neurotransmitter. In a similar system, gammarids (*Gammarus pulex*) are infected with a fish acanthocephalan, *Pomphorhynchus tereticollis*. Infected gammarids are positively phototactic, and the increased time they spend in exposed areas render them more visible to fish predators, which are the finals hosts for *P. tereticollis*. The serotonergic neurons of infected *G. pulex* show much stronger immunoreactivity when compared to the same neurons of uninfected individuals, indicating that infected gammarids have increased serotonin activity in their brains. Furthermore, infected individuals that display the strongest photophilia also have the highest serotonin immunoreactivity in their brains. It is unclear whether these acanthocephalans actively target the serotonergic metabolism of their hosts, or if altered serotonin metabolism constitutes part of the host response to parasitic infection. Nevertheless, if an altered host behavior somehow increases parasite fitness, any parasitic traits associated with those host behavior changes should be selected for.

Decapod crustaceans infected with rhizocephalan barnacles also display reproductive behaviors out of context. Rhizocephalan barnacles are completely parasitic and barely resemble their free-living counterparts – they lack a calcified shell and their bodies consist of little more than an absorptive, branchlike network extending throughout the body of their crab host. The only visible part of a rhizocephalan is a large, round bump called the externa, which protrudes from under the crab's abdominal flap (**Figure 2**). The externa is actually the female parasite's gonad. Female rhizocephalan larvae are released from the externa, locate and infect new hosts, and each develops inside a crab as a root-like network of tissue (called the interna), eventually producing an externa. Male rhizocephalans live the early part of their lives as free-living stages. Sexually mature males locate and enter the externa, remaining inside to permanently fertilize the female. The externa is located where the egg mass would be in an uninfected, ovigerous female crab. Infected female crabs still exhibit innate ovigerous behavior by aerating (fluttering abdominal flap) and grooming the parasite externa as if it were an egg mass. The behavioral modification is more dramatic for infected male crabs. The rhizocephalan feminizes the behavior and morphology of male crabs. An infected male exhibits the egg ventilating and grooming behavior normally limited to ovigerous females. Furthermore, the male's abdominal flap actually grows wider

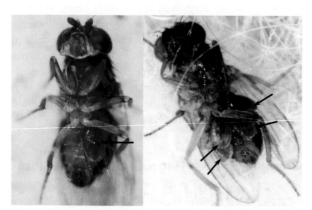

Figure 1 Desert flies (*Drosophila nigrospiracinla*) infected with the eccoparasitic mites *Mentochefes subhoridas* (induced by avirus) Photo courtesy if Michal Polak.

Figure 2 A European Green crab (*Carcinars mutex*) infected with the parasitic rihizocephalan bornacle. *Sacculini Carcini*. The rhizocepbalan's externa visibly bulges from under the erab's abdominal flap. Photo courtesy of Todd Husleui.

through subsequent molts, to resemble that of a female, to better house and protect the rhizocephalan externa.

The rhizocephalan benefits greatly from the altered behaviors of its host. The externa turns foul and necrotic without the crab's grooming, and the aerating behavior increases the parasite's reproductive success. The aeration performed by infected crabs likely helps to broadcast the female rhizocephalan's chemical cues, thereby increasing the number of potential males that will be attracted to the externa. Crabs have also been observed to ventilate the externa during the release of rhizocephalan larvae, which increases the dispersion of parasite larvae. The mechanisms of how rhizocephalans induce their hosts to nurture the externa remain unknown, although the feminization of males likely occurs through interference with the androgenic gland. Not only do rhizocephalans manipulate host reproductive behavior, they also parasitically castrate them, diverting energy from host reproduction toward parasite growth and reproduction (see section 'Parasitic Castration'). Rhizocephalans thus maximize their fitness by exploiting both the reproductive behavior and the physiology of their hosts.

Parasites have also been shown to alter host responses to sexual stimuli. This type of behavioral manipulation is referred to as behavioral castration, since the changes often prevent the host from reproducing successfully. For example, male flour beetles (*Tribolium confusum*) infected with the larval cestode *Hymenolepis diminuta* show a decreased response to female sex pheromones – an action critical for initiating the process of mating and mate recognition. Similarly, cockroaches (*Periplaneta americana*) infected with an acanthocephalan (*Moniliformis moniliformis*) also show decreased responses to sex pheromones. Infected cockroaches can still detect the pheromones, demonstrating that infection does not interfere with the host's ability to perceive scent stimuli. However, infected cockroaches were inconsistent in their responses to pheromones when compared to uninfected ones. An infected male cockroach that is unresponsive to female sex pheromones would almost certainly miss the opportunity to mate. Thus, the infected flour beetles and cockroaches experience reduced fecundity as a result of behavioral castration. In contrast, the parasite appears to reap some indirect benefit from its host's decreased interest in sexual cues. Male cockroaches fight for the right to mate with females, often sustaining injuries in the process. By preventing its host from engaging in such risky reproductive behaviors, the parasite increases its host's longevity and secures itself more time to mature (inside the host) and become infective to the next host.

Parasitic Castration

The most common and striking examples of parasite-modified host reproductive behavior involve the infamous phenomenon known as 'parasitic castration.' Parasitic castration has been classically defined as destruction or alteration of host gonadal tissue by parasites. Parasites can directly castrate hosts by damaging host gonads or by secreting neurochemicals that target host endocrine and/ or reproductive systems. Indirect methods of castration usually consist of gonadal atrophy due to energetic drain, as a result of infection. This destruction of reproductive organs is often accompanied by a change in the outward reproductive behavior of a host. Representatives of a diverse range of parasitic taxa castrate at least one host in their life cycles; these taxa include microsporidians; protozoans such as ciliates and gregarines; terrestrial, marine, and aquatic arthropods; and helminths of multiple phyla. This diversity suggests that castration may constitute a successful parasite strategy for gleaning host nutrients and energy while minimizing host mortality.

Larval stages (metacestodes) of the rat tapeworm *Hymenolepis diminuta* commonly infect the grain store beetles *Tenebrio molitor*. The metacestodes do not physically castrate their intermediate hosts, but infection does result in reduced fecundity of the infected beetles. This decreased fecundity may be due in part to the behavioral castration of infected male beetles, whereby they are less responsive to female sex pheromones. However, infection also results in morphological changes to host reproductive organs. The tapeworm larvae reside in the hemocoel (body cavity) of the beetle, and the lack of physical disruption to host reproductive organs points to a neurochemical mechanism by which they affect host fecundity.

Infected female beetles exhibit a marked decrease in the production of vitellogenin – the key component in egg yolk. Webb and Hurd isolated a peptide from *H. diminuta* metacestodes that directly impedes vitellogenesis, demonstrating that decreased host fecundity occurs as a possible parasite strategy and not simply as a by-product of infection. Metacestodes undergoing rapid development exerted the greatest vitellogenesis inhibition, suggesting that the worms may incur an energetic gain by diverting host resources toward parasitic growth.

The underlying strategy is less clear in the case of *H. diminuta*-infected male beetles. Rather than experience a marked depletion of a reproductive organ, infected male beetles have enlarged reproductive accessory glands, the primary function of which is the conversion of sugar to glucose. Enlarged glands retain higher protein contents, but there is yet no indication that this translates to increased host spermatogenesis or parasite growth. It remains unclear whether these changes benefit the host or parasite, or if they result as a pathological by-product of infection. Female beetles experience a greater impact of infection than do males, probably because ovarian development and egg production are more energetically costly compared to testicular growth and spermatogenesis. There may not have been much selective pressure for *H. diminuta* to fully castrate its intermediate host, since the tapeworm ultimately requires that a rat ingest its beetle host in order to complete its life cycle. However, it is possible that infected females exhibit some alteration in their reproductive behavior, perhaps their receptivity to breeding males, due to their overall decrease in reproductive output.

Tobacco hornworm caterpillars (*Manduca sexta*) serve as hosts for the parasitoid wasp *Cotesia congregata*. The female wasp oviposits multiple eggs inside the caterpillar, and the wasp larvae grow inside the caterpillar until they emerge from their host, spin and attach cocoons to the caterpillar's tegument (**Figure 3**, individual on left). Here, they remain for several days until they metamorphose. Caterpillar hosts cease feeding and movement during the period of cocoon formation and wasp metamorphosis, and eventually die after the adult wasps emerge. Infected male hornworms experience a reduction in testicular volume. Wasp larvae may benefit from this if they utilize the energy and nutrients diverted from developing host testes for their growth and maturation. Wasp larvae feed only on host hemolymph, so physical disruption of the testes (e.g., consumption) seems unlikely. Instead, methods such as neurochemical or hormonal signals, energetic drain, or both provide plausible explanations for castration in *M. sexta*.

Bopyrid isopods castrate their final crustacean hosts. These parasites are rather romantic (if one can refer to parasites as such), despite their host-sterilizing effects. Males and females live paired inside the host, and one or more dwarf males usually accompany the larger female (**Figure 4**, male on the left, female on the right).

Decapod crabs harbor bopyrids in their branchial chamber, whereas shrimps can have either branchial or abdominal infections (**Figure 5**, bopyrid infection underneath bulging carapace on right side of shrimp).

The methods of castration employed by bopyrids remain a mystery. There are strong indications for castration as a result of energetic drain, since bopyrids do not normally occupy the gonad region, and the parasite is large relative to host body size. This method of indirect castration seems especially plausible when considering castrated shrimps that are infected with abdominal bopyrids, where the parasite sits near the base of the host's pleopods, away from the reproductive organs. In contrast, branchial bopyrids are situated within the open circulatory system of the crab's hemocoel. It seems likely that these parasites could castrate their hosts by secreting neurochemicals that could induce atrophy or inhibit development of reproductive organs. The loss of reproductive organs may translate to a lack of circulating reproductive hormones, the physiological cues that would normally initiate and maintain reproductive behaviors in male and female crabs. Bopyrids

Figure 4   Dwarf male (left) and female (right) bopyrid isopods, removed from the branchial chamber of a mud shrimp (*Upegehta magniticorium*). Photo courtesy of Shane Anderson.

Figure 3   Infected (individual on the left) and uninfected (individual on the right) tobacco hardworm caterpillars (*Manduca secta*). The infected individual on the right has newly emerged larvene of the parasitoid wasp *Cotesra congregala*. Photo courtesy of Shelley Adasno.

Figure 5   Mud shrimp (*Upaghia magniticorium*) infected with bopyrid isopods in its branchial chamber. The carapace eventually bulges out, around the paired isopods, after several molt cycles. Photo courtesy of Shane Anderson.

also synchronize their molting cycles to those of their hosts, indicating that they have evolved means of reading host hormones. It is possible then that bopyrids are also able to secrete chemicals that communicate with host neuroendocrine systems and inhibit reproductive activity, such as sperm or egg production. Furthermore, effects of bopyrid castration can be reversed, with the host's gonad regenerating in some capacity (but not fully) after experimental removal or death of the bopyrid. Thus, bopyrids likely castrate through a combination of direct chemical or hormonal signals to the host reproductive system and indirect energetic drain on general host resources.

Rhizocephalan barnacles also castrate their crustacean hosts. In addition to modifying host reproductive behaviors, such as inducing egg-caring behavior in both females and males, the parasite also drastically alters the host's reproductive system. The rhizocephalan interna grows by branching out nutrient-absorbing processes that proliferate throughout the crab's body. Infected females usually have degenerate ovaries that are permeated by branches of the interna. The testes of infected males generally do not experience the same degree of atrophy, but the androgenic gland is often destroyed. This results in the cessation of sex hormone production responsible for spermatogenesis and secondary sexual characteristics. Compared to uninfected crabs, female crabs infected with rhizocephalans have higher levels of a gonad-inhibiting hormone (GIH), which inhibits ovarian growth and egg production. Although this hormone regulates the timing of reproduction in uninfected females, its continual circulation could suppress ovarian development while diverting potential energy toward use by the parasite. It remains unknown whether the parasite secretes GIH or another chemical that triggers the host to secrete it. Thus, it appears that rhizocephalans castrate their hosts through a combination of nutrient depletion and hormonal communication. The actions of GIH are remarkably specific when considering that the hormone reduces only the reproductive organ of an infected female and not the egg-caring behaviors displayed by gravid females.

Larval trematode parasites often infect the gonads of their molluscan intermediate hosts. In most cases, the hosts are castrated, and the trematodes completely fill the region formerly occupied by host gonad (see **Figure 6**, for cross-section of gonad). The California horn snail, *Cerithidea californica*, is a common benthic invertebrate in California and Baja California estuaries. The horn snail serves as first intermediate host to multiple species of trematodes, which go on to infect other molluscs, annelids, crustacea, and fishes as second intermediate hosts, and birds as final hosts. Many of these trematodes have asexually reproducing larval stages called rediae, which are mobile wormlike stages equipped with a mouth and pharynx. Rediae consume host tissue, including gonadal tissue, as the infection progresses, usurping host nutrients and space for parasitic growth. They have

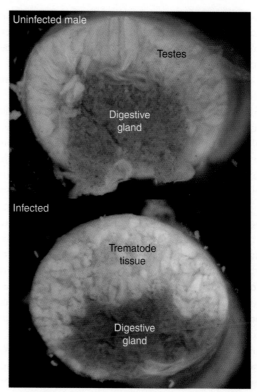

Figure 6   Cross-section of California horn snails (*Ceruhedia california*) uninfected and infected with larval tremalodes. Infected snails are parasitically castrated. Photo courtesy of Ryan Hechinger.

even been observed consuming larval stages of other trematode species in concurrent infections within the same snail. The sporocyst is another trematode larval stage that is similar to rediae in appearance but lacks a mouth, instead absorbing nutrients across the tegument. Depending on the trematode species, sporocysts may occur as precursors of rediae, or occur without producing rediae. Sporocysts are nonetheless associated with castration. Furthermore, there are also trematodes that infect only the snail's mantle region, the area above the snail's head. Snails with mantle infections – but not gonad infections – are also completely castrated. In addition to castration, these snails may also experience some slowness in retracting their heads in response to danger, due to the bulging mantle infection. Since the trematodes in the last two examples cannot castrate their host by ingesting reproductive tissue, they likely castrate through neurochemical alteration, energetic drain, or a combination of both strategies.

There are clear indications of energetic drain as a partial cause of castration in *C. californica*. Trematodes infecting *C. californica* undergo a tremendous amount of growth within the snail, reaching up to 2% of an individual host's biomass in some populations. The high energetic demands required for such growth likely usurp resources from host reproduction, resulting in castration even if the parasites are located elsewhere in the snail. Castration by manipulation of endocrine and/or neural systems remains to be identified in this host–parasite system; however, there is clear evidence that trematodes can castrate their snail hosts in the freshwater system described as follows.

Several studies document the complex mechanisms by which schistosome trematodes castrate their freshwater snail hosts through neuroendocrinological alteration. The avian schistosome *Trichobilharzia ocellata* manipulates both the energy budget and the reproductive system of its snail host (*Lymnaea stagnalis*) to maximize nutrient yield for parasitic growth and reproduction. The parasites infect the snail's gonad (ovotestis) and eventually castrate their hermaphroditic hosts. *Trichobilharzia ocellata* lacks rediae, so castration does not result from consumption of host reproductive tissue. Extensive work by Marijke DeJong-Brink and colleagues demonstrate that *T. ocellata* actively alters its host's neuroendocrine system using multiple strategies to successfully establish and advance infection.

Infected *L. stagnalis* upregulate a gene coding for a homolog to neuropeptide-Y (LyNPY). LyNPY is highly conserved in structure and function to that of vertebrate NPY, which is important in determining energy allocation toward growth and reproduction and can inhibit reproduction. Administration of synthetic LyNPY to uninfected snails, which mimics conditions of a *T. ocellata* infection, causes decreased or complete cessation of egg production and growth. This probably resulted from nutrient deprivation to the reproductive systems, because host feeding was not adversely affected, and LyNPY appears to prevent mobilization of energy to host reproduction and growth. The parasite also interferes directly with the development of ovotestes and gamete production. Neuroendocrine products excreted and/or secreted by the schistosome also directly inhibit cellular differentiation of the male copulatory organ. In later infection stages, the schistosome induces the host to produce and release schistosomin, a peptide with actions similar to those of vertebrate cytokines. Schistosomin simultaneously inhibits efficacy of gonadotropic hormones on reproductive organs and decreases the activity of central neuroendocrine cells that regulate ovulation, egg production, and associated behaviors. The intricate communication between host and parasite neuroendocrine systems suggests a highly coevolved system in which *T. ocellata* has successfully adapted to exploit host physiology in efforts to maximize parasite fitness.

Hosts Alter Reproductive Efforts in Response to Parasitism

Not all hosts passively succumb to the detrimental effects inflicted by parasites; some go down fighting. Hosts have demonstrated the ability to alter their life history strategies when faced with increased mortality or reproductive death (castration) due to parasitic infection. Several invertebrates actually ramp up their short-term reproductive efforts – generally defined as the energy and resources allocated to reproduction – in response to infection. Male desert flies (*D. nigrospiracula*) court females at a significantly higher rate after becoming infested with ectoparasitic mites (*M. subbadius*; see above). Infected males suffer from shorter life spans, probably due to loss of host hemolymph caused by feeding mites. This suggests that the increased courtship behavior may be an adaptive effort to compensate for potential long-term losses in reproduction. This may not always translate into actual reproductive gains, since infections of more than four mites can physically obstruct and prevent males from mating. Nevertheless, the plasticity of this behavior should still be selected for in fly populations if some infected males experience greater overall fitness by increasing reproductive efforts.

Biomphalaria glabrata are freshwater snails that serve as first intermediate host for the trematode *Schistosoma mansoni* – one of the major causative agents of human schistosomiasis. Schistosomes, like many other trematodes in their first intermediate snail hosts, completely castrate their hosts once they proliferate throughout the snail's gonad tissue. After Minchella and LoVerde exposed snails to *S. mansoni* in the laboratory, the exposed snails actually increased egg laying in the weeks immediately postexposure, before castration was complete. This phenomenon – termed 'fecundity compensation' – was even observed in snails that were exposed but did not become infected, indicating a persistent reaction to the initial threat of parasitism. The long-term egg output of these snails turned out to be less than that of their uninfected conspecifics, which probably resulted from the intense burst of reproductive activity early on. It may seem that the exposed but not infected snails ultimately sacrificed their maximum reproductive output for naught, but such immediate responses to infection may be genetically selected for in host populations that experience high rates of parasitic castration (see discussion on *Cerithidea californica* later). A short-term surge of reproductive activity may mitigate potential fitness losses for species that eventually become castrated.

Fecundity compensation has been observed in systems involving pathogens as well as parasites. *Daphnia magna*, a small aquatic crustacean, increases reproductive output in response to infection by the bacterium *Pasteuria ramosa*. Lab-infected individuals produce egg clutches earlier than uninfected ones, although they are eventually castrated. Interestingly, the number of clutches that an infected *Daphnia* produces before becoming completely castrated is inversely proportional to the reproductive output (number of spores) of the bacteria, where the increased reproductive efforts of the host compromise resources available for eventual use by the parasite. This indicates a struggle between host and parasite for energetic resources, where *Daphnia* struggles to maximize its reproductive output before it is sterilized, and *P. ramosa* races to castrate its host as quickly as possible in order to minimize its potential fitness losses.

There are indications that mammals can also compensate for fecundity losses due to parasitic infection. Mammals do not typically face the threat of parasitic castration, due to a number of variables including a large host-to-parasite body size ratio, which allows them to preserve resources for growth and reproduction even after some resources have been consumed by parasites. Instead, fitness losses may result from a shortened lifespan due to infection pathology. Described examples of mammalian fecundity compensation are restricted to short-lived species, which may indicate a selective condition for maintaining plasticity in reproductive timing and effort.

Wild deer mice (*Peromyscus maniculatus*) typically live about 6 months, whereas mice infected with schistosomatid trematodes (*Schistosomatium douthitti*) exhibit shorter lifespans. The schistosome infection is chronic, with worms living in the mouse for the duration of the mouse's life, and pathology can be severe. Adult worms live in the mesenteric vein system and pass hundreds of eggs a day through the venule walls to the intestine. Schistosome eggs then exit the host with the feces. However, many of these eggs become lodged in the host's liver and cause inflammation and eventually calcification. These symptoms are similar to those of certain types of human schistosomiasis (e.g., *Schistosoma mansoni*). Infected mice suffer from extensive liver damage and compromised thermoregulation, and they exhibit 30% higher mortality. Even though the schistosome worms do not castrate deer mice, female mice still respond to infection by increasing reproductive effort, likely due to the threat of a shortened lifespan. Experimentally infected females produce litters of significantly higher body mass compared to litters from uninfected females. This outcome can be viewed as a maternal fitness gain, since greater mass likely constitutes greater competitive ability and survival for an individual offspring. In contrast, longer-lived mammals can often overcome an infection and go on to reproduce in some capacity, which may explain the rarity of short-term fecundity compensation in mammalian hosts.

The examples above all confer the ability of individual hosts to alter short-term reproductive efforts when threatened with parasite-induced decreases in reproduction and, ultimately, fitness. However, host populations can also exhibit life history adaptations in response to high levels of parasitism. The California horn snail, *Cerithidea californica*, is found in high densities in many estuaries throughout California and Baja California. Most of these populations experience moderate-to-heavy rates of infection by multiple species of trematodes that use the snails as first intermediate hosts.

The trematodes infect both immature and mature individuals, grow via asexual reproduction, and gradually consume and/or occupy the snail's gonad, parasitically castrating the snail. Snails from populations with a high prevalence of infection matured at earlier life stages. Cross-transplant experiments, using individuals from both high and low prevalence populations, confirmed the genetic basis for earlier maturation. Larval and adult snails are relatively sedentary, so putative low rates of gene flow between sites – even within an estuary – further indicate that populations may adapt to local levels of parasitism. These findings are consistent with others that show individuals maturing earlier when lifespan and reproductive output are threatened, as with heavy predation pressure or harsh habitats.

Fecundity compensation and earlier maturation can be considered to be host responses to mitigate fitness losses caused by parasitism. These strategies are generally viewed as beneficial for the host and not parasite. In all the scenarios described earlier, the hosts do not evade infection or castration by altering their reproductive efforts in response to infection. Rather, hosts may be attempting to maximize their reproductive output once faced with an inevitable loss of reproductive ability.

Conclusions

Parasites that target host reproductive systems strike an optimal balance of resource consumption in hosts. The parasite gleans a nutrient-rich source of energy for their growth and reproduction, while taxing a host resource that is nonvital for host survival. Parasites can also exploit alterations in host reproductive behavior to further increase parasite fitness. Many alterations of host behavior benefit the parasite, and seem to constitute a parasite strategy for increased survival and reproduction. However, there are systems in which the host alters reproductive efforts in response to parasites, perhaps attempting to mitigate the damaging effects eventually induced by infection. Such struggles between host and parasite for host energetic resources drive the dynamics of host and parasite coevolution.

See also: Evolution of Parasite-Induced Behavioral Alterations.

Further Reading

Baudoin M (1975) Host castration as a parasitic strategy. *Evolution* 29(2): 335–352.

Beck JT (1980) Effects of an isopod castrator, *Probopyrus pandalicola*, on the sex characters of one of its caridean shrimp hosts, *Palaemonetes paludosus*. *Biological Bulletin* 158(1): 1–15.

Beckage NE (1997) *Parasites and Pathogens: Effects on Host Hormones and Behavior*. New York: Chapman and Hall.

de Jong-Brink M, Bergamin-Sassen M, and Soto MS (2001) Multiple strategies of schistosomes to meet their requirements in the intermediate snail host. *Parasitology* 123: S129–S141.

Helluy S and Holmes JC (1990) Serotonin, octopamine, and the clinging behavior induced by the parasite *Polymorphus paradoxus* (Acanthocephala) in *Gammarus lacustris* (Crustacea). *Canadian Journal of Zoology* 68(6): 1214–1220.

Hurd H (2001) Host fecundity reduction: A strategy for damage limitation. *Trends in Parasitology* 17(8): 363–368.

Kuris AM (1974) Trophic interactions: Similarity of parasitic castrators to parasitoids. *Quarterly Review of Biology* 49(2): 129–148.

Lafferty K (1993) The marine snail, *Cerithidea californica*, matures at smaller sizes where parasitism is high. *Oikos* 68(1): 3–11.

Minchella DJ and Loverde PT (1981) A cost of increased early reproductive effort in the snail *Biomphalaria glabrata*. *American Naturalist* 118(6): 876–881.

Webb TJ and Hurd H (1999) Direct manipulation of insect reproduction by agents of parasite origin. *Proceedings of the Royal Society of London Series B – Biological Sciences* 266(1428): 1537–1541.

Reproductive Behavior and Parasites: Vertebrates

H. Richner, University of Bern, Bern, Switzerland

Introduction

The Viewpoint of the Host

Host–parasite interactions during the reproductive period differ from those during other phases of the life cycle for several reasons. First, for many hosts, reproduction is one of the most demanding activities, and dealing with parasites at this time is much harder than it is during more relaxed phases of life. Second, many reproductive traits are linked to each other by trade-offs involving one or several behavioral activities that, under parasite pressure, require different optimization processes or strategies in order to maximize reproductive success. While most means of control of infection by hosts (e.g., parasite avoidance, preening, grooming, sun bathing, self-medication, or maternal effects) appear as obvious and often efficient in reducing parasite load, their frequency of occurrence will not be determined by their beneficial effects alone, but by the trade-offs incurred with other essential activities during reproduction, or by trade-offs with physiological functions. Third, many hosts are less mobile during reproduction, or are confined to a nest site, which makes them an easy target for parasites that require fixed locations for completing their own reproductive cycles. Fourth, newborns are naive with respect to parasites and have limited defenses, including their relatively immature immune systems. Protecting young from the damaging effects of parasites may then involve strategies that reduce exposure or alleviate damage if exposure occurs. An example of the latter strategy is the induction of maternal effects (e.g., immunoglobulins, hormones, antioxidants) that can be transmitted from mother to offspring via the egg, milk, or nutrients, and that protect young from the damaging effects of parasites. Fifth, parasitized young most often show compromised trajectories of growth and development that lead to permanent changes in phenotype and persist into adulthood. Thus, a control of the parasites' effects by behavior and other means during reproduction should be among an individual's highest priorities.

The Viewpoint of the Parasite

Fish, amphibians, reptiles, birds, and mammals are hosts to many parasite types of different taxonomic groups. These parasites are characterized by radically different life styles that pose different constraints for hosts and thus evoke different sets of host strategies. Such differences may be best understood if we adopt, for a moment, the viewpoint of the parasites. Protozoan blood parasites, for example, often depend on insect vectors for transmission to the next host, and may therefore attempt to manipulate hosts in a way that increases transmission to their specific vector. Tapeworms, as another example, live permanently inside the guts of their host; they produce vast numbers of eggs and compete with the host for the energy they both require for reproduction. The most successful tapeworms may then be the ones that hinder host reproduction. Flukes can occur in the blood, the guts, the bile ducts, the lungs, and other tissues of birds and mammals, and may therefore affect physiological functions that are important for hosts during the strenuous time of reproduction; they can thus induce strong changes in reproductive behavior and performance of hosts. The most successful parasites may not necessarily be the ones that win the competition with hosts for reproductive resources, but the ones that can best keep the host functioning for their own purpose.

In contrast, nest-based ectoparasites (e.g., mites, fleas, louse flies, blowfly, and botfly larvae) live off nestlings and have therefore no interest in reducing host fecundity. The parasites' interests are rather the other way round: more nestlings mean more resources for the parasites' offspring. Many ectoparasite species consume blood of nestlings and thus require an optimal synchronization of their own reproduction and larval growth with the reproductive phase of their hosts. Since larval growth of ectoparasites also depends on environmental temperature, the early breeding birds of temperate zones may not be optimal hosts and may constrain parasite reproductive success early in the season. In some ectoparasite species, the larvae develop underneath the host's skin and can thereby circumvent such constraints. The precise coordination of host and parasite reproduction is in most cases a tight coevolutionary arms race between hosts and such ectoparasites. Other ectoparasites such as ticks may live on hosts at any time of the year and their reproductive timing will therefore be less constrained. These are some of the examples that illustrate the diversity of parasite life styles and how they may set the scene for the more or less intimate relationships with their hosts.

Parasitism, Reproductive Traits, and Behavior

Correlations

A vast number of studies document correlations between parasitic infections of vertebrates and reproductive performance and behavior of hosts. Male three-spined sticklebacks (*Gasterosteus aculeatus*) naturally infected with a common cestode parasite, for example, showed weaker nuptial colors, lower courtship levels, and nesting activity than uninfected males. Pied flycatchers (*Ficedula hypoleuca*) naturally infected with blood parasites arrived later in the season at the breeding ground. Red Grouse (*Lagopus lagopus scoticus*) infected with round worms in their guts laid fewer eggs later in the season than uninfected individuals. American Kestrels (*Falco sparverius*) misplaced their eggs when infected with a nematode that lives in muscle tissue. Such correlations between host infection and behavioral changes, however, cannot in principle demonstrate a causal pathway from parasites to host behavior since it could just as well work the other way round, where phenotypes with particular traits and behaviors are more likely to contract parasitic infections. The few exceptions arise when parasites interfere directly with host organs required for reproduction, castrate hosts, or drastically reduce the levels of essential male hormones. In the white-footed mice (*Peromyscus leucopus*), the huge larvae of a botfly develop mainly in the groin region and thereby often prevent the descent of testicles and successful copulation. In male mice (*Mus musculus*), the common trematode (*Schistosomia mansoni*) reduces testosterone to castration levels. Following infection, the endogenous opiate system is activated and decreases the release of luteinizing hormones, which in turn reduces testosterone synthesis. Lowered testosterone levels can retard the growth and development of parasites and therefore be in the interest of the host, and also be initiated by the host.

These few examples demonstrate two important points regarding the cause–effect relationship in host–parasite interactions: First, the observed symptoms of infections may not be caused by the parasite, but instead may be a property of a phenotype that is correlated with a higher likelihood to become infected. Second, the observed symptoms are not necessarily adaptive for the parasite, but may be adaptive host responses to parasite infection. Since the cause–effect relationship is not obvious in most host–parasite relationships, it is important to discriminate between studies that simply draw conclusions from correlations, and studies that use experimental infections of randomly chosen host individuals in combination with properly designed controls. Furthermore, a careful interpretation of the observable correlations in host–parasite interactions with regard to the direct effects of parasites versus adaptive host responses is essential. Given the difficulty of disentangling parasite effects from effects due to the phenotype, nonrandom distributions over environments, or other nonrandom effects, the following examples are preferentially drawn from experimental studies that attempted randomization of infection.

A Life-History View

Understanding cause–effect relationships in host–parasite interactions also requires a life-history perspective. The commonly observed concurrence of reproductive activity and parasitic infections, for example, is expected if resource allocation to reproduction occurs at the expense of resource allocation to parasite defense or immune function. Such trade-offs will arise in any organism and, essentially, all functions that contribute to fitness can be involved in trade-offs. These trade-offs form the core of life-history theory.

For example, latent infections of vertebrate hosts with blood parasites are common, but many patent infections only arise during stressful periods, such as reproduction. What is the exact relationship then between reproductive effort and parasitism? The traditional view holds that parasitic infections will reduce reproductive effort and success, and thus interprets the lowered reproductive performance as a consequence of infection. The life-history view, in contrast, holds that reproductive effort will determine the chance of becoming infected or developing a patent infection, and thus views infection as the consequence of reproductive effort. An experiment by Heinz Richner and his students at the Universities of Bern and Lausanne on great tits (*Parus major*) infected by *Plasmodium* spp. illustrates the point: average prevalence of patent infection in reproducing females is around 20%. Females lay on average eight eggs. Richner and students removed the first two eggs in the laying sequence in one (experimental) group of breeding females; as a result, the experimental females laid one more egg but raised one offspring less than the females of the control group. The prevalence of infection in the control group was 20% but, strikingly, rose to 50% in the experimental group. This illustrates that reproductive effort and defense against blood parasites form a trade-off. Why, then, does an increase of 12.5% in reproductive effort lead to more than a doubling of the prevalence of parasitism? Relationships between life-history traits are expected to be nonlinear, and therefore an increase of reproductive effort at low levels of effort (**Figure 1**: from x_1 to x_2) will have small consequences for the resistance against parasites (from y_1 to y_2), whereas an equal increase at high levels of reproductive effort (from x_2 to x_3) will drastically reduce the resistance against parasites (from y_2 to y_3).

Behavioral Strategies for Coping with Parasites During Reproduction

Parasite Avoidance

The most obvious defense against parasites seems to be to avoid infestation in the first place. It implies being choosy

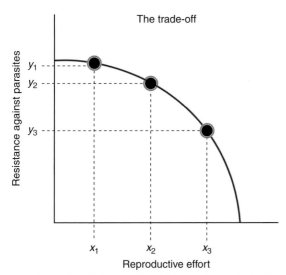

Figure 1 Trade-offs between life-history traits are typically nonlinear, with the consequence that equal changes in trait X lead to disproportionate changes in trait Y. An increase in reproductive effort at low levels of reproductive effort, for example, leads to a small reduction in resistance against parasites, whereas at higher levels of reproductive effort, the same increase leads to a more drastic reduction of resistance.

for mates, for breeding sites, and for places to roost and to forage. Gray treefrogs (*Hyla versicolor*), for example, laid fewer eggs in experimental pools containing snail vectors infected with a harmful trematode than in control pools with uninfected snails. Among great tit (*P. major*) females that were offered a choice between a flea-free and a flea-infested nestbox, almost 80% chose the flea-free nest site. When no such choice was given, 60% of females abandoned breeding at later stages. Female Eastern blue birds (*Sialia sialis*), given a choice between ectoparasite-infested and parasite-free nesting sites, preferred the ones without parasites. Cliff swallows (*Petrochelidon pyrrhonota*) used nesting sites that were lightly infested with cliff swallow bugs every year, but only used heavily infested sites every second year.

Since choice among options requires investment of time and energy, it becomes clear that parasite avoidance, although an intuitively obvious choice, may not always be the optimal strategy. While parasite avoidance per se is well documented, demonstrations that avoidance is the best behavioral response to potential infestation are rare. In the great tit example given earlier, the cost of deserting an infested brood must be offset by the benefits of breeding elsewhere.

Behavioral Control After Parasitism

Behavioral means for reducing the parasite load include preening, grooming, using chemical compounds of plants,

sun or dust bathing, anting, spacing of breeding sites, adjustments in social behavior, preening of conspecifics, and others. These behaviors, however, are not restricted to the period of reproduction and can occur at any time of the year. The frequency of occurrence of these behaviors throughout the life of an animal may be partly determined by parasite load, but probably equally important, by time constraints. Time constraints are especially relevant during reproduction, when demands for finding food, defending a nesting site, and providing a suitable environment for the developing offspring are peak priorities that leave little time for behaviors that control parental parasite load. In contrast, the control of parasites that target offspring may be given much higher priority. Birds that use plants with specific antiparasitic chemical compounds as nesting material provide a striking example. In an experimental study on starlings by Helga Gwinner, plants had no direct effect on the number of mites in bird nests but reduced the load of bacteria. Offspring in starling nests with added plant material fledged at higher body weights than did offspring in nests with plant material removed. Thus, the plants seem to compensate for the effect of the mites, which usually do have deleterious effects on growing young. Nestling Alpine swifts (*Apus melba*) can be heavily covered by the blood-sucking louse-fly (*Crataerina alba*). Nestlings often walk around the breeding colony and thereby become adopted by other parents. Experimentally infested nestlings sought adoption more often and earlier than noninfested controls, and their parasite load became smaller after adoption, due to a redistribution of louse flies among the members of the foster family.

An elegant illustration of the efficiency of parasite control by preening behaviors is provided by the work of Dale Clayton and his students at the University of Utah. In an ingenious experiment on rock pigeons (*Columbia livia*) ridden by feather lice, Dale Clayton prevented wild caught pigeons from preening efficiently by placing small metal bits between the upper and lower beak. The procedure had no effect on the condition or reproductive performance of the birds, but the load of two common feather-eating louse species increased 2–3 times. Also, rock pigeons with slight bill deformities had higher louse loads than pigeon without deformities. Interestingly, preening with normal bills selected for smaller lice that could escape preening, which illustrates the power of the coevolutionary process in host–parasite interactions.

Parasite Control by Maternal Effects

The discovery that the maternal environment before and during ovulation and pregnancy can permanently influence offspring development and behavior into adulthood is among the most important recent advances in understanding host–parasite interactions. Olivia Curno and collaborators at the University of Nottingham showed that mice housed in proximity to noncontagious conspecifics infected with the protozoan *Babesia microti* produced offspring that, as adults, showed a faster immune response to *B. microti* and lower aggression in social groups than offspring of mice housed close to uninfected mice. Such transgenerational effects appear to be common both in vertebrates and invertebrates. Birds are very well suited for experimental approaches to maternal effects since development of offspring occurs in the eggs and thus allows experimental separation of the prelaying and posthatching maternal influences on offspring. Technically, this can be done by first exposing mothers to a given parasite during egg formation in order to induce the maternal effect. This is then followed by a random cross-fostering of these eggs among females that have not been exposed. At same time, the females that have been exposed to parasites are given the eggs laid by the previously uninfested mothers. During the growth of the young, half of the nests are then exposed to the same parasite species that has been used to induce the maternal effects. Such a strict experimental protocol can show both the effect of an induction of a maternal response to a parasite on the rearing capacity of a female and the benefit of the maternal response for the development of offspring. Heinz Richner and his students at the University of Bern performed a series of such increasingly complex experiments in field studies on great tits infected by hen fleas. They found the following results: (1) females with a flea-induced maternal response have almost double the number of offspring compared to controls, (2) the maternal effect reduces survival times of fleas after they feed on nestlings, (3) the maternal effect can be beneficial to nestlings even if they are not exposed to parasites themselves, and (4) the maternal effect modulates natal dispersal of young. Birds whose mother has been exposed to fleas disperse shorter distances for breeding than birds with unexposed mothers. If infested with fleas, the birds with smaller natal dispersal distances are more successful breeders.

The mechanisms that mediate the transgenerational effects are varied. It has been shown that parasites can induce changes in the level of hormones (androgens or corticosteroids), antioxidants (e.g., carotenoids), and immunoglobulins deposited in the egg. In the great tit, exposure to fleas during egg formation leads to higher immunoglobulin concentrations and lower testosterone levels in eggs. A study where testosterone levels were directly manipulated in freshly laid eggs showed that lower testosterone levels led to shorter natal dispersal, thereby underlining the adaptive significance of the maternal response.

This type of transgenerational effect has been demonstrated in other bird species. For instance, Nicola Saino from the University of Milano and collaborators showed that female barn swallows (*Hirundo rustica*) exposed to an antigen (Newcastle disease virus vaccine, NDV) transfer significant quantities of anti-NDV antibodies to the egg yolk. Jennifer Greenstaff at the University of Indiana induced a maternal response in three groups of female Japanese quail (*Coturnix japonica*), one group as a control and the other two groups each with a different antigen. She demonstrated both elevated concentrations of specific antibodies in the offspring and higher growth rates if nestlings were immunized with the same antigen as their mother, compared to nestlings immunized with a different antigen. In general, growth and development are among the most important factors that determine the offspring's competitive ability, access to resources and other behaviors vital for survival both in the nest and as adults.

Another yet largely unexplored mechanism is genomic imprinting, that is, a change in the pattern of gene expression in the offspring, determined by the parent of origin, without an alteration of the underlying genetic code. Genomic imprinting is most commonly due to altered gene expression based on an induced change in DNA methylation or histone acetylation. The maternal environment induces such changes in genes passed on to offspring, and there is no reason to believe that parasites could not be an important inductor for genomic imprinting of genes.

Given that parasite-induced maternal effects can be highly beneficial in most situations where mother and offspring experience similar parasite environments, one would expect mothers to specifically seek exposure to these parasites before and during the time of egg formation, and also during pregnancy in species where young develop inside females and are essentially provisioned by the mother. There are hardly any studies that explore the occurrence of such behaviors as a consequence of expected parasite exposure and parasite type.

Some Current Questions

Despite the rather large number of studies that describe relationships between parasitic infections and reproductive traits and behaviors of hosts, the number of studies that have attempted to understand these interactions under strict experimental control is limited. Future studies should aim to achieve a proper randomization of parasitic infections among phenotypes and habitats in order to understand cause–consequence relationships, and perform experiments that can discriminate between pure parasite effects and adaptive and nonadaptive host responses. There should also be more studies that include investigation of longer-term trade-offs, such as linkages between the benefits for the current brood of behavioral adjustments due to parasitism and the cost of those adjustments for raising future broods.

Also, in studies of vertebrates there is a very strong bias toward birds, which limits our general understanding of host–parasite interactions. Life histories, ecology, and physiology of fish, amphibians, reptiles, birds, and mammals are in many respects fundamentally different, and predict different outcomes of host–parasite interactions and susceptibility to parasites.

The physiological mechanisms that mediate the effects of parasites on host behavior, and vice versa, are in most cases not well understood. Substances that are suspected to mediate these effects should be manipulated independent of parasites in order to assess their fundamental role in the mediation process. Recent developments in molecular biological techniques could be applied, for example, to understand parasite-induced changes in gene-expression in hosts, including imprinting of genes.

See also: Avoidance of Parasites; Ectoparasite Behavior; Intermediate Host Behavior; Maternal Effects on Behavior; Parasite-Induced Behavioral Change: Mechanisms; Parasites and Sexual Selection; Reproductive Behavior and Parasites: Invertebrates; Self-Medication: Passive Prevention and Active Treatment.

Further Reading

Combes C (2005) *The Art of Being a Parasite*, p. 291. Chicago, IL: The University of Chicago Press.

Hart B (1997) Behavioural defense. In: Clayton DH and Moore J (eds.) *Host-Parasite Evolution*, pp. 59–77. Oxford: Oxford University Press.

Hillgarth N and Wingfield J (1997) Parasite-mediated sexual selection: Endocrine aspects. In: Clayton DH and Moore J (eds.) *Host-Parasite Evolution*, pp. 78–104. Oxford: Oxford University Press.

Moore J (2002) *Parasites and the Behavior of Animals. Oxford Series in Ecology and Evolution*, pp. 315. New York, NY: Oxford University Press.

Oppliger A, Christe P, and Richner H (1996) Trade-off between clutch size and parasite resistance. *Nature* 381: 565.

Reproductive Skew

S.-F. Shen and H. K. Reeve, Cornell University, New York, NY, USA

Introduction

A major goal of evolutionary biology is to understand how the characteristics of animal societies emerge from the operation of natural selection on the social strategies employed by all members of the society. Of extreme interest is the possibility that variation in the major characteristics of societies can be understood within a single theoretical framework. Reproductive skew theory was born out of the desire to develop such a framework, one that could unify our understanding of social evolution in social insects, cooperatively breeding vertebrates, within groups of interacting plants. Skew models extend even among cells of a metazoan organism and among genes within a single organismal genome.

Reproductive skew is the unevenness of partitioning of reproduction among same sex individuals within social groups; in other words, some males or females may reproduce more than other males or females. Skew is an obvious major dimension along which all animal, including insect, societies vary. In high-skew societies, such as honeybee, yellow-jacket wasp, and many ant and termite societies, reproduction is monopolized by one or a few group members (i.e., the dominant breeders); in low-skew societies, such as communally nesting wasp and bee societies and some multiqueen ant associations, reproduction is shared almost equitably. Marked variation in skew occurs not only among species but also often within species, depending on ecological factors such as the time of year, competitive (e.g., body size) asymmetries among group members, and genetic relatednesses among group members.

Because reproductive skew is such a widely varying attribute of animal societies, the goal is a general theory that explains variation in skew among different ecological and genetic settings, and under distributions of different competitive asymmetries among group members. This would likely be a good candidate as a 'unifying theory' for understanding all of social evolution. To achieve this goal, skew theorists attempted to predict levels of reproductive skew by using a combination of game theory and kin selection theory. They asked what sorts of evolutionarily stable reproductive skews would result from games in which each of the group members was trying to maximize its own inclusive fitness.

In the initial development of skew theory, the connection between explaining reproductive skew and the problem of whether groups should form in the first place (and to what size these groups should grow) quickly became obvious. Consider the decision of an organism to join a group versus reproducing alone. The inclusive fitness payoff an animal receives for joining a group depends partly on how much of the total group reproduction that it receives by joining the group. However, this fraction depends on evolutionary games among group members that determine the reproductive skew. Thus, the problems of reproductive skew and group size are inextricably linked—one cannot be solved without solving the other.

More recently, it has become obvious that the problem of skew also strongly connects to the problem of how selfish and cooperative group members act within stable groups. In other words, what level of conflict will be observed within stable groups? The payoff for a selfish act depends on how much an individual increases its selfish share of the group reproduction. In turn, the potential for increasing its share depends on the current reproductive skew within the social group. How selfishly a group member acts in turn modifies its reproductive share and thus, affects the reproductive skew. As a result, the level of within-group conflict and the reproductive skew intertwine such that theory must solve both simultaneously.

In sum, skew models ultimately attempt to explain not only reproductive partitioning within societies but also whether groups should form, the evolutionarily stable sizes of groups, and degrees of intragroup conflict. These models apply, generally, across vertebrates and invertebrate species. Skew models unify our knowledge of reproductive skew across taxa by building on the twin theoretical pillars of kin selection theory and game theory. In addition, skew theory unifies our understanding of reproductive sharing, group size, and social conflict within taxa (formerly, these have been treated as almost separate problems).

In the next section, we describe the major classes of skew models and their major predictions. We then briefly discuss recent attempts to synthesize earlier models into a more comprehensive theory that accommodates the richly varying data on reproductive skew in animal societies, such as the social insects. Our discussion of the models roughly parallels their chronological development.

Models of Reproductive Skew

Transactional Models

The earliest transactional model of reproductive skew was proposed in 1979 by Sandra Vehrencamp and shortly thereafter by Stephen Emlen. The theory was later formalized explicitly into a game between dominants and subordinates by Reeve and Ratnieks, who also derived the conditions under which a group should form in the first place. The basic idea of the transactional model was simple: dominants always try to monopolize group reproduction, but if they are too greedy, the subordinates may get insufficient direct reproduction (offspring production) and respond by leaving the group to reproduce on their own. Thus, there are limits to how selfish a dominant can be; how much reproduction dominants must concede to (i.e., pay) subordinates is predictable from inclusive fitness theory: the lesser the share of reproduction for the subordinates, the greater their genetic relatedness to the dominant, the lesser their share of the reproduction. This is because the dominant can more exploit a closely related subordinate without making it favorable for the subordinate to leave. A subordinate receives greater indirect genetic benefits by helping a more closely related dominant, so such a subordinate does not require as large a direct reproductive share to gain by staying in the group. Also, a subordinate's reproductive share should decline as the success of solitary breeding decreases (since the dominant does not have to pay it as much to make it favorable for the subordinate to stay) and as the benefit of forming a group increases (because subordinates will accept a smaller share of a bigger total group pie).

Thus, skew, measured as the dominant's reproductive share, is particularly high under conditions of high relatedness, large group benefit, and harsh ecological constraints on solitary breeding. Above certain critical values of these variables, theory predicts complete reproductive skew: the dominant monopolizes the group reproduction, and the subordinate receives no reproductive share at all. Finally, the transactional model predicted that groups form under the simple condition in which the total reproduction of the group exceeds the sum of what the subordinates and the dominant would produce alone.

The transactional model assumes that dominants have essentially complete control over the subordinate's reproductive share. Dominants have the power to 'push' a subordinate's reproductive share to the point at which it almost 'breaks even' in inclusive fitness when staying is compared with leaving the group. Cant and Johnstone pointed out that an alternative model (a 'restraint' or 'eviction' model in contrast to the earlier 'concession' model) yields completely opposite predictions if the subordinate maintains complete control over reproductive shares. In this case, the subordinate increases its reproductive share up to the point at which theory favors eviction by the dominant. However, the predictions of the concessions and restraint come back into alignment if the dominant is defined as the individual controlling reproductive shares.

Another restrictive assumption of the transactional model is that the dominant and the subordinate have symmetrical genetic relationships with each other's offspring (e.g., as sisters or unrelated individuals). However, eusocial insect societies are, by definition, groups in which the dominants are mothers of the subordinates (workers). Reeve and Keller expanded the transactional theory to the latter situation and found that reproductive skew always is higher in mother–daughter associations than in sibling associations or among groups of nonrelatives. This occurs because the daughter can be as closely related to the dominant's offspring (i.e., siblings) as to her own offspring, so she has little genetic incentive to leave an association with a monopolizing dominant. In contrast, the dominant mother has great genetic interest in monopolizing the reproduction.

Transactional theories of skew were highly attractive; the models were simple (mostly algebraic) and had straightforward predictions. These models had to be tweaked for application to Hymenopteran systems (e.g., to cases where Hymenopteran relatedness asymmetries lead to split sex ratios) but nonetheless were straightforward. The major successful prediction was that mother–daughter groups should have generally higher skew (where workers are viewed as subordinates) than should groups of siblings or nonrelatives, as in queen-foundress associations or in communal colonies of wasps and bees.

Detailed tests of the theory in groups of symmetrically related individuals yielded mixed results, however. Bourke and coworkers showed that the theory nicely explains variation in skew in small queen associations in leptothoracine ant populations that experience varying ecological constraints on independent founding. The theory seems to explain why, in early summer colonies of the paper wasp *Polistes fuscatus*, reproductive skew in early season offspring production varies positively with both foundress relatedness and comb size (as a surrogate measure of group productivity). However, in other *Polistes* species such as *Polistes dominulus and P. bellicosus*, and in some stenogastrine wasp species, the skews are typically too high to test whether skew varies in correlation with other parameters. For example, relatedness does not significantly negatively correlate with skew in foundress associations in *P. bellicosus*. This finding does not necessarily contradict the transactional model, as some empirical researchers have erroneously asserted, because the transactional model predicts no relationship between skew and relatedness under some conditions. This happens when most or all colonies are essentially above the threshold for complete skew, and occurs when relatedness is high, group benefits are large, and/or the success for independent founding is low. Thus, the theory is at best difficult to test in species with typically (uniformly) high skews, because the theory predicts no strong relationships between skew and relatedness or group output. If most colonies within a population are above the complete skew threshold, subordinates also are no longer predicted to almost break even in inclusive fitness when the benefits of particularly the synthetic models, are still lagging. Particularly important will be determining whether variation in the outside options of group members modulates reproductive skew and intragroup conflict as predicted by skew theory.

See also: Colony Founding in Social Insects; Queen–Queen Conflict in Eusocial Insect Colonies; Reproductive Skew, Cooperative Breeding, and Eusociality in Vertebrates: Hormones.

Further Reading

Bourke AFG (2001) Reproductive skew and split sex ratios in social Hymenoptera. *Evolution* 55: 2131–2136.

Bourke AFG and Heinze J (1994) The ecology of communal breeding: The case of multiple-queen leptothoracine ants. *Philosophical Transactions of the Royal Society (London), Series B* 345: 359–372.

Buston P and Zink A (2009) Reproductive skew and conflict resolution: A synthesis of transactional and tug-of-war models. *Behavioral Ecology* 20: 672–684.

Cant MA (1998) A model for the evolution of reproductive skew without reproductive suppression. *Animal Behavior* 55: 163–169.

Field J, Solis CR, Queller DC, and Strassmann JE (1998) Social and genetic structure of paper wasp cofoundress associations: Tests of reproductive skew models. *American Naturalist* 151: 545–563.

Johnstone RA (2000) Models of reproductive skew: A review and synthesis. *Ethology* 106: 5–26.

Johnstone RA and Cant MA (1999) Reproductive skew and the threat of eviction: A new perspective. *Proceedings of the Royal Society (London), Series B* 266: 275–279.

Langer P, Hogendoorn K, Schwarz MP, and Keller L (2006) Reproductive skew in the Australian allodapine bee *Exoneura robusta*. *Animal Behavior* 71: 193–201.

Liebert AE and Starks PT (2006) Taming of the skew: Transactional models fail to predict reproductive partitioning in the paper wasp *Polistes dominulus*. *Animal Behaviour* 71: 913–923.

Reeve HK, Emlen ST, and Keller L (1998) Reproductive sharing in animal societies: Reproductive incentives or incomplete control by dominant breeders? *Behavioral Ecology* 9: 267–278.

Reeve HK and Ratnieks FLW (1993) Queen–queen conflicts in polygynous societies: Mutual tolerance and reproductive skew. In: Keller L (ed.) *Queen Number and Sociality in Insects*, pp. 45–85. Oxford: Oxford University Press.

Reeve HK and Shen S-F (2006) The bordered tug-of-war: A missing model in reproductive skew theory. *Proceedings of the National Academy of Sciences of United States of America* 103: 8430–8434.

Reeve HK, Starks PT, Peters JM, and Nonacs P (2000) Genetic support for the evolutionary theory of reproductive transactions in social wasps. *Proceedings of the Royal Society (London), Series B* 267: 75–79.

Vehrencamp SL (1983) Optimal degree of skew in cooperative societies. *American Zoologist* 23: 327–335.

Reproductive Skew, Cooperative Breeding, and Eusociality in Vertebrates: Hormones

W. Saltzman, University of California, Riverside, CA, USA

Introduction

Reproductive skew refers to the degree of asymmetry in the distribution of direct reproductive success among individuals within a social group. In high-skew societies, direct reproduction is monopolized within one or both sexes by one or a small subset of behaviorally dominant individual(s), whereas low-skew societies are characterized by a more equitable distribution of reproduction among all adult group members. Numerous theoretical models have been developed to explain the evolution and maintenance of reproductive skew. Simultaneously, considerable work has been devoted to determine the behavioral and hormonal mechanisms generating asymmetric reproductive success in high-skew societies.

High-skew societies are exemplified by cooperative breeding systems, including their most extreme manifestation, eusociality. Strictly speaking, cooperative breeding refers to any breeding system in which some individuals provide alloparental care for the offspring of other animals. These so-called helpers or alloparents may be male or female, related or unrelated to the breeding pair, and adult or immature. Among singular cooperative breeders – that is, species in which reproduction is limited to one female within each social group – helpers are often reproductively inactive, adult or subadult offspring of the breeding pair. Thus, additional common characteristics of singular cooperative breeding systems are delayed dispersal from the natal group and delayed or suppressed reproduction in helpers. Eusociality is a form of cooperative breeding characterized by extremely high reproductive skew and, according to some definitions, occurrence of irreversibly distinct behavioral castes. Cooperative breeding has been identified in a broad diversity of taxa, including insects, arachnids, crustaceans, fish, birds, and mammals, whereas eusociality has traditionally been considered to occur only among insects, especially some species of hympenoptera, isoptera, hemiptera, and thysanoptera. More recently, however, several species of sponge-dwelling shrimps (*Synalpheus*) and two mammalian species, the naked mole-rat (*Heterocephalus glaber*) and the Damaraland mole-rat (*Cryptomys damarensis*), have also been characterized as eusocial.

The mechanisms underlying both the suppression of reproduction and the performance of parental-like behavior by nonbreeding alloparents have generated considerable interest among behavioral endocrinologists and behavioral ecologists. Here, I focus on the hormonal aspects of these two hallmarks of high-skew vertebrate societies.

Hormonal Mechanisms of Reproductive Inhibition in Eusocial/Cooperative Breeding Systems

Failure of adult individuals to breed can be mediated by inhibition of reproductive behavior, suppression of reproductive physiology, or a combination of the two. Behavioral mechanisms may involve inbreeding avoidance in animals living with their natal families or interference in mating behavior. Physiological suppression typically involves dysfunction of the hypothalamic–pituitary–gonadal endocrine axis and may be manifest in impairments in gonadal endocrine function, gametogenesis, and pregnancy maintenance. Both behavioral and physiological mechanisms of reproductive failure can be conceptualized as being either imposed on subordinate individuals (i.e., helpers) by dominants, to the benefit of the dominant but at a cost to the subordinate, or self-imposed by subordinates as an adaptive response to specific organismal, social, or environmental cues.

The proximate mechanisms underlying reproductive inhibition have been investigated in numerous cooperatively breeding birds and mammals, and at least one cooperatively breeding fish. Most of these studies have compared activity of the hypothalamic–pituitary–gonadal axis between breeders and nonbreeders to discern whether or not nonbreeders are physiologically capable of reproducing. Briefly, in breeding adults, the hypothalamus secretes gonadotropin-releasing hormone (GnRH), which stimulates the anterior pituitary to secrete luteinizing hormone (LH) and follicle-stimulating hormone (FSH). These two gonadotropins, in turn, exert stimulatory effects on the gonads, promoting both gametogenesis and production of gonadal steroids (primarily testosterone in males, estradiol and progesterone in females) and peptide hormones (e.g., inhibin). Gonadal hormones feed back to the brain and pituitary to regulate secretion of GnRH, LH, and FSH. Reproductive impairments in subordinates can be caused, potentially, by dysfunction at the level of the gonads, pituitary, hypothalamus, or higher brain structures.

puberty (e.g., house mouse), inhibit sexual behavior (e.g., rhesus macaque, *Macaca mulatta*), delay conception (e.g., savannah baboon), block implantation (e.g., white-footed mouse, *Peromyscus leocopus*), induce spontaneous abortion or prenatal litter reduction (e.g., golden hamster, *Mesocricetus auratus*), or impair maternal care (e.g., ringtailed lemur, *Lemur catta*) in subordinate females. Such mechanisms can be activated by a variety of stressors such as agonistic interactions among females or reduced access to food or other resources.

Hormonal Mechanisms of Alloparental Care

A second aspect of cooperative breeding/eusociality of interest to behavioral endocrinologists, in addition to the mechanisms of reproductive curtailment in nonbreeding helpers, is the proximate control of alloparental behavior. Parental behavior in breeding individuals is activated by specific hormones (especially estrogen, progesterone, prolactin, and oxytocin) acting upon the brain while, in some cases, acting simultaneously upon peripheral structures (e.g., mammary glands, crop sac). In birds, the onset of parental behavior toward hatchlings is facilitated by the hormonal sequelae of incubation, whereas in mammals, the endocrine events stimulating the onset of maternal behavior are intimately linked to pregnancy, parturition, and lactation, processes that do not occur in nonbreeding helpers. Because cooperative breeding systems are characterized by nonbreeding individuals engaging in parental-like behavior, the question arises as to whether this alloparental behavior, like parental behavior, is activated by specific hormonal events. The identification of hormonal mechanisms regulating alloparenting, especially in birds, has also been considered crucial in determining whether alloparental behavior is evolutionarily distinct from parental behavior and results directly from natural selection, or whether it simply reflects a stereotyped response to cues from nestlings, which evolved in the context of parents provisioning their own offspring and which may not be dependent on hormonal priming. Thus, understanding the hormonal influences on alloparenting, if any, may elucidate both the proximate and ultimate causes of cooperative breeding.

Prolactin

Most studies of the endocrine correlates of alloparental behavior have focused on prolactin, a peptide released both within the brain and by the anterior pituitary into the general circulation. Prolactin has been demonstrated conclusively to play a key role in the onset of parental behavior in mammalian mothers and in avian mothers and fathers, and has been associated correlationally with paternal behavior in mammalian fathers in biparental species. Similarly, correlational evidence supports the hypothesis that prolactin promotes alloparental behavior in several cooperatively breeding birds and mammals: both inter- and intraspecific studies have found positive associations between alloparental behavior and prolactin concentrations in male and female helpers. In meerkats, for example, adult males have significantly higher circulating prolactin concentrations prior to a bout of 'babysitting' than prior to a bout of foraging, suggesting that prolactin might influence individual helpers' decisions to provide alloparental care. In Florida scrub jays and Harris's hawks, plasma prolactin levels in both breeders and nonbreeders rise during the incubation stage and correlate with individual differences in helping behavior. Prolactin secretion can be stimulated by incubation, however, or by contact with or exposure to nestlings/infants, and very few experimental studies have been conducted to determine whether prolactin plays a causal role in alloparenting. Thus, while prolactin remains the most obvious candidate for an 'alloparental hormone,' its role in the expression of alloparental care, if any, is not yet clear.

Testosterone

High circulating levels of testosterone have been shown to influence – usually, to inhibit – paternal behavior in a number of birds and mammals. Therefore, several studies of cooperative breeders have focused on testosterone in helpers as a possible determinant of alloparental care. No clear pattern has emerged in the relationship between individual alloparents' current testosterone levels and helping behavior. In male azure-winged magpies (*Cyanopica cyana*), endogenous circulating testosterone concentrations do not correlate with helping behavior, but treatment of male helpers with exogenous testosterone elevates their rates of feeding nestlings. In contrast, low endogenous testosterone levels in male Mongolian gerbils are associated with high alloparental responsiveness. More importantly, perhaps, exposure to androgens or other steroid hormones during the perinatal period is likely to permanently alter the propensity to provide alloparental care by exerting organizational effects on the central nervous system, especially in species that exhibit sex differences in alloparental behavior.

Pregnancy

In several cooperatively breeding mammals (e.g., common marmoset, meerkat, Mongolian gerbil), both dominant and subordinate females may commonly become infanticidal during pregnancy, presumably as a manifestation of female–female reproductive competition. The mechanisms underlying this pattern are not known but are likely to involve pregnancy-related hormonal changes.

Glucocorticoids

Few studies have addressed a possible relationship between glucocorticoid hormones and alloparental behavior. In meerkats, however, correlational evidence suggests that high glucocorticoid levels may increase pup-feeding rates by male helpers.

Future Directions

Clearly, much remains to be learned about the endocrinology of reproductive skew, cooperative breeding, and eusociality in vertebrates. While hormones undeniably play a role in limiting reproduction in at least some male and female cooperative breeders, the endocrine/neuroendocrine mechanisms underlying such effects, as well as the social or sensory cues activating these mechanisms, remain almost entirely unknown. Moreover, although rank-related differences in glucocorticoid concentrations have been reported in many cooperatively breeding vertebrates – often with dominant, breeding individuals exhibiting higher glucocorticoid levels than subordinate nonbreeders – nothing is known about the functional significance of these differences. Finally, our understanding of hormonal influences on alloparental care is rudimentary. Focused, experimental studies, ideally involving hormone manipulations, are needed to further elucidate the role of the endocrine system in determining reproductive and behavioral profiles of vertebrates living in high-skew societies.

See also: Caste Determination in Arthropods; Cooperation and Sociality; Fight or Flight Responses; Helpers and Reproductive Behavior in Birds and Mammals; Invertebrates: The Inside Story of Post-Insemination, Pre-Fertilization Reproductive Interactions; Mammalian Female Sexual Behavior and Hormones; Pair-Bonding, Mating Systems and Hormones; Parental Behavior and Hormones in Mammals; Parental Behavior and Hormones in Non-Mammalian Vertebrates; Reproductive Skew; Sexual Behavior and Hormones in Male Mammals.

Further Reading

Creel S (2001) Social dominance and stress hormones. *Trends in Ecology and Evolution* 16: 491–497.

Faulkes CG and Bennett NC (2001) Family values: Group dynamics and social control of reproduction in African mole-rats. *Trends in Ecology and Evolution* 16: 184–190.

Goymann W, Landys MM, and Wingfield JC (2007) Distinguishing seasonal androgen responses from male–male androgen responsiveness – Revisiting the Challenge Hypothesis. *Hormones and Behavior* 51: 463–476.

Hager R and Jones CB (2009) *Reproductive Skew in Vertebrates: Proximate and Ultimate Causes.* Cambridge, UK: Cambridge University Press.

Jamieson IG and Craig JL (1987) Critique of helping behavior in birds: A departure from functional explanations. In: Bateson PPG and Klopfer P (eds.) *Perspectives in Ethology*, vol. 7, pp. 79–98. New York: Plenum Press.

Russell AF (2004) Mammals: Comparisons and contrasts. In: Koenig W and Dickinson J (eds.) *Ecology and Evolution of Cooperative Breeding in Birds*, pp. 210–227. Cambridge, UK: Cambridge University Press.

Saltzman W, Digby LJ, and Abbott DH (2009) Reproductive skew in female common marmosets: What can proximate mechanisms tell us about ultimate causes? *Proceedings of the Royal Society of London B* 276: 389–399.

Schoech SJ (2004) Endocrinology. In: Koenig W and Dickinson J (eds.) *Ecology and Evolution of Cooperative Breeding in Birds*, pp. 128–141. Cambridge, UK: Cambridge University Press.

Solomon NG and French JA (1997) *Cooperative Breeding in Mammals.* Cambridge, UK: Cambridge University Press.

Ziegler TE (2000) Hormones associated with non-maternal infant care: A review of mammalian and avian studies. *Folia Primatologica* 71: 6–21.

times animals are exposed to predators) and/or intensity (e.g., degree of severity) of risk. Several studies have found a clear relationship between the risk history and the animal response to a given situation of risk, with lower antipredator responses (usually measured as the use of refugia or reductions in activity levels) after a history of frequent high-risk situations. Specifically, studies on cichlid fish by Maud Ferrari, Grant Brown, Patricia Foam, and others provide consistent support for the risk allocation hypothesis and highlight the predicted interaction between frequency and intensity of risk (see earlier). However, several other studies did not find clear evidence supporting the influence of the frequency of risk on prey antipredator responses.

These contradictory results were found even in closely related species, like in the studies conducted by Keith Pecor and Brian Hazlett on crayfishes, in which the behavior of *Oronectes virilis* matched the predictions of the risk allocation hypothesis, but the behavior of the closely related *O. rusticus* did not support the hypothesis. In a recent comparative study, Kate Boersma and collaborators suggest that different species of North Pacific flatfish may diverge in how they adjust to the risk allocation model depending on species-specific factors, such as the species ability to perceive risk changes, their preferred habitat (shallow and turbid waters that protect from most predator vs. deep and clear waters with higher predation risk), and their relative ability to avoid predation as a result of predator gape size. These species-specific factors could make some experimental designs unsuitable to detect patterns of risk allocation, as each species may have different risk thresholds before they start limiting antipredator responses. For instance, under a scenario with low frequency of high-risk events, small increases in the frequency of risky events may not elicit a reduction in antipredator behavior, since animals still have prolonged periods of safety to exploit resources. However, as the frequency of risky events continues to increase, it may reach a threshold that would start causing the reduction in antipredator effort predicted by the risk allocation hypothesis. Differences in this threshold among species may cause interspecific variation in risk allocation behavior. While this concept of risk threshold has only been suggested in the flatfish study, it could be present even between populations or individuals.

Risk thresholds may also explain the lack of support to the risk allocation hypothesis in other experimental studies. Usually, lab experiments include only two frequency of risky scenarios (low and high), assuming that at least one will fall above the risk threshold; however, the specific threshold level is largely unknown for the model species. Moreover, the physiology of the model species may affect the need to engage in foraging efforts. Josh Van Buskirk and collaborators, as well as Keith Pecor and Brian Hazlett suggested that given the low temporal

scale of many risk allocation experiments, some animals may not require foraging at all over the entire duration of the experiment, and thus invest nearly all their effort in antipredator behavior (e.g., animals are not sensitive to frequency of risk). Although this shortcoming could be solved by modifying food-deprivation schedules, the ability of prey to detect changes in predation risk is an implicit assumption of the risk allocation hypothesis.

Field Evidence Supporting the Risk Allocation Hypothesis

Some patterns observed under field conditions may actually be the result of risk allocation. For instance, the reduction in antipredator responses to humans in areas with a high rate of human visitation has been found in many vertebrate taxa, with habituation to humans usually proposed as the mechanism underlying this pattern. However, it is unclear whether habituation and/or risk allocation are involved. Humans do not directly prey on wildlife in many areas but nonetheless they are perceived as predators, causing antipredator responses similar to those elicited by real predators. Under the risk allocation hypothesis, this scenario of frequent high-risk situations (e.g., high frequency of human approaches) should result in a reduction in antipredator responses to each risky situation as compared to scenarios with less frequency of high-risk situations. This prediction has been tested by Iñaki Rodríguez-Prieto and collaborators using blackbird *Turdus merula* flight responses in urban parks. This study clearly differentiated for the first time the effects of risk allocation from those of habituation, finding that habituation complements risk allocation to produce the pattern of reduced antipredator responses in areas with a high frequency of human visitation. In this study, animals experienced different risk histories by park-specific daily patterns of human visitation. While the history of risk was naturally produced by human visitors, the tests on blackbird flight responses were performed by both human observers and novel predators (e.g., radiocontrolled vehicle) to help differentiate habituation from risk allocation.

That risk allocation is not predator specific is usually overlooked. For example, the risk history produced by frequent encounters with a snake species may lead to reduced prey responses not only to the snake but also to other types of predators, like raptors or mammals. Of course, the type of antipredator response may differ depending on the predator, but all antipredator responses are expected to be reduced as the frequency of high-risk situations from any predator increases.

While most of the known patterns of reduced antipredator responses in places with high density of predators come from areas with varying levels of human visitation, there are also some examples from areas with different densities of real predators. A recent study by Scott Creel and others on elk antipredator behavior in relation to wolf presence has tested the prevalence of risk allocation in relation to other models by studying several prey populations experiencing temporally and spatially variable levels of actual predation risk. The findings provide support for the risk allocation hypothesis, with wintering elks reducing antipredator vigilance in areas where they were more frequently exposed to wolf predation. Because predation risk was produced by the main predators of elks, habituation of elks to wolves was not likely a problem in this study.

Risk allocation and habituation predictions are difficult to distinguish. If animals habituate to risk cues that are not coupled with potential predation, some experimental designs may not be able to tell apart behavioral responses produced by risk allocation from those produced by habituation, as acknowledged by Reehan Mirza and others. A strategy to avoid this problem may be to study scenarios with real predation where actual predators almost always pose a threat and actually attack and kill prey, and thus may preclude habituation. However, if researchers cannot perform the study with real predators, a potential solution may be to use at least two types of different predator cues, one for creating the risk history and the other for testing antipredator responses. For instance, if a scenario of frequent high-risk situations is created by the intermittent addition of predatory fish odor cues in the water, animals may become habituated to these cues and thus reduce their responses to further predatory fish cues just by a process of habituation, but not by risk allocation. However, once the risk history is established, another type of predator stimulus, like an overflying raptor silhouette, could be used to test if the antipredator responses follow risk allocation, since the potential habituation to the previously used odor cue can be ruled out as responsible for patterns of reduced responses to the raptor. Another strategy may be to take advantage of the different temporal scales in which habituation and risk allocation could be acting. A bird cannot be expected to habituate and dishabituate periodically in response to short-term changes in the frequency of risk. However, risk allocation can predict increases and reductions in antipredator responses to high-risk situations following changes in the frequency of these situations, for example, between consecutive and cyclical periods such as mornings and afternoons.

Broader Implications of Risk Allocation

Risk allocation can alter some paradigms commonly used in ecological models. For instance, most studies on the ecological impacts of predators assume that all the effects of predation increase with increasing exposure to predation risk. Scott Creel and collaborators suggest that while direct mortality should increase with increasing attack ratios, the same does not apply to the costs of antipredator behavior since risk allocation suggests that the costs associated with antipredator behavior could be reduced or at least could remain relatively unchanged with increasing frequency of exposure to risk under the scenarios discussed earlier. Similarly, common ecological models assume that individual predators would suffer from reduced capture rates in areas with high density of competing predators since prey depletion and avoidance behavior may lead to a reduction in the per capita predator–prey encounter rate. However, Rodriguez-Prieto and collaborators suggested that the probability of prey being captured in any given predator–prey encounter would increase in areas with high density of predators, as animals would reduce their responses to predator approaches in those areas; hence increasing their vulnerability to predator attacks, which would potentially alter predator–prey dynamics.

The risk allocation hypothesis has opened new venues for behavioral research. While there have been some contradictory results, evidence is mounting in favor of risk allocation being an important force shaping the animal responses to temporal variation in predation risk. More research is needed, and there are some questions that seem particularly promising, for example, how and why different species (and probably different populations) vary in the thresholds at which they respond to predators, and how this variation affects predator–prey interactions.

See also: Ecology of Fear; Economic Escape; Predator Avoidance: Mechanisms; Trade-Offs in Anti-Predator Behavior; Vigilance and Models of Behavior.

Further Reading

Boersma KS, Ryer CH, Hurst TH, and Heppell SS (2008) Influences of divergent behavioral strategies upon risk allocation in juvenile flatfishes. *Behavioral Ecology and Sociobiology* 62: 1959–1968.

Brown GE, Rive AC, Ferrari MCO, and Chivers DP (2006) The dynamic nature of anti-predator behaviour: Prey fish integrate threat-sensitive anti-predator responses within background levels of predation risk. *Behavioral Ecology and Sociobiology* 61: 9–16.

Brown GE, Bongiorno T, DiCapua DM, Ivan LI, and Roh E (2006) Effects of group size on the threat-sensitive response to varying concentrations of chemical alarm cues by juvenile convict cichlids. *Canadian Journal of Zoology* 84: 1–8.

Creel S, Winnie JA, Christianson D, and Liley S (2008) Time and space in general models of antipredator response: Test with wolves and elk. *Animal Behaviour* 76: 1139–1146.

Ferrari MCO, Rive AC, MacNaughton CJ, Brown GE, and Chivers DP (2008) Fixed vs. random temporal predictability of predation risk: An extension of the risk allocation hypothesis. *Ethology* 114: 238–244.

Foam PE, Mirza RS, Chivers DP, and Brown GE (2005) Juvenile convict cichlids (*Archocentrus nigrofasciatus*) allocate foraging and antipredator behaviour in response to temporal variation in predation risk. *Behaviour* 142: 129–144.

Lima SL and Bednekoff PE (1999) Temporal variation in danger drives antipredator behavior: The predation risk allocation hypothesis. *American Naturalist* 153: 649–659.

(increased survivorship of offspring resulting in increased direct fitness), this section will focus on the costs of nest defense in the form of increased risk to the parent.

Birds defend their broods in many ways: flushing explosively, feigning injury (e.g., broken wing) to entice the predator to attack the parent and not its young, alarm calling, and individually attacking the predator with dives and strikes. Distraction displays take many forms, and Michael Gochfeld groups a large number of nest protective behaviors into this category, including running behaviors, injury feigning, tail flagging, and erratic flight. Most distractive behaviors make the performer appear to be easier to capture, and therefore an easier and more profitable meal than the parent's offspring, whether they have been detected or not. There is great variation in the form of these behaviors, and the type and frequency of behaviors used vary from individual to individual.

Although the most apparent cost of nest defense is increased risk of injury or death to the parent, there is little empirical evidence to suggest that this cost exists, and anecdotal evidence suggests that death is only occasional. Parents likely defend young in a manner that minimizes risk to themselves and may take fewer risks when the risk of predation posed by a particular type of predator is high. Additionally, parental survivorship costs of nest defense may be overstated because many defensive behaviors are performed in response to predators that pose little risk to the parents and only really threaten the offspring. The likelihood of nest defense in relation to costs may also depend on the parent's renesting potential (e.g., older parents, lower quality individuals, or parents nesting at the end of the breeding season when there is no chance of surviving the winter may take greater risks to protect their current offspring).

Mobbing

In many gregarious species, groups of animals sometimes advance toward a predator to inspect, follow, harass, or attack it until it leaves the area. Predators usually escape with just a few bites or pecks, but sometimes predators are captured or killed by mobbers. Mobbing is usually a group effort of offspring defense, varies considerably within and among prey species in form and intensity, and may take different forms specific to particular types of predators. Clear benefits of mobbing include reduced predation on gregarious species and increased reproductive success as a result of defending offspring. Eberhardt Curio outlined specific direct and indirect benefits of mobbing in birds; mobbing can harm or kill a predator (reducing risk), advertise detection of the predator and encourage a predator to leave an area sooner, dilute risk of any one individual attacker, attract larger predators that pose a risk to the attacker, alert other conspecifics to danger, or induce silence from offspring. Costs of mobbing, in addition to

temporal and energetic costs, should include decreased survivorship via the increased risk of reducing the distance between the predator and prey. Although anecdotes suggest that prey are occasionally captured while mobbing predators, the best evidence of the costliness of mobbing is indirect and contextual. Riskiness of mobbing may be inferred from observations of mobbers being more cautious (1) with more dangerous predators, (2) when closer to a dangerous location, (3) when terrain makes it difficult to escape, (4) when fewer mobbers are available, and (5) when prey are less nimble. Mobbing may also attract additional predators to the area, and mobbing parents may inadvertently leave their offspring unprotected and more vulnerable to predation.

Individual Attack

Members of some species show a willingness to confront a predator on their own. While many of these species also mob predators in groups, individuals are often capable and eager to take on approaching predators without help from conspecifics. Ungulate mothers sometimes confront predators to defend their offspring, attacking with their hooves and head ornaments (e.g., horns or antlers). Donald Owings, Richard Coss, Ron Swaisgood, and Matthew Rowe, among others, have studied the preemptive individual attacks by California ground squirrels to rattlesnake predators for many years. Squirrels may approach, dart toward, and retreat from the snake (**Figure 3**), tail flag, or bite or throw substrate (dirt) at the snake. This snake harassment behavior is undoubtedly influenced by the squirrels' resistance to rattlesnake venom and their interest in protecting more vulnerable offspring from predation: mothers are more likely than other squirrels to harass snakes.

Some prey launch counterattacks against predators that have attacked them. Species with noxious or toxic defenses are most famous for their counterattack strategies: skunks and striped polecats are exceptional in their ability to spray a target with anal gland secretion from several meters away. As these types of weapons are discussed in detail in another article, here I will focus on other forms of counterattack. Most nonnoxious prey that fight back against predators defend themselves with sharp teeth and claws. The small mustelid carnivores (e.g., weasels, martens, polecats) are excellent examples of small species with razor-sharp teeth that evolved to suit their voracious carnivorous appetite but that also are useful in defense. Many of these species can severely harm/maim an attacker in a cloud of dust and blood – the ratel (honey badger) is particularly well known for being fearless, pugnacious, and relentless in its response to harassment. Other animals may regurgitate oils onto their predators, which can lead to predatory birds becoming waterlogged and drowning and mammalian predators' fur

Figure 3 Illustration of a California ground squirrel (*Spermophilus beecheyi*) female jumping back after being struck by a rattlesnake. Ground squirrels recognize and harass rattlesnakes at close-range and throw substrate to drive them away; squirrels have fast reactivity and venom resistance. Painting by Richard G. Coss.

becoming foul-smelling. Some vultures regurgitate food when frightened; the vomit is highly acidic and can drive away predators. Clearly, the likelihood of counterattack depends on the costs and benefits of attacking versus flight and is based on the size and dangerousness of the predator, the size of the prey, the escape options on hand, and other environmental factors (e.g., availability of cover or refuge).

behaviors, carefully designed experiments and attention to environmental circumstances during the performance of the behavior can shed light on the benefits that performers might receive.

See also: Acoustic Signals; Alarm Calls in Birds and Mammals; Defensive Chemicals; Defensive Coloration; Defensive Morphology; Economic Escape; Group Living.

Conclusion

Clearly, when confronted by a predator, animals often do not flee immediately and behave in apparently risky ways that would appear to increase survivorship costs. However, in nearly all cases, closer examinations of the behaviors suggest that these 'risk-takers' are actually either optimizing fitness, so that fitness benefits exceed fitness costs, or are not behaving in a risky way at all, or both. Prey behave in risky ways throughout the attack sequence, whether they are alarm calling while the predator is still far away or purposely drawing a predators attention away from a nest of defenseless offspring. While some animals that accept increased survivorship costs at one stage of the predation sequence often compensate at a later stage by playing it more safely, other animals accept more risk in all contexts (i.e., a behavioral syndrome). While it is difficult to empirically assign functionality to risky antipredator

Further Reading

Caro TM (1994) Ungulate antipredator behavior: Preliminary and comparative data from African bovids. *Behaviour* 128: 189–228.

Caro TM (2005) *Antipredator Defenses in Birds and Mammals*. Chicago, IL: University of Chicago Press.

Conover MR (1994) Stimuli eliciting distress calls in adult passerines and response of predators and birds to their broadcast. *Behaviour* 131: 19–37.

Curio E (1978) The adaptive significance of avian mobbing: I. Teleonomic hypotheses and predictions. *Zeitschrift für Tierpsychologie* 48: 175–183.

Gochfeld M (1984) Antipredator behavior: Aggressive and distraction displays of shorebirds. In: Burger J and Olla BJ (eds.) *Behavior of Marine Animals,* vol. 5, pp. 289–377. New York, NY: Plenum Press.

Ioannou CC and Krause J (2008) Searching for prey: The effects of group size and number. *Animal Behaviour* 75: 1383–1388.

Krause J and Ruxton GD (2002) *Living in Groups*. Oxford: Oxford University Press.

Lima SL and Dill LM (1990) Behavioral decisions made under the risk of predation: A review and prospectus. *Canadian Journal of Zoology* 68: 619–640.

Lingle S, Rendall D, and Pellis SM (2007) Altruism and recognition in the antipredator defence of deer: 1. Species and individual variation in fawn distress calls. *Animal Behaviour* 73: 897–905.

Magurran AE (1990) The adaptive significance of schooling as an antipredator defense in fish. *Annales Zoologici Fennici* 27: 51–66.

Montgomerie RD and Weatherhead PJ (1988) Risks and rewards of nest defence by parent birds. *Quarterly Review of Biology* 63: 167.

Owings DH and Coss RG (2008) Hunting California ground squirrels: Constraints and opportunities for Northern Pacific rattlesnakes. In: Hayes WK, Beaman KR, Cardwell MD, and Bush SP (eds.) *The Biology of Rattlesnakes*, pp. 155–168. Loma Linda, CA: Loma Linda University Press.

Sargeant AB and Eberhardt LE (1975) Death feigning by ducks in response to predation by red foxes (*Vulpes fulva*). *American Midland Naturalist* 94: 108–119.

Sherman PW (1977) Nepotism and the evolution of alarm calls. *Science* 197: 1246–1253.

Taylor RJ, Balph DF, and Balph MH (1990) The evolution of alarm calling: A cost-benefit analysis. *Animal Behaviour* 39: 860–868.

Zuberbühler K, Noë R, and Seyfarth RM (1997) Diana monkey long-distance calls: Messages for conspecifics and predators. *Animal Behaviour* 53: 589–604.

Robot Behavior

E. S. Fortune and N. J. Cowan, Johns Hopkins University, Baltimore, MD, USA

Introduction

Behavior is typically defined as the activity of an organism, machine, or natural phenomenon, particularly in response to external stimuli. This definition is broad, and applies to robots and animals alike. A categorical difference between animals and robots is that the latter are typically designed by human beings to perform a set of highly constrained objectives, while animal behavior is typically rich and largely unconstrained.

The overwhelming majority of all robots manufactured worldwide are industrial robot manipulators. These are typically open kinematic chains – that is, a series of links and joints, whose first joint is mounted to a stationary base – with an end effector mounted to the final link. It is true that many industrial robot arms are at least approximately anthropomorphic, with six degrees of freedom: a waist joint, shoulder joint, elbow joint, and three degree-of-freedom wrist. However, the connection to biology ends there because the mechanical design, construction, control, and performance is unnatural: industrial robots are rigid, extremely precise, and some designs can produce forces on the order of tens of kilonewtons (thousands of pounds) while maintaining positioning error on the order of centimeters. In fact, some industrial manipulators can accelerate automobile engines at several times the acceleration due to gravity. These machines can produce a wide array of movements reliably, repeatably, and for much longer than animals can without fatigue. By many metrics, these machines already outperform animals, so a natural question arises: Why copy nature?

Once programmed, industrial manipulators are generally quite limited in their ability to respond to their environment and typically execute an even more limited repertoire of behaviors. These machines are rigid not only mechanically, but behaviorally as well. Animals, on the other hand, exhibit complex and nuanced responses to sensory stimuli, can perform a wide range of behaviors, and seem to deal with uncertainty and complexity more effectively than current robotic designs. This has motivated some research groups to study the principles of animal behavior to design robotic systems on the basis of these principles.

In this article, we turn our attention to biologically inspired robots: those that are designed to emulate or copy some set of morphological and/or behavioral characteristics from nature. In this context, what are the relations between robotic and animal behavior and how can they be a source of insights for both engineers and biologists? Engineers increasingly look to the animal world for inspiration, but, as described later, translating animal behavior into robotic systems is fraught with pitfalls. Nevertheless, with care, this approach has led to technological advances, several examples of which are reviewed here. Robotic behavior can also be used by biologists to test specific hypotheses. This opens up the door for productive synergies between engineers and biologists, seeking to advance the state of the art in both fields.

Briefly, robots can be divided into three categories. Type 1 robots are not directly inspired by, nor designed to share features with biological systems. Robots that operate in assembly lines or household robots such as the Roomba™ (iRobot Corporation) fall into this category. Type 2 robots are those that are biological inspired, incorporating specific behavior features observed in animals in order to achieve some set of functional objectives. Robots such as those envisioned by the popular science fiction writer Isaac Asimov fall into the Type 2 category. Type 3 robots are those that are built to test specific hypotheses about animal systems. These robots can be particularly helpful when certain manipulations of the actual animal are not possible. Here, we focus on Type 2 and Type 3 robots.

Type 2: Biologically Inspired Robot Behaviors to Service Human Needs

Even relatively simple animals can perform far more complex and nuanced behavior than can the most advanced robots. Take, for example, the case of winged flight. Fixed-wing aircraft can outperform animals in terms of velocity and lift, but cannot match flapping animal fliers in terms of maneuverability. In recent years, engineers have produced many biologically inspired flapping fliers, called *ornithopters*, in an effort to match the maneuverability of nature's designs. These Type 2 robots take their inspiration from the natural world to achieve a limited set of functional objections. Several of these machines are reviewed in the examples given later, but first some of the challenges in the process of bioinspiration are described.

Challenges in Translating Biological Behavior into Robotic Inspiration

The basic hypothesis underlying many efforts in bioinspiration is that animal behavior systems have been optimized over time by evolutionary processes. The notion is that organisms, through millions of years of evolution, have come up with fundamental design properties that solve problems associated with the natural constraints defined by the physical properties of the universe. The result is that animals can be high performing with respect to a variety of engineering metrics. Determining those fundamental design features for use in robotic design, however, is more difficult than simply copying specific features of an organism.

Take, for example, the differences among independently evolved vertebrate fliers such as pterosaurs, birds, and bats. Obviously, a pterosaur and bird and a bat do not look identical; in fact they are quite different in terms of the details of their design. If these animals were indeed 'optimal,' one might expect a greater convergence of materials, morphologies, and physiological properties between species. So, how does a roboticist determine those features of animal design that will contribute to better robots? As a start, the roboticist needs to consider that every behavior an animal performs must be interpreted in the context of the entirety of the organism. This is due to three features of animal behavior.

First, behavior is mediated by mechanisms that are shared by other behaviors and physiological functions of the animals. Therefore, the mechanisms for any particular behavior cannot be assumed to be optimal for that behavior. In fact, a better assumption is that the mechanisms for any particular behavior are actually suboptimal. For example, songbirds produce songs that are necessary for successful reproduction. Nevertheless, the mechanisms for song production use organ systems that are shared for grooming, breathing, eating, and thermoregulation. Ergo, birds are not optimized for producing song alone.

Second, the behavior of an animal occurs in the context of its evolution. The evolutionary history of an organism imposes a set of constraints on behavior and physiological mechanisms that may not be easily described or understood. A trivial example of this sort of phylogenetic constraint might be food manipulation in birds and mammals. Birds must use the hindlimbs to manipulate food items because the forelimbs have been co-opted for flight, whereas most mammals use the forelimbs. In other words, evolution makes use of the features that are available, rather than developing an optimal solution 'from scratch.'

Third, behavior must be interpreted in terms of the ultimate evolutionary goal of the animal, which is reproduction. Thus, although the proximate goal of a particular behavior might appear straightforward, such as prey capture, that behavior represents only part of the behavioral repertoire necessary for reproduction. In short, no animal is optimized to achieve any single behavior, but rather they are designed to carry out a suite of behaviors that are sufficient for survival and reproduction.

Design features can more reliably be determined via a comparative approach. Specifically, comparisons among species permit the identification and separation of those features that are unique to a particular organism or a clade of organisms, versus those features that are similar among organisms or clades. Of interest are those features that are similar among the widest array of animals: those features are most likely to represent the fundamental design constraints for a particular category of behaviors.

Evolutionarily speaking, such similarities can arise from two sources. Similarities across clades can be plesiomorphic – meaning that the feature arose in an evolutionary event that occurred long ago and that all the extant clades have inherited that feature. Indeed, there are genetic sequences that are found in very distantly related species, and can be shown, using quantitative analysis, to have first occurred well over 200 million years before present. Similarities can also be homoplastic – meaning that the feature arose independently in different clades. This process is also known as 'convergent evolution.' A simple example of convergence is the eyes of vertebrates, cephalopods, and cnidarians. Each of these clades independently evolved an eye with a single lens in front of a photoreceptive sheet.

Convergent strategies are of particular interest. Take, for example, the case of legged locomotion. Comparative studies that carefully account for dimensional scaling reveal remarkably similar mechanics and energetics across dramatic variations in, morphology (including 2-, 4-, 6-, and 8-legged runners) and body size (millimeters to meters). These similarities include mechanical and metabolic energetics, gait, stride frequency, and ground reaction forces. These ubiquitous scaling relations have been used to inform the design and implementation of biologically inspired legged robots such as the RHex robotic hexapod described below.

Examples of Type 2 Biologically Inspired Robots

There is a rich history of biological inspiration in mechanical systems, perhaps starting with the human fascination with avian flight. Leonardo da Vinci is widely known for his study of birds and he produced many conceptual designs of artificial flying machines, including an early design for a hang glider which was ultimately successful. Many of his designs were for ornithopter flapping flight. While it is true that for centuries, engineers have been inspired by flapping flight, ultimately engineers separated the mechanisms of thrust, lift, and, to some extent, control. This is a decidedly nonbiological approach, because for birds and other flapping fliers, flapping wings perform all three of these tasks.

Deciphering these integrated, shared mechanisms has proven a substantial challenge for engineers and has remained essentially unsolved for decades. In the last two decades, substantial progress has been made in this regard, encouraging researchers to return to the design and construction of ornithopters inspired by animal flapping flight. This has been facilitated by a multitude of technological and scientific advances. For example, improved fabrication technology has accelerated the rate at which new designs can be implemented and tested. Moreover, a wide variety of actuator technologies, such as electroactive polymers (EAPs), are becoming available for use as artificial muscles that offer many of the characteristics of animal muscle. From a scientific standpoint, there have been increasingly sophisticated studies into the mechanisms for lift and maneuvering in natural flapping fliers, including a wide range of new experimental and analysis techniques: high-resolution and high-speed videography, complex 3D fluid simulations, electrophysiological recordings during behavior, etc.

A recent example of the confluence of science and technology is the Micromechanical Flying Insect (MFI; http://robotics.eecs.berkeley.edu/~ronf/mfi.html) project, which began as a collaboration between biologists and engineers at UC Berkeley. These small-scale robotic flying machines are inspired by the discoveries in high-frequency flapping flight of flies. Biological investigations into the mechanisms of unsteady aerodynamics of flies, sensory integration from the haltere, and wing actuation have all fueled MFI technology. Recently, the Harvard Microrobotics Laboratory built a successor to the MFI that generated sufficient lift to take off.

In terrestrial locomotion, basic research at the PolyPEDAL Laboratory at the University of California, Berkeley, has inspired several engineering research labs to build hexapedal robots capable of rapid running and climbing, including the following projects:

- RHex (http://kodlab.seas.upenn.edu/RHex/Home): This is a collaboration between researchers at University of California, Berkeley, the University of Pennsylvania, and McGill University to build a dynamic hexapod capable of high-speed terrestrial locomotion.
- Robots in Scansorial Environments (RiSE; http://www.riserobot.org/): This is a large consortium of researchers from academics and industry, including University of Pennsylvania, University of California, Berkeley, Stanford University, Lewis and Clark College, and Boston Dynamics, Inc., to build biologically inspired climbing machines.

Type 3: Robots as Biological Research Tools

The Role of Robots as Physical Models

Biological hypotheses must often be tested using indirect methods, such as mathematical models and computer simulations. This is particularly true for fields like paleobiology,

where direct measurement of behavior is impossible, but it is also true for behavioral biology in general. For example, ethical considerations or technical challenges with small organisms can preclude many types of manipulations and measurements. In this case, computer models and numerical simulations can certainly be helpful, but there are cases when these are not sufficient either. This motivates the use of robotics as a tool for simulating animal behavior in order to test hypotheses.

In a certain sense, physical simulations using a robot are no different than those in a computer. In the context of behavioral biology, however, robot models have two potential advantages over computer simulations. First, computer simulations rely on approximations to the underlying physics of movement that cannot be independently verified. Second, robots are of special importance for investigations of social behavior, because robotic surrogates can be introduced into ecologically relevant settings; these surrogates can then be used to test specific hypotheses about the nature of social interactions in a way that may be impossible using animals alone.

Examples of Type 3 Robots to Model Biological Behavior

The mechanics of flapping flight of the fruit fly is challenging to study for many of the reasons noted earlier. First, the small scale of fruit flies makes instrumentation unfeasible using state-of-the-art technologies. Moreover, accurate mathematical simulations of the aerodynamics are currently an open research problem in the fluid mechanics community, so such simulations cannot, alone, be completely trusted as a test of a biological hypothesis. As a consequence, a team of engineers and biologists developed a Type 3 robot called 'RoboFly,' a large dynamically scaled robotic model of a fly wing, designed to flap in mineral oil. In a hybrid approach, the researchers measured the 3D flight kinematics for real fruit flies turning in free space and replayed the measured wing kinematics (appropriately scaled in time) through RoboFly. RoboFly's large scale facilitated instrumentation of forces and torques, enabling researchers to measure for the first time the forces required during maneuvering. Using the principle of similitude, the forces measured on RobotFly were used to determine the forces at play at the original scale of the fruit fly itself. In a synergistic collaboration, members of the same team have developed a series of Type 2 robots through the MFI project mentioned earlier. These small-scale robotic flying machines are inspired by the discoveries the team has made about flapping flight using RoboFly. This synergy between biologists and engineers – involving the development of both Type 2 (MFI) and Type 3 (RoboFly) robots – is increasingly common.

Another area where biological experimentation can be difficult is in the complex cues used in social communication.

how they interact; (2) examining the efficacy of the behavior in eliciting a response across a wide range of environmental conditions; and (3) assaying the characteristics of responding animals, for example, so that different responders can be compared by their aggressiveness, fearfulness, or responsiveness to a controlled stimulus. These types of experiments are discussed here, with examples for each.

Measuring the Response to Components of Complex Behaviors

Animal behaviors are often complex, and researchers may want to know which particular features of a complex behavior are necessary and sufficient to elicit a response from other animals. Robots can allow researchers to answer this question by experimentally mimicking a particular feature of the behavior and measuring the response. For example, there was a long-lasting controversy over whether honeybees communicate the location of food sources to other workers with elaborate waggle dances, or whether the dance serves only to focus the attention of other workers, which then use the scent of the pollen to find the food source. Alex Michelson and colleagues were able to separate these two components by creating a robotic honeybee that could dance in an experimentally controlled manner (**Figure 1(a)**). The robot was designed to mimic the waggle movement and sound-producing wing-vibrations of a dancing worker. The robot was scented like the food source, as a dancing worker would be, but it did not look like a bee – in fact, it was made of brass and had a vibrating piece of razor blade for wings. But looks were not important, since the dance occurs in the dark of a hive. By creating an experimental mismatch between the scent and dance signals, researchers were able to direct bees away from the scented target location, to the location indicated by the dance. This showed that the waggle dance independently conveys information about the distance and direction of food sources, though scent cues are clearly important as well.

In addition to being able to decouple and assess the response to different components of behaviors, researchers can examine how these components interact. Peter Narins and his colleagues did just this using a robotic dart-poison frog in a field study in Costa Rica. With the 'Robo-Rana,' Narins could compare the response to the acoustic signal alone, the visual signal of the inflating vocal sac on the throat, and the combination of both signal components (**Figure 1(b)**). They found that only the combination of acoustic and visual displays was effective at eliciting an aggressive response from males. Further, they were able to carefully adjust the timing and spatial position of the acoustic and visual signal sources (the speaker and the robot), to study how separation affected the integration of signal components by receivers (**Figure 2**). In another study examining how multiple signal components interact, Aaron Rundus, Don Owings, Sanjay Joshi, and colleagues

built a robotic ground squirrel (**Figure 1(c)**) that could produce both infrared (IR) and tail-flagging (i.e., side-to-side waving) signals used in interspecific communication with rattlesnakes. California ground squirrels heat up their tails while performing a tail-flagging display to rattlers, who have IR-sensing pit organs, but not to gopher snakes, who lack pit organs. Supporting the hypothesis that the IR increases the efficacy of the tail-flagging signal to rattlers, they found that the combination of the flagging and heat was more effective than tail flagging alone in repelling rattlesnakes.

In addition to studies of communication, robots have great promise for studies of social cueing. Social cueing is the use of cues from other individuals to make decisions about group formation, movement, settlement, foraging, and other social behaviors. In a recent study, José Halloy, Grégory Sempo, and colleagues used robots to study group decisions in cockroaches (**Figure 1(f)**). The robots were autonomous, and programmed with a model for how real cockroaches behave, including their response to other cockroaches, creating a closed behavioral loop between the robots and roaches. This allowed Halloy to use the robots as a test for alternative mechanisms of behavior (see section 'Biomimetic Robots') as well as to study how real cockroaches respond to social cues. The robots did not look like roaches, but were scented with roach pheromones. Real cockroaches followed the robots, even when the robots were programmed to lead them from their preferred dark nooks and crannies into unsafe, open areas.

Another form of social cueing is conspecific attraction, where gregarious animals are attracted to areas that have conspecifics present. Conservation biologists have exploited this tendency to positive effect by using model animals to promote settlement of animals into underutilized habitats. In some cases, static decoys may suffice, but robots may increase the effectiveness of this method. One example of this effect – in this case with detrimental effects to the attracted animals – is the use of powered decoys with spinning-wing by hunters to attract ducks. Joshua Ackerman, John Eadie, and colleagues found up to a 50% increase in kills with the use of motorized decoys, suggesting that the visual cue associated with wing movement may be used by ducks in conspecific cueing. The use of robotic models may thus help to determine which cues and signals are used in conspecific attraction, and may have important conservation and management implications.

Measuring Environmental Effects on Response

Not only are behaviors often complex, but they also occur in a complex world that varies in space and time; to be successful, animals need to produce behaviors that are effective and advantageous in the social and environmental context in which they are used. Robots allow researchers to examine how the efficacy of signals and other

Figure 1 Examples of robots used for playbacks experiments. (a) The robot honeybee used by Michelsen and colleagues to study communication; the robot can imitate the waggle dance and direct workers toward an experimental food source. Photo courtesy of A. Michelsen. Reproduced from Michelsen A, Andersen BB, Stom J, Kirchner WH, and Lindauer M. (1992) How honeybees perceive communication dances, studied by means of a mechanical model. *Behavioral Ecology and Sociobiology* 30: 143–150. (b) The 'Robo-Rana' dart-poison frog used by Peter Narins and colleagues to study male territorial signaling; to the left is a speaker for acoustic playbacks, centers is the model male with inflatable vocal sacs, right is a real male attacking the robot. Illustration courtesy of P. Narins. Reproduced from Narins PM, Grabul DS, Soma KK, Gaucher P, and Hodl W (2005) From the cover: Cross-modal integration in a dart-poison frog. *Proceedings of the National Academy of Sciences of the United States of America* 102: 2425–2429. (c) The robot squirrel used in playbacks of tail-flagging to rattlesnakes; the tail can be heated to mimic the infra-red signals produced by ground squirrels. Photo reprinted with permission from the National Academies of Science. Reproduced from Rundus AS, Owings DH, Joshi SS, Chinn E, and Giannini N (2007) Ground squirrels use an infrared signal to deter rattlesnake predation. *Proceedings of the National Academy of Sciences of the United States of America* 104: 14372–14376. (d) A robotic female sating bowerbird used by Gail Patricelli and colleagues to study male how male response to female signaling affected male success in courtship; the robots can look side to side – to appear more realistic – and perform female signals of interest in the courting male – crouching downward and spreading their wings. Photo courtesy of G. Patricelli. Reproduced from Patricelli GL, Uy JAC, Walsh G, and Borgia G (2002) Sexual selection: Male displays adjusted to female's response. *Nature* 415: 279–280. (e) The robotic female greater sage-grouse used by Gail Patricelli and Alan Krakauer to study how males adjust their displays in response to female proximity; the robot can look back and forth, move toward target males on model train tracks and rotate to face them; she is outfitted with an on-board microphone and video camera to measure male display from a female perspective. Photo courtesy of G. Patricelli. Reproduced from Patricelli GL and Krakauer AH (2010) Tactical allocation of effort among multiple signals in sage grouse: An experiment with a robotic female. *Behavioral Ecology* 21: 97–106. (f) The robotic cockroach used to study collective decision making and movements in cockroaches. The autonomous robot is programmed to interact with real cockroaches, and can elicit novel behaviors from them. Photo courtesy of J. Halloy.

behaviors varies in these different contexts. For example, Terry Ord and Judy Stamps tested how the presence or absence of an alert component (a push-up display) in an *Anolis* lizard affected the response of territorial males to the main part of the signal (the 'headbob') in varying light conditions. They used a robotic male anole, which could be programmed to produce different patterns and combinations of push-ups and headbobs, and deployed in the

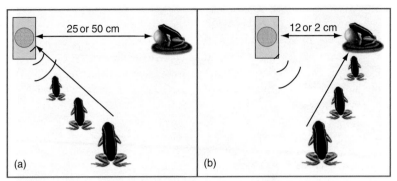

Figure 2 Results from Narins and colleagues, who used a robotic male dart-poison frog to study investigating cross-modal integration in dart-poison frogs. Trials in which the speaker and visual stimulus (the robot with moving vocal sac) were separated by 12 cm elicited significantly less aggressive behavior than with separation of 25 or 50 cm. Figure reprinted Narins PM, Grabul DS, Soma KK, Gaucher P, and Hodl W (2005) From the cover: Cross-modal integration in a dart-poison frog. *Proceedings of the National Academy of Sciences of the United States of America* 102: 2425–2429, with permission from the National Academies of Science.

Puerto Rican forest habitat of the anoles. They found that signals with an alert were detected more quickly in environments with poor lighting than those without an alert. In good lighting, however, the advantage of the alert disappeared. This supported the hypothesis that alerts are important for calling attention to the main part of the signal and that their effectiveness varies with environmental conditions. Using only observations of natural behaviors, such comparisons of the efficacy of the same behavior in different conditions would be impossible, since animals often change their behavior to match the background.

Assaying Responder Traits

Animal behaviorists often need to assay the traits of known individuals to determine how different aspects of behavior relate to each other or to life history traits, like fitness. For example, researchers may want to know how males differ in their courtship behavior, and how these differences relate to male success at convincing females to mate. These kinds of relationships can be difficult to measure in observational studies, since successful males are more likely to be found courting receptive females, and female receptivity behaviors may encourage males to court more enthusiastically. One of the most important uses of playback experiments is to assay the responses of different individuals to a controlled stimulus, allowing researchers to compare individuals on a level playing field.

For example, in collaboration with Gerald Borgia and colleagues, I used a robotic female satin bowerbird to ask whether male satin bowerbirds that display more intensely are more successful in courtship, or whether differences in display intensity are due to differences in the receptivity behaviors of courted females. The robot was able to mimic female signals of receptivity to courtship and could be deployed in the field where courtship would normally occur (**Figure 1(d)**). We found that successful males indeed displayed more intensely, even when female behaviors

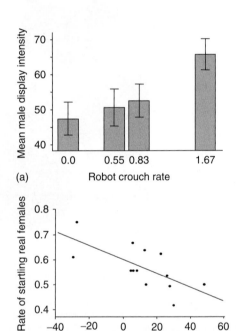

Figure 3 Results from Patricelli and colleagues, who used a robotic female satin bowerbird to assay male courtship behaviors. (a) Males increased their display intensity in response to increasing signals of interest (crouching) from the robot. (b) Males who adjusted their intensity more strongly (i.e., more responsive males) startled real females less often during courtship, and were thus more successful in courtship. Reprinted with permission from Nature Publishing Group.

were held constant using a robot. Moreover, we examined how males adjusted their display intensity in response to female crouching signals. We found that males increase the intensity of their displays in response to increased crouching (**Figure 3(a)**) and that males that adjust their display intensity more strongly (i.e., are more responsive) threaten real females less often (**Figure 3(b)**) and are thus more successful in courtship.

More recently, I worked with Alan Krakauer to examine how male greater sage-grouse adjust their display rate with proximity to females, and how this affects the quality of their displays. Using a robotic female sage-grouse that can be moved toward males on model train tracks (**Figure 1(e)**), we were able to control for the fact that real females tend to approach successful males more closely, which elicits faster strutting from the males. We found that successful males strut at a higher rate toward the robot and they adjust their displays more strongly with proximity to the robot, which allows them to produce higher-quality signals. The sage-grouse robot was also outfitted with an onboard audio and video recorder, so that it could be used as a data-collection tool as well as a playback stimulus; this allowed us to record male signals from the receiver's perspective – literally getting inside the head of a female during courtship. In both the sage-grouse and bowerbird studies, the use of a robot allowed us measure male responsiveness to controlled female behaviors, and to examine how this responsiveness varies with other male display traits and components of male fitness.

Practical Considerations for Robotic Playbacks

When is it appropriate to use robots in playback experiments? As discussed earlier, robotics is a powerful tool in playback experiments, but simpler tools may suffice in many cases. For example, if an animal will respond to video playbacks of the behavior of interest, then this may allow far more flexibility in manipulating the experimental stimulus (**Figure 4**). Using available video and animation software, video stimuli can be manipulated to change the form, color, size, behaviors, background, etc. Comparable manipulations with a robot would typically be more expensive and require more specialized training to construct, and in cases where behavior patterns are very rapid and complex, may simply be impossible with present technologies. But videos are not better in all cases – among other issues, they are two-dimensional and fail to elicit responses from some animals. Both robotic and video playbacks have pros and cons; researchers must choose which option will be cheaper, simpler, more reliable, and more flexible, considering the natural history of the study organism, the logistics of the study, and the questions being addressed.

In deciding whether to use a robot, researcher must also consider whether the focal species will respond to a robot. If the mimicked behavior is extremely complex, then it may be prohibitively expensive (or impossible) to build a robot that is sufficiently realistic to elicit a natural response. Thus far, robotic playbacks have been used primarily when behaviors are relatively simple, and/or the sensory systems or selectivity of the focal animal is sufficiently permissive. For example, researchers have elicited natural responses from male and female frogs, using robots that mimic male territorial and sexual signals. These robots look and move in a strikingly realistic way, with painted rubber or plastic models representing the body of the frog and inflatable vocal sacs (**Figures 1(b)** and **5**). In contrast, the sexual signals of male birds are often extremely complex, involving rapid movement and flight, such that even the best plastic or rubber bird models look far from real. Studies with robotic birds have therefore mimicked the simpler behavioral patterns of females, and have used jointed taxidermied mounts of the target species (**Figure 1(d)** and **1(e)**). Further, these studies have thus far been conducted on polygynous species, in which males are not very choosy about which females they court. No successful studies have yet focused on monogamous species, in which males and females are both choosy during courtship and in which the bar for a realistic model would likely be higher. Unfortunately, there is no way to predict whether a species will respond to a robot and how realistic the robot must be to elicit a natural response; trial and error is thus required in the early stages of the research.

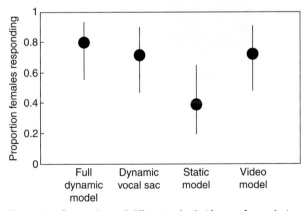

Figure 4   Comparison of different robotic túngara frog robots with video playback. Taylor and colleagues compared the proportion of females responding to a robotic model with a moving vocal sac, a moving vocal sac alone, a static model and a video playback of a displaying male. Females were given a choice between these stimuli and a speaker broadcasting the same call but lacking the visual stimulus of a model frog or video. Reprinted Taylor RC, Klein BA, Stein J, and Ryan MJ (2008) Faux frogs: Multimodal signalling and the value of robotics in animal behaviour. *Animal Behaviour* 76: 1089–1097, with permission from Animal Behaviour, Elsevier.

Biomimetic Robots

Mathematical and computational models have provided scientists with powerful tools for testing hypotheses for the mechanisms underlying behaviors. For example, scientists can test hypotheses about the neural and muscular mechanisms underlying locomotion by constructing models of these mechanisms and their interaction with the environment. They can then test whether their

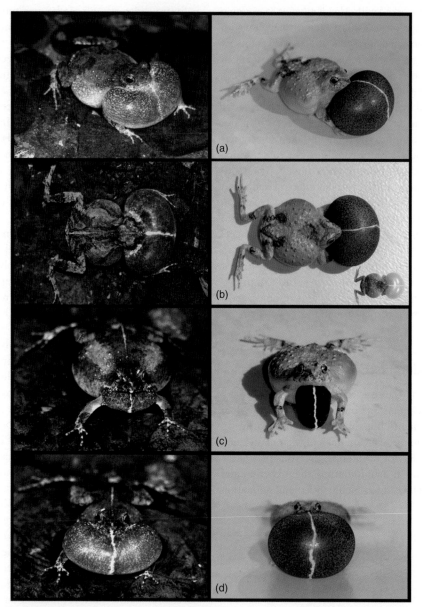

Figure 5 Photographs comparing real (left) and robotic (right) túngara frogs used by Taylor and colleagues to study female mate choice. The body of the robot is urethane, cast from a real túngara frog, and painted to match. The inflatable vocal sac is made from a latex catheter. Reprinted Taylor RC, Klein BA, Stein J, and Ryan MJ (2008) Faux frogs: Multimodal signalling and the value of robotics in animal behavior. *Animal Behavior* 76: 1089–1097, with permission from Animal Behavior, Elsevier.

models approach the behaviors observed in real animals. Robotics has provided another tool in this modeling toolkit and is often used in concert with mathematical and computational models. Here scientists test hypotheses about mechanisms by building physical models – robots – that use these proposed mechanisms and interact with stimuli in the outside world. This field is often called 'biomimetic robotics' or 'biorobotics' to distinguish it from other fields of robotics. Since the real world is typically more complex than even the most complex mathematical and computation models – even in the laboratory, where many factors are controlled – robotics

provides a robust test of hypotheses and a powerful tool for generating new predictions and refining models. These studies may also help researchers design more effective robots – thus animal behavior may contribute in turn to the development of robot technology.

Barbara Webb describes the process of robotic modeling as involving the following steps: 'identifying a target issue to be explained; offering an explanation; demonstrating that it accounts for observations; deriving further predictions and testing them.' Webb argues that the power of this approach is that it forces researchers to confront how the environment and body design affect behavioral capabilities, often causing researchers to reevaluate their assumptions. In some cases, robotic modeling approaches have demonstrated that the previous models were too simple to account for the complexity of observed behaviors in animals. In other cases, the opposite has been found – the simpler models can accomplish a task without the sensory or cognitive complexity that was assumed necessary. Researchers typically begin with the simplest possible mechanisms and add complexity as needed until the robot approaches natural behaviors.

The biomimetic robotics approach has thus far focused on simple behaviors, or simple aspects of complex behaviors, since there are more developed models of the underlying mechanisms for testing in these cases. The list of modeled behaviors includes escape behavior, locomotion, learning, recognition, social aggregation, collective behavior, and movement toward a stimulus source (taxis). A few examples of biomimetic research on taxis behavior are given in the following section.

Modeling Taxes

One of the most common behaviors addressed with biomimetic robots is taxis – directed movement toward a stimulus source, such as light (phototaxis), sound (phonotaxis), contact (thigmotaxis), and chemical cues (chemotaxis). For example, Frank Grasso, Jelle Atema, and colleagues have used robotic lobsters ('robolobsters') to examine the mechanisms that allow lobsters to efficiently locate the source of a chemical plume in turbulent water (**Figure 6(a)**). The robolobsters are designed to mimic the scale of a lobster in size, speed, maneuverability, and olfactory sampling, but not the biomechanics of movement (they roll on wheels). The robolobsters are programmed with several hypothesized mechanisms for localization, and the resulting behavior of the robolobster is compared with the behaviors of real lobsters facing the same task in the same conditions. Using this method, the researchers were able to reject hypothesized mechanisms for failing to reproduce lobster behavior, and direct future biological research toward more likely mechanisms.

Phonotaxis has been studied by Barbara Webb and colleagues in another invertebrate, the cricket. The robots created by Webb and her group model the sensory, cognitive, and biomechanical systems involved in phonotaxis (**Figure 6(b)**). Their robotic model of directional hearing suggested that female preference for a particular frequency (i.e., pitch) of male song may be a consequence of the physiology of the cricket ear, since only preferred frequencies can be localized by the robot. In addition to directional hearing, mate searching requires female crickets to recognize calls with the appropriate timing, and then orient and move toward the sound source. Webb and colleagues have modeled each step of this process. Currently, robots are being developed that integrate more

Figure 6 Examples of biomimetic robots. (a) A robotic lobster ('robo-lobster') used by Grasso and colleagues to test alternative mechanisms of chemotaxis, the localization and tracking of chemical signals, in this case in a current of water. Real lobsters were tested with an identical task for comparison. Reprinted Grasso FW and Atema J (2002) Integration of flow and chemical sensing for guidance of autonomous marine robots in turbulent flows. *Environmental Fluid Mechanics* 2: 95–114, with permission from Environmental Fluid Dynamics. (b) Three biomimetic crickets produced by Webb and colleagues to test hypothesized mechanisms of cricket behavior. Front, a robot that performs phonotaxis; middle, a robot that combines auditory and visual behaviors; rear, a robot that mimics insect-walking behaviors. Reprinted Webb B (2008) Using robots to understand animal behavior. In: Brockmann HJ, Roper TJ, Naguib M, Wynne-Edwards KE, Barnard E, and John CM (eds.) *Advances in the Study of Behavior*, pp. 1–58. New York: Academic Press, with permission from Advances in the Study of Behavior, Elsevier.

explicitly our increasing knowledge of the neurophysiology of the sensory and locomotor systems involved with phonotaxis. Ultimately, the goal is to develop a complete cricket model, which integrates all of the sensory modalities, as real animals do, to examine the interactions between sensory modalities and behaviors.

Studies of taxis behaviors can also lead to a more complete understanding of social aggregation and cueing. Christopher May, Jeffrey Schank, Sanjay Joshi, and colleagues developed robotic rat pups to study the apparently self-organizing and intentional behaviors of real rat pups in an arena. They programmed robots with either a random or a 'wall-following' response to contact with the edge of the arena (thigmotaxis), and they compared

Figure 7   Biomimetic robots using a random-movement algorithm reproduce the aggregation behaviors of real rat pups. Examples of aggregation patterns of seven (a) and ten (b) day-old rat pups (on the left) and robots (on the right) in experimental arenas. Reprinted May CJ, Schank JC, Joshi S, Tran J, Taylor RJS, and Esha I (2006) Rat pups and random robots generate similar self-organized and intentional behavior. *Complexity* 12: 53–66, with permission from Complexity.

the resulting behavior with the behavior of real rat pups in the same arena. They found that the simpler, random mechanism yielded behaviors strikingly similar to those of real rat pups (**Figure 7**). This does not suffice to prove that rat pups move randomly as well, but it clearly demonstrates that seemingly directed and complex behaviors can emerge from mechanisms far simpler than those previously assumed.

Biomimetic Playbacks: A Fusion of Approaches

Biomimetic robotics is distinguished from robotic playbacks, in that biomimetic robots are built to model the internal mechanisms of focal animals, rather than to mimic the outward behaviors in order to fool real animals. Since conspecifics are part of the environment experienced by animals, there are many cases where a combination of approaches – using biomimetic robots in interactions with real animals – is ideal. This approach can be used to test alternative hypotheses about mechanisms of response to social cues (biomimetic modeling), and may also be used to learn more about the signals and cues used by responding animals in mediating social interactions (playback experiments). The case discussed previously in which robotic cockroaches were used to study the cues used to coordinate group movements, is an example of this hybrid approach which yielded information about both the mechanisms and functional consequences of social cues. As technology improves and collaboration between animal behaviorists and engineers increases, this approach will undoubtedly become more common.

Limitations and Challenges

The use of robotics holds a great deal of promise for animal behavior research, but it is important to recognize the limitations and challenges to this approach as the field moves forward.

An important challenge for any behavioral study with robotics is how to validate the assay, demonstrating that the robot elicits or reproduces naturalistic behaviors. In playback experiments, this is often accomplished by testing whether an animal's response to the robot is correlated with its response to natural stimuli or by comparing the mean response of animals to the robotic and natural stimuli. In biomimetic robotics, the behavior of the robots is often compared with the behaviors of real animals. In both cases, researchers must decide what variable(s) to measure to assess validity. In addition, the strength of the relationships between robotic and natural cases will depend on many factors, and researchers must decide on a threshold to accept the assay as valid. In some cases, statistical significance may be used as a threshold, but in other cases, especially in biomimetic robotics, a closer relationship between the robot and real animal may be expected. Since it is difficult to imagine a single standard that would apply to all cases, devising methods and thresholds for validation will remain a challenge in future studies.

Even if a strong relationship between real and robotic behaviors emerges during validation, researchers must recognize that a similarity of outcome does not prove similarity of causation. While this concern is important in playbacks, it is a particular concern in biomimetic robotics, where explaining the mechanism is the primary goal. As discussed in the examples above, robotic modeling has the most power when it is used to eliminate untenable hypotheses and focus attention on more likely – and new – hypotheses for further study.

Researchers using robots in playback experiments must also consider the possibility of pseudoreplication – the use of the same stimulus in multiple experimental trials, which can artificially inflate the degrees of freedom. Pseudoreplication is problematic in any kind of playback experiment, when researchers use one or a few playback stimuli but generalize their results to all possible stimuli. Pseudoreplication can be minimized by the use of multiple playback stimuli or a stimulus that represents the average among multiple possible stimuli. These approaches may be feasible in some robot studies. But unfortunately, building multiple robots is far more difficult and expensive than recording or synthesizing multiple acoustic playback stimuli, and may not be possible or ethical when taxidermied mounts are used in building the robots. Many robotic playback studies to date have used only one or a few robots, and thus we do not have sufficient data to address whether different robots are likely to elicit different responses. Nonetheless, this issue must be addressed, and if not resolved, then at least acknowledged as a limitation of the study.

Despite the challenges and limitations discussed throughout this article, it is clear that robots will allow animal behaviorists to pursue questions that would be difficult or impossible to address otherwise. Used in combination with other approaches, robotics will allow us to generate new hypotheses and test existing ones to explain both the mechanisms and functions of animal behaviors.

See also: Acoustic Signals; Agonistic Signals; Alarm Calls in Birds and Mammals; Bowerbirds; Cockroaches; Collective Intelligence; Communication and Hormones; Communication: An Overview; Dance Language; Experiment, Observation, and Modeling in the Lab and Field; Group Movement; Herring Gulls; Honeybees; Insect Social Learning; Mate Choice in Males and Females; Mating Signals; Multimodal Signaling; Niko Tinbergen; Norway Rats; Playbacks in Behavioral Experiments; Robot Behavior; Sound Localization: Neuroethology; Túngara Frog: A Model for Sexual Selection and Communication; Visual Signals.

Further Reading

Göth A and Evans CS (2004) Social responses without early experience: Australian brush-turkey chicks use specific visual cues to aggregate with conspecifics. *Journal of Experimental Biology* 207: 2199–2208.

Grasso FW (2001) Invertebrate-inspired sensory-motor systems and autonomous, olfactory-guided exploration. *The Biological Bulletin* 200: 160–168.

Halloy J, Sempo G, Caprari G, et al. (2007) Social integration of robots into groups of cockroaches to control self-organized choices. *Science* 318: 1155–1158.

Martins EP, Ord TJ, and Davenport SW (2005) Combining motions into complex displays: Playbacks with a robotic lizard. *Behavioral Ecology and Sociobiology* 58: 351.

May CJ, Schank JC, Joshi S, Tran J, Taylor RJS, and Esha I (2006) Rat pups and random robots generate similar self-organized and intentional behavior. *Complexity* 12: 53–66.

Michelsen A, Andersen BB, Storm J, Kirchner WH, and Lindauer M (1992) How honeybees perceive communication dances, studied by means of a mechanical model. *Behavioral Ecology and Sociobiology* 30: 143–150.

Narins PM, Grabul DS, Soma KK, Gaucher P, and Hodl W (2005) From the cover: Cross-modal integration in a dart-poison frog. *Proceedings of the National Academy of Sciences of the United States of America* 102: 2425–2429.

Ord TJ and Stamps JA (2008) Alert signals enhance animal communication in noisy environments. *Proceedings of the National Academy of Sciences of the United States of America* 105: 18830–18835.

Partan SR, Larco CP, and Owens MJ (2009) Wild tree squirrels respond with multisensory enhancement to conspecific robot alarm behaviour. *Animal Behaviour* 77: 1127–1135.

Patricelli GL, Uy JAC, Walsh G, and Borgia G (2002) Sexual selection: Male displays adjusted to female's response. *Nature* 415: 279–280.

Patricelli GL and Krakauer AH (2010) Tactical allocation of effort among multiple signals in sage grouse: An experiment with a robotic female. *Behavioral Ecology* 21: 97–106.

Rundus AS, Owings DH, Joshi SS, Chinn E, and Giannini N (2007) Ground squirrels use an infrared signal to deter rattlesnake predation. *Proceedings of the National Academy of Sciences of the United States of America* 104: 14372–14376.

Taylor RC, Klein BA, Stein J, and Ryan MJ (2008) Faux frogs: Multimodal signalling and the value of robotics in animal behaviour. *Animal Behaviour* 76: 1089–1097.

Webb B (2000) What does robotics offer animal behaviour? *Animal Behaviour* 60: 545–558.

Webb B (2008) Using robots to understand animal behavior. In: Brockmann HJ, Roper TJ, Naguib M, Wynne-Edwards KE, Barnard E, and John CM (eds.) *Advances in the Study of Behavior*, pp. 1–58. New York: Academic Press.

S

Sea Turtles: Navigation and Orientation

C. M. F. Lohmann and K. J. Lohmann, University of North Carolina, Chapel Hill, NC, USA

Sea Turtle Life History

Among the sea turtles, orientation and navigation have been studied most thoroughly in loggerhead (*Caretta caretta*) and green turtles (*Chelonia mydas*). These two species have similar life histories and are highly migratory. In the typical case, females lay clutches of 100–150 eggs and bury them in the sand of the nesting beach. After ~ 2 months, the young turtles hatch below ground and slowly dig their way to the surface. The hatchlings emerge onto the surface of the sand, then scramble to the sea and migrate offshore, where they begin a prolonged period in the open sea, sometimes migrating across entire ocean basins. Later, juveniles recruit to various coastal feeding areas. Adults of most populations also exploit coastal feeding areas and then migrate from the feeding areas to the same breeding area in which they originally hatched.

Many of these migrations depend on the ability of turtles both to orient (i.e., maintain a course in a particular direction) and to navigate (i.e., to determine their position relative to their goal). The cues used for these tasks have been the focus of considerable research during the last 20 years.

Finding the Sea: Use of Visual Cues

When young turtles crawl from their nests, they almost immediately establish courses directly toward the sea. Recent research has confirmed the findings of studies published nearly a century ago: despite emerging at night, hatchlings use visual cues to set and maintain their courses to the sea. Apparently, the turtles are attentive to light levels near the horizon. They crawl toward low bright lights and veer away from dark silhouettes such as the outlines of vegetated dunes. The turtles ignore bright lights from overhead. The result is that, at night, turtles are drawn to the sea, which reflects starlight and moonlight far more than land does. Moreover, they are not often drawn to the brightest nighttime object, the moon, because it is too high in the sky overhead.

This reliance on light cues has undoubtedly served sea turtles well over the course of their 120 My history. On wilderness beaches, they are seldom misoriented toward the land. On beaches shared with modern humans, however, artificial lighting has been a serious problem. As development overtook the beaches of south Florida in the latter twentieth century, young turtles began crawling toward parking lots and tennis courts instead of toward the sea. Caught there at sunrise, the turtles often died from predation and desiccation before they could find their way back to the ocean. Happily, the understanding of their orientation mechanisms has informed conservation efforts throughout the world. In most populated areas, an effort is now made to reduce beach lighting on turtle nesting beaches and the incidence of misoriented hatchlings has declined.

The Offshore Migration

Migration by hatchlings in the ocean has been studied most thoroughly using loggerhead and green turtles from populations found along the Atlantic coast of Florida, USA. For these turtles, the first step of the migration is a journey from the beach out to the Gulf Stream current. This 'offshore migration' is accomplished during the first 2 or 3 days after emergence, when the turtles are nourished by yolk reserves and consequently do not feed. At this point in their lives, they are small swimming machines highly motivated to migrate.

Use of Wave Cues in the Ocean

When the ancestral turtles forsook the land and took up life in the sea, new orientation cues became available to them that their terrestrial brethren could not exploit. Among these is the information inherent in wave trains traveling across the ocean surface. The first indication that turtles might use wave cues came in studies of the offshore migration. Hatchlings in a floating orientation arena moored off the Florida coast swam in random directions when the ocean was still, but swam east when waves moved from the east toward the Florida shore. Subsequently, it was shown that when hatchlings first enter the water, they almost always swim into waves – even under unusual weather conditions when waves come from the landward direction.

Various anecdotal observations indicate that hatchlings cease swimming into waves after several hours in the ocean. These observations thus suggest that hatchlings use wave direction as an orientation cue only for a short time after entering the sea. Because waves in shallow water near the shore are refracted until they move nearly straight toward the shore, swimming into the oncoming waves would cause the turtles to orient straight toward the open ocean. This orientation allows turtles to avoid swimming parallel to the beach; instead, they head out to sea and quickly escape the nearshore, predator-filled waters.

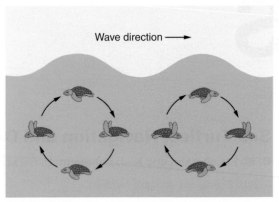

Figure 1 The motion of a hatching turtle swimming with and against the direction of wave propagation. For a hatchling oriented into waves (left), the sequence of accelerations during each wave cycle is upwards, backwards, downwards, and forwards. A turtle swimming with the waves (right) is accelerated upwards, forwards, downwards, and backwards. Modified from Lohmann KJ, Swartz AW, and Lohmann CMF (1995) Perception of ocean wave direction by sea turtles. *Journal of Experimental Biology* 198: 1079–1085.

Detection of Wave Direction

The direction of wave approach is fairly simple for a human to observe visually while standing high above the water in daylight hours. For a small turtle swimming below the surface of the water while migrating in the dark, the problem is more challenging. Under these conditions, wave direction cannot be seen from below, nor is the turtle able to raise its head high enough above the water to see the waves on the surface. In addition, as experiments in a laboratory wave tank demonstrated, turtles can swim into waves even in complete darkness.

Clearly, sight is not used by turtles to determine wave direction. Instead, it appears that turtles are able to monitor the accelerations that occur below the surface as waves pass overhead. Envision a waterlogged cork floating at a depth of a meter or so. When a wave passes through the ocean, the cork will describe a circle as it is accelerated around by the wave, eventually returning to its starting location (oceanographers refer to this as an orbital movement). Now, instead of a cork, imagine a turtle swimming at the same depth. Because the turtle provides its own forward accelerations, it is unlikely that its body will actually describe a circle, but it will feel the same accelerations that move a cork when a wave goes by. Moreover, the sequence of accelerations will appear to differ depending on how the turtle's body is oriented relative to the wave. For example, if the turtle is facing directly into the oncoming wave, the sequence will begin with the turtle perceiving an acceleration going up and back, then down and forward around to the starting point again. In contrast, a turtle swimming away from the wave would perceive a sequence in which it is accelerated upward and

forward, then downward and backward around to the starting point (**Figure 1**).

The hypothesis that hatchling turtles could distinguish between such patterns of orbital movements was tested in experiments that took advantage of the unusual nature of sea turtles. First, turtles are air-breathing reptiles so can be removed from the water allowing all hydrodynamic cues to be eliminated. Second, when suspended in air with flippers no longer in contact with the substrate, hatchlings behave as if in water and attempt to swim. When the suspended turtles were moved in gentle circles similar to those found beneath waves (**Figure 2**), the turtles responded by attempting to turn in a way that would cause them to experience the same accelerations they would feel if swimming into a wave. For example, when a turtle was subjected to simulated waves from the right, the turtle tried to turn right. Thus, turtles sense wave direction by monitoring the sequence of accelerations they experience beneath waves.

Use of a Magnetic Compass Sense

In some parts of Florida, the Gulf Stream current flows very close to the coastline and turtles can reach it in a few hours; in other areas, the current is several days away. In either case, however, wave cues are insufficient to guide turtles all the way to the Gulf Stream, because the wave refraction zone, where waves proceed directly toward shore, is relatively narrow. Once turtles pass through this zone, waves are no longer a reliable orientation cue.

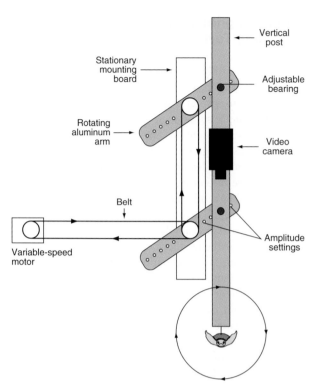

Figure 2 A machine designed to simulate wave motion by reproducing the sequence of accelerations that occur beneath a propagating wave. The responses of hatchling turtles to these simulated waves have been studied by placing turtles into cloth harnesses and subjecting them to orbital movements while they are suspended in air. Modified from Lohmann KJ, Swartz AW, and Lohmann CMF (1995) Perception of ocean wave direction by sea turtles. *Journal of Experimental Biology* 198: 1079–1085.

It seems likely that once beyond the wave refraction zone, turtles rely on their magnetic compass sense. Like other famous navigators (notably migratory birds and homing pigeons), turtles have the ability to sense Earth's magnetic field. This was demonstrated originally in a laboratory orientation arena (**Figure 3**) consisting of a pool of water in which a hatchling could be tethered. When the hatchling swam, it pulled a rotatable tracker arm. The tracker arm was in turn connected to electronics that recorded the direction toward which it was pointing and, by extension, the direction toward which the turtle swam.

The orientation arena was surrounded by a magnetic coil system which, when activated, reversed the magnetic field around the turtle. Thus, the turtle would perceive magnetic north in geographic south and so on. In initial experiments, turtles that swam in the unaltered Earth's field swam to geographic (and magnetic) northeast on average, while those in a reversed field swam to geographic southwest (which was now the new magnetic northeast). Thus, turtles could sense the magnetic field and use it to orient.

Subsequent work has shown that the direction turtles swim in such experiments is not the result of an innate directional preference. Rather, the turtles have the ability to acquire a magnetic preference based on other environmental cues such as the position of a light or the direction of waves. In nature, this flexible system presumably allows turtles to learn the direction of the open ocean from local cues. For example, after swimming eastward into waves, the Florida turtles on the east coast might acquire a directional preference for east and use the magnetic

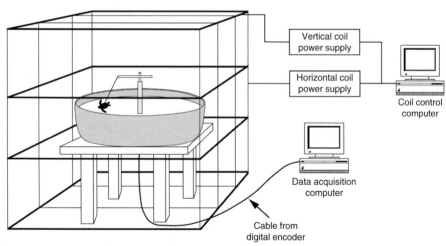

Figure 3 The orientation arena, coil system for generating earth-strength magnetic fields of different inclinations, data acquisition system and coil control system. Each hatchling was tethered to a rotatable arm mounted on a digital encoder (which was inside the central post of the orientation arena). The rotatable arm tracked the direction towards which the turtle swam in darkness. Signals from the digital encoder were relayed to the data acquisition computer. The coil control system regulated current running through the box-like magnetic coil system and was used to create specified magnetic fields inside the volume of the coil. Modified from Lohmann KJ and Lohmann CMF (1994) Detection of magnetic inclination angle by sea turtles: A possible mechanism for determining latitude. *Journal of Experimental Biology* 194: 23–32.

compass to maintain that course, whereas turtles on Florida's west coast might similarly set their magnetic compasses for west and then swim in that direction until reaching the Gulf Stream Current.

The Gyre Migration of North American Loggerheads

Loggerhead turtle hatchlings from the eastern coast of North America migrate offshore and then spend several years in the North Atlantic Gyre. This circular current system flows north along the southeastern United States, arches eastward across the Atlantic, turns south along the northern coast of Africa, and then flows back west toward the Caribbean.

Because small turtles cannot swim fast enough to progress against the current, researchers initially speculated that the turtles drifted passively in the gyre during the many years of their pelagic existence. Recent evidence suggests, however, that the turtles actively guide themselves in the gyre and adjust position to help stay within its confines.

Positional Information in the Earth's Magnetic Field

Human navigators have long known that the Earth's magnetic field provides compass information and the use of a magnetic compass sense by turtles and many other animals is now well-documented. A novel finding that arose largely from work on turtle navigation is that animals can also use the magnetic field to help them determine their position relative to a goal or habitat boundary. The use of positional information is based on the detection of magnetic features that vary more or less regularly across the Earth's surface.

Sea turtles are capable of detecting at least two such features: (1) magnetic intensity or strength, which increases steadily as one moves from the magnetic equator toward the magnetic poles and (2) magnetic inclination angle, which is defined as the angle that magnetic field lines make with the surface of the Earth; this angle becomes steadily steeper as one moves from the magnetic equator toward the poles.

The ability to detect either of these features would theoretically allow a turtle to roughly approximate its latitude. Detecting both might provide turtles with some information about longitude as well, because, while both inclination and intensity vary with latitude, they do so independently. Thus, most locations within an ocean basin have unique combinations of intensity and inclination.

Use of Magnetic Positional Information by Hatchling Turtles: Staying Within the Gyre

In an initial study of magnetic positioning, hatchling turtles were placed into an orientation arena (**Figure 3**) and exposed to several magnetic fields that differed only in inclination; a subsequent study involved fields that differed only in intensity. The swimming orientation of turtles in these magnetic fields clearly demonstrated that the turtles could detect both inclination and intensity. Moreover, the responses, when considered in the context of the hatchling's migratory route, would have had the effect, in nature, of keeping turtle hatchlings from straying out of the gyre. For example, when exposed to an inclination found near the northern edge of the gyre, the turtles swam south, but at an inclination found near the southern border of the gyre, they swam northeast.

In subsequent studies, the coil system was used to produce specific combinations of inclination and intensity that exist at three actual locations around the gyre. In all the three cases, the turtles swam in directions that would keep them within the gyre and progressing along their migratory route (**Figure 4**). It thus appears that hatchlings

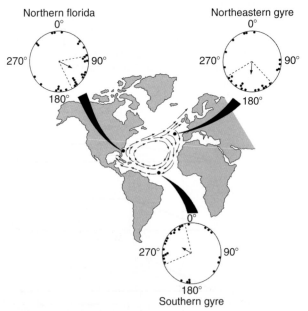

Figure 4 Orientation of hatchling loggerhead turtles in magnetic fields characteristic of three widely separated locations (marked by black dots on the map) along the migratory route. Generalized main currents of the North Atlantic gyre are represented on the map by arrows. In the orientation diagrams, each dot represents the mean angle of a single hatchling. The arrow in the center of each circle represents the mean angle of the group. Dashed lines represent the 95% confidence interval for the mean angle. From Lohmann KJ, Cain SD, Dodge SA, and Lohman CMF (2001) Regional magnetic fields as navigational markers for sea turtles. *Science* 294: 364–366. Reprinted with permission from AAAS.

employ a sort of magnetic waymark navigation to circumnavigate the gyre. When they encounter a particular magnetic field, they use it as a waymark that triggers a directional response that keeps them within the appropriate habitat.

Because hatchling turtles in these studies had no prior experience in the ocean, these studies strongly suggest that the turtles emerge from their nests programmed to respond to particular magnetic fields by swimming in certain directions. The possibility that these responses are innate has interesting conservation implications, because it might be that populations of loggerheads from different ocean basins have different inherited responses. If so, then it might be very difficult to reestablish endangered or eradicated populations by bringing in stocks from other areas; these introduced stocks would not possess the correct responses for the local migratory route. The failure to reestablish a nesting population of green turtles in Bermuda with eggs and hatchlings from Costa Rica may be the result of this difficulty.

Use of Magnetic Positional Information by Coastal Juveniles: Finding Specific Feeding Sites

After spending several years in the North Atlantic Gyre, young juvenile loggerheads move into coastal areas of the eastern United States and establish neritic feeding areas, to which they show long-lasting fidelity. In the northern part of the loggerhead range, for example, feeding sites are used seasonally. Tag returns indicate that the same individuals return to the same sites in successive years, even after migrating hundreds of kilometers in the interim. In other areas, turtles have been displaced from harbors to allow for dredging and other human activities, and the same turtles have subsequently returned to those areas.

These returns suggest an ability to navigate back to specific locations. Additionally, it seems unlikely that use of specific feeding sites is genetically programmed; rather, one would expect site selection on the basis of local conditions (i.e., food abundance and availability of shelter) that can easily change in response to storms. Thus, fidelity to particular feeding sites is unlikely to be based on innate responses. Instead, it appears that juvenile turtles learn the locations of their favored sites and can return to them from long distances.

The process of long-distance navigation in the ocean probably consists of two basic steps. To reach a specific site, turtles must first use long-range cues to guide themselves across the ocean and arrive in the general area. Then short-range cues might be used to identify the exact location. In sea turtles, the short-range cues have not been studied extensively, though visual, olfactory, auditory, and even wave cues are all possibilities. For the

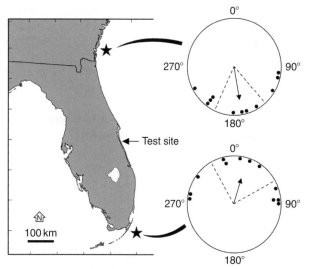

Figure 5   Evidence for a magnetic map in juvenile green turtles. Juvenile turtles were captured in feeding grounds near the test site in Florida, USA. Each turtle was exposed to a magnetic field that exists at one of two distant locations (represented by stars along the coastline). In the orientation diagrams, each dot represents the mean angle of a single turtle. The arrow in the center of each circle represents the mean angle of the group. Dashed lines represent the 95% confidence interval for the mean angle. Modified from Lohmann KJ, Lohmann CMF, Ehrhart LM, et al. (2004) Geomagnetic map used in sea turtle navigation. *Nature* 428: 909–910.

long-range cues, evidence suggests that sea turtles use magnetic maps based on inclination and intensity.

The pivotal experiment on magnetic maps was performed in Melbourne Beach, Florida, using juvenile green turtles with feeding sites roughly 100 m from the shore. These turtles were collected, brought to an onshore facility a few kilometers away, and placed in an orientation arena similar to that used for hatchlings except that it was much larger to accommodate the significantly larger size of the juveniles. When in the arena, the turtles were exposed to one of two magnetic fields: either the inclination and intensity of a location ~340 km to the north of the test site or the inclination and intensity of a location ~340 km south of the test site. Turtles exposed to the northern field swam south, whereas those exposed to the southern field swam north (**Figure 5**). Because the only difference between the two treatments was the magnetic field in the arena, the turtles were clearly detecting and responding to those fields. Moreover, the turtles behaved as if they had been displaced and were attempting to return to their feeding areas, apparently using magnetic positional information to determine whether they were north or south of their goal.

These results suggest that juvenile turtles have what is, in essence, a magnetic map. They have the ability to detect and remember magnetic features of their feeding grounds and have learned how these features vary across the surface of the Earth. When presented with a new set of magnetic features replicating a new location, they can determine where they are relative to their goal.

Use of Magnetic Positional Information by Adult Sea Turtles: Natal Homing

Like the juvenile turtles, adult turtles show fidelity to specific feeding sites. Males regularly migrate between these sites and courtship sites. Females migrate between feeding sites, courtship sites, and nesting beaches. Tagging studies indicate that females display fidelity to specific nesting beaches, sometimes returning to nest within just a few kilometers of their nest sites from previous years.

Recent genetic studies have confirmed an even more surprising ability. Apparently, sea turtles tend to nest in the same regions in which they themselves hatched. This behavior of natal homing is so strong that, for example, the population of loggerheads that nests in south Florida is genetically distinct from that nesting in north Florida.

How turtles accomplish natal homing is not known. It appears, however, that the task can be considered a special instance of long-distance ocean navigation. Thus, the turtle may first use long-range cues to navigate into the general location of the natal beach, and then use local, short-range cues to identify the appropriate nesting site. It seems likely that the initial long-range step is accomplished magnetically, and shorter-range navigation relies on other cues.

Recent evidence indicating the use of magnetic cues by adult turtles involves an experiment in which female turtles were displaced away from an island nesting beach. Some individuals were fitted with magnets that disrupted the ambient magnetic field and their homing ability was compared with controls fitted with nonmagnetic brass bars. The magnetically disrupted turtles found the nesting beach but took longer to do so and followed far less direct routes than controls. These results suggest that magnetic cues are indeed used by swimming turtles to find their nesting beach – but how do the turtles know which nesting beach is theirs?

The unique challenge in natal homing is that a turtle must quickly acquire, at a very young age, sufficient information that it can recognize its home region and return to it years later. One possible way that turtles might accomplish this task involves imprinting on the magnetic features of the natal area. If hatchlings detect and memorize the magnetic features of their home area – perhaps during the offshore migration, or even while in the nest – then as adults, turtles can potentially use their magnetic map sense to locate the general region of the natal beach.

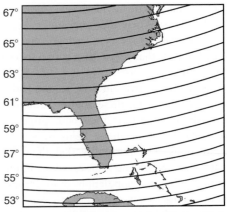

Figure 6 Diagram of southeastern United States with isoclinics (lines of equal magnetic inclination). The scale on the right shows the inclination angles in degrees. Each isoclinic intersects the Atlantic coastline only once; thus each area of coast is marked by a unique inclination angle.

Many major nesting beaches are located on continental shorelines aligned approximately north to south and in geographic areas where magnetic intensity and inclination also vary in a north–south direction. In such regions, different areas along the coastlines have unique magnetic signatures (**Figure 6**). A turtle could theoretically swim along the coastline until it finds a remembered magnetic intensity or inclination, or perhaps combination of the two, and then search until appropriate nesting habitat is discovered.

Not all nesting beaches are continental; some cases of turtle navigation involve navigation to remote, island nesting sites. It is interesting to note that these may be special cases of the more general process. For instance, a Brazilian population of green turtles nests on tiny Ascension Island, some 2000 km away from the Brazilian coast. To all appearances, gravid females find the 8-km long island with amazing pinpoint accuracy, far more accurate than what one would expect using a magnetic map. In this case, it seems likely that turtles use magnetic map information to arrive in the area of Ascension and then use local cues to find a nesting beach. Because the only suitable beaches in the area are the beaches of Ascension, all of the turtles eventually find and nest on the same tiny piece of land.

One difficulty with the hypothesis of magnetic imprinting of natal homing information is its potential vulnerability to problems with magnetic drift (secular variation). The magnetic field changes slightly each year and most species take 10–40 years to reach maturity; during this time, the field in the natal areas may change. If they relied solely on magnetic fields to find their nesting beaches, turtles might arrive many kilometers from their goals. Recent calculations suggest, however, that turtles in many geographic areas arrive within a region where short-range cues are available for use in identifying appropriate nesting sites. Kemp's ridley turtles, for example, nest in large aggregations on a 160 km beach near Rancho Nuevo, Mexico. If a turtle imprinted on the magnetic inclination at the center of that beach, calculations show that it would return to a location several kilometers away but usually within the limits of the nesting beach.

Future Directions

During the past two decades, tremendous progress has been made in unraveling the mechanisms that underlie orientation and navigation in sea turtles. Many questions remain unanswered. How do sea turtles sense the magnetic field? Do hatchlings indeed imprint on the magnetic field of their home beaches? What local cues are used to identify precise locations? What population differences exist? Nonetheless, the rapid progress in this field has provided a window into the world of one of the most remarkable animal navigators.

See also: Magnetic Orientation in Migratory Songbirds; Pigeon Homing as a Model Case of Goal-Oriented Navigation.

Further Reading

Bolten AB and Witherington BE (eds.) (2003) *Loggerhead Sea Turtles.* Washington, DC: Smithsonian Institution Press.

Carr AF (1979) *The Windward Road: Adventures of a Naturalist on Remote Caribbean Shores.* Tallahassee, FL: University Presses of Florida.

Carr AF (1986) *The Sea Turtle – So Excellent a Fishe.* Austin, TX: University of Texas Press.

Johnsen S and Lohmann KJ (2005) The physics and neurobiology of magnetoreception. *Nature Reviews Neuroscience* 6: 703–712.

Lohmann CMF and Lohmann KJ (2006) Quick guide to sea turtles. *Current Biology* 16(18): R784–R786.

Lohmann KJ, Cain SD, Dodge SA, and Lohmann CMF (2001) Regional magnetic fields as navigational markers for sea turtles. *Science* 294: 364–366.

Lohmann KJ and Lohmann CMF (1994) Detection of magnetic inclination angle by sea turtles: A possible mechanism for determining latitude. *Journal of Experimental Biology* 194: 23–32.

Lohmann KJ, Lohmann CMF, Ehrhart LM, et al. (2004) Geomagnetic map used in sea turtle navigation. *Nature* 428: 909–910.

Lohmann KJ, Lohmann CMF, and Endres CS (2008) The sensory ecology of ocean navigation. *Journal of Experimental Biology* 211: 1719–1728.

Lohmann KJ, Putman NF, and Lohmann CMF (2008) Geomagnetic imprinting: A unifying hypothesis of natal homing in salmon and sea turtles. *Proceedings of the National Academy of Sciences of the United States of America* 105: 19096–19101.

Lohmann KJ, Swartz AW, and Lohmann CMF (1995) Perception of ocean wave direction by sea turtles. *Journal of Experimental Biology* 198: 1079–1085.

Luschi P, Benhamou S, Girard C, et al. (2007) Marine turtles use geomagnetic cues during open-sea homing. *Current Biology* 17(2): 126–133.

Lutz PL and Musick J (eds.) (1997) *The Biology of Sea Turtles.* Boca Raton, FL: CRC Press.

Putman NF and Lohmann KJ (2008) Compatibility of magnetic imprinting and secular variation. *Current Biology* 18(14): R596–R597.

Spotila JR (2004) *Sea Turtles: A Complete Guide to Their Biology, Behavior, and Conservation.* Baltimore, MD: Johns Hopkins University Press.

Relevant Websites

http://www.unc.edu/depts/oceanweb/turtles/ – Sea Turtle Navigation.
http://www.seaturtle.org – Global Sea Turtle Network.

Seasonality: Hormones and Behavior

N. L. McGuire, R. M. Calisi, and G. E. Bentley, University of California, Berkeley, CA, USA

Introduction

The diversity of the regulation of hormones and behavior is vast. An organism's immediate environment (housing conditions, population density, time of day), social standing, social interactions, nutritional status, and even its mental state can all influence its endocrine system and consequently, its behavioral output. It must be remembered that behavior (either an individual's own behavior, or that of others around it) can also be processed by the brain and influence an organism's endocrine status. Thus, the amount of any particular hormone circulating in an individual's blood can vary rapidly during short periods of time in response to a great number of environmental variables.

On a different temporal scale, hormonal status also changes over longer periods of time (weeks and months) and influences changes in behavior that are appropriate for an individual's particular life-history stage. These changes in hormones and behavior thus occur on a seasonal basis, and it is the seasonality of hormones and behavior that this chapter addresses. Many organisms are adapted to live in seasonal environments and have evolved endocrine and behavioral mechanisms to predict the forthcoming seasons and either exploit them, endure them, or escape them – depending on the season. We have chosen a few key changes in hormones and behavior that are prevalent in seasonal environments and have illustrated them with specific examples.

The hypothalamo-pituitary-gonad and hypothalamo-pituitary-adrenal axes

Before describing key changes in hormones and behavior that occur as a result of adaptation to seasonal environments, we must first describe basic endocrine pathways that are present in all vertebrates and allow them to respond to the different seasons. The two main pathways that we will discuss are the *hypothalamo-pituitary-gonad* axis (HPG axis) and the *hypothalamo-pituitary-adrenal* axis (HPA axis). In broad terms, the HPG axis regulates reproduction and associated behaviors, and the HPA axis regulates the endocrine response to stress. These axes are present in all vertebrates, but respond differently to environmental and physiological cues in different organisms.

The HPG axis is pictured in **Figure 1(a)**. The HPA axis is depicted in **Figure 1(b)**.

The HPG axis seems to be present in all vertebrates studied, even in the Agnatha (jawless fishes: lampreys and hagfish), which are considered to be examples of primitive vertebrates. *Amphioxus,* a cephalochordate, appears to have an evolutionary precursor to the hypothalamo-pituitary system, with neurosecretory neurons projecting from a lobe of the brain to a rudimentary invagination on one side of the buccal cavity (roof of the mouth) that possibly secretes gonadotropins. Thus, hypothalamo-pituitary communication seems to have been established very early in vertebrate evolution. The HPA axis is also thought to exist in all vertebrates.

Next, we address how seasonality is regulated at different latitudes and discuss some of the key changes in endocrinology and behavior across the annual cycle.

Seasonality: Temperate Zone

Many animals have evolved to reproduce during specific seasons in order to optimize their reproductive success. Energetically demanding activities such as mating, gestation, and parental behavior are best conducted when the weather is clement and food is plentiful. In this way, the animals attain maximal reproductive fitness. For example, male white-tailed deer (*Odocoileus virginianus*) begin to secrete growth hormone (GH) from their pituitary gland in spring and summer, which stimulates the secretion of insulin-like growth factor (IGF) from the liver. These hormones induce antler growth, an energetically expensive process, while food is abundant in the summer. As fall approaches, the production of these hormones also decreases, allowing for calcification of the antlers in preparation for male–male combat during rutting in October, when females are reproductively receptive. By seasonally restricting antler growth and male–male combat, the white-tailed deer maximize their chances of reproductive success and get more 'bang for their buck' as it were. If house sparrows (*Passer domesticus*) in Minnesota mated, laid eggs, and hatched their chicks during the winter months when the temperature is cold and food is scarce, they would most likely lose those chicks to hypothermia and starvation, and perhaps they themselves would die as a result of the effort required to keep their chicks alive. Thus, these parents and their offspring would be naturally selected against. However, house sparrows that are able to time their reproduction during the warm spring months when food is ample are naturally selected for. Thus, there

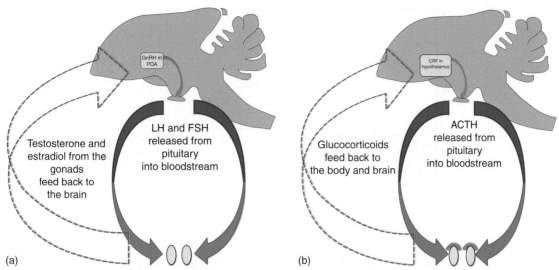

Figure 1 The HPG and HPA axes (simplified). (a) Environmental and physiological stimuli cause gonadotropin-releasing hormone (GnRH) in the preoptic area (POA) of the hypothalamus to be released to the pituitary gland (red arrow). The gonadotropins luteinizing hormone and follicle-stimulating hormone (LH and FSH) are released by the pituitary and carried in the blood to the gonads, causing gonadal activation. The gonads produce the steroid hormones testosterone and estradiol, which not only affect physiology and metabolism, but also feed back to the brain to influence behavior. (b) Environmental and physiological stimuli cause corticotropin-releasing factor (CRF) in the hypothalamus to be released to the pituitary gland (red arrow). Adrenocorticotropic hormone (ACTH) released by the pituitary is carried in the blood to the adrenal glands on the kidneys, causing release of the glucocorticoid cortisol (or corticosterone, depending on species). The glucocorticoids not only affect physiology and metabolism, but also feed back to the brain to influence behavior. N.B. the brain depicted is a 'typical' bird brain, but the axes are the same in mammals and other vertebrates.

is strong selection pressure for all animals to breed at the appropriate time of year for their species.

How do animals 'know' when to reproduce? What cues might they use to time their reproductive behaviors?

Day length is the most reliable environmental cue that animals can use to predict the forthcoming season and time annual changes in reproduction in temperate zones. The angle and rotation of the earth over a period of one year dictate how much light is received at any particular latitude (**Figure 2**).

At nontropical latitudes, days are shortest during the winter and longest during the summer. Changes in day length, or photoperiod, at these latitudes are therefore a better predictor for the timing of life-history events than other cues, such as temperature and rainfall. Thus, many seasonally breeding animals are what is termed '*photoperiodic*,' in that their reproductive systems are directly controlled by photoperiod. Temperature and rainfall can also correlate with the seasons, but they are often highly variable from year to year. Patterns of photoperiod remain constant annually, enabling animals to prepare their reproductive systems in advance of favorable conditions and exploit environmental factors such as suitable temperatures and rainfall for reproductive activities.

Many temperate zone animals that have short gestation periods, such as small mammals, mate in the early spring and rear their offspring in late spring and early summer. Most temperate zone birds exhibit this pattern of reproduction, too. Animals that breed during the spring

and the summer are often referred to as 'long-day breeders'. Generally, larger animals with longer gestation periods, such as sheep, goats, deer, and cattle, mate in the fall, gestate over the winter, and give birth in the spring. Animals that exhibit this type of reproductive strategy are called 'short-day breeders' because they mate when the days are shorter in the fall. This latter system most likely evolved to ensure that even with relatively long gestation periods, offspring are born at, or just prior to, a time of mild weather and sufficient food. Although mammalian young initially feed on mother's milk, other food sources are important for the health and well-being of the parents to ensure the energetic requirements needed for parental care.

How do animals time day length? Birds are one of many long-day breeding organisms that are photoperiodic, or use day length as an anticipatory cue to time reproduction. As short winter days become longer with the advance of spring in the northern hemisphere, photoreceptors that lie deep within the avian brain stimulate the HPG axis (**Figure 1**). In mammals, photoreception is exclusively by the eye. Light absorbed by the mammalian retina transmits information via the retinal-hypothalamic tract and suprachiasmatic nucleus (the body's circadian clock) to the pineal gland. The pineal gland is responsible for synthesizing and secreting the hormone melatonin. Generally, light inhibits melatonin production and darkness increases it. Thus, the body is able to measure day length according to the timing and duration of melatonin

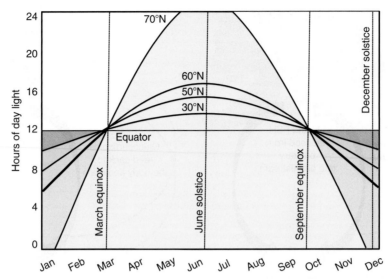

Figure 2   Hours of daylight received over the annual cycle at different latitudes. Note that change in daylength over the year can be very extreme (24 h light to 24 h dark at high latitudes), but it is always predictable for any given latitude.

secretion. This system differs from that of birds in that, in those species tested, neither the eyes nor the pineal gland are needed for birds to exhibit a reproductive response to a change in photoperiod. Melatonin does not appear to be as necessary for the avian photoperiodic response as it is in mammals, but there is some evidence for its involvement in seasonal processes. In fish, daily and seasonal rhythms of melatonin may be linked to reproduction, although a study on Atlantic salmon (*Salmo salar*) reported that melatonin was affected by the change of water temperature and not specifically by light. Experimentally modified seasonal photoperiods have been shown to affect the spawning time of many fish species, though effects are varied and complex. Bromage and colleagues review the details of the endocrine control of seasonal reproduction in fish (see Further Reading).

When do temperate animals 'turn off' their reproductive axis? Both birds and mammals eventually undergo a state of photorefractoriness, when the reproductive system's response to a particular day length qualitatively changes. Photorefractoriness in birds refers to a very different physiological process from photorefractoriness in mammals. In birds, the increase in photoperiod in spring will cause the photostimulation of the reproductive axis and growth of the gonads. However, late into the summer, when days are still long, a state of photorefractoriness will occur. Gonads then regress as the reproductive axis is turned off. These adaptations may have occurred to discourage breeding late into the season and thus having to raise offspring in harsher conditions in the fall and winter. After this period of photorefractoriness, birds will become photosensitive during the winter, meaning that their systems will once again be sensitive to an increase in photoperiod the following spring, and the

cycle will repeat. In general, birds must experience an increase in day length to become photostimulated and have their gonads fully recrudesce.

This phenomenon differs in long-day breeding mammals, such as seasonally breeding vole species (*Microtus* sp.). Unlike birds which become photorefractory to long summer days, these mammals continue to breed until day lengths decrease in the autumn. Their reproductive system then regresses in response to the increased duration of melatonin secretion in response to longer nights. After several weeks of exposure to short days (and long nights), many mammals become what is termed 'photorefractory' to short winter days, and their gonads will start to recrudesce as the reproductive axis switches on again. Experimental evidence suggests that when these animals are photorefractory, they are less sensitive to the nocturnal melatonin signal. In birds, photorefractoriness thus refers to an inhibition of reproduction after prolonged exposure to long days and in mammals, photorefractoriness refers to an activation of reproduction after prolonged exposure to short days.

Over 30 rodent species are characterized as 'long-day' breeders, yet subsets of many of these populations do not regress their reproductive systems under short, winter-like photoperiods. Approximately 30% of some of these species are classified as photoperiodic nonresponders based on laboratory experiments. This photoperiodic nonresponsiveness is heritable. Thus, it is likely that 30% of individuals in wild populations of these species retain the ability to breed year round, regardless of photoperiod. The fitness payoffs of photoperiodic nonresponsiveness in the wild have yet to be determined, but benefits must exist – otherwise this would not be an evolutionary stable strategy. A subset of photoperiodic nonresponders remains to be identified within any population of photoperiodic birds.

A depiction of these photoresponsive cycles in birds and mammals is given in **Figure 3**.

Seasonality: Arctic Zone

Dramatic seasonal changes occur in arctic climates, with the majority of the year being inhospitable to many animals. Changes in photoperiod range from 24 h dark per day in the winter to 24 h light in the summer (see **Figure 2**). Arctic animals can use these large changes in photoperiod to adjust their breeding and feeding schedules. Many arctic-breeding birds, such as the white-crowned sparrow (*Zonotrichia leucophris gambelii*), migrate to more favorable climates during winter and return to breed only during a short period in summer. In contrast, aquatic mammals, such as humpback whales (*Megaptera novaeangliae*), migrate to warmer climates in the summer to breed but take advantage of productive arctic feeding grounds in the winter. Reindeer (*Rangifer tarandus*) use photoperiod to adjust their food intake by decreasing levels of leptin, a hormone produced by adipose tissue, in the short days of winter. The decrease in levels of leptin during winter when food is scarce is thought to play a role in energy conservation by decreasing body temperature and inhibiting reproduction. Arctic charr (*Salvelinus alpinus*), which thrive in lakes covered with thick ice and snow, may be able to detect subsurface irradiance at very low intensities to measure day length, as they somehow receive photoperiod information to time the release of melatonin and thus provide a seasonally appropriate endocrine response.

Because of the extreme conditions in arctic zones, many species, particularly birds, migrate away from the Arctic during the nonbreeding season. Migration is discussed in a later section as well as in greater depth elsewhere in this encyclopedia.

Seasonality: Opportunism and Tropical Zones

Outside the temperate zone, seasons are often not defined by large changes in day length or temperature, but as rainy (abundant resources) and dry (limited resources). Tropical species that inhabit environments with predictable rainy and dry seasons often develop equatorial seasonality, while species inhabiting unpredictably dry and wet climates often develop opportunism.

It was long held that all equatorial species with annual cycles of reproduction must respond to nonphotic environmental cues that reliably change with the seasons, such as food availability, rainfall, presence of predators or conspecifics, or temperature. For example, Golden perch (*Macquaria ambigua*) and Crimson-spotted rainbow fish (*Melanotaenia fluviatilis*) of the Murray River in Australia spawn in the summer, when water temperature reaches 23 °C and 20 °C, respectively. Foraging behavior in squirrel monkeys (*Saimiri oerstedi*) of Costa Rica varies predictably across the yearly cycle of dry and rainy seasons. Foraging duration and choice depends on the food supply, with squirrel monkeys spending the greatest proportion of time on arthropods when this resource is limited in the late wet season. Rufous-collared sparrows (*Zonotrichia capensis*) of Equador regress and recrudesce their gonads,

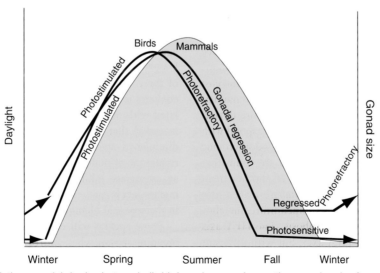

Figure 3   Changes in relative gonadal size in photoperiodic birds and mammals over the annual cycle of reproduction in nonequatorial regions. This cycle is explained in depth in the text given earlier (see section 'Seasonality: Temperate Zone').

testosterone, estradiol, and corticosterone concentrations, but this phenomenon varies across seasons. Not only do endocrine changes vary from one time of year to another, but the response to a specific hormone can vary over the same time frame – for example, there are marked seasonal changes in sensitivity to glucocorticoids in many seasonally breeding species. Even in humans (considered non-seasonal), social cues can elicit hormonal changes. The 'home-team advantage' in competitive sports is thought to be mediated by testosterone. Similarly in men, social cues tend to elicit a decrease in testosterone and an increase in prolactin at the time of their partner's late pregnancy and in the weeks just after childbirth – possibly to increase parental care. Thus, temporal changes in hormones and behavior are widespread, varied, and elicited by any number of environmental and behavioral variables. It is this variability and sensitivity to the environment that makes changes in hormones and behavior interesting, yet challenging to study and interpret in their natural environment.

See also: Aggression and Territoriality; Behavioral Endocrinology of Migration; Female Sexual Behavior and Hormones in Non-Mammalian Vertebrates; Hibernation, Daily Torpor and Estivation in Mammals and Birds: Behavioral Aspects; Immune Systems and Sickness Behavior; Male Sexual Behavior and Hormones in Non-Mammalian Vertebrates; Mammalian Female Sexual Behavior and Hormones; Molt in Birds and Mammals: Hormones and Behavior; Sexual Behavior and Hormones in Male Mammals; Stress, Health and Social Behavior.

Further Reading

Bradshaw WE and Holzapfel CM (2007) Evolution of animal photoperiodism. *Annual Review of Ecology, Evolution, and Systematics* 38: 1–25.

Bromage N, Porter M, and Randall C (2000) The environmental regulation of maturation in farmed finfish with special reference to the role of photoperiod and melatonin. *Aquaculture* 197: 63–98.

Calisi RM, Rizzo NO, and Bentley GE (2008) Seasonal differences in hypothalamic EGR-1 and GnIH expression following capture-handling stress in house sparrows (*Passer domesticus*). *General and Comparative Endocrinology* 157: 283–287.

Dawson A, King VM, Bentley GE, and Ball GF (2001) Photoperiodic control of seasonality in birds. *Journal of Biological Rhythms* 16: 365–380.

Dittami JP and Gwinner E (1990) Endocrine correlates of seasonal reproduction and territorial behavior in some tropical passerines. *Endocrinology of Birds: Molecular to Behavioral.* Tokyo/Berlin: Japan Scientific Societies Press/Springer.

Folstad I and Karter AJ (1992) Parasites, bright males, and the immunocompetence handicap. *American Naturalist* 139: 603–622.

Fusani L and Gwinner E (2005) Melatonin and nocturnal migration. *Annals of the New York Academy of Sciences* 1046: 264–270.

Paul MJ, Zucker I, and Schwartz WJ (2008) Tracking the seasons: the internal calendars of vertebrates. *Philosophical Transactions of the Royal Society B: Biological Sciences* 363(1490): 341–361.

Perfito N, Zann RA, Bentley GE, and Hau M (2007) Opportunism at work: Habitat predictability affects reproductive readiness in free-living zebra finches. *Functional Ecology* 21(2): 291–301.

Prendergast BJ, Kriegsfeld LJ, and Nelson RJ (2001) Photoperiodic polyphenisms in rodents: Neuroendocrine mechanisms, costs and functions. *Quarterly Review of Biology* 76: 293–325.

Soma KK, Scotti M-AL, Newman AEM, Charlier TD, and Demas GE (2008) Novel mechanisms for neuroendocrine regulation of aggression. *Frontiers in Neuroendocrinology* 29: 476–489.

Wingfield JC, Hegner RE, Dufty AM, and Ball GF (1990) The 'Challenge hypothesis': Theoretical implications for patterns of testosterone secretion, mating systems, and breeding strategies. *American Naturalist* 136: 829–846.

Seed Dispersal and Conservation

E. V. Wehncke, Biodiversity Research Center of the Californias, San Diego, CA, USA

The Seed Dispersal Process by Animals

A common approach used to study the whole seed dispersal process by animals has been to divide it into three different phases, a predispersal, dispersal, and a postdispersal phase. The predispersal phase has to do with the attraction of animals to fruits and the way they are influenced by several intrinsic and extrinsic factors. Intrinsic factors include tree fecundity, fruit size, fruit design and nutritional content, presence of secondary compounds, and fruit distribution, as well as availability in space and time. Many extrinsic factors not under the genetic control of the plant, such as the local abundance of other fruit resources or neighborhood effects, may also influence fruit attraction and its removal. Since considerable variation exists among seed dispersal agents in their responses to these factors, their movement and seed distribution patterns are likewise dependent upon a diverse array of disperser morphologies, physiologies, and behaviors that are commonly included in the dispersal phase. Thus, foraging behavior following fruit removal depends, at least in part, on intraspecific decisions (e.g., intragroup social relationships, mating), interspecific behaviors (e.g., competition, territoriality), environmental factors (e.g., abundance and availability of fluctuating food sources, site-specific habitat heterogeneity, predator pressure on seed dispersers), and abiotic factors that affect the habitat template where interactions take place. Finally, once seeds are deposited or discarded by dispersers (the postdispersal phase), the probabilities of seed germination and seedling establishment will depend not only on disperser defecation patterns and physical microenvironment conditions, but also on the likelihood of encountering seed predators or secondary dispersers. As a consequence, inconsistencies in the selection pressures exerted by seed dispersers upon plant traits are not unusual.

On the basis of the main topics emerging from the numerous studies that exist on plant–disperser interactions, here I intended to summarize and highlight some of the major animal behaviors intervening in each phase of the entire seed dispersal episode (from the time fruits are removed from parent plants, until seeds are deposited in the ground).

Fruit and Seed Removal: Animal Behavior Influencing the Predispersal Phase

Because often fruit must be removed so that the seeds can be dispersed, fruit removal may be considered the first step in the seed dispersal process and an important component of plant fitness. The pulp of fleshy fruits, consisting of a great variety of edible and nutritive tissues surrounding the seeds, provides the primary food resource for many frugivorous animals, especially mammals and birds, but also reptiles and insects. These animals may handle fruit and seeds in different ways. Thus, three types of frugivory modes can be identified according to their potential consequences for seed dispersal: (1) legitimate dispersers that swallow the entire fruit and defecate or regurgitate seeds intact and in viable conditions, (2) consumers that tear off and ingest only the pulp, sometimes allowing the seeds to come out, and (3) seed predators that crack and ingest the seeds, or consume the entire fruit and digest both pulp and seed. These categories define a wide gradient of seed dispersal activity provided by frugivores. The ability to handle, swallow, and process a given fruit efficiently depends on the relationship between animal size and fruit size. For example, ingested seeds must be able to fit into the mouth and throat of the animal to pass through unharmed. The number and the size of seeds determine the time needed by animals to process fruits, and the number and the size of fruits ingested determine the time needed to process them. Although several studies have reported that seed mass, the number of seeds per fruit, and the pulp mass per seed are important traits in fruit selection by frugivores, fruit size seems to be the main source of functional variation in fruit relative to the type of frugivores consuming it. Fruit also presents a great variety of secondary compounds that can be used to attract or repel frugivores (e.g., by reducing a fruit's digestibility). Some of these compounds may promote a high quantitative intake of fruit; while others have deterrent properties. Finally, whether frugivores behave as seed dispersers, pulp consumers, or seed predators may depend on diverse characteristics related to not just the frugivore and the plant but also the environment where the particular interaction occurs.

The influence of plant–frugivore interactions in the diversification of fruit structures and dispersal devices has been studied for several decades. The reliance of many fleshy-fruited plants, especially in the tropics, on birds and mammals for seed dispersal gave rise to the theory of a coevolved mutualism between fruiting plants and seed-dispersing frugivores. Fruit with pulp of high energetic and nutritive value, containing one or a few large seeds, would be one extreme of specialization by interacting with devoted frugivores which provide high-quality dispersal;

fruit with watery, carbohydrate-rich pulp and numerous small seeds occupies the other extreme by interacting with opportunist frugivores. There is empirical evidence that frugivores have a particular preference for one or the other type of fruit and exert selective pressure on fruit traits. For example, primates seem to prefer fruit with a high sucrose content, while birds prefer fruit with high fructose and glucose content. On the basis of the great variation of fruit traits and displays that exist in nature, it is not rare to find also a great diversity in frugivores' feeding behaviors and seed shadows.

Animals depend on different senses for the location of fruits and seeds. Those with color vision employ color as a primary cue for finding food; others, most of them mammals, reptiles, or nocturnal animals, use their sense of smell to locate seeds or fruits. Fish are known to congregate below ripening fig trees, presumably responding to the sound of the fruits hitting the water. Birds, for example, show preferences related to fruit color, nutritional content, presentation, accessibility, level of insect damage, and fruit size, even within a single individual plant and species. Mammals tend to select large, husked or protected, brown, green, orange, or yellow fruit, which are often odoriferous and low in protein content. Differences in the feeding ability or preferences between guilds of frugivores have been used to differentiate among fleshy-fruited species and have led to the description of fruit syndromes associated with particular frugivore guilds. Since animals differ enormously in their size, visual acuity, and ability to access and manipulate fruit, some studies suggest that a more constructive approach than the concept of syndrome may be necessary to identify frugivore traits that should be logically associated with certain fruit traits and vice versa. Fruit dispersed by terrestrial frugivores, for example, should fall at maturity; fruits dispersed by nocturnal frugivores are unlikely to be brightly colored but very likely to be odoriferous.

Fruits and seeds are also critical resources for a vast number of insect species (in the pre-and/or the postdispersal phases), and some of the prodigious radiation of insects is associated with these specializations. The size of fruit and seeds may influence the behavior of frugivorous insects. For example, the females of some species select determined seed sizes to lay their eggs. Likewise, frugivorous insects present a great diversity of behaviors that allow them to face periods of fruit scarcity and track the fluctuations in fruit abundance and distribution. On the other hand, insects may escape predation by frugivorous vertebrates because they may (1) make themselves toxic to frugivores by incorporating secondary seed compounds in their tissues, (2) complete their development before fruits ripen and become attractive to frugivores, or (3) make fruit unpleasant to frugivores since their excretion products change its flavor. Consequently, frugivorous vertebrates may avoid selecting fruits that contain insects. The activity

of insects and vertebrate frugivores on fruits and seeds seems, in some cases, to overlap in such a way that the spatial pattern of seed deposition should be considered as a result of the combined activity of both groups.

Animal Movement and Seed Transport: Animal Behavior Influencing the Dispersal Phase

By simply moving from one habitat patch to another, the dispersal of an individual has consequences that go beyond individual fitness. Because animal behavior plays a large role in determining patterns of movement, it is logical to expect that it would therefore also play an important role in determining dispersal effectiveness (defined as the contribution a disperser makes to the reproductive success of a plant, and measured by the quantity and quality components of seed dispersal). The quality of seed dispersal is characterized by the treatment that seeds receive from the disperser, the potential range of microsites and patches that those seeds will reach, and the spatial pattern in which they are deposited. While much emphasis has been given to quantity components, the quality component of seed dispersal may be the single most important factor determining the final fate of a seed.

One critical determinant of dispersal effectiveness is the distance of dispersal events, summarized by dispersal curves. In recent years, several models of seed dispersal, in particular inverse and mechanistic models, have played a prominent role in predicting the spatial distributions of seeds. However, mechanistic models of seed dispersal (based on seed passage times, displacement rates and speed, and direction of animal movement) have posed a greater challenge, primarily because animal movement patterns are complex, depend on many factors, and are difficult to quantify. Frugivores' influences on plant fitness do not end with seed transport. The seed shadow generated by frugivore species will depend on its identity, and in turn on the rate of movement and the pattern and quantity of seed deposition. Diverse external factors such as fruit patchiness in space and time, the presence of other competing frugivore species, predation pressure, and the distribution of alternative food sources and resources other than food (e.g., nesting sites, water holes) have been shown to influence disperser behavior, movement pattern, and the shape of seed shadows.

Seasonal fruiting patterns can have a great effect on the annual cycle of most frugivores and may cause dietary shifts in frugivorous animals which track the changes of fruit supply. For year-round resident frugivores, this implies having a highly diverse diet that provides a balance of nutrients. Reproduction, breeding, and migratory movements are commonly associated with seasonal patterns of fruit ripening. Predictable patchiness of fruit availability can influence the pattern of patch utilization by foraging frugivores and secondary dispersers, and their movements between patches have been suggested as a key mechanism for habitat connectivity and forest regeneration. Nonrandom seed deposition may result from movements that depend on the distribution of plant species fruiting at the same time of year. For example, the distribution of some parasitic and hemiparasitic plants (which need highly directed dispersal to particular plant hosts) shows consistency with the distribution of particular fruiting plant species that serve as recruitment foci for birds that use them as perches.

Social organization may influence patterns of seed deposition, since it is expected that home range size, dispersal distances, and the degree of clumping will increase as a function of group size. Animal dispersers can be solitary, forage in groups of 10–30 individuals, or congregate in hundreds of individuals, such as bats, squirrel monkeys, and some birds. Group foraging may increase the overall food intake rate, and foraging interference between individuals promotes longer foraging trips and the utilization of more food sources. Large group sizes may also increase clumping of seeds, because individuals will tend to disperse seeds together in space and time. While some animals produce small piles of seeds unevenly over the environment, others may concentrate large numbers of seeds under other fruiting species, in latrines, in places used for sexual displays, or roosting, nesting, or sleeping sites. Some bat species can deposit over 90% of the seeds they consume under roosts; and many species of ants carry considerable quantities of seeds to their nests, where they cut off the lipid-rich elaiosome attached to the seed and discard the rest in piles adjacent to the nest. Territoriality and/or resource defense is likely to be important in patterns of seed deposition, because they constrain the space used by an individual and may limit the number of potential dispersal agents visiting a given source tree. Some territorial birds regurgitate seeds within a close range of their feeding plant or perch, while elephants, tapirs, and large primates may use long, recurrent routes.

Mating and breeding behavior may also have potentially important consequences for seed movement and deposition. Both nesting and male display behaviors, as in lekking species, where individuals concentrate in a particular area, leading to some degree of seed clumping, though the quality of the microsites for germination and establishment in which seeds are distributed may vary. For example, some birds leave seeds inside nest caves where large numbers die in the absence of light, but others that nest along rivers may potentially promote long-distance dispersal of seeds taken downstream by the current. Males of some bird species prefer to sing perched in forest gaps, a behavior that leads to directed dispersal of seeds into forest gaps, which constitute favorable microhabitats for seedling recruitment in many plant species because of light and space availability.

Figure 1 Cebus capucinus consuming a fruit. Seeds will be deposited away from the fruiting tree, especially if the individual is not a dominant member of the group, and thus has to keep moving in the periphery of the foraging troop.

Intragroup social behavior of some primate species, such as capuchin monkeys (**Figure 1**), has the potential to affect diet diversity and seed deposition patterns. For example, members with low dominance rank avoid approaching trees with low fruit production until after the rest of the group has left. In the meantime, these individuals forage in the surroundings, exploring for new food items, and this prevents the clumping of seed depositions beneath fruiting trees. By turning over gut contents very fast, they may compensate for the low protein content of some foods and rid the gut rapidly of indigestible seeds. Consequently, short gut retention times can result in not only a high rate and diversity of tree visitation, but also more defecation events per day and fewer seeds per dung pile. The resulting scattered seed dispersal pattern may strongly influence postdispersal seed fate by reducing the probability of seed removal by secondary seed dispersers or predators. If gut seed retention times exceed the time frugivores remain in fruiting trees, defecation of seeds will surely occur at some distance away from them. Thus, the combined effect of the capuchin's social behavior and physiological traits may help to explain why occasional sequential selective foraging on favored fruiting species may not always result in dispersal under or near conspecifics.

Factors Determining the Final Fate of Seeds: Animal Behavior Influencing the Postdispersal Phase

Postdispersal seed germination and subsequent seedling establishment are affected by the effects of gut passage on seed viability and different types of defecation patterns. Though some studies show important gut passage effects

Sentience

L. Marino, Emory University, Atlanta, GA, USA

Introduction and Definitions

Sentience is a multidimensional subjective phenomenon that refers to the depth of awareness an individual possesses about himself or herself and others. When we ask about sentience in other animals, we are asking whether their phenomenological experience is similar to our own. Do they think about themselves the way we do? Do they ponder their own lives? Do they know that other individuals have feelings and thoughts? And, do they have an autobiographical sense of the past and future?

At its most cognitively sophisticated levels, sentience may be conceptualized in the context of three related psychological domains or capacities. It is becoming increasingly clear from the accumulating evidence that these three domains are not a cognitive 'package'; despite our still-limited knowledge, at this point, they appear to be separable related capacities. The first two have to do with one's awareness of self, physically and/or mentally. First, *self-awareness* is a sense of personal, particularly autobiographical, identity. Self-awareness may exist at a physical level, referred to as self-recognition, to more abstract levels of psychological continuity through time. Second, *metacognition* is the ability to think about, or reflect upon, one's own thoughts and feelings, and is clearly underwritten by self-awareness in the psychological realm but not necessarily by self-awareness in the physical realm (i.e., self-recognition). And third, *Theory of Mind (ToM)* comprises capacities, such as perspective-taking, modeling of others' mental lives, including empathy. ToM is others oriented, related to one's ability to take the physical and mental perspective of others, and is presumably underwritten by metacognition.

Sentience refers to any of these psychological phenomena. In normal adult human beings, all three of these capacities are found to some extent. The study of sentience in other animals is tantamount to determining how many of, and to what extent, these capacities are shared. Although we tend to view humans as having the full range and depth of sentience, it is important to acknowledge the possibility that other animals might have properties of sentience that humans lack. This possibility is difficult to assess.

History of Study

For the major part of human history, the question of sentience in other species remained a philosophical one. Potentially relevant observations could not be interpreted with any scientific rigor until research methods were developed that could provide strong tests of hypotheses. Thus, the domain of inquiry into animal sentience was typically considered beyond the reach of science, and was forever limited to 'thought experiments' and anecdotes. It was a very real and practical restriction on the ability to make scientific headway in this domain. There was also a more profound impediment to the serious study of sentience in other animals in the form of a pervasive anthropocentric world view of other animals. Whereas the question of awareness and cognition in other animals has always been great fodder for lively discussion and speculation, it has been burdened by an insidious restriction created by the tacit acceptance of the hierarchical nature of living beings and the inferiority of other animals to humans. This notion derives from the view of nature known as scala naturae, and its formalization is attributed to Aristotle (300 BC). The scala naturae, also known as the great chain of being, places humans at the top of a hierarchy of complexity, intelligence, and value. Furthermore, intrinsic to this scheme is the idea that there is a qualitative difference between humans and all other animals.

Our current understanding of nature, including, importantly, our own, has been laid bare by the insights of Darwin. Darwin provided the theoretical and mechanistic proof that all life on earth is related by descent with modification, showing that humans are a byproduct of the same natural nonprogressive and dispassionate processes that apply to all other living beings. Yet, while science, at large, no longer explicitly embraces progressive evolution in these post-Darwinian times, its signature remains engrained in many concepts and practices. And so it has been, historically, for such complex and abstract notions as sentience.

Post-Darwinian understanding of other animals has not always been consistent with the notions of continuity and nonhierarchical schemes of nature. In the twentieth century, the study of ethology and behaviorism emerged, but were still burdened by language that denied the psychological character of other animals. Therefore, the Darwinian revolution did not rid our thinking of scala naturae views. Rather, it presented a more realistic alternative that *could be* embraced. In the 1970s, the notion of psychological continuity across species finally began to be taken seriously. The publication of several seminal theoretical and experimental works opened minds to the idea of animal sentience as a serious possibility and topic of study. In 1970, the first evidence for mirror self-recognition in non-humans, chimpanzees, was published by psychologist Gordon Gallup in the prominent journal *Science*. This study demonstrated the empirical accessibility of animal sentience and changed views of human cognitive uniqueness, at least in some scientific circles, in one fell swoop. Also, in his books, *The Question of Animal Awareness* in 1976 and *Animal Minds: Beyond Cognition to Consciousness*, originally published in 1992, zoologist Donald Griffin provided a compelling set of observations and arguments for animal reasoning and sentience, including more sophisticated forms of self and other awareness that further set the stage for the serious study of these topics.

Sentience is, by its subjective nature, one of the most challenging psychological phenomena to study and, as we have seen, historical notions of human–non-human animal relationships have hindered serious consideration of this capacity in other animals up until recently. But the realization that not only are other animals sentient, but that sentience is amenable to investigation, has led to a revolution in our thinking about the psychological experience of other animals and a thriving scientific field of study. We turn now to current knowledge and understanding of the evidence for three domains of sentience (self-awareness, metacognition, and ToM) in other animals.

Self-Awareness

Self-awareness may be conceptualized as a sense of personal identity, that is, the subjective 'I.' At the bodily level, it is typically called self-recognition, the ability to become the object of one's own attention in the *physical* realm. At a more abstract level, self-awareness can take the form of robust *psychological continuity* over time (but it may also be more temporally fleeting in some cases).

Self-awareness has been identified with a variety of related terms that are fundamentally very similar in concept. These terms include phenomenal consciousness, self-consciousness, metacognition, and autonoetic consciousness. All of these designations ultimately converge on the broader concept of self-awareness, that is, awareness of one's own identity, one's own thoughts, and the consequences of one's own actions, through time.

Phenomenal consciousness refers to the subjective, experiential, or phenomenological aspects of conscious experience, sometimes identified with qualia. To consider animal consciousness in this sense, is to consider the possibility that, as expressed by Thomas Nagel in 1974, there might be 'something it is like' to be a member of another species. Phenomenal consciousness might be considered the basic minimum for more complex forms of self-awareness.

Self-consciousness refers to an organism's capacity for second-order representation of his or her own mental states. Because of its second-order character ('thought about thought'), the capacity for self-consciousness extends into the realm of metacognition ('cognition about cognition').

Another expression for self-consciousness that some might feel extends self-consciousness or metacognition is autonoetic consciousness. Autonoetic consciousness is the capacity to recursively introspect on one's own subjective experience through time, that is, to perceive the continuity in one's identity from the past to the present and into the future. Autonoetic consciousness has been likened to the concept of 'stream of consciousness' by William James.

Self-Recognition

Self-recognition has typically been tested through mirror self-recognition (MSR), but has also been probed in the context of other representations such as photographs and videos. Mirrors have the advantage of affording immediate direct and dynamic visual feedback. Normal humans do not show MSR reliably until they are 18–24 months of age. It is important to note that a demonstration of MSR in the current scientific paradigm requires not only the ability to correctly interpret information in a mirror as oneself, but also the motivation to use the mirror as a tool to view one's own body. It is this active self-investigatory nature of MSR that is tested in the mirror protocol. The protocol is thus quite stringent, and failure to pass the MSR test does not mean lack of self-recognition altogether.

The first experimental test of self-awareness in another species was conducted by comparative psychologist Gordon Gallup, when he presented evidence for MSR in chimpanzees in 1970. Gallup adapted the preexisting MSR test for children to the chimpanzees. The basic paradigm involves exposing mirror-naive individuals to a mirror. The chimpanzees in Gallup's study showed what has turned out to be a very typical pattern of behavior for chimpanzees (and human babies) in this situation. They initially react to their mirror image as though it is a social stimulus but they quickly (sometimes over a matter of minutes) learn that the mirror image is not another chimp. When social responses wane, chimpanzees begin to use the mirror to test the contingencies between the mirror image and their own body (e.g., making repetitive hand motions in front of the mirror) and to investigate parts of the body only visible in the mirror (e.g., sticking out the tongue).

The transition from social to self-directed responding at the mirror is compelling evidence for self-recognition. But Gallup moved the question into the experimental realm by following these observations with a hypothesis-driven test. Once the chimpanzees reliably showed self-directed behavior at the mirror, they were anesthetized, and a nontactile, nonolfactory, red dye was applied to parts of their face that were not visible without a mirror. Gallup argued and found that when his subjects awoke from the anesthesia and saw their altered image they, if so motivated, would use the mirror to touch and investigate the marks. Gallup showed that chimpanzees engage in vigorous mark-directed behavior at the mirror, providing conclusive evidence that they understand the mirror image as themselves.

Since this initial study, a substantial literature on MSR has accumulated on the phylogenetic distribution of self-recognition among primates. This literature converges on the finding that most common chimpanzees, bonobos, orangutans, and some gorillas, show convincing evidence of MSR. Individual differences and developmental trends have also been demonstrated among the great apes. Gallup was also the first to show that the presence of MSR is dependent upon the richness of early social experience. Chimpanzees raised in social isolation are impaired in MSR.

The phylogenetic range of self-recognition is now known to expand beyond primates with convincing positive findings on mirror tests for adult bottlenose dolphins (*Tursiops truncatus*), Asian elephants (*Elephas maximus*), and magpies (*Pica pica*). Each finding of MSR outside of the primate lineage marks an important advance in our understanding of the evolution of self-awareness; it shows that this capacity has evolved independently many times over. Put another way, it shows that there are many phylogenetic lineages converging on the same set of cognitive capacities.

Although MSR in great apes is relatively robust, gibbons, siamangs (the so-called lesser apes), and monkeys do not use mirrors to investigate their bodies, despite strong evidence that they understand the visual contingencies of mirrors. Moreover, individuals of some monkey species may show more sophisticated mirror responses than others. Capuchins, but not other monkeys, show robust evidence that they discriminate their own mirror image from that of a strange capuchin monkey. However, these intriguing findings fall short of providing unequivocal evidence for self-recognition in capuchins.

Other Forms of Self-Awareness

Besides mirror (and the conceptually equivalent photo and video) self-recognition, there are other ways that non-human animals demonstrate awareness of their own bodies. Dolphins and some primates have shown evidence for spontaneous vocal and behavioral mimicry. True mimicry requires awareness of one's own body and how it relates to that of another individual. Arguably, bottlenose dolphins show the most sophisticated range of capacities in this realm. For instance, bottlenose dolphins demonstrate awareness of their own actions by repeating a behavior just performed, in response to a 'repeat' command, or

performing a different behavior, if so instructed. They also understand the relationship between symbolic gestural references to their own body parts and the ability to use those body parts in whatever way requested by an experimenter. In the laboratory, the capabilities of bottlenose dolphins for both vocal and behavioral mimicry are well documented. Dolphins are capable of spontaneously imitating a wide variety of arbitrary, nonnatural, electronically generated sounds. Importantly, dolphins can develop a concept of mimicry – copying an observed behavior or sound when given symbol-based instruction to do so.

Metacognition

Metacognition refers to the ability to think about, or reflect upon, one's thoughts, perceptions, feelings, and memories. It is clearly underwritten by self-awareness (at least in the psychological domain), that is, it would be difficult to think about oneself without some minimal self-concept. Self-awareness in the physical sense, however, seems to imply less about the existence of metacognition. Furthermore, metacognition is introspective ('self' rather than 'other' oriented) and does not necessarily imply ToM capabilities.

Uncertainty Monitoring

One way to determine whether another animal is capable of metacognition is to experimentally assess whether that animal can indicate his or her subjective sense of certainty about information he or she needs, in order to perform a task. This kind of question was experimentally posed by David Smith and his colleagues a few years ago with startling results. Smith placed adult humans, bottlenose dolphins, and rhesus monkeys, in the same auditory discrimination task situation. All participants could optimize rewards by sometimes choosing a bail-out option, called an 'uncertain response,' on each trial that presented very difficult material. The human subjects reported verbally and the dolphin and monkey subjects reported by hitting buttons and paddles, respectively. What Smith and his colleagues found is that the overall pattern of response across the three species was essentially identical. Humans were able to respond optimally because they were able to introspect on their certainty or confidence about any given trial and opt out of those trials that were deemed too difficult to complete successfully. It turns out that dolphins and monkeys also responded in the same manner. In other words, dolphins and monkeys *know* when they do not know something. This is compelling evidence that dolphins and rhesus monkeys possess metacognitive abilities. Moreover, these findings are even more interesting in light of the fact that rhesus monkeys do not show evidence of mirror self-recognition, demonstrating the separable nature of the three domains of sentience.

Evidence for uncertainty monitoring also exists outside of the primate lineage. Rats have been shown to use a similar pattern of escape responses as above for duration discriminations as do pigeons in visual discrimination tasks.

Metamemory

Closely related to uncertainty monitoring is metamemory or memory awareness, which is the ability to accurately assess the availability or sufficiency of needed information in memory. Memory awareness can be assessed by determining the extent to which an individual seeks new or more information if the contents of his or her memory are insufficient for the given task. In these tasks, individuals with metamemory seek more information when needed, but do not do so when not needed. A common paradigm for testing metamemory involves the subject in a foraging task and assesses whether the subject will actively seek out the information, for example, by looking down an opaque tube, needed to complete the task. Several studies have shown that chimpanzees, orangutans, and rhesus monkeys, possess memory awareness. The evidence for this capacity in tufted capuchin monkeys is equivocal, as it is in a few other nonprimate species tested thus far, such as pigeons. It is interesting that pigeons show evidence for uncertainty monitoring, but are challenged when faced with a task that requires them to act more extensively upon their metacognitive perceptions, as is the case in the typical metamemory paradigm.

Theory of Mind

ToM refers to one's ability to take the physical and/or mental perspective of others. ToM is an enormously complex notion. Yet, there is evidence for various aspects of this capacity in other animals. This evidence falls primarily into three areas: perspective-taking (also referred to as perceptual and cognitive empathy), attribution, and emotional empathy. It is not clear that these subdomains of ToM are separable, but they will be presented as such for the purposes of this discussion.

Perspective-Taking (Perceptual and Cognitive Empathy)

Perceptual and cognitive empathy sit on the same continuum of capacities known as perspective-taking. Perspective-taking can refer to the strictly perceptual realm, for example, visual, as when we understand that someone who is turned in the other direction will not see an event that has taken place. Perceptual empathy shades into more cognitive realms when it affects our understanding of others' knowledge states and beliefs. So, in addition to knowing that someone did not witness something if they were facing the other direction, we also know that they are lacking certain information about the event because of this. And we know that this knowledge state may affect their attitudes and beliefs about the situation as well. When we are able to think about, and model, the mental state of others from their point-of-view, we call this cognitive empathy. Moreover, the dimension of perspective-taking shades into attribution ability.

Interestingly, the strongest evidence for perspective-taking in other animals comes from situations in which the subjects are put into a competitive situation with other individuals. This is because the more realistic the situation, the more meaningful it is from the animal's point-of-view, and the more it allows us to make fairly straightforward predictions about behavior that bears on the hypothesis of perspective-taking. Chimpanzees, rhesus monkeys, and corvids, use information about the perspective of others to devise effective social-cognitive strategies. This evidence for these complex abilities will be discussed below under attribution.

The study of visual perspective taking in non-human primates has been mottled with a wide disparity in results most likely resulting from the animal subjects' inability to extract meaning from certain highly artificial situations. However, there is now solid experimental evidence that non-human primates possess the capacity for visual perspective-taking, particularly in the domain of understanding gaze direction. Chimpanzees, mangabeys, and macaques, have been shown to be able to follow gaze direction of a conspecific to a target. Chimpanzees can take into account the properties of various barriers when tracking the gaze of others. Brian Hare of Duke University and colleagues have provided convincing evidence that chimpanzees know what a conspecific can and cannot see when placed in a competitive social situation, and can use this knowledge to develop effective strategies to win food competitions. Other studies indicate that chimpanzees understand the effect of blindfolds on humans.

Rhesus macaques have shown compelling evidence for visual perspective-taking in several contexts in addition to the above. For instance, in one study, when these monkeys were given the opportunity to steal a grape from a human competitor, they were more likely to do so if the human's eyes were directed away than toward them.

Bottlenose dolphins, not surprisingly, have shown convincing evidence of visual perspective-taking. Dolphins can reliably choose an object pointed to by a human informant or gazed at with a turn of the head. They can also understand spontaneously the use of pointing gestures substituted for symbolic gestures in language-like tasks. Additionally, dolphins can spontaneously produce pointing (using their rostrum and body alignment) to communicate desired objects to a human observer and appear to understand that the human observer must be present and attending to the pointing dolphin for communication to be effective.

Dogs also have knowledge about visual perception in others, including humans. Several studies have shown that dogs can use human gaze as a cue to locate hidden food. Primatologist Josep Call and colleagues have found that dogs are also sensitive to the attentional state of humans. He presented dogs with a situation in which a human placed a piece of food on the ground and forbade the dog to eat it. In one condition, the human left the room. In another, she turned her back on the dog or closed her eyes but stayed in the room. In yet another, she looked directly at the dog. Call et al. found that the dogs more often took the forbidden food in the conditions when the human could not see, including when the human's eyes were closed or back turned, than when the human had visual access to the dog.

An innovative body of work by Nicky Clayton and colleagues at the University of Cambridge has shed light on the perspective-taking abilities of birds, specifically scrubjays, members of the corvid family. Scrub jays are food-cachers and cache-thieves, and therefore engage in a number of cache-protection strategies. In a well-controlled series of experiments, Clayton and colleagues showed that scrub jays remember which individual watched them during particular caching events and alter their recaching behavior accordingly. These findings suggest that scrub jays remember who was present during prior caching events, and discriminate between specific individuals with different knowledge states.

In addition, Clayton and colleagues have shown that scrub jays who often steal food from the caches of other birds will go to elaborate lengths to hide and rebury their caches when other birds are present. Clayton and colleagues view this as evidence that the birds are using their own experience as thieves and projecting it onto other birds.

Attribution

As mentioned, the ability to attribute mental states to others depends upon perspective-taking; perspective-taking may be necessary but not sufficient for certain complex forms of attribution, such as intentionality. Attributions can become quite complicated when they become nested within one another in the form of higher-order intentions and beliefs, that is, I believe that you believe that I believe ..., etc. The ability to deceive (at least intentionally) also appears to be related to the capacity to understand and thus manipulate the mental states of others.

It is difficult to neatly separate many of the findings on perspective-taking and attribution into one or the other

category. It is likely that these capacities fall on a continuum. Moreover, although in the interest of scientific parsimony we typically do not extend interpretations of behavioral findings into the cognitive realm, it is reasonable to view some of the perspective-taking findings as indicators of attribution as well. Indeed, these processes may be equivalent.

In addition to the possibilities above, there is also evidence of attribution abilities in other animals from the domain of deception. While there are numerous instances of deception in the animal world, it is only intentional deception that is underwritten by perspective-taking and attribution. This is because it involves purposefully attempting to manipulate the psychological states of others. Hare and colleagues have shown that chimpanzees, when placed in a competitive situation for food with humans, actively conceal information that would lead the humans to the food. Eight chimpanzees, from their first trials chose to approach a contested food item via a route that was hidden from the humans' view. The fact that this was a 'first trial' response strongly supports the interpretation that the chimpanzees were engaging in the modeling of human mental states and not giving a learned trial and error response.

Emotional Empathy

Empathy generally refers to the capacity to recognize or understand another's state of mind or feelings. It requires the capacity to take the perspective of another. Emotional empathy is most popularly conceived of as sympathy. Emory University primatologist Frans de Waal defines emotional empathy as the capacity to be affected by and share the emotional state of another.

A series of creative experiments conducted at Yerkes National Primate Research Center by de Waal and his collaborators provide compelling, albeit somewhat indirect, evidence for emotional empathy in capuchin monkeys. The researchers exchanged tokens for food with eight adult female monkeys. Each capuchin was paired with a relative, an unrelated familiar female from her own social group or a stranger (a female from a different group). The capuchins were then given the choice of two tokens: the selfish option, which rewarded that capuchin alone with an apple slice; or the prosocial option, which rewarded both capuchins with an apple slice. The monkeys predominantly selected the prosocial token when paired with a relative or familiar individual, but not when paired with a stranger. De Waal interprets the fact that the capuchins predominantly selected the prosocial option as evidence that seeing another monkey receive food is satisfying or rewarding for them.

In another series of experiments by investigators at the Max Planck Institute, 12–18 semiwild chimps went out of their way to help an unfamiliar human who was struggling to perform a task (reach a stick), even when it required having to climb over an 8-ft rope barrier for no reward. Human toddlers will do the same thing. Chimps that had been taught how to unlatch a door also helped unfamiliar chimps struggling to get through the door – by unlatching the door for them. These findings support the interpretation that the assisting chimps were relating specifically to the situational intent of the other.

All of the above findings could be interpreted as evidence for cognitive empathy solely and, indeed, they depend upon the capacity. However, they seem to have an additional element of helping or choice that reflects the possibility that primates are not only capable of perspective-taking but also feeling sympathy and 'good will.'

In a classic experiment with rhesus monkeys, subjects were trained to make two different responses that delivered different amounts of food. The response that provided the larger amount of food, however, was paired with the sight of another monkey receiving a shock. After witnessing the shock, a majority of the subjects preferred the response that was not paired with shock but delivered less food. A number of the monkeys stopped making responses altogether and were literally starving themselves to prevent the shock to the conspecific. This response was enhanced by familiarity with the shocked individual.

There is also evidence for emotional empathy in rodents. When scientists at McGill University injected painful irritants into the stomachs and paws of mice, they noted the animals' 'writhing' responses. By placing a mouse in a neighboring cage, they observed that the witnessing mouse became significantly more sensitive to painful stimuli, but only if the neighbor was familiar with the writhing mouse. These mice showed clear indications that they recognized and responded to the pain of another mouse. Moreover, rats will actively work to avoid or alleviate distress on the part of a conspecific.

Another expression of emotional empathy or sympathy is consolation, which has been formally defined as contact initiation by a bystander who directs positive attention to a victim. Consolation has been convincingly demonstrated in chimpanzees, bonobos, and gorillas. Frans de Waal has recorded hundreds of instances of reconciliation and consolation in chimpanzees. His conclusions are based on analyses of postconflict observations that compare third-party contacts with baseline rates in a social group.

Recently, evidence suggestive of consolation behavior has been provided for dogs. Of 1711 observed conflicts between pairs of dogs in a fenced outdoor meadow, 36% were followed by affiliations (greeting, sitting/lying down together, anogenital sniffing, playing, and/or licking) from a third dog. Most of these involved the loser of the conflict, and most tended to be initiated by the third party rather than by a conflict participant. There is also some evidence for consolation in rooks, another member of the Corvid family. In these birds, third-party affiliations took the form of a distinctive behavior: bill twining, in which two birds interlock the mandibles of their beaks.

Discussion and Implications

The emerging evidence shows that sentience is not only distributed across the animal kingdom, but it is also multidimensional and highly complex. This situation demands that we consider ourselves on a psychological continuum with other animals. The implications for science, philosophy, and even the ethics of human–non-human animal relationships, are powerful.

Moreover, the fact that many different phylogenetic lineages show capacities related to complex high-level sentience is intriguing evidence for cognitive convergence in the animal kingdom. It may be that sentience is ubiquitous, but the more complex levels extractable from experimental paradigms, like the ones above, are a result of any species' brain obtaining a certain level of elaboration and complexity. Further research will offer a more nuanced understanding of sentience and its distribution in the animal kingdom.

See also: Deception: Competition by Misleading Behavior; Empathetic Behavior; Mental Time Travel: Can Animals Recall the Past and Plan for the Future?; Metacognition and Metamemory in Non-Human Animals.

Further Reading

Call J, Brauer J, Kaminski J, and Tomasello M (2003) Domestic dogs are sensitive to the attentional state of humans. *Journal of Comparative Psychology* 117: 257–263.

Clayton NS, Dally JM, and Emery NJ (2007) Social cognition by food-caching corvids: The western scrub-jay as a natural psychologist. *Philosophical Transactions of the Royal Society of London* 362: 507–522.

Cools AKA, Van Hout AJ-M, and Nelissen MHJ (2008) Canine reconciliation and third-party-initiated postconflict affiliation: Do peacemaking social mechanisms in dogs rival those of higher primates? *Ethology* 114: 53–63.

de Waal F, Leimgruber K, and Greenberg A (2008) Giving is self-rewarding for monkeys. *Proceedings of the National Academy of Sciences USA* 105: 13685–13689.

Foote LA and Crystal JD (2007) Metacognition in the rat. *Current Biology* 17: 551–555.

Gallup GG, Anderson JR, and Shillito DJ (2002) The mirror test. In: Bekoff M, Allen C, and Burghardt GM (eds.) *The Cognitive Animal: Empirical and Theoretical Perspectives on Animal Cognition*, pp. 325–333. Cambridge, MA: MIT Press.

Hare B, Call J, Agnetta B, and Tomasello M (2000) Chimpanzees know what conspecifics do and do not see. *Animal Behaviour* 59: 771–778.

Hare B, Call J, and Tomasello M (2006) Chimpanzees deceive a human competitor by hiding. *Cognition* 101: 495–514.

Herman LM (2002) Vocal, social, and self-imitation by bottlenosed dolphins. In: Dautenhahnm K and Nehaniv CL (eds.) *Imitation in Animals and Artifacts*, pp. 63–108. Cambridge, MA: MIT Press.

of observation of a same subject, or even of different subjects, are available, it is possible to aggregate all the data into a same contingency table, hence the same transition matrix is used. Obviously, we should not do that when the different sequences represent clearly different types of behaviors. In that case, a different matrix should be used for each sequence, or the hidden Markovian approach described later in the article should be considered.

The concept of transition matrix can be extended in two ways. First of all, it is possible to compute a matrix between two nonadjacent periods. For instance, we may be interested in the probability of observing a particular activity at time t conditionally to the activity observed at $t-2$. The rows of the matrix then represent the activities at time $t-2$, when the columns still represent t. Another possibility is to use the information of several past periods to explain the present. For instance, we can build a matrix, using any combination of the activities observed at times $t-1$ and $t-2$ to explain the activity at time t. Going back to our example with activities (P)lay, (E)xploration, and (R)est, each row or column of the transition matrix will be defined as one of the possible combinations between two successive observed activities:

$$(PP, EP, RP, PE, EE, RE, PR, ER, RR)$$

The first letter always corresponds to the first observed activity. The transition matrix estimated from our dataset then reads

Activities at time $t-1$ and t

	PP	EP	RP	PE	EE	RE	PR	ER	RR
PP	0.94			0.04			0.02		
EP	0.55			0.31			0.14		
RP	0.80			0.10			0.10		
PE		0.53			0.44			0.03	
EE		0.28			0.62			0.10	
RE		0.36			0.55			0.09	
PR			0.07			0.14			0.79
ER			0.00			0.00			1.00
RR			0.03			0.03			0.94

Activities at times $t-2$ and $t-1$

Notice that the majority of transitions are in fact impossible (they have been left blank in the matrix). Such nonexistent transitions are called structural zeros. For instance, the third row of the matrix means that we observed activity R at time $t-2$ and activity P at time $t-1$. From this situation, it is only possible to go to a column also indicating that activity P was observed at time $t-1$. Columns 1, 4, and 7 are then the only possibilities. The position of structural zeros being perfectly defined into the transition matrix, it is often better to rewrite the matrix in a collapsed form in which only the possible activities at time t appear in column:

Activity at time t

	P	E	R
PP	0.94	0.04	0.02
EP	0.55	0.31	0.14
RP	0.80	0.10	0.10
PE	0.53	0.44	0.03
EE	0.28	0.62	0.10
RE	0.36	0.55	0.09
PR	0.07	0.14	0.79
ER	0.00	0.00	1.00
RR	0.03	0.03	0.94

Activities at times $t-2$ and $t-1$

Lag-Sequential Analysis

Computing the probability of observing an activity in function relation to past activities is of interest, but the researcher can find it more useful to identify the most frequent patterns of successive activities occurring in the data. The basic method designed for that purpose, the lag-sequential analysis, was introduced by Sackett. Its principle is to choose a starting activity, Exploration for instance, and to compute the probability of observing another target activity such as Play after 1, 2, 3, etc. periods. Each of these probabilities is an element of a transition matrix computed between two times spaced by 1, 2, 3, etc. periods. The probabilities clearly exceeding their expected value are identified on the basis of a statistical test based on a normal distribution. Probabilities are then computed also for the same starting activity, but with each of the other possible target activity. The result is a view of the most probable activities occurring 1, 2, 3, etc. periods after the starting activity is observed.

In **Table 1**, the X indicates the transitions identified as very likely to occur. Here, a pattern of the form Exploration followed by Play, Exploration, Rest, and Rest appears very common. By repeating the procedure starting now from the activity identified as the most probable one at time $t+1$ (Play), we can check whether a transition between Play at time $t+1$ and Exploration at time $t+2$ also exceeds expectation, whether a transition between Play at time $t+1$ and Rest at time $t+3$ also exceeds expectation, and so on. If all transition probabilities exceeding expectation coincide, then the pattern Exploration–Play–Exploration–Rest–Rest is really a frequent pattern. The same process can then be repeated starting from another activity to identify other frequent patterns.

Notice that the activities composing a frequent pattern do not have to follow each other exactly. For instance,

Table 1   Example of a lag-sequential analysis

Starting activity at time t	Target activity	Statistically significant probabilities				
		$t+1$	$t+2$	$t+3$	$t+4$	$t+5$
Exploration	Play	X	–	–	–	–
	Exploration	–	X	–	–	–
	Rest	–	–	X	X	–

From the starting activity Exploration observed at time t, the probability of the three possible activities Play, Exploration, and Rest are investigated from time $t+1$ to time $t+5$. In the table, an X denotes the activities having appeared clearly more often than expected by chance only.

a pattern such as Play Play Play m m Play Play Play, in which m means that no particular activity was identified as very likely to occur, is possible. In that case, three periods of Play, followed by two periods spent in any activity, followed again by three periods of Play is a very common pattern in the data. Another example is E m m m R. Here, exploration is very often followed four periods later by a rest.

Log-Linear Models

The lag-sequential analysis focuses on the identification of particular patterns of successive activities, rather than on the global relation between all possible activities at different times. When the researcher is more interested by this higher level of analysis, Agresti suggests that the log-linear approach should be preferred. It is also possible to reformulate a lag-sequential analysis in log-linear terms. A log-linear model is designed for the analysis of multiple-way contingency tables. Consider for instance a situation where we would like to identify the possible relations between activities observed during four successive periods, from t to $t+3$. In addition to the single effect induced by the independent distribution of activities observed during each period, we can also consider the six possible two-way contingency tables computed between two different periods among four, the four possible tri-way contingency tables, and the unique four-way contingency table, using the four periods simultaneously. The starting point of a log-linear analysis, the so-called saturated model, uses all possible relations between the four periods, to exactly reproduce the observed data. This model is not really interesting by itself, since it does not reduce in any manner the complexity of the initial observed situation. However, on the basis of successive statistical tests, it is then possible to suppress one by one the nonsignificant relations, starting from the highest order relations, until reaching the simplest structure of association between the four periods being still statistically similar to the saturated model. For instance, we can obtain a final structure like this, each star indicating a significant relation between periods:

$$[t^*t+1, t^*t+2, t+1^*t+2, t+1^*t+3, t+2^*t+3,$$
$$t+2^*t+4, t+3^*t+4]$$

In this example, no relation between more than two periods simultaneously remains significant. Among the two-way relations, only relations between observations separated by, at the most, two periods are significant. We conclude that an activity observed at a given time can influence the activities observed one or two periods later, but that no influence remains later.

Markov Chains

Markov chains introduced in the early twentieth century by the Russian mathematician Andreï Andreïevitch Markov are another way of using the concept of transition matrix. The basic assumption of a Markov chain is that the value taken by a variable at time t is fully explained by the values observed for the same variable from times $t-k$ to $t-1$. In the most restrictive case, the first-order model ($k=1$), the immediately preceding observation is supposed to resume the entire past of the target observation. When the two immediately preceding observations are used ($k=2$), we have a second-order model, and so on.

Probability Distribution of Future States

Let Y_1 be a random variable observed from time $t=0$ to time $t=n$. At each time t, the variable Y takes its value in a given set of values, this set being often finite. In the example used throughout the article, this set is made of the three possible activities (P)lay, (E)xploration, and (R)est. In a first-order Markov chain, these values are identified as the states of the model. Let X_1 be a row vector giving the probability distribution of the different activities at time t:

$$\chi_t = (P(Y_t = P)\ P(Y_t = E)\ P(Y_t = R))$$

Let Q be the first-order transition matrix between two successive behaviors, as previously defined. If we know thelytokous parthenogenesis in eusocial insects. Thelytoky may be more difficult to detect in eusocial insects than in other insects because its presence is sometimes concealed or suppressed by social structure; the mechanisms that cause this are described in the next section. Indeed, findings by L. Keller of the conditional use of asexual reproduction in ants arose by pure serendipity. Future studies may reveal thelytoky in many more eusocial species. Possibly the rarity of thelytoky in eusocial insects is attributable to the importance of genetic diversity among colony members. While asexual reproduction by a queen would increase within-colony relatedness, the resulting reduction in genetic diversity within colonies would lower homeostasis in colonies. For example, low genetic diversity colonies are more afflicted by disease than genetically diverse colonies in bumblebees, honeybees, and leaf-cutting ants. In honeybees, high-diversity colonies maintain more uniform temperatures in their brood nests than did the uniform ones because of a system of genetically based task specialization. Thus, reduction in genetic diversity would render the colonies less resilient to environmental perturbation. A comprehensive survey of species capable of thelytokous parthenogenesis is a necessary next step in the effort to understand the costs and benefits of sexual and asexual reproduction for eusocial insects.

The genotypes of parthenogenetic offspring depend on the mode of parthenogenesis (**Figure 1**). Thelytokous parthenogenesis can be categorized into two major cytological divisions, 'apomixis (ploidy stasis)' and 'automixis (ploidy restoration).' In apomictic parthenogenesis, known as clonal reproduction in aphids, the features of meiosis are either entirely or partially lacking. Only one maturation division takes place in the egg and this division is equational. The offspring retain the genetic constitution of the mother excluding mutations, and heterozygosity is maintained in subsequent generations.

Automixis with Central Fusion

In Hymenoptera, females are usually produced by sexual reproduction and are diploid, while males develop from unfertilized eggs and are haploid. In some social Hymenoptera, however, females can produce female offspring from unfertilized eggs. In general, thelytoky in eusocial Hymenoptera, such as the ants *Pristomyrmex pungens* and *Cataglyphis cursor*, and the honeybee *Apis mellifera capensis*, is automixis with central fusion, in which two haploid pronuclei that segregate at meiosis I fuse to restore diploidy. Therefore, the offspring have the same genotype as their mother for the loci that did not recombine, whereas an offspring is homozygous for one of the two maternal alleles if recombination occurred. The recombination rate and thus the frequency of transition from heterozygosity to homozygosity vary across loci, depending on the distance to the centromere (**Figure 2**). Compared to other forms of ploidy restoration, the reduction of heterozygosity is slowed under central fusion.

In social Hymenoptera, the maintenance of heterozygosity at the sex locus is essential for successful long-term propagation of parthenogens because the locus must be heterozygous for the expression of female traits. Central fusion gradually increases homozygosity over time because of recombination. Reduced rates of recombination are predicted to evolve for the maintenance of genetic diversity in parthenogens propagating thelytokously with central fusion. In the little fire ant *Wasmannia auropunctata*, the absence of recombination in the central fusion presumably evolved as a mechanism for maintaining heterozygosity.

Automixis with Terminal Fusion

In contrast to eusocial Hymenoptera, thelytoky in termites, such as *Reticulitermes speratus* and *R. virginicus*, is accomplished by automixis with terminal fusion, in which two haploid pronuclei that divide at meiosis II fuse (**Figure 2**). Thus, offspring are homozygous for a single maternal allele at all loci that did not crossover, whereas offspring have the same genotype as their mother at loci where crossover occurred. This causes a rapid reduction of heterozygosity. The near-total homozygosity caused by terminal fusion requires that all recessive lethal genes be eliminated in the course of evolution prior to the appearance of parthenogenesis. Otherwise the species will be unable to produce viable parthenogens. Inbreeding should preadapt a population to parthenogenetic modes that promote rapid homozygosity, but this inbreeding preadaptation hypothesis has not been supported by empirical studies. In termites, however, genetic purging through repeated inbreeding might have been a necessary preadaptation for the evolution of thelytoky.

Conditional Use of Sexual and Asexual Reproduction in Eusocial Insects

In social insects genetic diversity at both the individual and colony level is important for colony growth, survival, and reproduction. Perhaps the best solution to the dilemma over the costs and benefits of sexual and asexual reproduction is to use both modes of reproduction conditionally and to therefore experience the best advantages of both. Indeed, recent studies have uncovered unusual modes of reproduction in ants and termites, in which they take advantage of the social caste system to use sex for somatic growth and parthenogenesis for germ line production.

Conditional Use of Sex in Ants

Reports on *Cataglyphis cursor*, a common ant in the dry forests of Europe, suggested that unmated workers

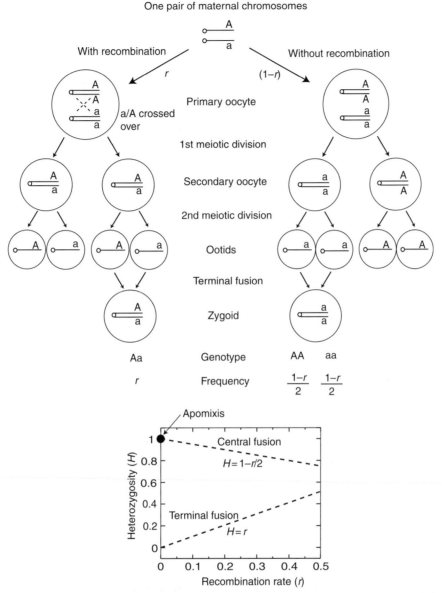

Figure 2 The cytological mechanism of ploidy restoration in automixis with terminal fusion (above) and the relationship between offspring heterozygosity and recombination rates in different modes of thelytoky (below).

produce both new queens and workers by thelytokous parthenogenesis in colonies that have lost the mother queen. Recently, M. Pearcy and colleagues discovered that both unmated workers and mated queens can use thelytokous parthenogenesis. The genotypes of queens and workers within colonies revealed that they consistently had different genotypes. Furthermore, a detailed analysis of 35 monogamous colonies showed that 96% of the queens (54 of 56) were produced by parthenogenesis. In contrast to queens, 97% of the workers (476 of 489) had been produced by normal sexual reproduction. Thus, queens of this species use alternative modes of reproduction to produce queens and workers.

The use of asexual reproduction allows queens to maximize their reproductive success by increasing the transmission rate of their genes to their reproductive female offspring. On the other hand, males normally archive direct fitness only through diploid female offspring because males usually develop from unfertilized maternal eggs in haplodiploid species. Therefore, thelytokous parthenogenesis by queens, which potentially reduces male reproductive success to zero, is a grave threat to males, implying major conflict between sexes.

In the little fire ant *Wasmannia auropunctata*, the evolutionary arms race between sexes has impelled a very unusual mode of reproduction. Similar to *C. cursor*, queens are produced parthenogenetically and workers sexually. Surprisingly, *W. auropunctata* males are also clonally produced by their fathers, whereby all males exhibit the same genotypes as their fathers. A plausible mechanism underlying this mode of reproduction is elimination of the maternal genome in eggs, and thus the resulting haploid males have genotypes identical to the sperm stored in the queens' spermathecae. Most recently, a mode of reproduction identical to that of *W. auropunctata* was found in the Japanese queen-polymorphic ant *Vollenhovia emeryi* with clonal reproduction by both queens and males. Because genes are transmitted only between individuals of the same sex, there is no gene flow between the male and queen gene pools. As a result of the obligate clonal reproduction by both sexes in these ant species, the queens' and the males' genomes are completely separated and form two distinct genetic lineages within the species.

Conditional Use of Sex in Termites

Termites have often been compared with ants. However, termites are basically social cockroaches, whereas ants evolved from wasps. Their different ancestries provided both groups with different life history preadaptations for social evolution. These phylogenetically divergent insects differ in fundamental ways. Unlike the haplodiploid Hymenoptera, both sexes of termites are diploid. Termites are hemimetabolous, whereas ants are holometabolous. Nevertheless, the structural elements of social organization in ants and termites are highly convergent, suggesting common selection factors in their social evolution. One of the most amazing convergences can be seen in the fact that conditional use of sexual and asexual reproduction evolved not only in ants but also in termites.

Opening the black box

In most termite species, one king and one queen usually found colonies. In termites, especially in lower termites, it has long been believed that the inbreeding cycles of generations of neotenic reproductives (offspring of the colony) propagate the colony after the death of the primary king and queen. Evidence for inbreeding depression in termites is mounting, such as higher mortality in inbred incipient colonies and lower lifetime fecundity in inbred colonies. Like most subterranean termites, *Reticulitermes* species have cryptic nesting habits with transient, hidden royal chambers underground or deep inside wood, making it difficult to reliably collect reproductives. Therefore, the breeding system of subterranean termites has been primarily estimated by genotyping workers or culturing laboratory colonies rather than censuses of field colonies.

R. speratus is the most common termite in Japan. To obtain reproductives from a sufficient number of natural colonies, we collected more than 600 nests in the field. We successfully found the royal chambers, where reproductives and young broods were protected, of 30 colonies. In nearly all cases, primary kings were continuously present but primary queens had been replaced by an average of 55.4 secondary queens. These results indicate that primary kings live much longer than primary queens; replacement of the primary king is rare, whereas replacement of the primary queen is the rule at a certain point in colony development. In addition, secondary reproductives always differentiate from nymphs but never from workers in natural colonies (**Figure 3**).

The paradox of the king-daughter inbreeding hypothesis

Sexual reproduction can lead to important conflicts between sexes and within genomes. In monogamous termites, conflicts between the primary king and queen can arise over parental investment and genetic contribution to offspring. Our finding that primary queens are replaced much earlier than primary kings in *R. speratus* leads to a paradox if the secondary queens are the daughters of the primary king. King-daughter inbreeding should result in uneven genetic contribution to the secondary offspring (offspring of secondary queens) by the primary king and queen. The life-for-life relatedness of a primary king to the secondary offspring produced by king-daughter inbreeding is 5/8, while relatedness of the primary queen to the secondary offspring is 1/4 when the primary king and the primary queen are unrelated. Because male primary reproductives are 2.5 times more related to offspring than is the primary queen under this system, colonies are expected to bias alate (new primary reproductives that disperse) production in favor of males. Contrary to this prediction under this breeding system, the alate sex ratio is slightly but significantly female-biased in this species (numerical ratio of male $= 0.43 \pm 0.02_{SE}$). Because of the larger size of females relative to males, the biomass sex ratio was even more biased toward females (investment ratio of male: $0.415 \pm 0.02_{SE}$). This inconsistency between king-daughter inbreeding and sex investment ratio in alate production suggests that there is a different breeding system in which the king and queen have more equal genetic contributions to offspring.

Queen succession through asexual reproduction

While examining genotypes within nests of *R. speratus*, we uncovered an extraordinary mode of reproduction. Secondary queens are almost exclusively produced parthenogenetically by the founding primary queens, whereas workers and alates were produced by sexual reproduction (**Figure 3**). By using parthenogenesis to produce secondary queens, primary queens are able to retain the transmission rate of their genes to descendants while maintaining genetic diversity in the workers and new primary reproductives even after the primary queen is replaced. The relatedness of the primary queens to workers ($r = 0.49$, $SE_{jackknife} = 0.04$) and to alate nymphs ($r = 0.58$, $SE_{jackknife} = 0.079$) is not significantly different from 0.5, the value expected between a female and her sexual offspring. This is twice what the expected genetic contribution queens would

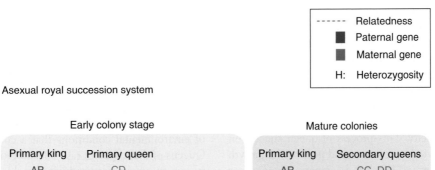

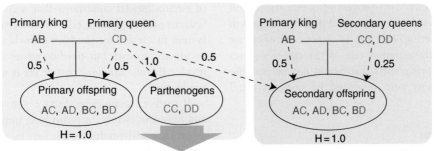

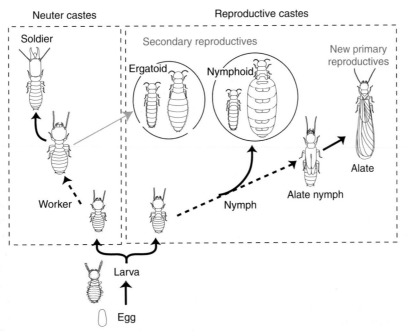

Figure 3 Asexual royal succession system in termites (above) and differentiation pathways of primary and secondary reproductives (below). This breeding system enables the primary queen to maintain her full genetic contribution to the next generation of primary reproductives (alates) while avoiding any loss in genetic diversity from inbreeding. Regression relatedness is shown in the figure. Life-for-life relatedness of the primary queen to parthenogens is 0.5. Secondary reproductives always differentiate from nymphs but not from workers in nature.

make to colony members under king-daughter inbreeding ($r_{\text{primary queen to king-daughter offspring}} = 0.25$).

Parthenogenetic production of secondary queens allows *R. speratus* to undergo queen succession without inbreeding. Heterozygosity of workers in colonies headed by secondary queens was as high ($H_{\text{o}} = 0.733$) as that expected for offspring produced by outcrossing of the primary king and the primary queen ($H_{\text{e}} = 0.736$). Likewise, there was no significant reduction of heterozygosity in nymphs produced in colonies with secondary queens. Further evidence of the lack of inbreeding in *R. speratus* colonies is provided by the low inbreeding coefficient of workers, which did not differ significantly from zero ($F_{\text{IT}} = 0.014$, $SE_{\text{jackknife}} = 0.048$, over all loci). The lack of consanguineous mating in this breeding system may also benefit primary kings. The offspring produced by outcrossing between the king and parthenogenetic queens may have greater fitness than those produced by king-daughter inbreeding.

The production of secondary queens through conditional parthenogenesis effectively extends the reproductive life of the primary queen, greatly expanding her reproductive capacity. This process of queen succession allows the colony to boost its size and possibly its growth rate without suffering any loss in genetic diversity or diminishing the transmission rate of the queen's genes to her grand offspring, feats that would not be possible if secondary queens were produced by normal sexual reproduction.

Purging selection

A faster rate of accumulation of deleterious mutations is a major cost of asexual reproduction. In haplodiploid organisms, deleterious alleles are directly exposed to selection each generation in the haploid males, and there is no masking effect of dominance. Therefore, purging selection will cause a more rapid decrease in the frequency of deleterious alleles at haplodiploid loci than at comparable loci in diploid organisms.

In termites, paradoxically, asexual queen succession can function to purge deleterious mutations. Parthenogenetic offspring are homozygous for a single maternal allele at all loci because of terminal fusion. Therefore, deleterious recessive genes are exposed to selection in homozygous parthenogens. Parthenogens carrying homozygous recessive deleterious alleles should not be able to survive or develop into functional secondary queens. The obligate occurrence of parthenogenesis in the normal life cycle of this species can eliminate recessive deleterious genes in every generation, much like the genetic purging that haploid males of Hymenoptera undergo, eliminating the transmission of deleterious recessive alleles to the sexual offspring.

Comparison of the Conditional Use of Thelytoky Between Ants and Termites

Asexual royal succession in *R. speratus* is in some ways analogous to the conditional use of sex found in the ants *C. cursor*, *W. auropunctata*, and *V. emeryi*. In these ant and termite species, queens do not require sperm from mates to produce diploid (female) offspring. Nevertheless, they retain sexual reproduction for production of workers and thus the genetic diversity in the worker force is maintained. By using alternative modes of reproduction for the queen and worker castes, genetic diversity in the worker population can be maintained.

As discussed earlier, reduced genetic diversity in the worker force may be detrimental to the colony because it leads to reduced defense against parasites, less efficient division of labor, and a decreased range of environmental conditions that a colony can tolerate. These costs are akin to those thought to lead to the instability of parthenogenetic reproduction in nonsocial organisms. Thus, sexual reproduction might have important benefits for colony function through increased defense against parasites, more efficient division of labor, and an increased range of environmental conditions that a colony can tolerate. Queens of these species take advantage of the social caste system to use sex for producing workers, which amounts to somatic growth but parthenogenesis, which does not involve the evolutionary cost of sex is used for germ line production.

If workers retain reproductive potential (totipotency) and have the chance to reproduce, queens may increase direct fitness by producing workers by parthenogenesis to some extent. In *C. cursor*, a small portion (2.5%) of workers is produced by parthenogenesis, and unmated workers have the chance of reproduction in colonies that have lost the queen. Under complete worker sterility, however, queens gain no fitness benefit by using parthenogenesis for worker production. In *W. auropunctata*, which exhibits complete worker sterility, all workers are produced by sexual reproduction. In the termite *R. speratus*, workers are always produced by sexual reproduction. We found that all of the 1660 secondary queens collected from field colonies were nymphoid, that is, neotenic reproductives with wing buds differentiated from nymphs. This indicates that workers have no chance to develop into secondary queens in nature (**Figure 3**). This suggests that worker sterility relaxes the sexual conflict over worker production and thus favors the use of sexual reproduction for worker production.

Another commonality is that the parthenogenetic queens produced in both systems are cared for their entire lives by workers. In ants, the conditional use of parthenogenesis for new queen production primarily occurs in dependent-founding species, most likely because the presence of workers compensates for the negative effects of parthenogenesis. Automictic parthenogenesis with central fusion gradually increases homozygosity over time because of crossing-over. By increasing the levels of homozygosity, parthenogenesis should result in reduced queen survival and fitness. In line with this prediction, queens of *Cataglyphis sabulosa*, in which colony reproduction proceeds through flight dispersal and independent colony founding, produce new queens by sexual reproduction although they are capable of thelytokous parthenogenesis.

In *R. speratus*, only the nondispersing neotenic (secondary) queens that develop within established colonies are produced asexually, whereas alates (adult primary queens) are produced sexually. Parthenogenesis with terminal fusion results in near-complete homozygosity, which should reduce the viability and fitness of the secondary queens. However, the consequences for secondary queens in this termite species may be minimal, because secondary queens stay in the natal nest protected and cared for by the existing worker force, unlike independent colony founding by the primary king and queen. Contrary to neotenic queens, alates of termites disperse and found colonies independently, and are thus subjected to a number of environmental contingencies in which genetic diversity is likely to be advantageous.

Genetic Influences on Reproductive Division of Labor

Eusocial insects, by definition, have a reproductive division of labor, which is often associated with a pronounced queen–worker polymorphism. Though caste differentiation is mostly controlled by environmental and social factors, recent studies have documented genetic influences on queen–worker differentiation. Queens in certain populations of *Pogonomyrmex* harvester ants are always homozygous, whereas workers are heterozygous hybrids. A very similar genetic caste determination associated with interspecific hybridization is also known in *Solenopsis* ants. Among multiple queens within colonies of *S. xyloni*, queens that mate with a conspecific male produce only new queens, while queens that mate with a *S. geminata* male produce only workers.

Genetic influences on queen–worker differentiation are essential to the conditional use of sexual and asexual reproduction. In the termite *R. speratus*, parthenogens are strongly biased to develop into secondary queens, suggesting that differentiation into this caste is genetically influenced, possibly by whether individuals are heterozygous or homozygous at certain loci. This asexual royal succession system can work only if parthenogens have priority to become secondary queens. Without any genetic influence on caste differentiation, it seems impossible to do this. The genetic system of homozygous advantage to be secondary queens makes it possible. Our latest analysis of the relationship between reproductive dominance of secondary queens obtained from experimentally orphaned colonies and their genotypes suggested that homozygosity at least two independent loci influenced the priority to differentiate into secondary queens, thus suggesting the existence of a multilocus queen determination system.

Future Perspectives

Recent findings of extraordinary modes of reproduction in phylogenetically divergent eusocial groups show that eusociality with its attendant caste structure and unique life histories can generate novel reproductive and genetic systems with mixed modes of reproduction that can provide important insights into the advantages and disadvantages of sexual reproduction.

Although conditional use of sexual and asexual reproduction is currently known only in three ants and a single termite species, breeding systems involving conditional parthenogenesis may occur in other eusocial insects. To date, the ability to reproduce parthenogenetically has been reported in unmated females in seven termite species from four different families, in which the production of neotenic replacement reproductives is common. This raises the possibility that the conditional use of sex and parthenogenesis could be widespread within this ecologically and economically important group of social insects. In addition, the genetic system of homozygous advantage to secondary queens may provide an ideal opportunity to identify the queen determination gene in termites. Indeed, molecular tools including RNAi may help reveal the queen determination gene in the near future.

See also: Kin Selection and Relatedness; Queen–Worker Conflicts Over Colony Sex Ratio.

Further Reading

Cremer S, Armitage SAO, and Schmid-Hempel P (2007) Social immunity. *Current Biology* 17: R693–R702.

Foster KR, Wenseleers T, and Ratnieks FLW (2006) Kin selection is the key to altruism. *Trends in Ecology & Evolution* 21: 57–60.

Fournier D, Estoup A, Orivel J, et al. (2005) Clonal reproduction by males and females in the little fire ant. *Nature* 435: 1230–1234.

Hamilton WD (1964) The genetical evolution of social behaviour. I, II. *Journal of Theoretical Biology* 7: 1–52.

Hughes WOH, Oldroyd BP, Beekman M, and Ratnieks FLW (2008) Ancestral monogamy shows kin selection is key to the evolution of eusociality. *Science* 320: 1213–1216.

Keller L (2007) Uncovering the biodiversity of genetic and reproductive systems: Time for a more open approach – American Society of Naturalists E. O. Wilson award winner address. *American Naturalist* 169: 1–8.

Matsuura K, Vargo EL, Kawatsu K, et al. (2009) Queen succession through asexual reproduction in termites. *Science* 323: 1687.

Maynard Smith J (1978) *The Evolution of Sex*. Cambridge: Cambridge University Press.

Oldroyd BP and Fewell JH (2007) Genetic diversity promotes homeostasis in insect colonies. *Trends in Ecology & Evolution* 22: 408–413.

Pearcy M, Aron S, Doums C, and Keller L (2004) Conditional use of sex and parthenogenesis for worker and queen production in ants. *Science* 306: 1780–1783.

Schmid-Hempel P (1998) *Parasites in Social Insects*. Princeton, NJ: Princeton University Press.

Thorne B and Traniello J (2003) Comparative social biology of basal taxa of ants and termites. *Annual Review of Entomology* 48: 283–306.

Wilson EO (1971) *The Insect Societies*. Cambridge, MA: Harvard University Press.

Relevant Websites

http://www.agr.okayama-u.ac.jp/LIECO/englishpage.html – Laboratory of Insect Ecology, Okayama University.

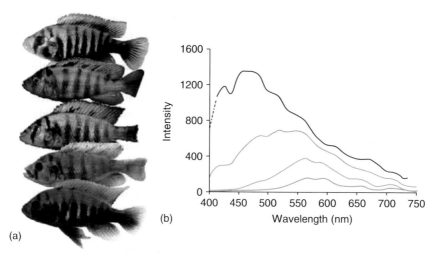

(b)

(a)

Figure 1   (a) Variation in male nuptial coloration in sympatric *Pundamilia* spp. cichlids of Lake Victoria, from the blue form of typical *Pundamilia pundamilia* (top) to the red form of typical *P. nyererei* (bottom). Females of both species have cryptic yellowish coloration. The red species *P. nyererei* is found at greater depths than the blue species *P. pundamilia*. (b) An illustration of how the spectrum of ambient light changes with depth in Lake Victoria, from the surface (blue) through three successive depths: 0.5 m (green), 1.5 m (orange), and 2.5 m (red). Modified and reprinted by permission from Seehausen O, Terai Y, Magalhaes IS, et al. (2008) Speciation through sensory drive in cichlid fish. *Nature* 455: 620–626, with permission from Nature Publishing Group. Copyright Macmillan Publishers Ltd.

have proposed a complex verbal model of the speciation process that includes sensory drive, Fisherian selection, gene flow, and reinforcement.

Hawaiian crickets in the genus *Laupala* have the highest speciation rate on record for anthropods. Closely related *Laupala* species are morphologically and ecologically indistinguishable and can produce viable hybrids. The only conspicuous difference between sympatric species is that they differ in the pulse rate of male courtship song. Tamra Mendelson and Kerry Shaw found that female *Laupala* can discriminate between conspecific and heterospecific song from a distance and are more likely to approach conspecific males. This suggested that correlated divergence in song and song preferences drove speciation, but further research by the same researchers revealed a more complex story. In a laboratory study of two allopatric species, conspecific courtship sequences usually went to completion, while heterospecific courtship rarely proceeded to the stage where males provide spermatophores. However, when females of the same species were paired with F2 hybrid males, which vary widely in song pulse rate, the song pulse rate of the males did not predict whether courtship proceeded to completion. A possible explanation is that chemical or tactile cues are exchanged between the sexes during courtship and that divergence in such cues, not courtship song, is responsible for the breakdown in heterospecific courtship. In support of this explanation, the researchers found evidence for rapid divergence between *Laupala* species in cuticular hydrocarbon (CHC) profiles. Whether species differences in CHCs contribute to reproductive isolation remains to be determined.

Three-spine sticklebacks provide an example in which sexual selection and ecological character displacement both appear to have played integral roles in speciation. In the lakes of British Columbia, sticklebacks occur in two ecologically and morphologically distinct species pairs: a larger benthic ecotype that forages in the littoral zone, and a smaller limnetic ecotype that forages in open water. Limnetic and benthic ecotypes within a lake are more closely related to each other genetically than they are to fish of the same ecotype in different lakes, probably because each lake was colonized independently by the marine ancestor (*Gasterosteus aculeatus*). Nevertheless, benthics and limnetics within a lake are reproductively isolated, while fish of the same ecotype from different lakes are not. This suggests that the same prezygotic isolating barriers arose independently in different lakes. Indeed, Janette Boughman and colleagues have shown that male coloration and female sensitivity to red light differ between ecotypes in the same direction in three different lakes. Compared to benthics, limnetic males have more red and less black coloration, and limnetic females are more sensitive to red light. In each lake, reproductive isolation between ecotypes appears to be maintained by female choice based on male size and color. The consistent direction of the differences between ecotypes in male color and female sensitivity to red light suggests that they are caused by habitat differences (albeit in the opposite direction as seen in African lake cichlids). Water color is red-shifted at greater depths in these lakes, and male benthics raised in red-shifted water in the laboratory develop less red and more black coloration than those raised in clear water. It would be informative to know whether this response to ambient light is present in the marine stickleback (i.e., the presumed ancestor). If so, and if the sensitivity of females to red light is similarly affected, then reproductive isolation between the ecotypes might have arisen as a byproduct of plastic responses to the environment (process 8). Alternatively, or in addition, the ecotype differences in color and red sensitivity might have evolved in response to selection, favoring increases in the visibility of males against the background water color (process 6). Reinforcement seems unlikely in this case because ecotypes differ only in the strength and not in the direction of the female color preference.

Research on brood-parasitic indigobirds (*Vidua* spp.) by Michael Sorenson, Robert Payne, and colleagues illustrates how within-generation shifts in the development of secondary sexual traits and mate preferences could cause rapid speciation (process 8). Indigobird nestlings, which are invariably raised by foster parents, imprint on the songs of their host species. Males later attract females reared by the same host species by mimicking host songs (**Figure 2**), and females preferentially lay eggs in the nests of their host species. Normally, this process of sexual imprinting maintains host-specificity between generations. But when females lay eggs in the nest of species other than their natal host, sexual imprinting may result in the sudden formation of new host races, or hybridization between existing host races, depending on whether the novel host already has its own host race of indigobirds. Molecular genetic data support this model of sympatric speciation with occasional hybridization. As a possible example of speciation in action, one of the ten recognized parasitic indigobird species occurs in two morphologically indistinguishable host races. Males sing host-specific

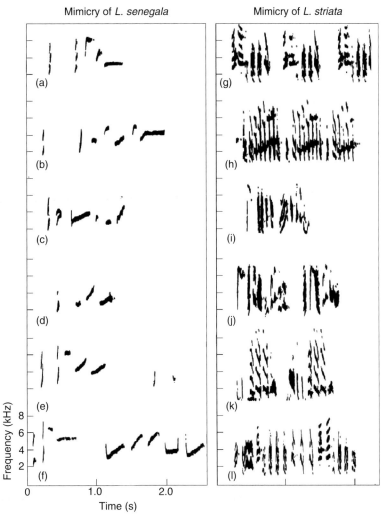

Figure 2 Sonograms showing host mimicry by male indigobirds (*Vidua chalybeata*). Males represented in the left column were reared by the normal firefinch host, while those represented in the right column were reared by Bengalese finches. Reprinted with permission from Payne RB, Payne LL, Woods JL, and Sorenson MD (2000) Imprinting and the origin of parasite–host species associations in brood-parasitic indigobirds, *Vidua chalybeata*. *Animal Behaviour* 59: 69–81, with permission from Elsevier. Copyright Elsevier.

Figure 4 A free swimming juvenile blacktip shark just after release from tagging (yellow tag is somewhat visible). (Photo: C. Simpfendorfer).

use, remain in, and return to a home site if displaced, and they display the ability to return if required.

On larger and longer-term scales are the migratory movements of sharks which often return to specific regions at specific times of year. Long-term studies have revealed that individuals can migrate over 100s of kilometers from a site and then later return to the exact region used previously. Mounting evidence for philopatric behavior in blacktip, blacknose, bull, nurse, and bonnethead sharks suggest that more pelagic species such as hammerhead, white, and tiger sharks also routinely return to specific sites for feeding, mating, or some other purpose. How sharks navigate and find their way to and from specific sites is still largely unknown and is an area of shark behavior that requires serious future research before any solid understanding can be gained. This is especially interesting, since in some species, naïve sharks less than 1-year-old undergo migrations of hundreds of kilometers with no parental guidance. These individuals then find their way back to summer habitats the following year, suggesting a sophisticated navigation mechanism and a strong drive to use specific regions. Despite our current lack of understanding, this is clearly a crucial component of shark behavior and ecology.

Behavior as a Physiological Aid

Several studies have reported shark and ray species behaving in ways to help cope with changes in the physical environment or to meet physiological needs. A broad example of behavior as an aspect of physiological tolerance is found in migratory species of many animals, including sharks. Once water temperatures decline in winter months, migratory shark species, such as blacktip

and sandbar sharks, living along the US east coast begin to travel south to where water temperatures remain warm. Temperatures in the northern extent of the species range are below the thermal tolerance for the species, so individuals must migrate. In this instance, movement/migration behavior is a key to survival.

A more specific example of environmental conditions driving movement is the report that bat rays along the California coast move into warmer waters during the day and out of these regions at night when water temperatures decline. It was suggested that this movement pattern is a form of behavioral thermoregulation where the rays are selecting specific environmental conditions to help with some physiological process. In this case, it was assumed that rays sought out warm water to aid in digestion. A similar case of behavioral thermoregulation was reported for leopard sharks on Catalina Island, California. In this instance, pregnant female sharks moved into shallow, warm waters during the day and out of the region during the colder night hours. The presence of only pregnant females displaying this behavior suggests that females use the warmer water temperatures to help speed gestation. A similar strategy has been reported for reef sharks in the Hawaiian Islands.

Other environmental factors may cause behavioral responses in sharks and rays. Recently, research into the movement and behavior of juvenile bull sharks within a highly variable river environment suggested that individuals choose habitat based on salinity preference (**Figure 3**). Results of this research revealed that sharks moved nearest to the river mouth (the area of greatest marine influence) during the wet season when the river approached fully freshwater conditions. During the dry season, when the river was well mixed, young bull sharks used the entire river region. This behavior could be a means of behavioral osmoregulation, in which individuals select specific salinity regimes to reduce the need to regulate their internal salt balance. Movement to remain in an optimal salinity may allow individuals to use energy for growth rather than salt regulation. Changes in flow and salinity conditions have been reported to similarly affect other species using estuarine regions including bonnethead sharks, cownose rays, bat rays, and leopard sharks.

A more extreme example of behavioral response to environmental conditions is the reported response of sharks to passing tropical storm systems. Long-term acoustic tracking data have revealed that as tropical storm systems come ashore, sharks move out of inshore regions into areas of deeper water. The best documented case of this behavior involves Tropical Storm Gabrielle making landfall on the gulf coast of Florida. The storm made landfall 50 km south of an inshore bay where juvenile blacktip sharks were tracked by an acoustic monitoring array. All monitored individuals left the study site 7 h prior to the storm's landfall. Analysis of storm conditions revealed that sharks left the inshore region during the period of greatest decrease in barometric pressure. This suggests that sharks are so finely tuned to their environment that subtle changes can cause them to leave a region. The change in barometric pressure is thought to have triggered a flight response, possibly to prevent individuals from getting caught in storm surge in shallow regions, or to avoid exposure to salinity declines that may accompany tropical weather systems. After passage of the storm, all of the monitored individuals returned to the study site and resumed movement patterns similar to those prior to the event.

Movement in response to environmental conditions may provide physical or physiological benefits, and in some cases, may be a means of avoiding mortality. This reinforces how important movement is as a behavioral response in mobile populations, and how finely tuned species can be to the conditions and changes within their environment.

Feeding and Predator–Prey Relationships

Feeding ecology of sharks is a large subject area that involves many aspects of shark ecology, physiology, and behavior. This includes sensory systems, jaw mechanisms, feeding strategy (grasping vs. suction), and biological aspects such as feeding periodicity. This discussion is limited simply to feeding activities in relation to behavior.

The majority of shark species feed on fish, crustaceans, and other sharks. Various feeding strategies are employed, with many species feeding in turbid water or areas of high prey density. Few studies attempt to define feeding behaviors of sharks and most of the available information is based on large predatory species, such as white or tiger sharks. From what is known, these species may use more of a stealth approach to feeding and either attack from below for surface prey, such as marine mammals and sea turtles, which must come to the surface to breathe, or attack from above to prey on bottom-dwelling fish and rays. The coloration of sharks, dark on the top and light on the bottom, is thought to help camouflage them when feeding on surface or bottom oriented prey who may find it difficult to see approaching sharks that blend in with surface or bottom coloration. Limited predation event data from tiger sharks suggests that they do not chase prey items, but rather take advantage of prey when they are vulnerable or not alert. Sharks may take advantage of situations when prey are not vigilant and a quick attack can occur. An extreme example of stealth feeding behavior is that of cookie cutter sharks. These small individuals are almost ectoparasitic as they swim up to large fish and marine mammals and bite a chunk of flesh out of their prey. These sneak attacks leave the prey with a melon-ball shaped, nonlethal wound. Whether small

(cookie cutter) or large (white, tiger), stealth appears to be a key component of shark feeding behavior.

Most shark species hunt as solitary individuals, regardless of their feeding behavior. However, data on sevengill sharks revealed group hunting for fur seals. Sevengill sharks group together and swim in circles around a fur seal until one or more of the sharks rush in to bite the seal. After the initial bite the rest of the group attacks and feeds upon the seal. This suggests that social feeding behaviors may occur in shark populations, although this has not been documented in any other species.

As well as being predators, sharks often fall prey to other, larger, sharks. Studies of juvenile blacktip sharks in a coastal estuary in Florida revealed that juvenile sharks tended to use habitat more in relation to predation risk than prey abundance. This suggests that sharks are not always top predators in the system and have predation concerns similar to those of other prey species. Further support for the use of habitat for protection rather than feeding comes from juvenile hammerhead sharks in Hawaii. Young hammerhead sharks remained within a nursery area despite the fact that it did not provide adequate food resources and a portion of the population actually starved while resident in the region. These populations presumably stay in the area to avoid predation by larger sharks that may not enter the shallow water regions these sharks inhabit. This reinforces the concept that large sharks can alter the behavior patterns of their prey species, and that prey species, including small sharks will use habitats in ways to avoid predation by these individuals. As these large predators move into a region, fish, small sharks, and other prey species, such as marine mammals, may change their distribution and movement patterns to avoid interactions and reduce their predation risk. As predators and prey, shark behavior clearly plays a key role in feeding activities and predator avoidance.

Social Aspects of Shark Behavior

Sharks are typically thought of as solitary individuals that do not associate with others, unless for very specific purposes such as mating. Given this assumption, very little research has been conducted to define social aspects of shark behavior. In one of the earliest studies (1954), dominance within a community of smooth dogfish was reported; larger individuals were dominant over smaller individuals, suggesting the existence of a social hierarchy. A similar study conducted in the 1970s looked at a group of bonnethead sharks in a penned enclosure. The movements and interactions of sharks in the pen were observed and recorded to determine if any social hierarchy existed within the group. Results of the research revealed that smaller bonnethead sharks moved out of the way of larger individuals or followed along behind a large animal.

Figure 4   A dangerous eavesdropper and deceptive signaler, a female firefly of the genus *Photuris* that has lured a male *Photinus* firefly to her; the male has paid the ultimate price for his mistake. Courtesy of James E. Lloyd.

Figure 3   A male thynnine wasp has grasped the deceptive female decoy of the elbow orchid *Spiculaea ciliata*. Note the orchid pollinia stuck to the wasp's thorax, a result of a previous wasp–orchid interaction.

immune to predation because of its unpleasant, unpalatable, or downright poisonous properties. The mimic communicates falsely to its predators that it is a member of a protected inedible class of prey. Predators that have experienced the bad-tasting or toxic protected species may avoid the Batesian mimic as well, treating it as it were unpalatable and should be left alone. In doing so, the predatory signal receiver loses food that it would benefit from having.

Then there are the acoustical deceivers, like those birds that give alarm calls warning others nearby of an impending danger that does not exist. By doing so, the alarm callers scare neighbors temporarily away, which enables them to forage for food without the competition provided by those they have frightened off. Olfactory, as well as visual and acoustical, deception has also evolved. Take the bolas spiders, which release a pseudosex pheromone similar to that emitted by certain female moths. When male moths of these species sense the odor, they respond as they would to a real female moth – by flying upwind toward the source of the scent. The spider takes advantage of this reaction to catch the male moth when he comes close enough to be struck and stuck to a ball of glue at the end of a silken line. The captive moth can then be reeled in, killed and eaten by the predator.

Why Communicate If Eavesdroppers and Deceivers Can Exploit the System?

The fact that eavesdropping and deception are widespread is puzzling, given the logic of the evolutionary argument that communication systems should not persist if the signaler or the receiver are damaged, reproductively speaking, by giving or responding to a signal. Biologists have, however, produced two different kinds of solutions to this puzzle. One answer can be labeled novel environment theory in which the maladaptive signal-giving, or signal-receiving, occurs because modern conditions are so different from those that shaped the evolution of the behavior in the past, and because there has not been sufficient time for advantageous mutations to occur that would 'fix the problem.' So, for example, certain male buprestid beetles respond to the color pattern signals provided by orangeish beer bottles and telecommunication signs (**Figure 5**) in Australia by flying to these objects and attempting to mate with them, just as they would attempt to mate with an orange-colored female of their species. Beer bottles and orange-colored signs have only very recently become part of the environment of the beetles and selection has had no time to favor individuals that happen to avoid these novel objects as they search for receptive female beetles.

Most cases of eavesdropping and deception, however, involve eavesdroppers and deceivers that have long been

Figure 5 A male buprestid beetle attracted to a sign because of its orange color, a maladaptive response due to a novel environment. Prior to recent human activity in the area, males that responded to the color orange were likely to locate potential mates efficiently.

part of the exploited signal-giver or signal-receiver's environment. Novel environment theory cannot easily explain these instances of communication gone awry. Instead, we can apply what might be called net benefit theory to most of these cases, arguing that although it does not pay to be exploited by an eavesdropper or deceptive signaler, the loss of reproductive success imposed by these exploiters is on average less than the benefits gained by remaining as part of a mutually beneficial communication system. In other words, even though there is some chance that a signaler will be harmed by an illegitimate receiver, or a receiver will be damaged by a deceptive signaler, giving the signal or responding in a particular way to a given signal yields a net gain in reproductive success on average.

For example, a male cricket that calls for females or a male firefly that approaches the signals associated with females of his species derives a benefit, the opportunity to pass on his genes, from being part of a communication system with females of his own species. If the risk of exploitation is reasonably low, then even though parasites and predators of other species reduce the average reproductive benefits associated with the legitimate communication system, the benefits could still outweigh the disadvantages of signal-giving or signal-receiving, leading to the persistence of these behaviors. Imagine that a mutant male *Photinus* firefly appeared that did not respond to the signals given by deceptive *Photuris* femme fatales. This male would probably live longer on average than his fellow males, which respond to these signals, but since the females of its own species also give the same

signals, the mutant male would be very unlikely to find a mate and pass on the genes for his safe but sterile behavior.

If this argument is correct, then eavesdroppers and deceivers should usually exploit a system that has clear adaptive value under most circumstances. This seems to be the case as can be judged from the fact that most of the examples given earlier have to do with communication between males and females of the same species that lead to reproductive payoffs for both participants. Túngara frogs often gain mates by calling for them, even if they occasionally attract a fringe-lipped bat instead. Thynnine wasp males that respond to the sex pheromones released by females have a chance to copulate with these individuals, even if they sometimes are fooled into 'mating' with an orchid flower. Likewise, it can pay a bird to always dash for cover when hearing an alarm call because the response may save its life, a major benefit, even though occasionally the alarm is false, which comes with the modest cost of losing a bite to eat.

Reducing the Costs of Dealing with Damaging Eavesdroppers

Even though it may still pay to communicate with others despite the risks posed by exploitative eavesdroppers and deceptive signalers, we might expect communicators to evolve tactics that at least reduce the costs imposed by these parasites and predators. If such tactics were possible, their use would increase the net benefit associated with participation in an adaptive communication system. Ameliorative tactics of this sort have evolved in many species. Take the túngara frog, which combats the fringe-lipped bat by altering its calls under some circumstances. When a male is calling with relatively few other individuals nearby, and so is at greater risk of being singled out by a frog-hunting bat, the frog is especially likely to drop one component of the call labeled the 'chuck.' Calls with a chuck are more readily located by bats, so by eliminating this feature, male frogs make themselves harder for their predators to find. Frogs giving chuck-less signals are almost certainly also harder for females to find as well, but this price will be worth paying when the risk posed by an eavesdropping bat predator is particularly high.

Likewise, small songbirds give soft, high-frequency 'seet' alarm calls when they spot deadly daytime hunting hawks rather than the louder, lower-frequency 'mobbing' call that they use when they encounter a sleepy owl during the daytime. The 'seet' call does not travel far and so is not as effective in alerting the fellow flock members to nearby danger, but the weaker signal is also harder for hawks to detect and use to target an alarm caller.

The evolution of counteradaptations to signal parasites is apparent in the visual displays of the northern swordtails. These fish use their flashy tails to attract mates but

The NREMS-promoting effects of GHRH are independent of GH release as they persist in hypophysectomized rats. Microinjection studies revealed that the medial preoptic area (MPA) is a likely target for GHRH to stimulate NREMS. It seems that stimulation of GH secretion and promotion of NREMS are two parallel outputs of hypothalamic GHRHergic neurons: one is a (neuro)-hormonal action through the hypothalamic-hypophyseal portal circulation, and the other is due to neuromodulator activity in the preoptic region. It remains unknown if two sets of GHRH neurons under the control of a common regulator or the same population of neurons are responsible for the sleep and GH-stimulatory effects. In addition to NREMS, GHRH also weakly stimulates rapid-eye-movement sleep (REMS) in rats. This effect is abolished by hypophysectomy suggesting that this REMS-promoting action of GHRH may be mediated by GH. Acute sleep loss is followed by compensatory rebound increases in sleep, a phenomenon called 'homeostatic sleep regulation.' GHRH may be an important hormone in this homeostatic response. Sleep deprivation enhances hypothalamic GHRH release, and rebound sleep that usually occurs after sleep loss, is greatly reduced in transgenic mice and rats with impaired GHRH activity. GHRH receptor antagonists or anti-GHRH antibodies suppress spontaneous sleep as well as inhibit rebound sleep increases after sleep deprivation.

Somatostatin

The other hypothalamic hormone in the somatotropic regulation pathway, somatostatin, was one of the first peptides tested for somnogenic actions. SST and other SST receptor agonists rapidly suppress NREMS and stimulate REMS. In rats, NREMS suppression is followed by a large increase in NREMS intensity, a phenomenon similar to what is seen during recovery responses after sleep deprivation. These NREMS changes parallel the effects of SST on GH in that GH secretion was suppressed when NREMS was suppressed and surges of GH reoccurred when the duration of NREMS normalized and NREMS intensity was enhanced. In humans, potent SST agonists impair sleep in both young subjects and in the elderly. These findings are consistent with the expectation that a hormone, SST, that inhibits GHRH release, also suppresses NREMS and GH secretion.

Feeding/Adiposity-Related Hormones

A growing body of evidence supports the existence of a strong interaction between sleep/vigilance and feeding. Several hypothalamic areas, such as the suprachiasmatic nucleus, lateral hypothalamus (LH), and ventromedial hypothalamic nucleus, have long been implicated in the regulation of both sleep and food intake. Changes in the amount and/or content of food greatly affect sleep–wake activity in rodents. For example, there is a significant correlation between meal size and the subsequent duration of sleep. Starvation induces a marked sleep loss in rats, while refeeding after food deprivation induces increased sleep in rats. The calorie-rich 'cafeteria diet' induces hyperphagia and increases the amount of sleep in rats. Intravenous administrations of nutrients affect sleep in rats differently. The diurnal distributions of NREMS and REMS in rats are significantly altered when food access is restricted to the light period. Feeding-related peptides and hormones also affect sleep. In this section, we focus on the sleep-modulating effects of the hunger signal ghrelin, the satiety signal cholecystokinin (CCK), and the adiposity signal leptin.

Ghrelin

The preproghrelin gene (Ppg) encodes two distinct peptides, ghrelin and obestatin. Ghrelin was first isolated from the stomach and later was found in the brain and in several other tissues such as the distal parts of the gastrointestinal system, thymus, gonads, and adrenal gland. In the brain, ghrelin-producing neurons are present in the arcuate nucleus (ARC), and in the hypothalamic area between the LH, paraventricular (PVN) and dorsomedial hypothalamic nuclei. Ghrelin is the endogenous ligand of the ghrelin receptor (GHS-R1a) which is widely distributed in the brain and also expressed by various peripheral tissues. Ghrelin secretion by the stomach and circulating plasma ghrelin levels inversely correlate with feeding. Plasma levels are elevated before a meal and rapidly decrease in response to eating in both humans and rodents. Ghrelin peptide levels in the rat hypothalamus also have a diurnal rhythm; it is low during the light phase when the animal is inactive, gradually increases toward the end of the light period, and peaks at the beginning of the dark phase.

It is well established that ghrelin promotes increased feeding and contributes to metabolic regulation in fasting states. Recent results indicate that ghrelin may also play a role in sleep regulation. For example, intracerebroventricular administration of ghrelin at the beginning of the inactive period of rats stimulates wakefulness and simultaneously suppresses NREMS and REMS for 2 h after injections. Local microinjections of ghrelin into different hypothalamic areas such as the LH, PVN, or MPA promote wakefulness at the beginning of the inactive period of rats. The LH appears to be the most sensitive site of ghrelin's wakefulness-promoting and sleep-suppressing effects. Intravenous (i.v.) injection of ghrelin decreases slow-wave sleep and increases wakefulness for 30 min after injections in rats. Increased hunger and feeding activity lead to increased wakefulness. Ghrelin, however, promotes wakefulness in those rats which have no access to food, indicating that the increased wakefulness is not simply the consequence of increased eating activity. Ghrelin is a part of a well-characterized hypothalamic neuropeptide circuit that also involves neuropeptide Y (NPY) and orexin. Ghrelin-containing neurons innervate orexinergic cells in the LH and NPY-containing neurons in the ARC and stimulate the release of NPY from presynaptic axon terminals in the PVN and LH. Orexin is well known for its key role in maintaining wakefulness. NPY also promotes wakefulness when injected into a lateral cerebral ventricle or the LH in rats. We hypothesize that the hypothalamic circuit formed by ghrelin, orexin, and NPY has two parallel outputs, increased wakefulness and increased food consumption. These behaviors, as well as increased locomotor activity, are characteristic of the beginning of the dark period in rodents and collectively comprise the 'dark onset syndrome.' The ghrelin-NPY-orexin mechanisms that are responsible for the dark onset syndrome play an important role in integrating the regulation of sleep and feeding.

Although exogenous administration of ghrelin causes robust increases in wakefulness, Ppg KO animals do not show any major change in their sleep–wake pattern. Recently, it became evident that the Ppg also codes for another biologically active peptide, called 'obestatin.' Interestingly, while ghrelin has strong wake-promoting activities, obestatin facilitates sleep. Such dissociation between ghrelin and obestatin was described for food intake as well. The fact that Ppg knockout animals exhibit normal feeding and sleep at thermoneutral ambient temperature could be explained by the simultaneous lack of two peptides which have opposite effects on sleep and feeding. When challenged by cold exposure (17 °C for 3 days), Ppg KO mice show increased cold sensitivity indicated by suppressed body temperature and sleep compared to controls. Ppg KO mice are even more sensitive to the combined challenge of cold exposure and fasting. Ppg KO mice enter hypothermic bouts associated with reduced sleep culminating in a marked drop in body temperature to near ambient levels. Prior treatment with obestatin attenuates the hypothermic response of preproghrelin knockout mice. This further supports the role of obestatin in the coordinated regulation of metabolism and sleep.

In humans, the effects of ghrelin and other ghrelin receptor agonists on sleep are less clear. Repeated intravenous bolus injections of ghrelin during the night increase NREMS, particularly stage 4 sleep, and suppress REMS in males, but not in females or when given in the early morning. Hexarelin, a more potent ghrelin receptor agonist, suppresses sleep. Intravenous administration of growth hormone-releasing peptide-6 increases stage 2 sleep but has no effect on SWS and REMS, while growth hormone-releasing peptide-2 has no effect on sleep in humans.

CCK

CCK is secreted by the endocrine cells of the small intestines and produced by neurons in various brain regions. Intestinal CCK serves as a gastrointestinal hormone and paracrine agent while neuronal CCK is a neuromodulator. Post-translational processing of prepro-CCK, a 115-amino acid peptide, is cell specific. In the brain, the predominant form is CCK octapeptide while in the circulation, longer forms also exist. Two G-protein-coupled CCK receptor subtypes have been identified. CCK-1 receptors dominate in the gastrointestinal tract, and in the brain, the most common form is the CCK-2 receptor.

Two of the best characterized behavioral effects of CCK are its satiety- and sleep-inducing effects. In fact, CCK was the first peptide implicated in the common regulatory mechanisms of sleep and feeding/metabolism. Single bolus injection of CCK elicits the complete behavioral sequence of the satiety syndrome, that is, the cessation of feeding, increased sleep, decreased motor activity, and social withdrawal. Systemic CCK injections promote sleep in cats, rats, mice, and humans. Central administration of CCK does not suppress feeding or promote sleep indicating that these effects are mediated by peripheral targets. Injection of selective CCK-2 receptor agonists does not have sleep-promoting effects, but CCK-1 antagonists suppress CCK-induced sleep, as well as, sleep increases after eating. The feeding-suppressive effects of CCK are mediated by the vagus nerve; however, vagotomy does not prevent the sleep-inducing effects of CCK in rats. Also, sleep induction is independent of the pancreatic insulin, another somnogenic hormone stimulated by CCK. OLETF rats that lack CCK-1 receptors have normal baseline sleep–wake activity suggesting that in the absence of CCK signaling through the peripheral receptor subtype normal sleep–wake activity can be maintained.

Leptin

Leptin is a 16-kD peptide, the product of the *ob* gene. Its major source is the adipose tissue and the epithelial cells of the stomach. The best characterized effects of leptin are the suppression of feeding and the stimulation of metabolism; these actions together lead to the reduction of fat stores. Leptin is secreted proportional to the size of the adipose tissue and is thought to serve as an adiposity signal to the brain. Gastric leptin is released from the stomach after eating and may play a role in the short-term regulation of feeding as a satiety signal. Circulating leptin is transported into the brain where leptin receptors are expressed in various hypothalamic nuclei and in the brainstem. Leptin suppresses food intake through the stimulation of melanocyte-stimulating hormone secretion and suppression of NPY-producing cells.

Relationships between obesity on sleep have long been acknowledged. For example, obesity is the most common metabolic abnormality in sleep-related breathing disorders; gains in body weight reliably predict the incidence of obstructive sleep apnea. Increased incidence of obesity and the different primate species do not interpret a human's gazing behavior toward one of the locations as a communicative gesture produced to inform them about the location of the food. Instead, they choose mainly randomly between the different locations, presumably ignoring the human's gesture. The reason for this lack of understanding is currently a highly debated issue, much more so because domestic dogs, a species very distantly related to humans, are very flexible in using different kinds of communicative gesture from humans. Dogs' skills in this domain are outstanding in the animal kingdom and are most likely the result of selection pressure during domestication and an adaptation to life with humans.

Perspective Taking

To test whether chimpanzees understand something about other individuals' perspective, Hare and colleagues designed a paradigm based on chimpanzees' natural tendency to compete for food. Two individuals, one dominant over the other, were placed in a situation in which they had to compete over two pieces of food. The subordinate chimpanzee, which would normally not have had a chance to gain food with the dominant present, had an advantage. While it had visual access to both pieces of food, the dominant individual could see only one piece, the other piece being hidden by a wooden barrier. The subordinate chimpanzee preferred to approach the piece of food behind the barrier, the one the dominant could not see, to the piece in the open and visible to the other individual. The chimpanzee's preference for the hidden piece was not merely based on a preference for eating behind an obstacle, because when they are alone, subjects choose randomly between both food locations. Neither did subordinates preferentially approach the piece of food behind a barrier if the barrier was transparent, even if it potentially protected them from the competitor physically. Chimpanzees also prefer to reach through an opaque rather than a transparent tunnel for food if in competition with a human whose eyes they can not see while reaching but who can potentially see their hand reaching through the transparent tunnel. Taken together, these results suggest that chimpanzees based their behavior on some sensitivity to the visual perspective of others.

Another group of species that seem to possess a flexible understanding of others' visual perspective are birds, in particular, members of the crow family (Corvidae). Most members of this family cache (hide) food for future consumption, which requires not only a good spatial memory to relocate the caches at a later date, but also certain strategies to reduce the probability of others stealing those caches. This work has focused on the cache protection strategies of western scrub-jays and ravens. If confronted with a situation in which scrub-jays have a choice of where to cache while a conspecific is observing, cachers use distance to reduce the

visual acuity available to observers, by preferring locations relatively far from the observer to those nearby. They also prefer to cache behind an opaque barrier or in a tray that is located in the shade to caching out in the open or in a tray located in the light. These results suggest that scrub-jays also adjust their behavior, based on an understanding of others' visual access to their actions.

Knowledge Attribution

Different animals understand something not only about others' current, but also about others' past, visual access. Povinelli and coworkers conducted a series of experiments in which they wanted to test whether chimpanzees could take into account what a human had seen in the immediate past. To test this, they confronted chimpanzees with a situation in which they had to distinguish between two human experimenters who informed them about the location of hidden food. One of the experimenters (the knower) witnessed food being placed in one of several containers while the other experimenter (the guesser) waited outside the room. After the guesser reentered the room, the humans, the guesser and the knower, pointed to different containers. The chimpanzee was then allowed to choose between the containers and could potentially base her choice on the information coming from the most reliable source, the knower. In this setting, chimpanzees could only differentiate between humans after several hundred trials, which was most likely the result of discriminating between whether the human was present or absent during baiting. However, one general critique of this paradigm is that it is rather unnatural for chimpanzees. A human indicates the location of food in a very cooperative manner, something that would not occur in a group of chimpanzees. It is highly unlikely that one chimpanzee would indicate the location of food to another chimpanzee with the intention of letting her have it.

In a paradigm also based on chimpanzees' natural tendency to compete over food, Kaminski and colleagues allowed two chimpanzees to compete over two pieces of food. Subject and Competitor sat opposite one another, between them was a sliding board, which the human could slide back and forth between both individuals. The task began with a hiding event, in which food was hidden under one of three cups while both participants were watching. Another piece of food was hidden under a second cup while only the Subject was watching. Hence, while the locations of both pieces of food were known to the Subject, only one of them was known to the Competitor. Now the Competitor was given the first choice with the Subject unable to see this choice being made. After the Competitor had made its choice, it was the Subject's turn. The chimpanzees in this situation preferred the piece of food unknown to the Competitor presumably because they understood that the other piece, the one the Competitor had information about, was likely to be gone by the time of her choice. This finding supports previous studies showing that chimpanzees may take into account what others have seen in the immediate past.

Scrub-jays, like chimpanzees, may base their cache protection strategies on an understanding of others' knowledge states. In a recent study, Dally and colleagues presented scrub-jays with a situation in which they had to decide which tray to recover their caches from. Earlier, the birds were allowed to hide food in tray A in the presence of observer A with a second tray (B) present, but covered so that caching that tray was not possible. After a short delay, the same bird was allowed to cache in tray B in front of observer B, with tray A now covered. After 3 h, the birds were given the opportunity to recover their caches from both trays in one of four conditions: to recover in the presence of observer A, or in the presence of observer B, or in front of a naïve bird (C) that had not witnessed caching in either tray, or to recover in private. Interestingly, the birds specifically recovered the caches that observers had seen them make. For example, the birds selectively recached items from tray A when recovering in the presence of observer A, but did not recache any items from tray B. By contrast, the birds did not recache items from either tray when recovering in the presence of the naïve bird, suggesting possible attribution of ignorance as recaching in front of the naïve bird would have provided information to the naïve bird that they previously did not have. Finally, the birds recached items from both trays when they recovered in private, as observers A and B had seen caches being made in these trays, so recaching them in private would move the caches to new places that potential pilferers had not seen. When recaching in front of a 'knowledgeable' observer, the cacher moved their caches around up to six times using previous cache sites, suggesting that the cacher was attempting to 'confuse' the observer as to the final location of the cache, possibly through memory interference.

Whether any of these studies can show that animals attribute mental states to other individuals is still a highly controversial issue. One criticism of all the studies mentioned above is that the animals in those studies simply base their strategies on associations formed during the experiment or in earlier life or simply read others' behavior and act based on that information. Instead of having some concept of seeing, animals may simply learn to (e.g., in the chimpanzee example) associate the eyes of their competitor with one piece of food and not the other. The stimulus 'eye' may be seen as an aversive stimulus which the subject then associates with the food and therefore avoids that particular piece (the so-called 'evil eye hypothesis'). In the examples with scrub-jays, the observing jays could have formed an association with two individuals with two trays (Bird A caching in Tray A, and Bird B caching in Tray B). As the cachers interacted only with one tray each, the observer may have formed associations based on the former events. Another, nonmentalistic interpretation of the results is that animals in these settings do not form concepts of other individuals' mental states but concepts about others' behavior and that this is sufficient to succeed in all paradigms used with animals so far. This concept of behavior is formed based on previous experience with conspecifics and may lead to certain rules like 'Every time I do X, the other individual does Y' or 'every time my competitor behaves like X, I am safe to do Y.' The fact that this area of research is still controversial and the results remain debatable suggests that more research is needed to answer the question of whether animals have an understanding of others' mental states rather than their behavior.

Understanding Others' Beliefs

One ability seen as a benchmark for mental state attribution and therefore theory of mind is the understanding that others have beliefs and that those beliefs can be true or false. Having an understanding that another individual's belief is false requires an understanding that another person's mental states can be contradictory to one's own mental states and, more importantly, contradictory to reality. So far there is no evidence that animals can make this distinction. In one version of a false belief task, chimpanzees were again confronted with a situation in which a subordinate and a dominant chimpanzee had to compete over food. In this setting, one piece of food was hidden from the dominant behind one of two opaque barriers. The subordinate chimpanzee approached the food if the dominant was uninformed (i.e., did not witness the placing of the food) but did approach less when he was informed (i.e., did witness placing of the food). Again the low-level explanation is that the subordinate, instead of attributing knowledge to the dominant, simply prefers food with which he does not associate the dominant's eyes.

In another version of the test, the dominant always witnessed the initial placement of the reward, but then the reward was moved to a second location. The dominant sometimes witnessed the moving of the food such that they were informed about its final location, but sometimes the dominant did not witness the moving, such that they were misinformed (potentially having a false belief). Interestingly, the subordinate chimpanzees did not distinguish between both situations and approached the food equally often.

Summary

Certain social cognitive skills like reading others' attentional state and following others' gaze direction seem to be relatively widespread throughout the animal kingdom. Such skills appear to be reflexive and possess a high survival value, for example potentially aiding in the rapid location of predators or avoiding conflict. Other skills, such as the ability to take anothers' perspective or understand what others' have seen in the immediate past do not seem to be so widespread and thus may be based on more complex cognitive operations. Although there is some evidence, albeit controversial, that apes and corvids form representations of other's knowledge states, such as what they may have seen in their immediate past, no non-human animal has, so far, demonstrated the ability to attribute false beliefs to others. This suggests that a truly representational theory of mind may be a uniquely human cognitive capacity.

Box: Ape and Corvid Social Cognition: An Example of Convergent Evolution

Two families, apes and corvids, seem comparable in their social cognitive capacities and superior to other animal species. One example is that chimpanzees (as a member of the apes) as well as scrub-jays (as a member of the corvids) have a very flexible understanding of what other individuals can and cannot see. Chimpanzees, as well as scrub-jays, seem to understand when a competing individual's vision is or is not blocked (e.g., by a barrier) and when their opponent cannot see a target object or them caching food items. In addition, and perhaps more impressively, chimpanzees and scrub jays also appear to understand and remember what potential competitors have seen in the immediate past. They memorize what another individual saw and develop their counter strategies, based on this knowledge accordingly. So far, chimpanzees, scrub-jays, and ravens (another corvid) are the only non-human species which seem to have quite flexible skills in this domain.

From an evolutionary perspective, this is interesting as chimpanzees and scrub-jays are very distantly related species, separated by around 300 million years of evolution. Therefore, their comparable cognitive skills suggest convergent rather than homologous evolutionary events. This means that those distantly related species evolved similar traits independently, but via similar evolutionary processes and most likely as adaptations to similar socioecological challenges, such as competition with conspecifics over food.

That apes and corvids have similar cognitive skills as a result of convergent evolution is supported by the fact that the morphology of the mammalian (ape) brain is completely different from the avian (corvid) brain. While the mammalian brain is laminated, with connections between and within layers, the avian brain is nucleated with connections between nuclei. It remains unclear how information is processed in this very differently structured neural system or how avian brains that are much smaller in absolute size than mammalian brains can process similar complex forms of information.

One reason why apes' and corvids' cognitive skills may be similar is that both groups of species face similar socioecological challenges, such as recognizing individuals, predicting their future actions, and using such statistical regularities of their behavior to deceive them. This may therefore be support for the social intelligence hypothesis formulated in 1976 by Humphrey.

See also: Apes: Social Learning; Conflict Resolution; Cooperation and Sociality; Culture; Deception: Competition by Misleading Behavior; Decision-Making: Foraging; Imitation: Cognitive Implications; Mammalian Social Learning: Non-Primates; Monkeys and Prosimians: Social Learning; Punishment; Referential Signaling; Sex and Social Evolution; Social Recognition.

Further Reading

Call J and Tomasello M (2008) Does the chimpanzee have a theory of mind? 30 years later. *Trends in Cognitive Sciences* 12: 187–192.

Dally JM, Emery NJ, and Clayton NS (2006) Food-caching western scrub-jays keep track of who was watching when. *Science* 310: 1662–1665.

Emery NJ (2000) The eyes have it: The neuroethology, function and evolution of social gaze. *Neuroscience & Biobehavioral Reviews* 24: 581–604.

Emery NJ and Clayton NS (2004) The mentality of crows: Convergent evolution of intelligence in corvids and apes. *Science* 306: 1903–1907.

Emery NJ and Clayton NS (2009) Comparative social cognition. *Annual Review of Psychology* 60: 87–113.

Hare B, Call J, Agnetta B, and Tomasello M (2000) Chimpanzees know what conspecifics do and do not see. *Animal Behaviour* 59: 771–785.

Heyes CM (1998) Theory of mind in nonhuman primates. *Behavioral and Brain Sciences* 21: 101–114.

Humphrey NK (1976) The social function of intellect. In: Bateson PPG and Hinde RA (eds.) *Growing Points in Ethology*, pp. 303–317. Cambridge, UK: Cambridge University Press.

Jarvis ED, et al. (2005) Avian brains and a new understanding of vertebrate brain evolution. *Nature Reviews Neuroscience* 6: 151–159.

Kaminski J, Call J, and Tomasello M (2008) Chimpanzees know what others know but not what they believe. *Cognition* 109: 224–234.

Miklósi Á, Topál J, and Csányi V (2004) Comparative social cognition: What can dogs teach us? *Animal Behaviour* 67: 995–1004.

Penn DC and Povinelli DJ (2007) On the lack of evidence that non-human animals possess anything remotely resembling a 'Theory of Mind'. *Philosophical Transactions of the Royal Society of London B* 362: 731–744.

Povinelli DJ and Eddy TJ (1996) What young chimpanzees know about seeing. *Monographs of the Society for Research in Child Development* 61: 1–152.

Premack D and Woodruff G (1978) Does the chimpanzee have a theory of mind? *Behavioral and Brain Sciences* 1: 515–526.

Whiten A and Byrne RW (1988) The Machiavellian intelligence hypotheses: Editorial. In: Byrne RW and Whiten A (eds.) *Machiavellian Intelligence: Social Expertise and the Evolution of Intellect in Monkeys, Apes, and Humans*, pp. 1–9. New York, NY: Clarendon Press/Oxford University Press.

Social Evolution in 'Other' Insects and Arachnids

J. T. Costa, Western Carolina University, Cullowhee, NC, USA; Highlands Biological Station, Highland NC, USA

Introduction

The 'other' social insects and arachnids comprise a heterogeneous group that includes three primary forms of social structure. Simple parental care (maternal, paternal, and biparental care) involves a range of parent–offspring associations but no division of labor. Resource-defense societies are built around protection of a valuable commodity necessary to the survival of the family – typically a nesting site that also provides sustenance, as in galls. These societies consist of simple family groups with a defense-based division of labor among siblings. Finally, herd-like colonies consisting of either all juveniles (larvae or nymphs) or a combination of adults with juveniles are found in many insect taxa; these societies often consist of extended families or mixtures of unrelated family groups.

This diversity poses a challenge to any simple classification of these societies, and they likely represent multiple pathways of social evolution (discussed later). The treatment of these groups in an 'other' category in the present work reflects the history of our efforts to understand social evolution. These societies have traditionally been classified relative to the most complex social forms, the eusocial species, and as such their primary commonality is the lack of eusociality, rather than any uniquely defining characteristics. This approach to the classification of noneusocial societies was adequate as long as their social diversity remained undiscovered. Beginning in the 1970s, however, and continuing through the 1990s, a number of unusual and new social forms were described from several insect orders and other arthropod taxa, and at the same time other such societies, long known, were found to possess a greater degree of complexity than previously realized. Today, there is no universally accepted classification for these 'other' societies. There have been several proposals to redefine some of the categorical terms used to describe them, but to date no one approach has gained general acceptance.

Classification

Social behavior in insects and their relatives is taxonomically widespread, particularly under the definition of sociality given by Edward O. Wilson in his 1971 book *The Insect Societies* suggesting that social species exhibit 'reciprocal communication of a cooperative nature.' This general definition includes the full range of social forms that have traditionally (i.e., through much of the twentieth century) been arrayed along a scale from *solitary* (nonsocial) to *eusocial*, the latter being the term applied to the most sophisticated insect societies (**Table 1**). 'Eusocial' was coined in 1966 by entomologist Suzanne Batra, in a study discussing sociality in halictid bees, a behaviorally variable group. Eusociality has traditionally been defined by three criteria: co-occurrence of parents and offspring in a colony (overlapping generations), care for offspring performed by both parents and nonparents of those offspring (cooperative brood care), and unequal reproduction among colony members, such that most colony members forego some or all personal reproduction in favor of caring for the offspring of others, often termed 'queens' (reproductive division of labor). Traditionally, this term was applied strictly to termites (Dictyoptera, suborder Isoptera), ants, and some bees and wasps (Hymenoptera, suborder Apocrita). The noneusocial forms of sociality were given terms such as *subsocial, quasisocial, parasocial,* and *communal,* depending on the degree of parent–offspring interaction exhibited relative to the eusocial state. Some of these terms, such as 'subsocial,' are still commonly used, but most are not.

Wilson's general definition of sociality is useful in differentiating ephemeral or incidental groups of animals – *aggregations* – from social groups per se. Aggregations might form, for example, by the common attraction of individuals to a common resource such as food or a breeding area; spawning salmon is one such aggregation, and another is that of the bears that gather to fish them. Individuals in these circumstances aggregate for reasons other than interacting cooperatively with each other, forming 'selfish herds.' (Note that groups that simply gather to breed or display are not considered societies, which is not to say the behavior exhibited is not a form of social interaction. The common application of the word 'colony' to both breeding aggregations and social groups can create confusion on this point.) The 'selfish herd' concept was introduced by evolutionary biologist William D. Hamilton in 1971. Hamilton pointed out that animals may benefit in several ways from grouping, largely as a result of increased vigilance and a reduction in per capita predation probability. In some cases, it behooves individuals in selfish herds to jockey for the optimum position to either avoid predation or gain information on resources or mates.

Social groups, in contrast to aggregations or selfish herds, exhibit some degree of behavioral integration or coordination directed toward a common goal; these might include, for example, shelter or nest building and hygiene,

Table 1 Summary of the traditional social insect classification that emerged following the formal definition of eusociality in 1966

Category	Defining traits
Subsocial	Parental care of eggs or young
Communal	Members of the same generation co-occur in colony
	No brood care
Quasisocial	Members of the same generation co-occur in colony
	Cooperative brood care
Semisocial	Members of the same generation co-occur in colony
	Cooperative brood care
	Reproductive division of labor
Parasocial	Used in place of Communal + Quasisocial + Semisocial
Presocial	Used in place of Subsocial + Communal + Quasisocial + Semisocial
Eusocial	Overlapping generations
	Cooperative brood care (offspring and parents engage in brood care)
	Reproductive division of labor

Adapted from Wilson EO (1971) *The Insect Societies*. Cambridge, MA: Belknap/Harvard University Press.

finding and/or retrieving food, defending the group, caring for young, or physiological regulation (notably thermoregulation). What makes such behaviors cooperative, however? In this context, the word 'cooperative' refers to individuals acting in common, with or without mutual modulation of behavior such that, as a group, tasks are executed more readily than can be achieved by single individuals acting alone, or produce an end product (such as a nest) that benefits the entire group, that cannot be produced by individuals acting alone.

With this in mind, the significance of 'reciprocal communication of a cooperative nature becomes clearer: communication is the means by which individuals in social groups coordinate or direct their behavior. In social insects, this is most often achieved through chemical communication (chemotactile, trail-based, or airborne pheromones), or acoustic communication (substrate-borne vibrational signals) – think of communication as the glue that holds such societies together.

Three Social Structures, Four Evolutionary Pathways

The three forms of social structure found in the 'other' insect societies – parental care, resource defense, and herd-like colonies – represent four distinct evolutionary pathways of sociality. Parental care as a behavioral phenomenon is sufficiently general in that it can be achieved via different evolutionary routes. In this case, it is thought that exclusive maternal and biparental care proceeds along an evolutionary pathway distinct from that of exclusive paternal care (**Table 2**).

Maternal and Biparental Care

Maternal and biparental care (**Figures 1** and **2**) is extremely widespread among insects and other arthropods, and has been hypothesized to arise in several ecological circumstances. Edward O. Wilson first discussed this in his 1975 book *Sociobiology*, when he proposed 'prime environmental movers' that favored the evolution of increased parent–offspring interaction. These prime movers include (1) stable, structured habitats, (2) unusually stressful physical environments, (3) scarce, specialized food sources, and (4) predation pressure. All four promote some degree of prolonged parent–offspring coexistence and intimacy (care in the form of defense, nest building, provisioning, etc.), albeit for quite different reasons. For example, stable, structured habitat and scarce resources are thought to promote *k*-selected strategies that require extended cohabitation; rotting wood, for example, is an abundant but nutrient-poor resource that requires gut symbionts for exploitation, a condition that may favor extended parental interaction in wood specialists. Ephemeral but high-value resources such as dung or carrion may necessitate parental care for different reasons, in this case for rapid exploitation and to compete effectively with rivals for utilization of the resource. Finally, the risk associated with searching for a patchy resource could favor central-place nesting and parental provisioning behavior.

Entomologists Douglas Tallamy and Thomas Wood were the first to suggest that maternal and biparental care behaviors arose convergently in many insect taxa primarily due to resource use. Their *resource hypothesis* posited that the quality, persistence, and spatial distribution of resources such as plants and plant derivatives (foliage, fruit, flowers, wood, sap, pollen) as well as fungi, carrion, and dung play an important role in the evolution of parental care behavior, notably maternal and biparental care (see **Box 1** for a description of these resources and the social structures associated with them).

An alternative view of the ecology and evolution of maternal care behavior (and to a lesser extent biparental care) is based on reproductive strategy. Many species exhibiting these forms of sociality are *semelparous*, where females reproduce at more or less a single period in their lifetime, producing a single large clutch of eggs. This contrasts with *iteroparous* reproduction, with multiple smaller clutches spread out over a period of time. The *semelparity hypothesis* of maternal care evolution posits that since semelparous species have a great deal invested in their one clutch, they can best maximize their reproductive success by remaining with their eggs, and later, defending sheltering, and/or feeding the young and so

Table 2   Expressions of sociality by order and family in the 'other' insect societies and some noninsect arthropod groups

Order/family	Maternal care	Biparental care	Paternal care	Resource defenders	Larval/nymphal colonies	Mixed family colonies
Dermaptera (Earwigs)						
Labiduridae	X					
Forficulidae	X					
Anisolabididae	X					
Spongiforidae	X					
Orthoptera (Grasshoppers and Crickets)						
Romaleidae					X	
Acrididae					X	X
Pyrgomorphidae					X	X
Gryllide	X					
Gryllotalpidae	X					
Embiidina (Webspinners)						
Oligotomidae	X					X
Embiidae	X					
Anisembiidae	X					X
Clothodidae	X					X
Mantodea (Mantids)						
Mantidae	X					
Blattodea (Cockroaches)						
Blaberidae	X	X				X
Cryptocercidae		X				X
Psocoptera (Barklice)						
Peripsocidae	X					
Psocidae						X
Archipsocidae						X
Zoraptera						
Zorotypidae						X
Sternorrhyncha (Aphids)						
Pemphigidae				X		
Hormaphididae				X		
Aphididae						X
Auchenorrhyncha (Tree-, Plant-, and Froghoppers)						
Membracidae	X	X			X	X
Heteroptera (True Bugs)						
Acanthosomatidae	X					
Pentatomidae	X					
Cydnidae	X					
Coreidae	X		X		X	
Phloeidae	X					
Reduviidae	X		X			
Tingidae	X				X	
Belostomatidae			X			
Gerridae						X
Veliidae						X
Thysanoptera (Thrips)						
Phlaeothripidae	X			X		X
Coleoptera (Beetles)						
Scarabaeidae	X	X				
Passalidae		X				
Staphylinidae	X					
Silphidae		X				X
Curculionidae	X	X			X	X
Chrysomelidae	X				X	
Erotylidae	X				X	
Tenebrionidae	X	X				
Diptera (Flies)						
Syrphidae					X	
Mycetophilidae					X	

Continued

Table 2   Continued

Order/family	Maternal care	Biparental care	Paternal care	Resource defenders	Larval/nymphal colonies	Mixed family colonies
Lepidoptera (Moths and Butterflies)						
Eupterotidae					X	
Saturniidae					X	
Lasiocampidae					X	
Notodontidae					X	
Lymantriidae					X	
Arctiidae					X	
Limacodidae					X	
Zygaenidae					X	
Tortricidae					X	
Geometridae					X	
Pyralidae					X	
Yponomeutidae					X	
Papilionidae					X	
Pieridae					X	
Lycaenidae					X	
Riodinidae					X	
Nymphalidae	X				X	
Hymenoptera						
Halictidae	X					
Pergidae	X				X	
Argidae	X				X	
Diprionidae					X	
Tenthredinidae					X	
Pamphiliidae	X				X	
NONINSECT ARTHROPOD GROUPS						
ARACHNIDA (spiders and allies)						
Araneae (Spiders)						
Theridiidae	X					X
Dictynidae						X
Eresidae						X
Oxyopidae						X
Thomisidae						X
Agelenidae						X
Sparassidae						X
Amblypygi (Tailless Whipscorpions)						
Amblypygidae	X					
Opiliones (Harvestmen)						
Cosmetidae	X					
Cranaidae	X					
Stygnopsidae	X					
Gonyleptidae	X		X		X	
Manaosbiidae			X			
Assamiidae			X			
Podoctidae			X			
Pseudoscorpionida (Pseudoscorpions)						
Chthonidae	X					
Atemnidae						X
Scorpionida (Scorpions)						
Vaejovidae	X					
Scorpionidae	X					X
Ischnuridae						X
Diplocentridae						X
Acari (Mites)						
Tetranychidae	X					X
Cheyletidae						X
CHILOPODA (Centipedes)						
Scolopendridae	X					
Geophilidae	X					

Continued

Table 2   Continued

Order/family	Maternal care	Biparental care	Paternal care	Resource defenders	Larval/nymphal colonies	Mixed family colonies
Mecistocephalidae	X					
DIPLOPODA (Millipedes)						
Andrognathidae			X			
Platydesmidae						X
CRUSTACEA: MALACOSTRACA (Shrimp, Crabs, and Allies)						
Isopoda	X	X				
Decapoda	X			X		X
Amphipoda	X					

The table includes taxa in which the indicated forms of social structure have been well documented in the scientific literature; it is not exhaustive, and sociality is likely to occur in many insect and noninsect groups not listed.

Figure 1 Brood guarding in the rare Australian beetle *Pterodunga mirabile* (Chrysomelidae: Chrysomelinae). Females are live bearing (viviparous), with young produced over a period of time leading to mixed-age cohorts that she guards. This species feeds on plants in the family Proteaceae. Photograph courtesy of Mr. Jack Hasenpusch, Innisfail, Queensland, Australia.

increase their chances of survival. She has little to lose since it is unlikely she will produce another clutch, and much to gain by helping ensure the success of the brood she produced.

Paternal Care

Examples of exclusive paternal care are rare in comparison with maternal and biparental care (**Table 2**). Among insects, this behavior is found in giant water bugs (Heteroptera: Belostomatidae), a few genera of assassin bugs (Heteroptera: Reduviidae), and a single genus of leaf-footed bug:

the 'golden egg bug' *Phyllomorpha* (Heteroptera: Coreidae) found in the circum-Mediterranean region. Paternal care is also known from a few noninsect arthropods, including two millipede genera in the family Andrognathidae and a few species of harvestmen (Opiliones) representing several families (**Figure 3**). Paternal care in these groups typically takes the form of egg guarding, though some, such as brooding giant water bugs, must also care for the eggs so as to prevent drowning or desiccation. In one giant water bug subfamily (Belostomatinae), females cement their eggs to the back of the male, who must forgo flying while thus encumbered. These male water bugs rhythmically pump their abdomen and thorax to keep the eggs moist and aerated. In another subfamily (Lethocerinae), clutches are deposited on emergent aquatic vegetation, and males periodically crawl to the eggs to 'water' them.

Evolutionarily, exclusive paternal care arises due to enhanced reproductive opportunities for males. Sexual selection favors female choice of males that exhibit parenting prowess, in this case demonstrable success in defending broods, securing a territory or resources, etc. – behaviors that may reflect honest signals of male fitness. That is, females compete for mates with demonstrable parental abilities. This dynamic should result in males caring for eggs or young regardless of whether they sired them; if larger clutches or broods under a male's care lead to more subsequent matings for that male, he should tolerate or even seek to acquire broods to care for even if they are not his. This expectation is borne out for some groups: brooding male assassin bugs and harvestmen have been shown to tolerate clutch contributions from females they have not mated with and can usurp the clutches of rival males. Brooding giant water bugs, on the other hand, do not show such tolerance. In cases such as this where males have limited ability to care for eggs – having only so much space on their backs, say – they have no tolerance for eggs of females he has not mated with. The male-care phenotype may arise more via natural than sexual selection in this group.

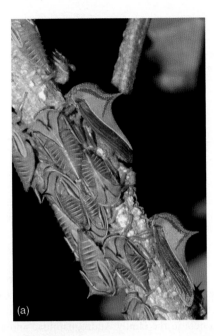

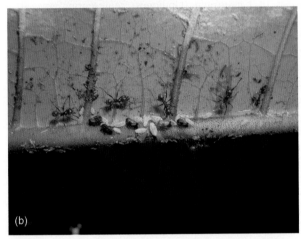

Figure 2 Treehopper colonies (Auchenorrhyncha: Membracidae). (a) *Umbonia spinosa*, a 'thorn bug' treehopper from South American that forms family groups lasting a single generation. This species, which does not associate with ants, initially forms parent–offspring colonies. Parental care takes the form of defense, though in some treehopper species, the mother will also facilitate nymph feeding by perforating the host plant with her ovipositor. *Umbonia spinosa* nymphs continue to associate to adulthood after the death or departure of the mother. Groups of adults and nymphs as seen here thus occur together only during eclosion, when some individuals have reached the adult stage and some have not yet done so. Colonies are maintained through the use of vibrational signals, detected through the hostplant, produced in synchronized bursts by drumming the substrate. (b) *Aphetea inconspicua*, another neotropical treehopper, seen here feeding along the midrib on the underside of a leaf while being tended by ant mutualists. Colonies consist of mixed groups of adults and nymphs, likely representing a merger of multiple families. Colonies accordingly persist for multiple generations and so are often found on the same hostplant over a period of years. Photographs courtesy of Dr. Rex Cocroft, University of Missouri.

Resource ('Fortress') Defenders

Three distantly related arthropod groups have converged on a most remarkable social lifestyle. These are gall thrips (Thysanoptera: Phlaeothripidae), gall aphids (Sternorrhyncha: Hormaphididae, Pemphigidae, and a few Aphididae), and snapping shrimp (Decapoda: Alpheidae), groups in which specialized soldier or defender morphs have evolved in the defense of a static and critical resource, often a nest site that doubles as food source. All three taxa exhibit a range of social structures, the most striking of which involve a morphological division of labor into soldiers and nonsoldier forms. Thrips and aphid soldiers have appendages that are enlarged and thickened relative to those of nonsoldiers in the same colony; these are often armed with spines. Similarly, snapping shrimp soldiers bear enlarged chelae. These social groups do not exhibit parental care or much general social interaction; rather, sociality centers on defense of a valuable and largely irreplaceable resource: the colony domicile. For this reason, this form of sociality is termed *resource-defense sociality*. In some cases, other forms of cooperation are evident. Some social aphids, for example, engage in nest hygiene, expelling honeydew globules that accumulate as a result of the way these insects feed on sap and excrete excess water.

The domicile of social thrips and aphids (**Figure 4**) is produced by a foundress that induces development of a hollow gall on young stem or leaf tissue of the host plant. Her brood colony will develop within the safety of the hollow gall, feeding on the plant tissue from within. If the gall is breached, however, the defender morphs will emerge and patrol the surface, aggressively attacking any intruder they encounter. These species are beset by a host of predacious insects such as syrphid fly larvae; gall thrips also face severe pressure by usurping kleptoparasitic thrips species that kill the foundress and make the gall their own. The domicile of snapping shrimp consists of crevices in sponges, a resource that is less ephemeral than galls but also highly susceptible to usurpation. All of these domiciles represent a valuable, if not critical, commodity that is not easily replaced if lost.

Mortality stemming from predation and nest usurpation is likely the most important ecological pressure in the evolution of defender morphs, though it is important to point out that kin selection, in the case of thrips and aphids, may play a role as well. Soldier-producing thrips species, which like all members of the order Thysanoptera are haplodiploid, often exhibit high levels of inbreeding. Aphid colony mates are clonally produced by the foundress, a mode of asexual reproduction called 'parthenogenesis' that is found in virtually all aphids and related groups. In social species of both thrips and aphids, there is, accordingly, a high degree of genetic relatedness between colony mates: an average of 0.75 in social thrips colonies, and 1.0 in colonies

---

Box 1 Nutritional Strategies Commonly Associated with Maternal and Biparental Care Behavior

Herbivores such as foliage, fruit, and sap feeders that feed exposed are subject to heavy predation and parasitism. Maternal guarding behavior (protecting eggs and/or young) has arisen convergently in many taxa, including treehoppers (Membracidae), stink bugs (Heteroptera: Pentatomidae), leaf beetles (Coleoptera: Chrysomelidae), and several sawfly taxa (Hymenoptera: Symphyta). Others sequester their young in sheltered nests or burrows and either provision the nest or guard the young while foraging. Examples include earwigs (Dermaptera), burrower bugs (Heteroptera: Cydnidae), and mole crickets (Orthoptera: Gryllotalpidae).

Wood feeders often live in galleries excavated within their food source. Some can produce cellulolytic enzymes endogenously, while others rely on symbiotic organisms. Examples of biparental care include beetles of the family Passalidae, the adults of which masticate wood for larval feeding and help construct pupal chambers for their offspring, and wood roaches of the family Cryptocercidae and Blaberidae, which inoculate their young with gut symbionts. Bark beetle colonies (family Curculionidae) are typically maternal; these insects inoculate the walls of their wood galleries with fungi, which serves as the derivative nutritional source for the young.

Dung and carrion feeders often sequester their rich but ephemeral resource in subterranean nests, reducing the chances of losing the resource to usurping competitors. Some dung beetles (family Scarabaeidae) carve out dung balls and roll them some distance before burying them by undermining. Carrion-specializing beetles (family Silphidae) also bury their resource by undermining. The perishable meat is then processed with antimicrobial and digestive salivary secretions to condition it for larval feeding and prevent it from rotting.

Figure 4 The North American gall aphid *Pemphigus obesinymphae* (Sternorrhyncha: Pemphigidae), which induces galls on the petiole of *Populus fremontii*. Note soldiers patrolling gall surface. Photograph courtesy of Dr. Patrick Abbot, Vanderbilt University.

Figure 3 The neotropical harvestman *Ampheres leucopheus*, one of several Opiliones species with exclusive paternal care. This male's clutch could be the product of one or more females, which compete for access to him. Exclusive paternal care is a rarity in arthropods best developed in certain harvestmen, millipedes, and true bugs. Photograph courtesy of Glauco Machado, Sao Paulo University.

of social aphids. This may favor kin selection in these groups, in which indirect fitness – realized through the successful reproduction of relatives – more than compensates the loss of direct fitness that results from altruistic sacrifice (soldier morphs reproduce less and have a greater chance of dying as a result of colony defense).

Herds

'Herding' insect societies come in two forms: juvenile-only *larval/nymphal herds* and mixed *adult–juvenile herds*, reflecting the demographic makeup of the colony and its temporal persistence. In both cases, the colony may consist of family members or may be mixed-family groups of unrelated individuals formed by colony merging. Social interaction in juvenile herds is limited to the immature stage with strictly solitary adults, while mixed adult–juvenile herds are more like their ungulate counterparts with one or more adults and juvenile cohorts co-occurring in the herd. By definition, societies of immatures are more ephemeral than those that include adults, in that the group dissolves at adulthood. Larval/nymphal herd social structure is found in many species of caterpillars (Lepidoptera), sawflies (Hymenoptera: Symphyta), leaf beetles (Coleoptera: Chrysomelidae), short-horned grasshoppers (Orthoptera: Acrididae), and even some syrphid flies (Diptera: Syrphidae). What these groups have in common is that they are all phytophagous (even the herding syrphids, a fly family that is more typically predacious) (**Figures 5–7**).

Exposed feeding by immatures presents a suite of challenges – ultimately, the need to procure sufficient

Figure 5 The 'Ugly Nest Caterpillar' *Archips cerasivoranus* (Lepidoptera: Tortricidae), a patch-restricted forager of North America that feeds on cherry (*Prunus*). This species imparts a trail pheromone to the silk as it is extruded from the spinnerets. Their colonies sometimes grow to thousands of individuals as a result of merging of multiple independent family groups. Photograph courtesy of Dr. Terrence D. Fitzgerald, SUNY College at Cortland.

(a)

(b)

Figure 6 'Spitfire' sawflies, *Perga affinis* (Hymenoptera: Pergidae), found throughout eastern Australia. (a) These sawflies form large nomadic colonies on their *Eucalytpus* host trees. They are well defended, sequestering sap and chemicals derived from their hostplant (in this case, eucalyptus oil) in gut structures called 'diverticulae,' and ejecting the liquid when threatened. Colony structure is maintained through the use of substrate-borne vibrations that the larvae generate by tapping the tree with their sclerotized anal plate. (b) Early instars exhibit cycloalexic 'circle-the-wagons' behavior, thought to be a defensive formation. Photographs courtesy of Dr. Lynn Fletcher, Franklin College.

resources for development while avoiding predation. Grouping in these insects is accordingly thought to have positive effects on group defense, thermoregulation, foraging, and/or pupation success. Aposematic coloration and antipredator displays, for example, may be more effective in groups by amplifying signals, rendering groups more apparent or threatening, and grouping also results in a dilution effect whereby per capita predation rate declines. Larvae working in groups may be more efficient at shelter construction, expending less energy per capita and making a more robust structure en masse. Similarly, the thermal benefits of group basking, both in terms of the rate of heating and the conservation of body heat by lateral transfer, is in part a function of grouping and group size, and success at excavating subterranean pupation chambers in some group-pupating sawflies is also related to group size.

Expressions of sociality in larval/nymphal societies often relate to foraging strategy: *patch-restricted foragers* typically nest in or on a food patch, in many cases constructing a silk shelter that encompasses the patch. *Nomadic foragers* are ranging herds, typically constructing no shelter and often feeding exposed as they move among patches. Species in this category often have aposematic coloration or sport defenses such as spines and hairs. *Central-place foragers* nest in a more or less fixed location and venture forth solitarily or in groups to procure food, returning to the shelter after each foraging bout. These species, too, tend to be well defended chemically and mechanically (see **Box 2** for representative species exemplifying each foraging strategy).

Degree of social complexity, as gauged by communication and the suite of collective behaviors contributing to the colony, increases from patch-restricted to nomadic

to central-place foraging. Among social caterpillars and sawflies, for example, patch-restricted foragers appear to engage in little social interaction beyond employing trail or arena-marking pheromones that promote group cohesion. Nomadic foragers often employ trail pheromones for group cohesion as well, while some groups (e.g., sawflies of the family Pergidae) employ substrate-borne vibrational cues to accomplish cohesion. A rare and remarkable form of locomotion apparently restricted to nomadic foragers is processionary movement, characterized by head-to-tail trains of individuals that remain in continuous or near-continuous physical contact. Species exhibiting processionary movement include the neotropical silkmoth *Hylesia lineata* (Saturniidae), the processionaries

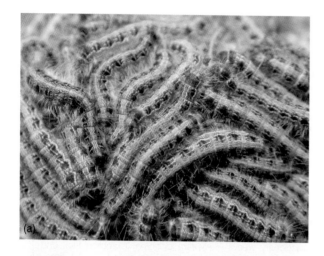

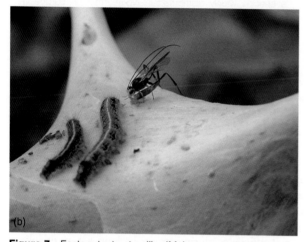

Figure 7 Eastern tent caterpillar (*Malacosoma americanum*, Lasiocampidae), a central-place forager that employs trail-based recruitment communication to cooperate in locating high-quality feeding sites. (a) Closeup of caterpillars basking on the tent surface. Group basking can raise caterpillar body temperatures considerably above ambient even under freezing or near-freezing conditions. (b) Stragglers left on the tent surface are targets for parasitoids such as this ichneumonid wasp. The tent provides protection from predators and parasitoids, and there is also safety in numbers when the colony is grouped in this manner: isolated individuals are more easily picked out by predators and parasitoids. Photographs courtesy of Susan Roberts, Haywood Community College.

moths (*Thaumetopoea* spp., Notodontidae: Thaumetopoeinae) of central and southern Europe, and the neotropical weevil *Phelypera distigma* (Curculionidae: Hyperinae). Processionary behavior has been reported from just two species outside the Insecta, the western Atlantic spiny lobster *Panuliris argus*, and an early Cambrian crustaceanomorph arthropod. Maintenance of processionary columns involves chemotactic cues associated with setae at the tip of the abdomen. Given the widespread occurrence of trail following behavior among nomadic foragers, it is not clear why this form of locomotion has arisen only rarely.

Group antipredator displays are common among nomadic foragers, but whether or how alarm signals are spread among colony mates (as opposed to being individually directed toward threats) is poorly understood. Some group-displaying species such as diprionid sawfly larvae achieve remarkable synchrony in their rapid body-flicking display, and it would be interesting to learn if larvae are rapidly responding to signals from sentinel colony mates, or to external threats. Such displays may serve to intimidate visual predators such as birds, or provide a moving target to parasitoids flies and wasps, making oviposition difficult.

Central-place foragers, which also exhibit group displays, employ a mode of foraging that sets the stage for recruitment communication – a form of cooperation found in very few larval societies. This behavior is best studied in tent caterpillars of the north-temperate lasiocampid genera *Malacosoma* and *Eriogaster*, caterpillars of which use trail-based recruitment trails to communicate the location of profitable patches to colony mates. *Malacosoma* remains the only social caterpillar for which a trail pheromone has been characterized; this species employs the steroid 5β-cholestane-3,24-dione, perhaps derived from a phytosterol produced by the host plant. *Malacosoma americanum* employs its trail pheromone in a two-tiered system consisting of exploratory and recruitment phases, with strength of overmarking recruitment trails related to the nutritional quality of the food source. Though the caterpillars produce silk trails, the silk alone does not elicit trail following but must be overmarked with pheromone secreted from a site between the anal prolegs. These caterpillars are sensitive to pheromone concentrations as small as 10^{-11} g l^{-1}.

Mixed adult–juvenile herds tend to be temporally stable colonies and often consist of amalgamations of multiple family units. This form of sociality is found in a diverse array of arthropods, including species of barklice (Psocoptera), treehoppers (Hemiptera: Membracidae), true bugs (Heteroptera), thrips (Thysanoptera), short-horned grasshoppers (Acrididae), webspinners (Embiidina), and many spiders (Araneae) (**Table 2**). Maintenance of social structure in most of these groups is poorly studied. Trail pheromones have been shown to play a role in maintaining colonies of barklice, and treehoppers communicate in the same manner as pergid sawflies by sending vibrational signals through their host plant. In many treehoppers, this form of communication plays a dual role, serving to both promote group cohesion and signal their presence, or alarm, to ants. (Treehoppers, such as many sap-sucking insects excrete sugar-rich honeydew highly prized by ants. Mutualistic interactions between the honeydew secreting insects and ants have evolved in many such groups. The ants aggressively defend the sap-sucking insects as a resource and thereby protect them from predators.) Social spiders, too, likely communicate via silk-borne vibrations. Sociality in these mixed-family groups often takes the form

---

Box 2   Representative Species of 'Herding' Caterpillars, Sawflies, and Beetles

• **Patch-restricted foragers**

Site-limited, constructing shelters of leaves or enveloping leaves in silk and feeding on leaves from within shelter. Patch-shelter is expanded to incorporate more leaves as they are consumed. The best studied patch-restricted foraging caterpillars include the fall webworm *Hyphantria cunea* (Arctiidae), cherry scallop-shell moth *Hydria prunivoranus* (Geometridae), ugly nest caterpillar *Archips cerasivoranus* (Tortricidae), and the common ermine moth *Yponomeuta cagnagellus* (Yponomeutidae). The best sawfly examples come from the web-spinning sawfly family Pamphiliidae, such as *Neurotoma* spp. There are no known examples of patch-restricted larval beetles.

• **Nomadic foragers**

Wandering groups that move from patch to patch, typically feeding exposed and constructing no shelter; the most widespread form of larval/nymphal society. Common and well-studied nomadic foraging caterpillars include the North American forest tent caterpillar *Malacosoma disstria* (Lasiocampidae), oakworms of the genus *Anisota* (Notodontidae), the arsenurine tropical silkworm *Arsenura armida* (Saturniidae), and processionary caterpillars of the genera *Hylesia* (Saturniidae) and *Thaumetopoea* moths (Notodontidae). Sawfly examples include the red-headed pine sawfly *Neodiprion lecontei* (Diprionidae) of North America, and Australian spitfire sawflies of the genus *Perga* (Pergidae). Beetle examples include the neotropical processionary weevil *Phelyptera distigma* (Curculionidae), and the imported willow leaf beetle *Plagiodera versicolora* (Chrysomelidae), a species widely distributed in North America, Europe, and Asia. A good orthopteran example is the nymphal colonies of the lubber *Romalea guttata* (Romaleidae), common in the southeastern United States.

• **Central-place foragers**

Static, long-term nest site is established (typically constructed of silk and/or leaves) from which larvae emerge to forage, often in synchronized bouts. This rarest foraging strategy of social larvae appears to be found only in the Lepidoptera. The best-studied groups include several Lasiocampidae, among them the tent caterpillars *Malacosoma americanum* and *M. neustria*, the small eggar moth *Eriogaster lanestris*, and the Australian bag-shelter moth *Ochrogaster lunifer*. The only known central-place foraging butterfly is the Mexican species *Eucheira socialis* (Pieridae).

---

of cooperative prey capture, with multiple individuals joining forces to subdue prey much larger than themselves, and parental care in that juveniles share in the food captured by the adults.

Both larval–nymphal herds and mixed adult–juvenile herds can arise in two ways: (1) *communal oviposition*, where two or more females oviposit or less simultaneously into a common batch (e.g., *Peripsocus quadrifasciatus* (Peripsocidae), and several species of *Heliconius* butterfly (Nymphalidae)); and (2) *group merging*, where broods hatched from initially separate clutches coalesce to form larger, combined family groups (e.g., the red-headed pine sawfly *Neodiprion lecontei* (Diprionidae), and eastern tent caterpillar *Malacosoma americanum* (Lasiocampidae)). Both these phenomena are likely to be more common than is currently thought. This is particularly true of group merging, based on reports of colony sizes that exceed, sometimes considerably, the size of those expected based on knowledge of typical clutch size.

Note that even a modest degree of family merging would reduce intracolony relatedness in the herding social insects. This suggests that maintenance of family structure in these societies may be unimportant relative to gains realized by other correlates of group merging, especially group size. There are few empirical tests of this. Studies with eastern tent caterpillars demonstrated a strong group size effect, where larvae reared in groups differing in size but controlled for genetic background attained significantly larger body mass developing in

larger versus smaller colonies. Caterpillars also exhibited less individual variation in growth in larger colonies. Likely, group size has important consequences for several aspects of larval biology: nest (tent) building rate and thermal efficiency during group basking are positively affected by group size, as is foraging efficiency, or the efficacy of group defense. The potential costs are greater apparency to predators, greater likelihood of disease transmission, and competition for food resources. Insofar as many of the herding societies show evidence of group merging or communal oviposition by unrelated females, the ecological benefits of 'supercolony' formation are considerable. Moreover, the ready merging of unrelated family units suggests that kin selection plays little-to-no role in the evolution of sociality in these groups, insofar as ecological circumstances may promote grouping and cooperation regardless of relationship among colony mates.

A range of biotic and abiotic ecological pressures, including predation, parasitism, disease, resource dispersion, and thermal regime and its effects on assimilation and growth, likely play a role in the evolution of sociality in the herding societies, albeit in varying combinations and degrees of importance. It is important to stress that it is very difficult to test the assumption that sociality enables these larvae to meet the challenges and opportunities of their environment. Consider that many nonsocial congeners or confamilials co-occur in the same environment with social ones. The historical factors that lead to sociality in some lineages and not others are perhaps impossible to assess, and the empirical work aimed at elucidating the pressures faced by social species may only tell us about the possible maintenance, not origin, of sociality in this or that group. However, a series of studies undertaken by Sillén-Tullberg and colleagues does point the way to a historical approach to some relevant questions. These investigators employed used phylogeny to assess the relationship between aposematism and sociality (gregariousness) of caterpillars, and found that aposematism appears to evolutionarily precede sociality in many groups, a finding that supports the idea that caterpillar sociality has arisen in the context of amplifying warning signals.

Conclusions

The remarkable diversity in *expressions* of sociality in the 'other' societies of insects and their allies summarized in this review is itself a strong argument for multiple pathways of social evolution. Whether simple parent–offspring groups, extended families, cohorts of unrelated juveniles, or complex mixtures of adults and offspring varying in familial affinity, what these societies fundamentally have in common is more or less reciprocal and significant influence of colony members on one another's fitness. This is something that ephemeral aggregations lack. The ultimate evolutionary currency is reproductive success, and it is instructive to consider sociality as a strategy, albeit one that reflects the circumstances of ecology and constraints of history in ways that are often difficult to identify. Studies of 'other' social arthropods have progressed considerably beyond the simplistic assumption that social evolution proceeds in simply ladder-like fashion from solitary through eusocial. Yet a complete understanding of the ecology and evolution of these 'other' societies is critical to fully understanding eusociality as well. The study of the noneusocial societies has lagged considerably behind that of the more complex eusocial groups; although much is known about a handful of them, the field is still very much in a natural history phase where fundamental knowledge of life history, behavior, and ecology has yet to be realized for the vast majority of groups. Empirical studies of communication mechanisms, life history, colony genetic structure, behavioral

repertoire, and more are needed to fill in the gaps in our knowledge of most of the 'other' social insect groups.

See also: Ant, Bee and Wasp Social Evolution; Cockroaches; Colony Founding in Social Insects; Communication: An Overview; Cooperation and Sociality; Crustacean Social Evolution; Foraging Modes; Group Living; Group Movement; Kin Recognition and Genetics; Olfactory Signals; Parent–Offspring Signaling; Recognition Systems in the Social Insects; Social Information Use; Social Insects: Behavioral Genetics; Social Recognition; Spiders: Social Evolution; Subsociality and the Evolution of Eusociality; Termites: Social Evolution; Vibrational Communication; Vigilance and Models of Behavior; Wintering Strategies: Moult and Behavior.

Further Reading

Choe JC and Crespi BJ (eds.) (1997) *The Evolution of Social Behavior in Insects and Arachnids.* Cambridge, UK: Cambridge University Press.

Cocroft RB and Rodriguez RL (2005) The behavioral ecology of insect vibrational communication. *BioScience* 55: 323–334.

Costa JT (2006) *The Other Insect Societies.* Cambridge, MA: Belknap/Harvard University Press.

Costa JT and Fitzgerald TD (1996) Developments in social terminology: Semantic battles in a conceptual war. *Trends in Ecology and Evolution* 11: 285–289.

Costa JT and Fitzgerald TD (2005) Social terminology revisited: Where are we a decade later? *Annales Zoologici Fennici* 42: 559–564.

Eickwort GC (1981) Presocial insects. In: Hermann HR (ed.) *Social Insects,* vol. 2, pp. 199–280. New York: Academic Press.

Fitzgerald TD (1995) *The Tent Caterpillars.* Ithaca, NY: Cornell University Press.

Hamilton WD (1971) Geometry for the selfish herd. *Journal of Theoretical Biology* 31: 295–311.

Reed RD (2005) Gregarious oviposition in butterflies. *Journal of the Lepidopterists' Society* 59: 40–43.

Tallamy DW (2001) Evolution of exclusive paternal care in arthropods. *Annual Review of Entomology* 46: 139–165.

Tallamy DW and Brown WP (1999) Semelparity and the evolution of maternal care in insects. *Animal Behavior* 57: 727–730.

Tallamy DW and Wood TK (1986) Convergence patterns in subsocial insects. *Annual Review of Entomology* 31: 369–390.

Trumbo ST (1996) Parental care in invertebrates. *Advances in Study of Behavior* 25: 3–51.

Whitehouse MEA and Lubin Y (2005) The functions of societies and the evolution of group living: Spider societies as a test case. *Biological Reviews* 80: 347–361.

Wilson EO (1975) *Sociobiology: The New Synthesis.* Cambridge, MA: Belknap/Harvard University Press.

Sernland E, Olsson O, and Holmgren NMA (2003) Does information sharing promote group foraging? *Proceedings of the Royal Society of London B* 270: 1137–1141.

Stamps JA (1988) Conspecific attraction and aggregation in a territorial species. *American Naturalist* 131: 329–347.

Valone TJ (2007) From eavesdropping on performance to copying the behaviour of others: A review of public information use. *Behavioral Ecology and Sociobiology* 62: 1–14.

Ward P and Zahavi A (1973) The importance of certain assemblages of birds as ''information-centres'' for food-finding. *Ibis* 115: 517–534.

Social Insects: Behavioral Genetics

B. P. Oldroyd, University of Sydney, Sydney, NSW, Australia

Introduction

Insect colonies do amazing things. Leaf-cutting ants cultivate fungus gardens in massive subterranean galleries that you could easily stand in. Some termite species construct earthen towers several meters high that air-condition the nest beneath. Honeybee colonies locate and exploit the best sources of food over an area of 100 km^2 using a symbolic language that coordinates the efforts of 10 000 foragers. These remarkable feats require a degree of organization that rivals that of the most successful multinational corporations. Yet, each individual worker has a brain less than the size of a pinhead, and there is no managerial control of a colony's behavior. In this article, we explore the emerging field of 'sociogenomics' – understanding how genetically determined rules followed by workers result in colony-level behavior.

Social insects are those that live in groups in which some group members rear offspring that are not their own. This definition excludes insect aggregations and swarms, which may also show group-level behavior, but lack the key element of cooperative brood care. Social insect species include all ants and termites; some wasps, bees, and ants; a few gall-forming thrips and aphids; and a bark beetle (Costa, 2006). Insect societies range from simple family groups with no significant physical differentiation between reproductives and nonreproductives, to extremely sophisticated societies in which queens and workers are strongly differentiated, expressing very different genes and behavioral and physical phenotypes. The latter societies usually comprise large numbers of specialist sterile workers that build and maintain well-defended homeostatic nests and forage over a broad area.

Individual- and Colony-Level Phenotypes

In discussing social insect behavioral phenotypes, it is important to understand that there are different levels of biological organization in a colony (**Figure 1**). For example, the amount of pollen a honeybee colony collects and stores in its brood nest is a colony-level behavioral phenotype that is readily quantified. Beekeepers can recognize high pollen-hoarding colonies, and high and low pollen-hoarding strains can be produced by selection on the colony-level phenotype. But at the same time, a colony is comprised of individual workers which also have phenotypes. Studies on a pair of selected strains showed

that the average worker from a high pollen-hoarding strain had a set of behavioral characteristics that was different from that found in an average worker from the low pollen-hoarding strain (Hunt et al., 2007). Not surprisingly, these included the size of typical nectar and pollen loads collected by individual foragers and a preference for pollen collection in the high pollen-hoarding strain. But the age at onset of foraging and various physiological characteristics, including the size of the ovaries of workers, was also changed by the colony-level selection. This example illustrates how subtle differences in behavior at one level of biological organization (the behavior of individual workers) can result in significant shifts in behavior at another level (the kind of food stored by a colony).

Task Stimulus, Task Threshold, and Task Specialization

The work of a social insect colony needs to be divided among the available workers in a way that maximizes colony productivity. This requires that all necessary tasks are done and that the most important and labor-intensive tasks are allocated to the most workers. The allocation of workers to tasks needs to change dynamically in response to the changing needs of a colony and the available foraging opportunities. How do colonies do this, for, unlike human factories, there is no top-down management? The answer is that in insect colonies, division of labor is self-organized.

The most important explanatory model for the organization of work in insect colonies is the Fixed Threshold Model (Bonabeau et al., 1998). This model holds that each worker has a fixed probability that it will engage in a particular task and that workers differ in this probability (i.e., they differ in their task threshold). Workers with different task thresholds will behave differently in response to the same task stimulus – some measure of the colony's need for a task to be performed. This results in the often-reported phenomenon of task specialization, in which particular workers are more likely to engage in particular tasks than other ones, given a particular level of task stimulus. Variance in task thresholds among workers arises from both genetic and environmental causes.

Many of the more complex insect societies do not have a simple family structure of a single once-mated queen and her worker offspring. Indeed, as societies become

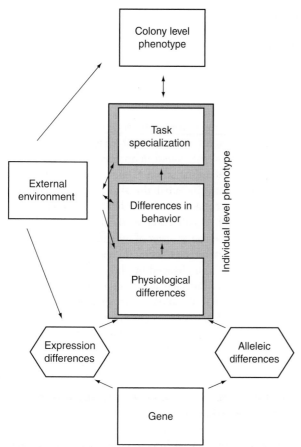

Figure 1 Levels of biological organization in a social insect colony. Genotypes are represented at the level of genes in individual workers, collective genes of the combined workers, and the collective queen and workers – the colony-level genotype. Worker phenotypes arise from the genes carried by individual workers, and the environment that they experience. The environment workers' experience is influenced by the genotypes of sister workers and their social environment. Colony-level phenotypes are an emergent property of the collective phenotypes of workers and their interactions with each other. Redrawn from Oldroyd BP and Thompson GJ (2007) Behavioral genetics of the honeybee, *Apis mellifera. Advances in Insect Physiology* 33: 1–49.

more complex, there seems to be a strong tendency for the genetic architecture of the colony to become more baroque. Either the queens would have mated with many males (e.g., honeybees, harvester ants, driver ants, leaf cutter ants, and army ants) or there will be multiple queens per nest (e.g., sugar ants). (Exceptions to this general trend are the stingless bees and many termites, where a single once-mated queen seems to be the rule despite their often large and complex colonies.) Multiple fathers or multiple queens mean that these societies are comprised of multiple subfamilies – either the half-sister daughters of the different males that have mated with the single queen, or the matrilines that are the daughters of the different queens in the colony.

In colonies with multiple subfamilies, we often see the emergence of genetic task specialization (GTS) – workers of different subfamilies have different task thresholds that are inherited from their different parents. Thus, if we go to a colony and collect the workers performing a particular task like nest ventilation, we often find that a few subfamilies are greatly overrepresented in the sample relative to a random sample of workers or workers performing a different task. (Subfamily membership can be determined using the techniques of forensic genetics and DNA microsatellite markers. Each male and female parent passes its own genetic signature to offspring workers. Parentage is determined by examination of a worker's microsatellite genotype and inference of its parent from this profile.) The phenomenon of GTS has been demonstrated for a large number of tasks across a broad range of taxa (**Table 1**).

Why Have Task Specialization?

Modeling studies show that diversity in task thresholds within a colony enhances colony-level well-being by producing a more homeostatic environment. Imagine a colony trying to regulate the interior of its nest at the optimal temperature for rearing brood. If all the workers have the same task threshold for heating and cooling the nest, the colony is likely to swing precipitously between all the workers engaging in the task and all the workers disengaging. (Old-fashioned air-conditioners have the same problem. They are either on (and you freeze) or off (and you swelter). The temperature fluctuates around the optimal while never being at the optimum.) In contrast, a colony where the workers have a range of thresholds there will have a more modulated response to changes in temperature. Empirical evidence now shows that genetically diverse honeybee colonies, with a range of genetically based task thresholds, make more honey, have higher survival, and do a better job of regulating their internal nest environment than do genetically uniform colonies (Jones et al., 2004; Mattila and Seeley, 2007). Similarly, studies of harvester ants in the field have shown that genetically diverse colonies have larger populations and produce more offspring than more uniform ones, presumably because (among other reasons) they regulate their internal nest environment better.

Physiological and Genetic Basis of Task Thresholds

In many advanced ant and termite species, a worker's role in the nest is determined by its subcaste: soldiers are involved in defence of the nest, major workers in foraging, and minor workers in in-nest work such as tending fungus

Table 1   Examples of genetically based task preferences in social insects

Taxon	Description	Source of variance[a]
Honeybees	Preferred forage (pollen, nectar, water, or floral location)	A
Apis	Scouting for food or nest sites	
	Guarding	
	Feeding and tending larvae	
	Grooming other individuals	
	Feeding other individuals	
	Fanning in response to high temperatures	
	Corpse removal	
	Worker reproduction	
Ants	Worker caste determination has a genetic component	A
Acromyrmex	Tasks associated with tending fungus gardens	A, B
	Undertaking	B
Camponotus consobrinus	Worker caste determination has a genetic component	B
Pogonomyrmex badius	Worker caste determination has a genetic component	A
Eciton burchelli	Worker caste determination has a genetic component	A

[a]A = Multiple fathers, B = Multiple queens.

gardens or nest cleaning. It has generally been assumed that the development of an ant larva into a worker or a soldier is environmentally determined, particularly the amount of food it receives as a larva: the best-nourished larvae develop as large soldiers, whereas poorly nourished larvae develop as minor workers. Yet, evidence is mounting that larval feeding is not the whole story. In the Australian sugar ant and in army ants, some matrilines and patrilines respectively are overrepresented in the soldier caste (**Table 1**), providing evidence for a genetic influence on worker caste development and ultimately worker behavior.

In contrast to ants, in social bees and wasps, there is limited morphological variation among workers, and the variation that does exist is unlikely to be of functional significance. In these societies, the primary determinant of a worker's task threshold is its age. Young workers tend to perform tasks within the nest, but as workers age, they begin foraging tasks. This age-based task ontogeny is accompanied by fundamental changes in the exocrine and endocrine systems so that workers of a particular age are equipped with the appropriate glandular secretions that allow them to undertake tasks typical for that age. The age-based ontogeny is adaptable depending on the colony's needs. So if all the foragers are lost in a storm or at the hands of predators, young workers will switch to foraging precociously.

Sitting on top of this basal behavioral ontogeny is a secondary genetically based variance in perception of the environment, and this modulates a worker's task thresholds and thus the work that she is likely to perform. A well-studied example of this comes from honeybees. A honeybee forager carries nectar in her crop and pollen on 'baskets' on her hind legs. She can barely fly when they are both full, so foragers tend to specialize in either pollen collection or nectar collection. Individuals that specialize in nectar collection have a low 'sucrose threshold.' This is

the minimum concentration of sugar in nectar that is needed for the worker to recognize nectar as being sweet and worth collecting. Workers with a high sucrose threshold find most types of nectar unattractive and end up specializing in collecting only very high-quality nectar. Researchers can predict the foraging specialism of even preforaging age bees by measuring the minimum concentration of sucrose that the bees will respond to by sticking their tongue out to feed (**Figure 2**).

Behavioral Genetics of HoneyBees

In honeybees, there has been considerable progress toward identifying the actual genes that influence task threshold and the behavior of workers. Honeybees are particularly suited to behavioral genetic studies: first, a complete genomic sequence is available at http://www.ncbi.nlm.nih.gov/mapview/map_search.cgi?taxid=7460 and the sequence helps researchers locate functional genes; second, the honeybees have a very high recombination rate, allowing researchers to map genes more finely than in species like fruit flies where the recombination rate is lower; and third, honeybees show an amazing array of behavior that is easily measured and genetically variable.

To locate behavioral genes that influence task threshold, researchers cross two strains that differ strongly in a colony-level behavior. They then backcross a daughter queen of this F_1 to both parental strains. The assumption here is that one strain is pure-breeding for alleles for the high task threshold (*AA*), whereas the other is pure breeding for alleles for low task threshold (*aa*). The F_1 queen will therefore be *Aa* at all loci that affect the task threshold. Thus, the backcross to the *aa* parent will produce a colony in which there will be a Mendelian segregation producing workers that are either *Aa* or *aa*. Under an

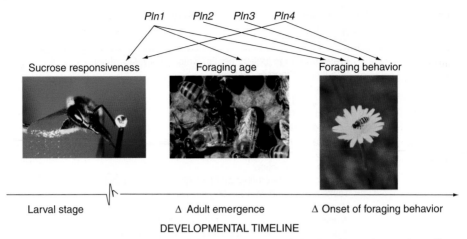

Figure 2 Four quantitative trait loci (*Pln* 1–4) thought to influence the foraging behavior of honeybee workers. Genes near these QTL are candidates for the regulation of honeybee foraging behavior. Solid bars represent linage groups. Genetic markers nearby the QTL are shown on the left of the linkage groups. Genes near the QTL are indicated. Most of these are orthologues of fly genes. Dashed lines indicate 97% confidence intervals of the QTL. Adapted from **Figure 2** of Hunt GJ, Amdam GV, Schlipalius D, et al. (2007) Behavioral genomics of honeybee foraging and nest defense. *Naturwissenschaften* 94: 247–267 with permission.

assumption of complete dominance, the *Aa* workers will have a high threshold and the *aa* workers a low one.

After the creation of the backcross, about 300 workers are scored for the behavior of interest to identify 150 low-threshold and 150 high-threshold individuals. The 300 workers are then genotyped at 350 functionless genetic markers that are evenly distributed along the honeybee's 16 chromosomes, using the online genomic sequence as a guide. If the markers are evenly spaced throughout the genome, at least one of them will be genetically linked to the *A* locus which affects the task threshold. This marker, call it M, will reveal itself by statistical association between genotype (*Mm* or *mm*) and behavioral phenotype (high or low task threshold).

After the identification of M, the genome sequence is again consulted to locate additional markers nearby M. The workers are genotyped at these additional loci to locate the marker with the strongest statistical association to behavioral phenotype. The location of this marker identifies the region of the genome that contains the *A* locus. Bioinformatic searches of the genome in this region will reveal several genes, in theory as few as 5–10. These are then candidate regulators of the task threshold. This list of genes may be narrowed further by consideration of their likely function (e.g., genes coding structural proteins are unlikely to be relevant but a pheromone-binding protein might be). Final identification requires a demonstration of change in the task threshold of workers following experimental manipulation of gene expression by RNA interference.

So far, such crosses have been made for colony defensiveness, nest cleaning, pollen hoarding, and worker reproduction. Over a dozen Quantitative Trait Loci that affect aspects of these colony-level behaviors have now been identified – that is, relatively small areas of the genome that likely contain genes that regulate behavior.

As yet, no gene affecting task threshold has been definitively identified by the final step of RNA interference. Nonetheless, some candidate genes that are thought to affect stinging behavior were shown to be differentially expressed between an Africanized (defensive) strain and a European (nondefensive) strain, strongly suggesting that these candidate genes identified by QTL mapping are indeed the genes that affect stinging behavior.

Genes Affecting Foraging Behavior in HoneyBees

Greg Hunt and colleagues performed the pioneering experiment designed to locate QTL affecting foraging behavior (Hunt et al., 1995). This first experiment located 3 QTL that affected the amount of pollen stored by colonies. These QTL named *Pln* 1–3 have been confirmed in different crosses in different populations using different markers. Furthermore, an additional QTL, *Pln* 4, was located during these subsequent crosses (**Figure 2**). These QTL, located via a cross between high and low pollen-hoarding lines, have effects on characters associated with individual bee behavior including the age at which they start foraging, their sucrose responsiveness, and their foraging preferences (**Figure 3**). *Pln* 4 is particularly interesting. This QTL was not located via mapping. Rather, an ortholog of a gene known to affect foraging behavior in the fruit fly *Drosophila melanogaster* was investigated in honeybees. The *Forager* gene encodes a cyclic GMP-dependent protein kinase. This gene is now known to influence foraging behavior in such diverse organisms as fruit flies, nematode worms, harvester ants, and honeybees, apparently by stimulating the desire for movement. In honeybees, for example, it is involved in the rate at which workers mature to foraging tasks from in-nest tasks. A genetic marker within *Amfor* (the honeybee equivalent of *Forager*) shows a statistical association with foraging behavior, suggesting that *Amfor* affects a bee's task threshold for pollen collection. However, this inference remains to be confirmed because *Amfor* may be linked to another gene that is the actual cause of the differences in behavior.

Another aspect of honeybee foraging behavior known to be genetically influenced is the dance language. Upon discovery of a desirable food source, a forager will tell her nestmates where it is by means of a symbolic dance. In the dance, the worker strides forward while vigorously waggling her abdomen from side to side. She then takes

a sharp turn to the left or right and returns to her starting point. Here she completes another waggle phase, this time ending the waggle with a turn in the direction opposite to the previous run. Thus, the dancing bee traces out a pattern that looks a bit like the figure of 8 (**Figure 4**).

Other workers trip along behind the dancer and learn the direction and distance to the food from the colony. The direction the recruit should fly is indicated by the angle of the waggle run, and the distance by the dance 'tempo.' The closer the food is the shorter the duration of the waggle run and the number of circuits completed per unit time. At some point, the dancer has to turn so quickly that the waggle runs do not line up on top of each other. This form of the dance is known as a transition dance. Still closer

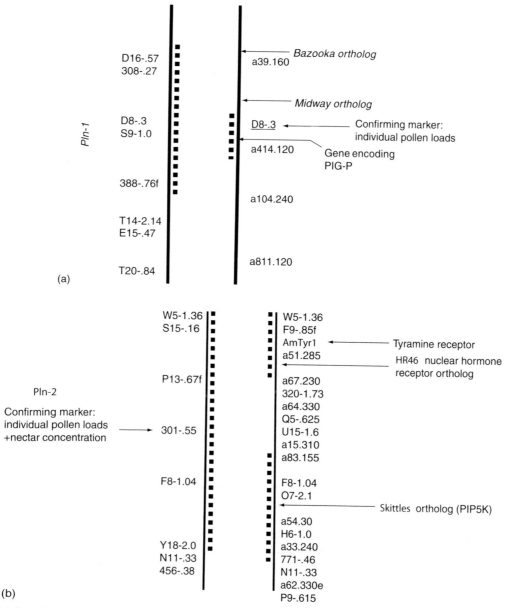

Figure 3 Continued

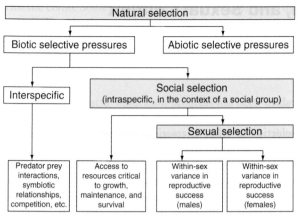

Figure 1 A flow chart illustrating the position of social selection and sexual selection within the broader category of natural selection. Note that our definition of sexual selection also includes abiotic selective pressures and biotic selective pressures such as predator prey interactions if they act differently on one sex than another.

Figure 2 Some examples of monomorphically showy species. (a) Collared araçari (*Pteroglossus t. torquatus*), photo by P.A. Gowaty. (b) Broad-billed motmot (*Electron platyrhynchum minor*), photo by J.P. Drury. (c) Keel-billed toucan (*Ramphastos sulfuratus brevicarinatus*), photo by J.P. Drury.

should look carefully at the context in which individuals perform social behaviors.

Bizarre and elaborate juvenile traits

Parental choice leads to another important case of nonsexual social selection: bizarre and elaborate traits of juveniles. Coots (*Fulica* spp.), for example, are relatively

Figure 3 An adult coot (*Fulica atra*) with its brightly colored young. The bright coloration in young animals is due to social selection for access to parental care. Photo © Böhringer Friedrich, licensed under Creative Commons.

unremarkable as adults, yet coot chicks have bright collars, which influence parents to bias parental care toward the more ornamented young (**Figure 3**). Likewise, the natal pelage of dusky leaf monkeys (*Trachypithecus obscurus*) and many other primates is bright and conspicuous compared with adult pelage, so that natal pelage is likely involved in attracting care from alloparents. Other similar examples include begging frequencies in birds, the sounds some insect larvae make, and the elaborate patterns found in avian nestlings' mouths.

Other examples and future directions

West-Eberhard mentioned several other examples of selection acting in social contexts such as the evolution of eusociality in Hymenopterans, the increased size of the human brain, large size in dinosaurs, and the elaborate competitive behaviors in insects as having evolved via social selection.

In spite of the high number of citations for West-Eberhard's works on social selection, few investigators have attempted to develop further hypotheses about social selection. We know of no examples of quantitative models of the effects of social selection. Further investigation into the role of selection on showy traits in social species, even outside of mating contexts, should be a priority for biologists interested in the evolution of such traits.

Sexual Selection

Darwin's Challenge

In response to the critics of *On the Origin of Species*, Darwin elaborated his earlier ideas on sexual selection, which solved the challenge from his critics who argued that natural selection could not explain traits that lowered the survival probabilities of their bearers. In 'Principals of Sexual Selection,' the first chapter of Part II of *The Descent of Man and Selection in Relation to Sex*, Darwin defined sexual selection as that type of natural selection that "... depends on the advantage which certain individuals have over other individuals of the same sex and species, in exclusive relation to reproduction." Darwin distinguished sexual selection from natural selection as that process which results from some individuals having gained a reproductive advantage over other same-sex, conspecific individuals, not through different habits of life, but from reproductive competition with rivals. The definition we employ here, following Juan Carranza, and discuss further later, however, does include differences arising from habits of life, because sex-dependent selection encompasses all selection that acts differently on the different sexes. We justify our decision further by noting that 'habits of life' affect variables, such as individuals' survival probabilities or their encounter rates with others with whom they may potentially mate, as well as their encounter rates with rivals – all variables intimately associated with narrow-sense sexual selection. Thus, it seems scarcely possible to parse habits of life to those affecting only natural or only sexual selection. In practice and in life, natural selection and sexual selection may be the same thing; thus, our definition of sexual selection as within-sex selection is conservative and inclusive.

The evolution of bizarre and elaborate traits

In his 1871 volume, Darwin focused mostly on the effects of reproductive competition on secondary sexual organs and ornaments, sex-limited traits that are not directly connected with copulation, spawning, or gamete transfer or even secondary sexual characters that he said 'were not directly connected with the act of reproduction: for instance in the male possessing certain organs of sense or locomotion, of which the female is quite destitute, or having them more highly-developed, in that he may readily find or reach her; or again, in the male having special organs of prehension so as to hold her securely.' Yet, Darwin clearly appreciated that within-sex reproductive competition could have favored the evolution of such traits, including the ones he explicitly set aside from his 1871 discussion. He clearly knew that 'contrivances,' as he called them, including especially those related to finding mates and, once found, holding on to them, or even the variation in copulatory parts (in males of intromittent organs and in females sperm storage or sequestration organs) could have evolved via sexual selection. He put them aside from discussion possibly because further discussion would confuse his readers, most of whom were completely new to the idea of sexual selection. Nonetheless, Darwin's use of the term 'sexual' in 'sexual selection' referred to the within-sex nature of selection occurring from 'within-sex reproductive competition' and not simply to sex or copulation. Notably, modern investigators, armed with new empirical methodologies and new theoretical tools, are not so constrained, and consequently they consider as sexually selected many of the organs

that Darwin would not discuss or was unwilling to label as sexually selected.

Bateman's Insight

In 1948, A.J. Bateman published what was to become the most cited empirical study in sexual selection. This work solidified the view that variation in number of mates predicts variance in reproductive success. Setting aside for a moment the flaws in Bateman's methods and conclusions, the historical importance of his review cannot be set aside because it put the study of sexual selection on a more quantitative footing, so that modern scholars associate sexual selection acting through variation in number of mates with within-sex variance in a measure of fitness such as fecundity or productivity. However, other reasonable scenarios for the operation of within-sex selection focus on variation in other measures besides number of mates, such as the quality of mates.

Those who emphasize within-sex variation in number of mates as the only mechanism of sexual selection make the assumption that this very proximal component of fitness is strongly correlated, rather than just correlated (as it must be) with more distal fitness components. Number of mates often is correlated positively with the numbers of eggs laid and offspring born, and also, but more weakly with the number of offspring surviving to reproductive age. The quality of mates is sometimes a better predictor of variance in number of offspring surviving to reproductive age than number of mates. Thus, some investigators describe sexual selection more broadly than just variance in number of mates (see section 'Moving Past Narrow-Sense Sexual Selection: Modern Insights').

Parental Investment and Anisogamy Theory

In 1972, Robert L. Trivers hypothesized that sexual selection via numbers of mates variance worked more strongly on the sex that provided the least parental investment. His argument makes intuitive sense and was that the sex that provided the most parental investment (usually females) was limited by access to resources for reproduction and would experience selection to be choosy about mating as mistakes could be very costly. In contrast, the sex providing the least parental care (usually males) should experience selection to mate with as many potential mates as possible, because what limits their reproductive success is access to mates. This idea more than any other modern idea fueled studies that have provided so much information about the lives of non-human animals. It predicted the evolution of further sex differences on the basis of existing sex differences and led to many studies of narrow-sense sexual selection.

Similarly, the strong argument of Geoff Parker and colleagues explaining the evolution of anisogamy, gametes of two different sizes, also published in 1972, contributed importantly to narrow-sense sexual selection. His idea was that disruptive selection acted on the normal distribution of ancestral gametes, because of the benefits that accrued to very large, resource obtaining gametes and very small, but highly active gametes. Parker's insight explained that the sexes are what they are because of the size of the gametes they carry. Like Trivers's idea, anisogamy theory predicted further sex differences on the basis of the evolution of pre-existing sex differences. Like Trivers's idea, it sparked additional interest in narrow-sense sexual selection.

As a result of discussions and investigations fueled largely by these two studies, modern investigators redefined sexual selection more narrowly than Darwin; they confined their attention to selection favoring sex differences and almost entirely to fancy traits that are associated with within-sex variance in number of mates (**Figure 4**).

Moving Past Narrow-Sense Sexual Selection: Modern Insights

In attending largely to the bizarre and elaborate traits of males in some species, Darwin was responding to his critics, but his text points to general principles that extend

Figure 4 A visual representation of narrow-sense sexual selection: the bright, elaborate plumage of the male violaceous trogon (*Trogon violaceus concinnus*) is in full focus, while the female remains out of focus in the background. Photo by P.A. Gowaty.

to other traits than those involved in males obtaining mates. His discussion included many cases of female traits, especially in humans, that he said evolved via male choice. Notably, these traits are neither bizarre nor elaborate. Narrow-sense sexual selection changed and constrained Darwin's definition (see section 'Darwin's Challenge'), which invoked within-sex variance in reproductive success and did not rely on, but did explain, the existence of bizarre and elaborate traits. Nor did Darwin's ideas rely on strict sex differences to create reproductive success variance that was sexual selection.

Two seminal studies began the effort to redefine sexual selection as sex-dependent selection. In the mid-1980s, William Sutherland published a paper showing that the variances in reproductive success reported by Bateman could be due entirely to chance. Shortly after that, Steve Hubbell and Leslie Johnson described a quantitative model of the effects of demographic stochasticity on variances in fitness; they concluded that the opportunity for sexual selection should be redefined not as the observed variance in fitness but as the residual variance after accounting for chance effects on observed fitness variances. Thus, when comparing male and female fitness variances, it is important to limit comparisons to residual variances after accounting for chance effects on observed fitness variances.

Hubbell and Johnson also showed quantitatively that demographic stochasticity plus the existence of potential mates of only two qualities (better and worse) could favor the evolution of choosy and indiscriminate behavior independent of the sex of an individual. Much later, in 2005 and 2009, Gowaty and Hubbell extended these ideas to show that demographic stochasticity plus quality differences among all potential mates (i.e., if there are say 100 potential mates, there could be 100 different qualities of mates) would favor the evolution of adaptively flexible individuals, not sexes. Adaptively flexible individuals are those able to modify their reproductive decisions to accept or reject potential mates moment-by-moment as their social and ecological circumstances change. Gowaty and Hubbell's work suggests that the habits of life of individuals can affect the parameter's associated effects on fitness variances. So, if an individual's encounter rate with potentially mating individuals, for example, is changed by chance alone or by chance in combination with deterministic effects (e.g., selected traits), their quantitative model predicts that these individuals will change their behavior to enhance their fitness (i.e., individuals are adaptively flexible). This model is an alternative hypothesis to narrow-sense view that coyness in females and eagerness in males are fixed, genetic traits. Thus, if individuals of different sexes have different habits of life, within-sex fitness variances may differ. This could happen if individuals of different sexes inhabit different environments, have different life histories, or face different predators or pathogens. Since differences in habits of life of individuals of different sexes can affect within-sex variances in fitness, these differences matter in sexual selection.

Mechanisms of Sexual Selection

As we emphasized earlier, although sexual selection is measured as within-sex variance in reproductive success, the mechanisms of sexual selection include behavioral and physiological interactions between individuals, whether intrasexual or intersexual, that result in within-sex variance in some component of fitness. Here, we treat several major mechanisms of sexual selection: mate choice, within-sex contests, and in a separate section below, sexual conflict.

Mate Choice

Although Darwin discussed the efficacy of male choice of females, he emphasized female choice as one of the main mechanisms of sexual selection favoring the evolution of elaborate male traits. Because males can also be choosy, as Darwin reported and many recent investigations have also borne out, the term 'mate choice' is more appropriate. Mate choice occurs when a choosing individual assesses more than one potential mate and when variation in some trait of potential mates influences the chooser's reproductive decisions, such as whether or not to copulate, whether or not to inseminate or allow insemination, whether or not to allow fertilization, or whether or not to invest in parental care.

Selection from the perspective of choosers
For selection to act on mate assessment, the fitness consequences of choices are what matter. Many biologists conceptually divide the fitness consequences of preferences into ones that arise via the genotype of offspring, known as 'indirect benefits,' and those that do not, known as 'direct benefits.'

Two classical models of indirect (genetic) benefits that address the evolution of exaggerated secondary sexual traits are Fisherian runaway selection and Zahavi's handicap principle. R.A. Fisher hypothesized that a genetic correlation between a (male) trait and the (female) preference for that trait leads to rapid exaggeration of that trait. Selection on such correlated traits is such that over many generations, selection should 'runaway' so that the trait would become more and more exaggerated until countered by selection in another context. The fitness benefit that choosing individuals obtain is predicted to be offspring that produce the exaggerated trait (sexy sons) and are thus more likely themselves to obtain high reproductive success. The other classical model, known as 'Zahavi's handicap principle,' posits that

exaggerated secondary sexual traits act as indicators of mate quality, since only high-quality mates will be able to develop an exaggerated trait in the face of the survival costs that they entail. In other words, individuals are handicapped by these traits such that they become potentially honest signals of some aspect of mate quality. Thus, individuals who choose mates with an exaggerated trait will pass on this quality to their offspring. The exact genetic underpinnings of the exaggerated traits in these two models are rarely explicitly examined, nor does the quality of the assumed choosing individuals – usually females in these models – get mentioned.

Indirect benefits to choice do not require that a secondary sexual trait be involved. Individuals can choose mates based on the presence or absence of certain alleles in the potential mate. These alleles may be of an additive effect, acting regardless of the composition of the chooser's genome, or of a nonadditive effect, where their effects are dependent on the alleles present in the choosing individual. One oft-cited example of this class of benefit to choice is that of choice for alleles involved in immunity. This famous hypothesis by W.D. Hamilton and Marlene Zuk posits that individuals choose mates such that they confer specific alleles against specific diseases to their offspring.

Selection from the perspective of potential mates
Selection should favor traits in potential mates that increase the probability of being accepted by an opposite-sex individual. Signals may evolve in this context, as can other traits, such as qualities associated with an increased probability of encountering potential mates, metabolic efficiency, or parenting.

One set of adaptations among potential mates includes traits that manipulate choosing individuals by exploitation of pre-existing sensory biases, or sensory traps. Examples of sensory traps include bees (*Xylocopa* spp.) with sex pheromones mimicking floral odors, characid (*Corynopoma riisei*) males with appendages resembling a small prey insect that they wave in front of females prior to copulation, and guppy (*Poecilia reticulata*) females preferring females of the color of a common food source.

The handicap principle (see section 'Selection from the Perspective of Choosers') also suggests an important selective pressure on potential mates. Namely, there should be selection for individuals to recognize and adjust the development of a trait to the highest degree of exaggeration possible, such that the indicator trait becomes a potentially honest signal of some index of quality. Selection should favor individuals able to maximize the signal without suffering survival costs.

Finally, selection should favor those individuals who maximize contributions that confer direct fitness benefits choosers, such as nuptial gifts, parental care, territory, etc.

It takes two: thinking beyond the classical framework

Perhaps as an artifact of thinking of females as coy and males eager, theoretical treatment of mate choice makes the assumption that the chooser is for the most part critical of potential mates; that is, will reproduce only with those few mates that have some agreeable trait, and the potential mate is indiscriminate; that is, will mate with any choosing individual that deems him/her acceptable. As recently suggested by Gowaty and Hubbell, if we move beyond an imagined sex difference and imagine instead that both individuals in a mate choice interaction are both assessing each other as potential mates, while simultaneously being assessed, other opportunities for selection to act become clearer to investigators.

One example of such an extension of mate choice theory would be a re-examination the genetic models that are often employed that focus on the alleles of potential mates. These good gene models posit that alleles of additive effect present in potential mates somehow benefit the choosing individuals, yet they do not address how the presence or absence of this additive allele in the chooser will influence reproductive decisions. Nor do these models offer solutions for alleles of nonadditive effect, where the alleles present in the potential mate act differently depending on other alleles of the gene or genes in question. Thus, future theoretical and empirical work on the genetic benefits of mate choice should incorporate an understanding of the underlying genotypes of both choosers and potential mates.

According to the choosy-chooser and indiscriminate-other logic, classical models assume that in all pairings that in fact occur, both individuals accept one another. There are other possible dyadic interactions, however, including situations where the female accepts and the male rejects, where the male accepts and the female rejects, and where both individuals reject each other. Social and ecological constraints on mating decisions occur such that individuals may reproduce with individuals that they do not prefer.

The timing of choice

As stated earlier, mate choice occurs whenever an individual assesses the relative benefits of reproducing with one versus another mate. These initial reproductive decisions likely affect subsequent decisions. For example, pre-touching mate assessments likely affect the cascade of reproductive decisions that follow, such as how long to copulate, what components besides sperm to ejaculate, what sperm females should save, store, use, sequester, or kill, and what resources parents should contribute to eggs, zygotes, neonates, and juveniles. With the general definition of mate choice, it is possible for mate choice to happen at many points during a mating encounter: prior to copulation, during or after copulation and prior to insemination, after insemination, and even after birth.

Within-Sex Contests

Same-sex interactions include dominance interactions that can control access to mates, access to breeding resources, or even access to nonbreeding resources that might later affect an individual's competitiveness relative to reproduction. Such dominance interactions can be subtle and hard to observe such as those that occur through pheromonal signaling. Or, dominance interactions can be dramatic such as same-sex fights prior to, during, or even after the mating and reproductive seasons. These interactions are subject to sexual selection, because they create variance in reproductive success.

Same-sex fights

Most commonly, investigators report fights between males. Aggressive and dramatic male–male fights occur in many animals including mammals, birds, and snakes. In some cases, it has been relatively easy to attribute these fights to competition over access to females, or competition over resources that males can later broker to females in mating negotiations. Some experimental studies in birds, such as those in eastern bluebirds, *Sialia sialis*, have demonstrated that male–male aggression is temporally variable occurring most reliably during the breeding season only when males are guarding females from the possibility of mating with other males. Thus, male–male fights are most dramatic on the territories of males whose social mate is fertile, nearing, or in laying. These fights defend a male's genetic paternity, and may have a profound effect on a given male's reproductive success, so that these fights no doubt contribute in the end to variance in male reproductive success.

Fights also occur between females, but female–female fights are reported less frequently than fights among males, which might be because females are less noticeable than males or because fights between females occur less frequently. In contrast, females commonly compete through dominance interactions, something that Sarah Blaffer Hrdy's 1981 book *The Woman That Never Evolved* emphasized. Hrdy's book is a broad review of non-human primates, and one of its central points was that these females compete sometimes ferociously for access to the resources they need for reproduction. Her book made it clear that female–female fights result in variation in reproductive success and even survival of females. Some experimental studies in birds, such as those in eastern bluebirds, have demonstrated that the ferocious fights one often observes in the early stages of nesting attempts are most exaggerated during egg laying when females' nests are vulnerable to conspecific nest parasitism, that is, the threat that another female will lay eggs in her nest.

Same-sex clubs

Not all variance in within-sex reproductive success needs to be the result of antagonistic within-sex interactions; sexual selection may result from same-sex affiliative behavior, though admission to such same-sex clubs may be the result of intense competition.

One famous example of same-sex clubs is that of bonobo females, where nonkin groups maintain affiliative interactions. Richard Wrangham posited in 1993 that the forward placement of the clitoris in female bonobos and their exuberant female–female genital rubbing is a trait that mediates the interactions in these same-sex clubs. Bonobo females are in charge of access to food and social networks, and as such female bonds are an important source of sexual selection.

Other examples of same-sex clubs are leks, where many males gather to perform elaborate mating displays, groups of same-sex individuals that are aggressive to opposite-sex individuals, such as male red-cockaded woodpeckers (*Picoides borealis*) that provoke dispersal of female kin, and affiliative same-sex behavior mediated by kin selection, such as juveniles that help their parents raise future offspring.

Sperm competition

In internally fertilizing species where females mate multiply, intramale competition does not end at copulation. The male side of postcopulatory competition involves the various mechanisms by which sperm compete with other males' sperm for access to eggs, otherwise known as 'sperm competition'. Sperm competition can be either passive or active. With passive sperm competition, the relative concentration of sperm of males predicts fertilization success, thereby selecting for mate-guarding behavior, increased testicle size in many mammals, and physiological chastity belts such as sperm plugs. With active sperm competition, on the other hand, there is some mechanism wherein sperm from one male competes with sperm from another male by either blocking admission of another males' sperm to the female reproductive tract, killing another males' sperm, or displacing it from sperm storage sites inside the female reproductive tract. Active sperm competition can be either nonspecific, with mechanisms acting whether or not the female being mated has mated previously (e.g., apyrene sperm of Lepidopterans as sperm that displace previous males' sperm), or specific, wherein mechanisms act only when a female has mated multiply (e.g., the killer sperm hypothesis of Baker & Bellis).

Most work on sperm competition has been carried out under the assumption that female reproductive tracts are passive playing fields. Since it is now clear that this is not true, much further research is needed to illuminate how pretouching assessment by females influences the outcome of sperm competition.

Sexual Conflict

Differences in the definitions of sexual conflict reflect the goals of investigators. Those interested in patterns of trait evolution across taxa, usually say sexual conflict occurs when the evolutionary interests of the sexes differ. Others, such as Locke Rowe and Goren Arnqvist, who are primarily interested in genomic conflict, say sexual conflict occurs when the genetic interests of the sexes conflict. Less gene-centric definitions focus additional attention on many mechanisms of heredity not just genes. A broader definition is more appropriate for those who wish to include flexible phenotypes such as behavior. Generally speaking, sexual conflict can occur whenever the survival or reproductive interests of the sexes are at odds.

Sexual conflict can occur before copulations; during copulation, after insemination; after pair bonding; or over compensatory reproduction. A famous example of sexual conflict involves a mismatch between paternal and maternal alleles involved with the distribution of resources to developing fetuses in humans, leading to ailments such as high blood pressure and diabetes in many pregnant women. Another example of sexual conflict is a more dramatic example of differing fitness interests of males and females: infanticide by males. This generally happens when females form social groups between which males disperse. When a new male arrives to a group, he may threaten or kill the suckling infants in the group to bring mothers into estrous again.

Sexually antagonistic selection pressures are the main source of sexual conflict. For example, when potential mates encounter one another and one individual accepts but the other rejects a copulation, sexual conflict occurs. There are many opportunities for sexual conflict over reproduction, including manipulation of the reproductive decisions of one partner by the other, for example, how much to allocate to offspring, or how much to compensate for offspring viability deficits. How sexual conflict is resolved depends on the fitness differentials – what one individual may gain and what the other may lose if the conflict comes out in one's favor or not, and of course, also on the power differential (size differences, metabolic differences, resource or wealth differences) between the individuals. These topics all involve reproductive decisions and the behavior of individual females and males.

Sexual conflict is undoubtedly common, just as sexual cooperation is. Quantifying the force and magnitude of sexual conflict is conceptually simple. All one needs to know is what the relative within-sex, fitness outcomes for individuals experiencing a conflict are. In practice, it is a bit more difficult. Consider a male and a female who do not agree over whether to mate. If the female says 'no,' but the male says 'yes' to a potential copulation, the outcome of the conflict is seemingly obvious: if they do not copulate, the female 'wins'; if they do copulate, the male 'wins.'

But, this is far from the end of the calculations one must make to evaluate the fitness outcomes of this single conflict. Remember that fitness is relative; and in this case the fitness of each individual is relative to other individuals of the same sex. Thus, what is needed next is an evaluation of how the outcome of the contest affected each individual's fitness relative to others: for the male to other males, and for the female to other females. These considerations make it clear that sexual conflict is a mechanism of sexual selection. In other words, we should not ask the question 'Are males who force copulation upon females "winning" in a coevolutionary sense?' but rather 'Do males who force copulation have higher fitness than males who do not force copulate?' and 'Do females who are able to resist forced copulation have higher fitness than those females who are unable to resist it?'

Sexually antagonistic selection pressures can result, but not necessarily, in sexually antagonistic alleles (alleles that confer a positive fitness effect to one sex and a negative fitness effect to the other); yet, sexually antagonistic selection pressures – whether due to ecological forces, social forces, or genomic forces, will result in fitness variation among individuals. The recent book by Locke Rowe and Goren Arnqvist describes current scholarship on sexually antagonistic genes.

See also: Aggression and Territoriality; Bateman's Principles: Original Experiment and Modern Data For and Against; Compensation in Reproduction; Cryptic Female Choice; Differential Allocation; Flexible Mate Choice; Forced or Aggressively Coerced Copulation; Helpers and Reproductive Behavior in Birds and Mammals; Infanticide; Invertebrates: The Inside Story of Post-Insemination, Pre-Fertilization Reproductive Interactions; Kin Selection and Relatedness; Levels of Selection; Mate Choice in Males and Females; Maternal Effects on Behavior; Monogamy and Extra-Pair Parentage; Pair-Bonding, Mating Systems and Hormones; Parasites and Sexual Selection; Sex Change in Reef Fishes: Behavior and Physiology; Sperm Competition.

Further Reading

Andersson MB (1994) *Sexual Selection*. Princeton, NJ: Princeton University Press.

Arnqvist G and Rowe L (2005) *Sexual Conflict*. Princeton, NJ: Princeton University Press.

Bateman AJ (1948) Intra-sexual selection in *Drosophila*. *Heredity* 2: 349–368.

Carranza J (2009) Defining sexual selection as sex-dependent selection. *Animal Behaviour* 77: 749–751.

Clutton-Brock T (2007) Sexual selection in males and females. *Science* 318: 1882–1885.

Darwin C (1871) *The Descent of Man and Selection in Relation to Sex*. London: J. Murray.

Fisher RA (1930) *The Genetical Theory of Natural Selection*. New York: Dover Publications.

Gowaty PA and Hubbell SP (2009) Reproductive decisions under ecological constraints: It's about time. *Proceedings of the National Academy of Sciences* 106: 10017–10024.

Hamilton W and Zuk M (1982) Heritable true fitness and bright birds: A role for parasites? *Science* 218: 384–387.

Hrdy SB (1981) *The Woman That Never Evolved*. Cambridge, MA: Harvard University Press.

Hubbell SP and Johnson LK (1987) Environmental variance in lifetime mating success, mate choice, and sexual selection. *American Naturalist* 130: 91–112.

Jones AG (2009) On the opportunity for sexual selection, the Bateman gradient and the maximum intensity of sexual selection. *Evolution* 63: 1673–1684.

Lande R (1981) Models of speciation by sexual selection on polygenic traits. *Proceedings of the National Academy of Sciences* 78: 3721–3725.

Parker GA, Baker RR, and Smith VGF (1972) The origin and evolution of gamete dimorphism and the male–female phenomenon. *Journal of Theoretical Biology* 36: 529–553.

Sutherland WJ (1985) Chance can produce a sex difference in variance in mating success and account for Bateman's data. *Animal Behaviour* 22: 1349–1352.

Trivers RL (1972) Parental investment and sexual selection. In: Campbell B (ed.) *Sexual Selection and the Descent of Man*. Chicago: Aldine Publishing Company.

West-Eberhard MJ (1983) Sexual selection, social competition, and speciation. *The Quarterly Review of Biology* 58: 155–183.

Zahavi A (1975) Mate selection: A selection for a handicap. *Journal of Theoretical Biology* 53: 205–214.

Sociogenomics

C. M. Grozinger, Pennsylvania State University, University Park, PA, USA
G. E. Robinson, University of Illinois at Urbana-Champaign, Urbana, IL, USA

Introduction

Sociogenomics is the study of social life at the molecular level. This field seeks to characterize the molecular mechanisms that regulate both the production and evolution of social behavior, including the molecular pathways that are influenced by social information, and thereby adjust the behavior of organisms to their social context.

Animals perform many activities during the course of their lives to survive and reproduce: they find food and mates; defend themselves; and in many cases, care for their offspring or other relatives. These activities become 'social' when they involve interactions among members of the same species in a way that influences immediate or future behavior.

Unlike behavioral genomics and neurogenomics, sociogenomics focuses primarily on the genes regulating the behavior of animals that live in social groups. Furthermore, a key feature is that many sociogenomic studies are performed in an ecologically relevant context, in order to be able to understand how the social environment influences behavior and related molecular processes in the brain. Studies of 'model social' species have become increasingly feasible because of the development of new high-throughput sequencing and functional genomics techniques. However, the focus on animals that live in social groups is not exclusive, because it is thought that many behaviors performed by solitary animals provide the substrates for the evolution of social behavior. Thus, a great deal of information can be gained by understanding the molecular pathways that regulate basic behaviors such as courtship, mating, foraging, and aggression in species with mostly solitary lifestyles.

This article begins by outlining general issues, mechanistic and evolutionary, associated with sociogenomics. This is followed by select examples of sociogenomic studies to give the reader a sense of the current status of the field. The article ends by outlining future directions for research in sociogenomics.

Mechanistic Perspectives

Sociogenomics deals with both mechanistic and evolutionary questions. At the proximate level, the focus is on identifying genes that regulate or are associated with specific social behaviors. These behaviors can include courtship and mating, parental care, nest construction, foraging and provisioning of the offspring or communal group, dominance hierarchies, and division of labor, in which individuals within a group specialize on different tasks. The genes involved in these processes can function at the level of an individual cell (i.e., an olfactory receptor in a sensory neuron), a neural network (a neuropeptide that alters response thresholds), physiology (a gene involved in regulating metabolic processes), development (a particular gland or sensory structure), or in more than one of these contexts.

From a mechanistic standpoint, genes associated with social behavior fall into two broad categories. First, social information can lead to changes in behavior by causing changes in gene expression patterns, particularly in the brain. These 'socially responsive' molecular pathways are exquisitely tuned to the social environment and serve to change neuromodulatory, neuroendocrine, or physiological systems, thereby helping to change the animal's response to new conditions.

The second category of genes relate to how genetic variation between individuals causes differences in social behavior. Genetic variation can modify social responsive pathways, thereby resulting in differences in how an individual will respond to a particular social situation. In addition to modulating these acutely responsive pathways, genetic variation can also result in changes in development, metabolism, and neural circuitry that influence the production of social behavior.

There is also increasing evidence for genotype by environment interactions, and the effects of social environment can be strikingly different on individuals from different genetic backgrounds. Genetic variation between individuals also of course forms the basis for evolutionary differences in social behavior.

Evolutionary Perspectives

Many theories have been developed to explain the evolution of social behavior, especially the most altruistic forms that involve strong differences in reproductive success between members of social groups. These theories include *kin selection, mutualism,* reciprocal altruism and other cost–benefit analyses of group living. Identification of the genes and molecular pathways that regulate the production of various types of altruistic behaviors should eventually help ascertain the likelihood that specific theories pertain to the evolution of a particular social system.

From an evolutionary standpoint, one line of inquiry is how much and what type of genetic variation is necessary for the evolution of social behavior from solitary systems, including the evolution of new genes or new functions for conserved genes. Similar questions have been addressed by the field of evolutionary and developmental biology (evo-devo). It is reasonable to assume that similar molecular mechanisms underlie the evolution of morphology, physiology, and behavior.

New genes can evolve to coordinate social behavior. The evolution of new genes may be particularly critical for detection of species-specific sensory information. For example, specific pheromone receptor genes have evolved from olfactory receptor genes to detect sex and social pheromones in a variety of species.

It seems likely that for most other central neural or physiological processes, existing genes are utilized in novel ways, by modifying when, where, and how much they are expressed. These 'toolbox' genes are functionally conserved, but undergo evolutionary changes so that they can be utilized in different contexts. On a large scale, expression of networks of genes may be modified to shift expression of behaviors or physiological states to different contexts. Thus, the genes in these networks may remain essentially functionally intact, but the signaling pathways that activate the networks or the transcriptional factors that regulate them may change.

Two distinct types of genetic changes are involved in the modification of existing genes and the evolution of new traits. Differences in the nonprotein coding, *cis-regulatory regions* of genes can alter the ability of transcriptional regulators to bind to promoter regions, or change the types of transcriptional regulators that can bind. Thus, *cis*-regulatory mutations can modulate the timing, location, and level of expression of the RNA transcripts. In addition, mutations in protein-coding regions can structurally alter the resulting protein, causing changes in stability, activity, and the ability of the protein to bind to ligands or other proteins.

At this point, it is not clear which type of genetic change is a more common or critical driver of evolutionary change for animal phenotypes in general, and there are many empirical examples of both. Proponents of *cis*-regulatory evolution argue that changes in these regions are less likely to cause deleterious *pleiotropic* effects because of the modular nature of transcription factor binding sites in the promoters of genes. According to this view, new regulatory modules can be added or subtracted without substantially altering existing functions. However, proteins can be modular as well, and complete domains can be included or removed by *alternative splicing*. Even the addition or deletion of short protein sequences can alter a protein's ability to interact with signaling pathways and binding partners in specific cellular contexts. Indeed, there is evidence that transcription factors themselves are modular and can evolve without deleterious pleiotropic effects.

Selected Examples of Sociogenomic Studies

In the following section, we highlight examples of sociogenomic studies that have identified 'socially responsive' genes, have shown how variation in gene expression is associated with differences in social behavior, or have discovered that genetic variation modulates responsiveness to social environment. These accounts are by no means exhaustive, and the reader is encouraged to consult the suggested reading list for more information.

Regulation of Genes by Social Environment

Perception of information from the social environment can cause changes in brain gene expression and behavior on a surprisingly short time scale. Expression of the transcription factor *early growth response factor 1* (*Egr-1*; also known as *Krox 24*, *NGFI-A*, *Zif268*, *Tis8*, and *ZENK*) is significantly upregulated, within seconds, in the auditory forebrain of male zebra finch songbirds (*Taeniopygia guttata*) upon hearing novel conspecific songs. But this expression is not triggered by hearing familiar songs or songs from other species; thus, this gene may be involved in priming birds for learning new songs. *egr-1* also helps orchestrate changes in dominance hierarchies in cichlid fish (*Astatotilapia burtoni*). When the dominant male is removed from a group, *egr-1* is induced specifically in the hypothalamic anterior preoptic area of some subordinate males. This response is tightly associated with the perception of new social opportunity; a subordinate male assumes dominance within minutes, displaying dramatic changes in body coloration and behavior. Many of these changes are also associated with upregulation of the gene encoding gonadotropin releasing hormone (GnRH).

Social information also can cause long-term effects on brain gene expression and behavior. Differences in maternal care in Norway rats (*Rattus norvegicus*) lead to differential methylation at several locations in the genome, including the promoter region of the *glucocorticoid receptor* gene. Increased maternal care leads to increased activity of Egr-1 and enhanced binding of this transcription factor to the promoter region of the *glucocorticoid receptor* gene. This results in reduced promoter *methylation* and increased glucocorticoid receptor expression, especially in the hippocampus of rat pups. The differences in glucocorticoid receptor expression lead to altered expression of downstream genes involved in mediating responses to stress. Individuals reared by mothers that display high levels of maternal care themselves have high glucocorticoid receptor expression, even into adulthood. They also display reduced anxiety, reduced stress responses, and high maternal care, compared to rats reared by mothers that show low maternal care.

There are a growing number of examples in which genes involved in behavioral regulation in solitary species apparently have been co-opted for use in social species. In honeybees (*Apis mellifera*), these include genes originally discovered in *Drosophila melanogaster* such as *period* (a gene involved in circadian rhythms) and *foraging*, a gene that causes differences in foraging style. Both are involved in the regulation of age-related division of labor in honeybees, in which bees work in the hive when they are young and then shift to foraging outside when they get older. Expression differences of *period* are associated with changes in temporal rhythms associated with division of labor: round-the-clock activity while taking care of the brood in the hive and a stricter diurnal schedule while foraging. Expression differences of the *foraging* gene help regulate the age at onset of foraging. Brain expression of *foraging* is higher in foragers than in nurses, just as it is in 'rover' flies, which are more active in food collection than 'sitter' flies. Treatment that upregulates the molecular pathway associated with the *foraging* gene causes precocious foraging in honeybees.

These examples focus on single genes, but the regulation of social behavior involves the coordinated actions of many genes. It is thus no surprise that the responses to social stimuli can be massive, involving hundreds or thousands of genes and perhaps many different brain regions at once. Exposure of worker honeybees to queen pheromone changes the expression of hundreds of genes in the brain, in a behaviorally relevant manner. Queen pheromone slows the transition from hive work to foraging, thus ensuring adequate care of the brood. This occurs because queen pheromone downregulates genes that are typically upregulated in foragers. In the swordtail fish, *Xiphophorus nigrensis*, suites of genes are turned on in the brains of females as they interact with attractive males, but are off when they interact with other females, and vice versa. Individual swordtail females that are more 'choosy' show a stronger genomic response, suggesting a connection between individual differences in behavior and gene expression profile. Using transcriptional profiling to study the roots of individual differences in social behavior will undoubtedly prove to be a fertile line of research in the future.

Genetic Variation in Genes Regulating Social Behavior

Genetic variation between individuals within a species is widespread and has been linked to differences in a broad array of social behaviors. Even seemingly minor sequence changes in individual genes can have profound impacts on gene networks, brain function, and social behavior. A mutation in the *FoxP2* gene is associated with a disorder in speech and language skills in a human family. FoxP2 is a transcription factor, and the mutation results in a single amino acid change that disrupts the DNA binding ability of this protein, thus altering expression of a large number of downstream target genes. FoxP2 appears to be associated with orofacial movements and thus may be important for regulating subtle muscle movements necessary for producing language in humans and vocalization in nonprimate species, including mice, zebra finch, and bats. Comparison of *FoxP2* gene sequences from different species reveals that while this gene is generally highly conserved across mammals, there are much larger differences between chimpanzees and humans, and there is evidence in humans for a '*selective sweep*' associated with the evolution of human speech.

Genetic variation between vole species is associated with differences in the activation of specific neural networks. Prairie voles (*Microtus ochrogaster*) are monogamous and engage in extended biparental care, while montane voles (*Microtus montanus*) are promiscuous and do not form long-lasting pair bonds, nor do they engage in extended biparental care. Sequence variation in the promoter region of the *vasopressin receptor* gene (*V1aR*) causes differences in where this gene is expressed in the brain and in mating preferences. The prairie vole promoter has an expanded *microsatellite* region of ~500 bp compared to the montane vole. Transgenic mice expressing a copy of the prairie vole *V1aR* gene (including the promoter region with the microsatellite) have similar brain expression patterns of *V1aR* as prairie voles, and infusion of vasopressin increased affiliative behavior in these transgenic mice. Similar effects occur in montane voles that are given the prairie vole promoter. Based on the location of the expression differences between prairie and montane voles, one hypothesis to explain these differences is that the pattern of *V1aR* expression in monogamous voles results in a tighter coupling of 'social memory' and 'reward' circuits, facilitating the formation of long-term male–female bonds. However, the 500 bp expansion is not perfectly associated with monogamous and polygamous behavior across all vole species, suggesting that other elements of genetic variation are also involved. It will be interesting to see whether they also target circuits associated with social memory.

With genome sequence information becoming increasingly available for model social species, genome-wide scans of different populations are being used to search for genetic variation that might underlie variation in social behavior. This approach has been used to compare different subspecies of honeybees, which differ in numerous social traits, including colony defense, foraging, and the allocation of resources to (colony-level) somatic growth versus reproduction. There is evidence for positive selection acting on the protein-coding regions of ~10% of all honeybee genes when comparing the infamous 'killer' bees (the African *A. mellifera scutellata* subspecies) and more temperate evolved subspecies. These provide attractive candidates for future research.

Effects of Genotype–Environment Interactions on Social Behavior

The effects of social environment on brain function and social behavior differ among individuals as a result of individual genetic variation. Evidence of genotype–environment interactions influencing both social behavior and gene expression has been found in studies of the fire ant *Solenopsis invicta*. Fire ants, like honeybees, live in colonies with thousands of workers, but while honeybee colonies have just a single queen, fire ant colonies can have one or more. Variation in queen number has a genetic basis in fire ants, and one genetic locus has been identified. *General Protein 9 (Gp-9)* is involved in regulating a key aspect of fire ant social organization, namely, the treatment of queens by the workers. Queens homozygous at the *Gp-9* locus (*BB*) are larger and more fecund than *Bb* queens, and *BB* workers will only accept a single *BB* queen, resulting in one-queen colonies. *Bb* workers will accept multiple *Bb* queens, resulting in larger multiqueen colonies that are ecologically more invasive. Cross-fostering showed that *BB* workers in a *Bb* colony become tolerant of multiple *Bb* queens, and also take on a *Bb* gene expression profile. It appears that worker gene expression profiles are more strongly affected by colony genotype than their own genotype. In other words, the genotype of the worker population determines colony organization, which in turn regulates brain expression patterns of the workers. Unfortunately, the function of *Gp-9* is not clear. There is some sequence similarity to odorant binding proteins in other insects, but functional analyses have not been performed.

Another example of a genotype–environment interaction comes from research on rhesus macaques (*Macaca mulatta*). Levels of serotonin (5-HT) are associated with individual differences in behavior, particularly aggression, depression, and anxiety in a variety of vertebrate species. Both humans and rhesus macaques have two alleles of the gene encoding the serotonin transporter (5-HTT) gene, which differ in their promoter regions. The 'short' allele of the human ortholog is associated with reduced expression of this gene in cell culture studies relative to the 'long' allele, resulting in decreased 5-HT uptake. Rhesus macaques that possess the short allele display delayed neurobiological development, impaired serotonergic functioning, excessive aggression and fearfulness, and increased alcohol consumption. However, these traits are only displayed when individuals are reared in peer groups, without access to their mothers. Peer-group reared animals tend to have increased fearfulness and aggression into adulthood compared to animals reared with their mothers, suggesting that peer-group rearing is a more stressful environment. Rearing environment and early life experiences strongly interact with genotype, resulting in significant differences in adult behavior. Similarly, stressful life events were shown in one study to be able to predict episodes of depression in human individuals possessing the short allele, but not for individuals that were homozygous for the long allele. These studies suggest that genotype can 'buffer' individuals from environmental conditions and vice versa. It is hypothesized that genetic variation in the 5-HTT gene may have evolutionary significance, by better adapting certain individuals to certain types of physical and social environments.

Conclusions and Future Directions

Sociogenomics is an integrative and multidisciplinary field, requiring a synthesis of genetic, physiological, neural, behavioral, social, and evolutionary information to understand social life in molecular terms. Sociogenomics is also a very young field, and important challenges remain. We highlight a few of them here.

Systems biology and sociogenomics

Most studies to date have focused on understanding the contributions of individual genes or identifying set of genes by microarray analysis. However, genes do not act alone, and instead are part of genetic networks. Understanding what networks of genes regulate which behaviors will allow researchers to better determine how behavior is controlled, both mechanistically and evolutionarily. Are there specific networks that control certain behaviors? How do networks interact? Are correlated behaviors controlled by the same genetic networks? Are there master regulators that control multiple behaviors? How plastic are these networks? Answering these questions requires development of systems biology approaches to social behavior.

How Does Social Behavior Evolve from Solitary Behavior?

An emerging insight is that social behavior often seems to arise from modifications of similar behaviors performed by individuals in typically solitary species. Will this prove to be the dominant trend in social evolution? Are networks associated with certain genes more likely to be co-opted than others?

How Do Genes and Environment Interact to Modify Behavior?

Are the interactions between genes and the environment infinitely plastic, or are there evolved mechanisms to canalize specific environmental conditions into specific behavioral responses? Are there certain types of behavior or genetic pathways that are more sensitive to environmental conditions? Addressing these questions will require large-scale studies of different types of social behaviors in ecologically relevant environmental contexts, and a detailed understanding of how the genes associated with these behaviors shift expression patterns under different conditions.

How Does Individual Experience Modify Gene Expression and Behavior?

In the context of social behavior, how common is long-term modification of gene expression via epigenetic mechanisms? Can multiple types of experiences cause long-term changes in behavior? If species are 'primed' to respond to certain experiences in this way, how did this evolve?

Continued advances in genomic, proteomic, and metabolomic techniques will make it feasible to study an even wider array of species, as well as study the differences between populations and individuals within a species. This will allow these and other questions to be addressed, leading to a deeper understanding of social life in molecular terms.

See also: Ant, Bee and Wasp Social Evolution; Behavioral Ecology and Sociobiology; Caste in Social Insects: Genetic Influences Over Caste Determination; Development, Evolution and Behavior; *Drosophila* Behavior Genetics; Genes and Genomic Searches; Kin Selection and Relatedness; *Nasonia* Wasp Behavior Genetics; Nervous System: Evolution in Relation to Behavior; Social Insects: Behavioral Genetics.

Further Reading

Grozinger CM (2010) Genomic approaches to behavioral ecology and evolution. In: Westneat DF and Fox CW (eds.) *Evolutionary Behavioral Ecology*, pp. 488–505. New York, NY: Oxford University Press.

Hoekstra HE and Coyne JA (2007) The locus of evolution: Evo devo and the genetics of adaptation. *Evolution* 61: 995–1016.

Hofmann HA (2003) Functional genomics of neural and behavioral plasticity. *Journal of Neurobiology* 54(1): 272–282.

Hudson ME (2007) Sequencing breakthroughs for genomic ecology and evolutionary biology. *Molecular Ecology Resources* 8: 3–17.

Mackay TF and Anholt RR (2007) Ain't misbehavin'? Genotype-environment interactions and the genetics of behavior. *Trends in Genetics* 23: 311–314.

Robinson GE and Ben-Shahar Y (2002) Social behavior and comparative genomics: New genes or new gene regulation? *Genes, Brain and Behavior* 1: 197–203.

Robinson GE, Fernald RD, and Clayton DF (2008) Genes and social behavior. *Science* 322(5903): 896–900.

Robinson GE, Grozinger CM, and Whitfield CW (2005) Sociogenomics: Social life in molecular terms. *Nature Reviews Genetics* 6(4): 257–270.

Smith CR, Toth AL, Suarez AV, and Robinson GE (2008) Genetic and genomic analyses of the division of labour in insect societies. *Nature Reviews Genetics* 9(10): 735–748.

Toth AL and Robinson GE (2007) Evo-devo and the evolution of social behavior. *Trends in Genetics* 23: 334–341.

Wagner GP and Lynch VJ (2008) The gene regulatory logic of transcription factor evolution. *Trends in Ecology & Evolution* 23: 377–385.

Wray GA (2007) The evolutionary significance of cis-regulatory mutations. *Nature Reviews Genetics* 8: 206–216.

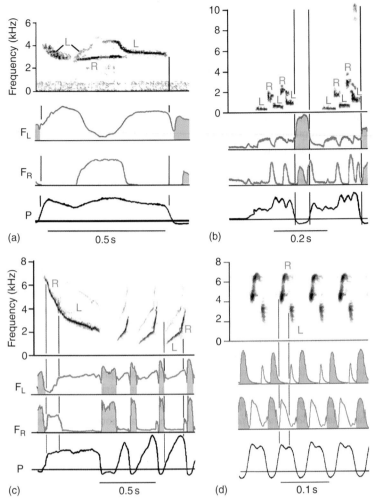

Figure 6 Species differences in the pattern of syringeal lateralization. (a) Independently modulated two-voice syllable of brown thrasher. (b) Two note clusters from song of brown-headed cowbird. Abrupt frequency steps without a silent interval between notes is achieved by alternating note production between opposite sides of the syrinx. Final whistle is not shown. (c) Frequency-modulated sweeps from northern cardinal song with smooth transition of phonation from one side of syrinx to the other in the middle of the sweep. (d) Part of a 'sexy' phrase from a domestic canary, showing the high repetition rate, wide bandwidth and two-note structure of the syllables. In all species studied the right side of syrinx tends to produce higher frequencies than the left. See legend of **Figure 4** for abbreviations. (a–c) Adapted from Suthers RA and Zollinger SA (2004) Producing song. The vocal apparatus. *Annals of the New York Academy of Sciences* 1016: 109–129, with permission.

Acoustic Importance of the Vocal Tract

The source–filter model of the vocal tract

The sound produced by the vocal organ is further modified as it passes through the vocal tract. In birds, the vocal tract includes the trachea. The source–filter theory assumes that the vocal tract filter is not strongly coupled to the source (the vibrating vocal folds or labia). In other words, the column of air oscillating in the vocal tract at the resonant frequency does not control or significantly affect the frequency at which the source vibrates. The following discussion assumes independence between the source and the filter. Although this theory is widely accepted in human speech and in bioacoustics, there are almost

certainly some kinds of vocalizations or some circumstances in which source–filter coupling plays an important role (see howler monkey in this article).

Resonance filters

The broadband sound produced in the larynx or syrinx must pass through the supra-laryngeal vocal tract before it is broadcast into the environment as an acoustic signal for vocal communication. In birds, the vocal tract includes the trachea. The air-filled passages of the vocal tract act as resonance chambers that pass some frequencies better than others, depending on the relationship between the dimensions of the chamber and the wavelength of the

Table 1 Patterns of song lateralization: Motor implications for vocal performance

Independent bilateral phonation: Brown thrasher and gray catbird

Advantages

    Two independent voices increase spectral and phonetic complexity.

Disadvantages

    Expensive in use of air supply. Best suited for a low syllable repetition rate.

Unilateral dominance: Waterslager canary[1]

Advantages

    Conserves air, favoring shorter minibreaths and longer Phrases with pulsatile expiration. Separation of phonatory and inspiratory motor patterns to opposite sides of syrinx. Both of these may facilitate higher syllable repetition rates.

Disadvantages

    Use of one voice limits frequency range and certain kinds of spectral and temporal complexity. Minibreath may be smaller than tracheal deadspace.

Alternating lateralization: Brown-headed cowbird and domestic canary (each side of syrinx produces separate note)

Advantages

    Enhances spectral contrast between notes. Efficient use of air supply. Extended frequency range for overall song.

Disadvantages

    Two-voice complexity limited to note overlap.

Sequential lateralization: Northern cardinal (phonating side of syrinx switched without interrupting note)

Advantages

    Extended frequency range for continuous FM sweeps. Conserves air supply.

Disadvantages

    Lacks spectral complexity of two voices.

Source: Modified from Suthers R (1999) Motor basis of vocal performance in songbirds. In: Hauser M and Konishi M (eds). *The Design of Animal Communication*, pp. 37–62. Cambridge, MA: MIT Press.

[1]Extreme unilateral left dominance of song production by inbred Waterslager canary has probably evolved because this strain is deaf above about 2 kHz and cannot hear sounds produced by right side of syrinx.

sound. As sound from the vocal organ travels through the vocal tract, some of it is reflected back toward the vocal organ because of impedance changes at the mouth or changes in the dimensions of the vocal tract along its length. If the positive pressure peaks of the reflected sound waves coincide with the positive peaks of the wave traveling toward the mouth, they undergo constructive reinforcement to produce a relatively high-amplitude standing wave. The frequencies at which this occurs are the natural or resonance frequencies of the vocal tract. The reflected frequencies in the source spectrum that are not close to the resonance frequencies will be out of phase with the primary sound pressure waves and will be subjected to destructive reinforcement that damps (i.e., attenuates) their amplitude in the vocal output.

The vocal tract's resonance frequencies appear in some vocalizations as bands of relatively high-amplitude sound called formants. Formants are often visible as dark horizontal bands in spectrograms of broad band vocalizations containing nonperiodic noise or a periodic signal with multiple higher harmonics (**Figure 7**). These broad band nonperiodic vocalizations of some birds contrast with the more tonal, periodic vocalizations of other species which lack prominent overtones and usually have most of their energy concentrated in the fundamental frequency. The formant frequency may coincide with the fundamental frequency but in the absence of

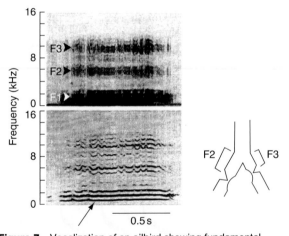

Figure 7 Vocalization of an oilbird showing fundamental frequency at about 800 Hz (arrow) and its higher harmonics (lower spectrogram with narrow band filter) and same vocalization with wide band filter (upper spectrogram) showing three formants. F1 is tracheal resonance, F2 is resonance of left bronchus between syrinx and trachea and F3 is from corresponding portion of right bronchus. Length of bronchus between each semi-syrinx and trachea varies between individuals so each individual has unique F2 and F3 formant frequencies that could potential be used for individual recognition. Modified from Suthers RA (1994) Variable asymmetry and resonances in the avian vocal tract: A structural basis for individually distinct vocalizations. *Journal of Comparative Physiology A* 175: 457–466.

substantial sound energy at other frequencies, multiple formants are not apparent. This is the case, for example, in some bird song. Vocalizations of this type are better designed to carry biologically relevant information in the frequency or modulation patterns of their dominant frequency, for example a large repertoire of different syllable types. The perceived pitch of a vocalization depends on the fundamental frequency of the vibrating sound source in the vocal organ. Formants, on the other hand, contribute to the timbre of the vocalization.

A tracking filter in songbirds

Most birdsong contains frequency-modulated notes or 'syllables' that change in pitch. For these songs, a vocal tract filter could be maladaptive if it is tuned to a fixed, narrow frequency band, since frequencies outside this band will be attenuated. This problem can be avoided if the bird can adjust the tuning of his filter to match the changing frequency of the song.

Evidence that songbirds can control the tuning of their vocal tract was obtained from birds singing in light gas (a mixture of helium and oxygen). Since sound travels faster in light gas than in air, the wavelength for any given frequency is longer. This shifts formant frequencies upward and alters the filter properties of the vocal tract. By observing how formants change in light gas, one can deduce the resonance frequencies and how they change during normal song in air.

What is the physical basis of the tracking filter and how does the bird change the frequency to which it is tuned? Birdsong is usually accompanied by beak movements in which the beak opening, or gape, tends to be large at high frequencies and small at low frequencies. These observations have led to the hypothesis that birds use changes in beak gape to vary the effective acoustic length of their vocal tract and tune its resonance to the changing fundamental frequency of the song. Recently, X-ray cinematography of singing birds has revealed that songbirds actively adjust the dimensions of their vocal tract between the

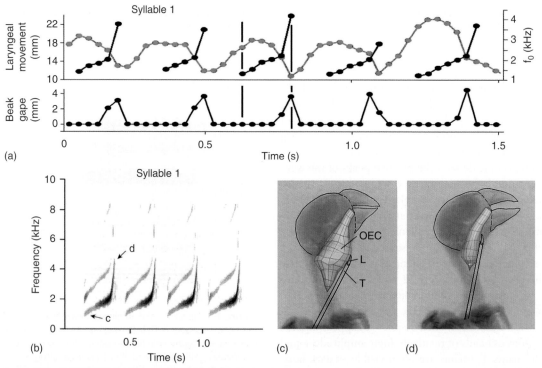

Figure 8 Northern cardinal tunes the major resonance of its vocal tract to the changing fundamental frequency of its syllables. (a) Volume of oropharyngeal cavity (red curve) increases at low fundamental frequency of upward sweeping song syllable (superimposed black lines). At high fundamental frequency cavity volume decreases and beak opens. Each dot is measurement from successive frames of X-ray movie. Red dots indicate movement of larynx relative to anatomical reference. Upward movement of red line corresponds to increased volume of oropharyngeal cavity. (b) Four song syllables in which the fundamental sweeps upward from about 1 kHz to about 5 kHz. (c) and (d) X-ray images at beginning (1 kHz) and end (5 kHz) of syllable 1 (c). Pink polygon shows expanded oropharyngeal cavity and cranial end of esophagus (OEC) at beginning of syllable. (d) At end of syllable esophagus has collapsed and volume of oropharynx is much reduced. A computational acoustic model based on volume of upper vocal tract estimated from a three-dimensional model indicates the primary resonance is about 1 kHz in (c) and 5 kHz in (d). Trachea (T) forms tube slanting upward from syrinx (not in image) to larynx (L) which opens into the mid-ventral region of the oropharynx. Adapted from Riede T, Suthers RA, Fletcher NH, and Blevins WE (2006) Songbirds tune their vocal tract to the fundamental frequency of their song. *Proceedings of the National Academy of Sciences of the United States of America* 104: 5543–5548, with permission.

larynx and beak so that its primary resonance frequency matches the fundamental frequency of their song. At low fundamental frequencies, the hyoid apparatus moves the larynx downward and forward, greatly enlarging the vocal tract by expanding the oropharynx and opening the cervical end of the esophagus (**Figure 8**). A computational acoustic model of the songbird's vocal tract suggests that beak gape contributes to the overall filter properties of the vocal tract, but the relative importance of beak gape versus changes in the dimensions of the oropharynx and esophagus remains an area of current research. In any case, by tuning its vocal tract to the dominant frequency component of the song, the bird increases the vocal efficiency and intensity of the emitted song. Higher harmonics that do not coincide with a resonance peak are suppressed, resulting in a more tonal vocalization.

Vocal tract morphology and acoustic diversity

Many animals have evolved morphological features of their vocal tract that could convey useful acoustic information. For example, the frequency spacing of formants in vocal signals can potentially convey important information regarding such things as individual identity, body size, dialect, etc. A number of different birds, including some cranes, swans, and members of several other taxa have evolved greatly elongated tracheas which form loops or coils inside the thorax or under the skin (**Figure 9**).

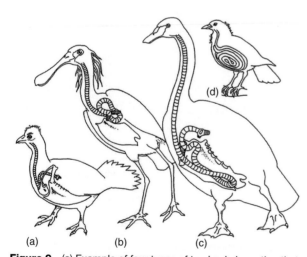

Figure 9 (a) Example of four types of tracheal elongation that may have evolved as a way of exaggerating body size by altering the formant frequency dispersion. Species shown are
(a) intraclavicular coiling, crested guinea fowl *Guttera edouardii*;
(b) intrathoracic coiling, European spoonbill *Platalea leucorodia*;
(c) intrasternal coiling, trumpeter swan *Cygnus buccinators*;
(d) subdermal coiling, trumpet manucode *Manucodia keraudrenii*. Reproduced from Fitch WT (1999) Acoustic exaggeration of size in birds via tracheal elongation: Comparative and theoretical analyses. *Journal of Zoology London* 248: 31–48, with permission.

This tracheal elongation might have evolved to make the sender sound larger than he really is and perhaps thereby increasing his attractiveness to a potential mate.

Many tetrapods have sac-like outpouchings of the vocal tract at various locations along its length from the trachea to nasal chambers. Their function in sound production and communication is not completely understood. In some cases, these structures may function as Helmholtz cavity resonators that amplify their resonant frequency while suppressing other frequencies. In other cases, they may provide a low-impedance pathway to facilitate sound transmission across the wall of the sac from the vocal tract to the external environment. Inflatable vocal air sacs are present in a number of other vertebrates in which an impedance-matching function has been experimentally verified in some cases.

Adult male Howler monkeys (*Alouatta*), whose name derives from their loud, low-pitched vocalizations have a hypertrophied larynx with long vocal cords that produce a low fundamental frequency. They also have laryngeal sacs that are surrounded by an enlarged bony hyoid bulla under the chin and open into the vocal tract near the glottis. These sacs are thought to act as Helmholtz resonators with their resonant frequency coupled to the vibration of the vocal cords. If this hypothesis is correct, they represent an exception to the source-filter model, which assumes no significant coupling.

Vocal sacs in anura appear to have multiple functions. The calls of many anura are accompanied by the inflation of a balloon-like subgular vocal sac. Nearly all anura have at least one vocal sac and some have laterally paired sacs. Air enters the vocal sac through slits in the floor of the buccal cavity as buccal pressure increases during phonation. When fully inflated, the vocal sac in some species is almost as large as the frog.

Since anura vocalize with their mouth and nares shut, the sound has to pass through the body wall to leave the animal. The thin-stretched wall of the inflated vocal sac probably provides a relatively low-impedance path that increases the amount of sound energy that passes through the wall of the sac from inside to outside the frog, but does not change the frequency composition of the sound – i.e., does not filter the sound. Experiments indicate that the vocal sacs of frogs are not Helmholtz cavity resonators in which the air inside the chamber vibrates. They might, however, be a different kind of Helmholtz resonator in which the tissue composing the wall of the chamber amplifies sound by vibrating at a resonant frequency. A similar mechanism probably occurs in the wall of the inflated esophagus of doves, which like anura, coo with their mouths and nostrils closed.

Anuran vocal sacs probably also serve nonacoustic functions. For example, vocal sacs store elastic energy and use it to quickly reinflate the lungs. This eliminates the need for buccal pumping before each call, thus increasing the energetic efficiency of call production and allowing higher call repetition rates. The vocal sac also provides a visual, as well as auditory, cue in bimodal communication in some species.

Underwater Sound Production by Marine Mammals

Marine mammals include several diverse taxa. Some are amphibious, like seals, walruses, and polar bears, which spend part of their lives on land. Others are obligatory marine species like whales and dolphins that live entirely in the water. Many species produce sounds underwater that are presumably used for acoustic communication and some are known to use echolocation. Acoustic signals allow marine mammals to exploit deep dimly lit or dark portions of their environment where visual information is severely limited or absent. Although the songs of baleen whales, for example, humpbacks and bowheads, have received the most popular attention, the mechanism by which they are produced is not well understood. Here I will focus on sound production by toothed whales and dolphins (suborder Odontoceti), since the mechanism by which they produce sound is better understood than it is in other obligate marine mammals. These toothed whales produce a wide variety of social whistles, broadband burst pulse sounds, and clicks.

Phonic Lips: A Nasal Sound Source

The speed of sound in water is nearly five times that in air and sound energy travels more efficiently through water than through air, greatly increasing the distance over which acoustic signals can be heard. In terrestrial mammals, vocalization is powered by airflow through the larynx, but a marine mammal is limited to the air it carries from the surface in its respiratory system. In order to solve this and other problems of submarine acoustic communication, toothed whales have evolved a completely new mechanism for sound production using a series of chambers and valves in their nasal passages. Their larynx is not used for sound production, but retains its important function as a valve that prevents food from entering the trachea. Since the vocal tract is not involved in Odontocete sound production, it is technically incorrect to refer to these sounds as 'vocalizations.'

Only the main features of Odontocete sound production will be described here. Instead of terminating in a pair of external nares at the tip of their nose as they do in most terrestrial mammals, Odontocete nasal passages have moved posteriorly where they converge in a chamber on top of their forehead that contains their single blowhole. The left and right nasal passages originate in the pharynx from which they travel dorsally on each side of the midline. As each nasal passage approaches the blow hole, it gives rise to several anatomically complex diverticula, the nasal air sacs,

that can be expanded or compressed by associated muscles. A short distance below the blow hole transverse ridges, the phonic lips (previously called 'monkey lips'), are present on the wall of the nasal passages. These ridges protrude into the lumen of the naris and are intimately associated with a fatty body or bursa (**Figure 10**).

Prior to phonation, the larynx is inserted into the proximal end of the nasal passages where they open into the pharynx. Muscles press the phonic lips together, on opposite walls of the naris, in a valve-like action that blocks airflow through the distal naris. Studies of vocalizing dolphins (*Tursiops truncatus*) show that sound is preceded by an increase in nasal air pressure, believed to be caused by the muscular compression of nasal chambers while the larynx is closed. Although the exact mechanism of sound production is not clear, it appears that sound is produced when air is forced across the phonic lips, pushing them apart and causing them to vibrate. Endoscopic observations of the phonic lips show that they oscillate in synchrony with the acoustic pulses. Air used to produce vibration is rarely exhausted through the blowhole, but is instead conserved by recycling it in the nasal system.

Acoustic Lenses

The acoustic energy of sounds initiated by airflow across the phonic lips of odontocetes must be efficiently

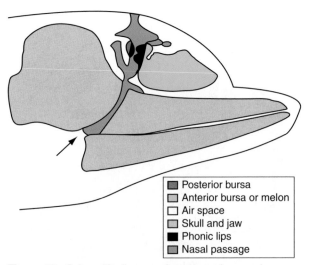

Legend:
- Posterior bursa
- Anterior bursa or melon
- Air space
- Skull and jaw
- Phonic lips
- Nasal passage

Figure 10 Schematic diagram of structures for sound production in the forehead of dolphins. Anterior bursa = melon. During sound production the larynx is inserted into the opening of the internal nares (arrow) but glottis is closed during sound production, which occurs at the phonic lips. Modified from Cranford TW, Amundin M, and Norris KS (1996) Functional morphology and homology in the odontocete nasal complex: Implications for sound generation. *Journal of Morphology* 228: 223–285 and Frankel AS (2009) Sound production. In: Perrin W, Wursig B, and Thewissen J (eds.) *Encyclopedia of Marine Mammals*, 2nd edn, pp. 1056–1070. Academic Press, with permission.

transmitted through their body tissue and then from this tissue into the surrounding water. Since the density of water is almost 800 times that of air, there is a severe impedance mismatch between air and body tissue (which is mostly water). Odontocetes have solved this problem by evolving masses of fatty tissue adjacent to the phonic lips. Vibration of the phonic lips is transmitted directly to these lipid-filled bursae, which resonate at the frequency of the phonation.

These fatty tissues also focus the radiated sound energy into a forward-directed beam. This is important because the similarity in the density of body tissue and water would otherwise cause sound to radiate in all directions. A large anterior bursa, called the melon, located in the dolphin's forehead just in front of the sonic lips (**Figure 10**), acts as an acoustic lens focusing sound energy in a forward direction. The speed of sound changes in a systematic way from the middle of the melon to its outer edge, because of corresponding differences in the chemical composition of its lipids, which have been referred to as 'acoustic fats.' The speed of sound in the middle of the melon is lower than that in sea water, but increases as one moves toward the outer edge of the melon where it is greater than in sea water (**Figure 11**). These peripheral sound waves are thus bent back toward the middle of the melon, to form a forward-directed beam. Concentrating the sound energy into a relatively narrow beam reduces the spreading losses that would otherwise occur if sound radiated from the body in all directions and increases the distance over which the animal can echolocate prey or communicate with conspecifics.

See also: Communication: An Overview; Dolphin Signature Whistles; Mating Signals; Túngara Frog: A Model for Sexual Selection and Communication; Vocal–Acoustic Communication in Fishes: Neuroethology

Further Reading

Au WWL and Hastings MC (2008) *Principles of Marine Bioacoustics.* New York, NY: Springer.

Bass AH and Clark CW (2002) The physical acoustics of underwater sound communication. In: Simmons AM, Popper AN, and Fay RR (eds.) *Acoustic Communication,* pp. 15–64. New York, NY: Springer.

Bradbury JW and Vehrencamp SL (1998) *Principles of Animal Communication.* Sunderland, MA: Sinauer Associates, Inc.

Cranford TW and Amundin M (2004) Biosonar pulse production in odontocetes: The state of our knowledge. In: Thomas JA, Moss CF, and Vater M (eds.) *Echolocation in Bats and Dolphins,* pp. 27–35. Chicago, IL: University of Chicago Press.

Feng AS and Narins PM (2008) Ultrasonic communication in concave-eared torrent frogs (*Amolops tormotus*). *Journal of Comparative Physiology A: Neuroethology Sensory Neural and Behavioral Physiology* 194: 159–167.

Fitch WT and Hauser MD (2002) Unpacking 'honesty': Vertebrate vocal production and the evolution of acoustic signals. In: Simmons AM, Popper AN, and Fay RR (eds.) *Acoustic Communication,* pp. 65–137. New York, NY: Springer.

Fletcher NH (1992) *Acoustic Systems in Biology.* New York, NY: Oxford University Press.

Gerhardt HC and Huber F (2002) *Acoustic Communication in Insects and Anurans.* Chicago, IL: University of Chicago Press.

King AS (1989) Functional anatomy of the syrinx. In: King AS and McLelland J (eds.) *Form and Function in Birds,* pp. 105–192. London: Academic Press.

Suthers RA (1999) The motor basis of vocal performance in songbirds. In: Hauser M and Konishi M (eds.) *The Design of Animal Communication,* pp. 37–62. Cambridge, MA: MIT Press.

Suthers RA and Zollinger SA (2008) From brain to song: The vocal organ and vocal tract. In: Zeigler HP and Marler P (eds.) *The Neuroscience of Birdsong,* pp. 78–98. Cambridge: Cambridge University Press.

Titze IR (1994) *Principles of Voice Production.* Englewood Cliffs, NJ: Prentice Hall.

Walkowiak W (2007) Call production and neural basis of vocalization. In: Narins PM, Feng A, Fay RR, and Popper AN (eds.) *Hearing and Sound Communication in Amphibians,* pp. 87–112. New York, NY: Springer Verlag.

Zollinger SA and Suthers RA (2004) Motor mechanisms of a vocal mimic: Implications for birdsong production. *Proceedings of the Royal Society of London B* 271: 483–491.

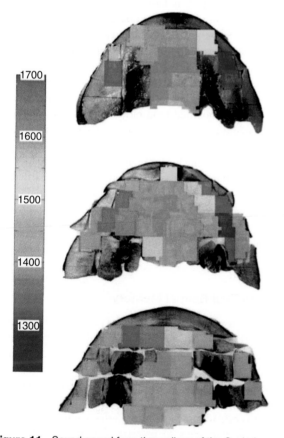

Figure 11 Sound speed from three slices of the Cuvier's beaked whale melon. Acoustic fats at center of melon have low sound speed compared to periphery and contribute to melon's function as an acoustic lens. Color bar indicates sound speed in m s^{-1} Mean sound speed in seawater is about 1500 m s^{-1}. Modified from Soldevilla MS, McKenna MS, Wiggins SM, Shadwick RE, Cranford TW, and Hildebrand JA (2005) Cuvier's beaked whale (*Ziphius cavirostris*) head tissues: Physical properties and CT imaging. *The Journal of Experimental Biology* 208: 2319–2332, with permission.

Spatial Memory

S. D. Healy, University of St. Andrews, St. Andrews, Fife, Scotland, UK
C. Jozet-Alves, University of Caen Basse-Normandie, Caen, France

Introduction

What Is Spatial Memory?

Most animals move around their environment at some point in their lives. Some of that movement simply involves moving away from their current location while other movement involves returning to a place, which may be a nest, a roost, a burrow, or a territory. Successful return to a specific place can be achieved in different ways using varying degrees of sophistication and resulting in more or less reliability. For example, among the simplest means of navigation is the use of a beacon, or a prominent landmark near to home, toward which the animal heads. Depending on the size of the landmark and how accurately it denotes the home location, this kind of homing constrains an animal to a relatively restricted range. An animal could move further away by leaving a visual or chemical trail such as the mucus trail left by limpets as they forage away from their 'home scar.' However, as in the case of Hansel and Gretel, who dropped breadcrumbs to help lead them back home, such trails are open to interference, damage, and loss, in their case by small animals eating the crumbs and removing enough of their trail to make it impossible for the children to find their way home. Another common, superficially simple, method of getting home is that of 'dead reckoning' in which even when the outward path is a circuitous one, the animal is able to head directly home because it continually updates its current position by taking into account the movements it has made. A swathe of work has been done on the use of this mechanism in invertebrates (although vertebrates use it too), especially in certain desert ants of North Africa.

Many animals appear to prefer to rely on an internal capability for accurate return, collectively called 'spatial memory.' Broadly, this is when an animal remembers information about a location that enables it to know where the location is. The animal is not simply remembering how objects at a location look, sound, smell, or feel, although sensory information of all these kinds, and more, may be put together to build spatial memory. Knowing where a place is means the animal can return to it from within familiar surroundings and, sometimes, even from unfamiliar surroundings.

Spatial memory has been demonstrated in a wide taxonomic range of species, from insects and cephalopods, to fish, amphibia, reptiles, birds, and mammals, including humans, with most of the evidence coming from laboratory data. Spatial memory in rodents, especially, is of both academic and industrial interest: rodents are a model species for memory drug testing by pharmaceutical companies and memory is of special interest. Data from the 'real' world are, thus far, surprisingly limited. Even for migration, where it seems plausible that migrants remember the locations of breeding and overwintering grounds, there is little evidence to demonstrate that these animals use memory to do so, precise as their relocation often is. On the other hand, it has taken laboratory evidence to demonstrate any involvement of spatial memory in some behaviors. Some birds, for example, may hoard hundreds, even thousands of food items, and not return for this food for days, weeks, or months. It took laboratory experiments by Krebs and coworkers to demonstrate convincingly that food-storing birds do, indeed, use spatial memory to retrieve their hoarded food rather than simply searching for it, as unlikely a feat as it may be. Additional elegant work by Clayton, Dickinson, and others has shown that some food-hoarding birds remember not only the location of the food items but also when and what they hoarded. As a result of this latter work, investigation into spatial memory has more recently become associated with interest in episodic (humans) and episodic-like memory (non-human animals), in which memories for the 'what,' 'where,' and 'when' of an event are recollected as one unit.

How to Test Spatial Memory

The way in which spatial memory is investigated depends on two things: firstly, the kind of spatial memory that is of interest, and secondly, the species to be tested. Although we may consider that spatial memory in the 'real' world concerns the ability of an animal to navigate around its world and return to specific locations, for logistical reasons it is rare to test this ability in that context. One of the most common ways to investigate spatial memory has been to use some kind of maze to examine memory for rewarding locations.

There are two kinds of maze that are used most often. One of these, the radial arm maze, traditionally consists of a small central area with eight arms fanning out from it, along which the animal walks to find food (or not). By observing whether the animal can remember both previously rewarded and emptied arms, both reference and working memory aspects of spatial memory can be assessed.

For example, in some versions of the task, the animal learns across days that some of the arms are rewarded and some never contain food. Running down an arm that is never rewarded then constitutes a reference memory error. Running back down an arm from which the animal has taken the food on that particular trial constitutes a working memory error. Radial arm mazes can have different numbers of arms, the common component being that the animal must return to the central arena after each choice.

An innovative version of the radial arm maze was developed for use with birds (which do not perform well in traditional mazes), whereby the 'arms' were shafts of light coming from each end of an 'arm.' In the middle of the 'maze' was a central perch where the bird was required to sit at the beginning of the trial. Once the animal was perched in the center, the room lights were switched off and the bird offered eight alternative equidistant perches, each lit with a small light and a feeder. Once the bird flew to one feeder (which may or may not have contained food) and fed, the lights on each of these perches were switched off and that on the central perch switched on. In this way, the bird flew to and from a central place, in a way that was analogous to the terrestrial version of the radial arm maze.

Other kinds of maze require the animal to move from a start point to a goal, learning to avoid arms or routes that do not lead to the reward, in a manner similar to mazes used to test humans. This kind of maze has often been used for animals that are not terrestrial, as, while the radial arm maze works well when rodents and other terrestrial mammals are to be tested, it is less effective for other animals that use different modes of movement. Aquatic animals and invertebrates, for example, have rarely been tested with radial mazes but rather with unidirectional mazes. These unidirectional mazes still allow examination of a single or a series of choices that an animal makes as it progresses through the maze. A very simple version of such a maze is the T-maze, which is also used in testing rodents. The T-maze is literally a maze in the shape of a T and the animal is placed in one arm of the T (sometimes the arms are all the same length). The animal, therefore, has a choice between two options.

The final 'maze' that is extremely commonly used is, in fact, not a maze at all. It is the Morris water maze (MWM), developed by Richard Morris in 1982, and is now probably the most typical method for testing spatial memory in rodents. An MWM consists of a large container of opaque water (usually 1.5–2 m in diameter), deep enough that the animal must swim to move around in the pool, in which there is a platform hidden just below the surface of the water. The animal is trained to find, and climb out onto, the platform. Both reference and working memory can be tested in this apparatus: in a reference memory version, the platform remains in the same place across multiple swims (usually one or two a day for

multiple days), while in a working memory version, the platform stays in place for several swims each day but is moved across days. In this test, the animal has no choice but to locate the platform and the measure of performance is the time and path length that the animal takes to reach the platform.

Although human spatial memory can readily be tested by means that approximate navigation, such as map reading, or giving and following navigational instructions, mazes have been used only relatively recently and in a virtual context, that is, computer-presented. Most tests of spatial ability in humans can appear rather remote from navigation. One such test is that of delayed-matching-to-sample (DMTS) or delayed-nonmatching-to-sample, in which a sample (usually an image on a screen) is presented, to which the subject may be required to respond, followed by a delay (often only a few seconds) and then the subject is offered a choice between the sample and one or more alternatives. The correct choice depends on whether the subject is expected to choose the matching or the nonmatching image/object. This method is used to test spatial memory by requiring the subject to use the location of the image/object to make the correct choice and can be very useful logistically as multiple trials can be presented in a short space of time, images/objects and their locations can be readily varied and, in many cases, the data can be collected automatically by computer. There are similar tasks that can also be used to test spatial memory such as spatial alternation in which the subject must simply alternate between two locations to receive reward.

In nearly all these test situations, the subject must be trained to find the rewarded location, but there are a few contexts in which this is not the case. Tests of spatial memory in food-hoarding animals involve the animal hoarding food items and returning to them, just once. Homing in pigeons requires the homing ability to be initially conditioned, but once this is acquired the birds can be released anywhere (sometimes hundreds of miles away) and they will return home.

Components of Spatial Memory

In the kinds of task described earlier, an animal's ability to remember a location can be tested. In the same tasks, it is also often possible to determine what information the animals are using to make their decision. For example, by manipulating the visual landmarks within and outside a maze, one can see whether the animal uses proximal (nearby cues) or distal (distant) cues, and whether the animal uses a single cue as a beacon that points the animal directly to the goal location, or whether it uses the spatial arrangement of cues. From these studies, it appears that animals learn distances and directions as well as the visual attributes of landmarks to find goal locations, with males and females relying differentially on the two types of

Spatial Orientation and Time: Methods

K. E. Mabry, New Mexico State University, Las Cruces, NM, USA
N. Pinter-Wollman, Stanford University, Stanford, CA, USA

The study of animal space use and movement is a rapidly growing area in animal behavior. New technologies such as GPS (global positioning system) tags allow researchers to obtain unprecedented amounts of information about how animals move through and interact with the landscape, and with each other. Animal movement and spatial behavior is an exciting research area, and one that often bridges the fields of animal behavior, population ecology, and conservation biology. For example, animal behavior researchers may be interested in the influence of mechanistic and ecological factors on the dispersal behavior of individual animals. The data gathered in such a study of dispersal behavior may also apply to questions about population redistribution across the landscape, or the prediction of how endangered species will respond to habitat fragmentation and modification. The number of tools and techniques available for analyzing animal spatial data is rapidly expanding; in this article, we present a subset of the available methods. We assume that the data for each tracked animal consist of (at minimum) a series of X, Y coordinates and associated time stamps; an almost infinite number of additional measurements, such as habitat type, elevation, and distance to water or roads, may be associated with each location.

Data Collection Considerations

Before spatial analyses can be conducted, the data must be collected in a manner appropriate to answer the study question. For this reason, and because it is often costly (in terms of both equipment and researcher time) to conduct a tracking study, we recommend that researchers think carefully about several issues before beginning data collection. There are matters to consider both in tag deployment and in tag reading. Tag deployment issues include which, and how many, animals to tag. Tag reading considerations are how frequently to collect locations, at what time of day to track, how many locations to obtain, and location error. Because studies of spatial behavior are so diverse in terms of both objectives and study species, decisions made about tag deployment and reading will depend greatly upon the specific study.

Which animals to track will depend on the study question: a researcher interested in natal dispersal will track juvenile animals, and a study of sex differences in home range areas will require tracking both male and female adults. The number of animals that can be tracked is often restricted by financial concerns, but despite financial constraints, one must ensure that a large enough sample size will be obtained to adequately test hypotheses. Ad hoc power analysis may be used to determine the required sample size.

Data collection frequency will depend on both tag constraints and the question of interest. Small tags (for small animals) often have limited battery life, requiring that locations be collected frequently over a relatively short period of time. In addition, obtaining detailed movement paths requires a high frequency of data collection, and the timing of data collection will depend on the movement speed of the study animal. The time of day for data collection will depend upon the goal of the study – if it is to characterize the inclusive space use of an animal, data must be collected at all hours of the day and night. However, many studies require data collection only during the animal's active period (i.e., during the day for diurnal animals or at night for nocturnal creatures). How many locations to collect per individual will also depend upon the goals of the study; for example, when analyzing home ranges of resident animals, a good rule of thumb is that at least 30–50 locations should be collected per individual. Finally, researchers should consider and measure location error. Location error may be due to a variety of factors, including the distortion of VHF radio signals due to 'signal bounce' off physical obstacles (such as hills) and errors in GPS locations when GPS tags are located in densely vegetated areas. If the error is large relative to the size of the study area or the animal movements, the data collected may be of limited use.

Categories of Spatial Analyses

We address two primary types of analyses in this article: (1) space use by resident animals (home ranges) and (2) the analysis of movement paths. The spatial behavior of resident and nonresident animals may be quite different from one another and will often require different methods of data collection and analysis. For example, studies of the spatial behavior of resident animals often focus on delineating and quantifying the home range area (the area typically used by an animal in its daily activities) or the territory size (the area defended by an animal). However, the home range concept does not apply to nonresident animals – in fact, the space used by a migrating animal may change on a daily basis. It is important to carefully consider the objectives (i.e., describe home ranges of adults, or follow the movement paths of dispersers) of a study of spatial behavior before data collection begins, to ensure that the study design will result in data appropriate for the analysis method.

Analysis of Space Use by Resident Animals

Many methods, consisting primarily of home range estimators, are available to describe space use by resident animals. Because of the large number of home range estimators, which are highly variable in their data requirements, ease of use, and suitability, we focus our discussion here on the two most widely used methods: the minimum convex polygon (MCP) and the kernel density estimator (KDE). MCPs and KDEs share some assumptions about the manner in which data are collected, the most important of which is the assumption of temporal independence of locations. For many years, achieving a lack of temporal autocorrelation between the locations of successive points was considered a primary goal of data collection for home range studies. The focus on temporal independence was due to the fact that in addition to violating the assumption of independence of data points, temporally autocorrelated points also yield less information about space use than do independent points. More recently, it has been realized that complete temporal independence is not necessary for most studies, but it is still true that data points should be collected across the animal's active period, with a reasonable amount of time between locations. To illustrate why the temporal spacing of data points is recommended, consider that a researcher will obtain much more information about the activity of a diurnal animal if ten locations are collected throughout the daylight hours than if all the ten locations are recorded between 3 pm and 4 pm. Home range analysis methods can be implemented in a variety of computer programs, and the most commonly used programs by researchers in animal behavior and wildlife ecology are ArcView 3.x (using the Animal Movement Extension), RANGES, and CALHOME, although other programs, including ArcGIS, GRASS, and MapInfo are also used.

Minimum convex polygon

The oldest, and perhaps the most widely used, home range estimator is the MCP. The MCP is calculated by connecting the outermost points at which an animal was located, which results in a two-dimensional polygon representing the home range (**Figure 1(a)**). In practice, researchers often discard 5% of point locations when calculating the home range, which results in a 95% MCP (**Figure 1(b)**). The rationale behind this approach is that discarding 5% of locations should eliminate outliers. However, the 5% figure is arbitrary, and researchers may calculate MCPs, using any percentage of the available data. For example, 50% MCPs may be used to represent 'core areas,' and 100% MCPs may be used by those who do not want to miss any space used by an animal (but note that 100% MCPs may include area that is not used by the animal at all; see **Figure 1(a)**). MCPs were developed before the widespread availability of computers and are thus computationally quite simple – a researcher may even physically plot locations on a map and draw the borders of the home range by hand, but most now use computer programs that quickly plot locations and calculate home range areas.

Because of the intuitive nature of the MCP, its ease of calculation, and widespread historical use, this method is still employed by many researchers. However, there are quite a few drawbacks to the MCP. Technically, MCPs may be drawn with as few as three unique locations, but recent simulation studies have suggested that as many as 100–300 locations may be required to achieve an accurate representation of the home range, using MCP estimators. Because the MCP method results in a two-dimensional polygon with no indication of the relative use of different areas within the home range, too much weight may be assigned to points on the edge of the range. That is, the presence of a few locations occurring far from the bulk of the points may result in a substantially larger range than would otherwise be obtained (compare **Figure 1(a)** and **1(b)**). Finally, MCPs do not generate estimates of centers of activity. Some researchers now suggest that home range studies report the results of both the MCP and at least one additional home range estimator, the rationale being that the MCP is comparable among studies and allows comparison with results from the older literature. However, other researchers disagree, and suggest that the disadvantages of the MCP method are so great that this method should not be used at all. We leave this decision up to the researcher.

Kernel density estimators

The use of KDEs has become quite widespread in the past decade. Unlike MCPs, KDEs allow researchers to determine the animal's relative use of different areas of the home range, and allow for the determination of multiple centers of activity. KDEs estimate the intensity (or probability) of use at particular locations (the utilization distribution, or UD). KDE home ranges may be represented in three dimensions as map contours, with higher 'peaks' in areas with a greater probability of use (**Figure 2**). Software packages generate KDEs by evaluating a probability density function: a mathematical formula that incorporates a kernel function, the number of point locations, the smoothing parameter, and the coordinates at which the animal was located. The probability density function estimates the animal's probability of using a particular point on the landscape. Researchers often implement KDEs with the most-intensely used 50% of locations to estimate core areas, and 95% KDEs

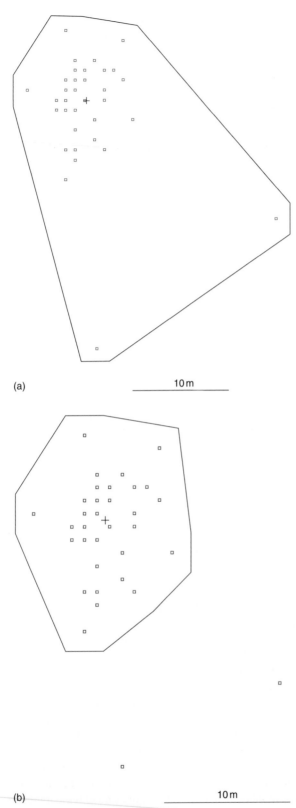

(a)   10 m

(b)   10 m

Figure 1 (a) 100% MCP home range of a male prairie vole. Note the large influence of the two points at the bottom of the range. The area of this home range is 625.5 m². (b) 95% MCP home range, generated from the same data shown in (a). Without the inclusion of the two distant points, the area of this home range is 194.5 m².

to estimate the larger home range area. A critical issue when calculating KDEs is the choice of smoothing parameter h (also called bandwidth; see **Figure 2(b)–2(d)**). If the value chosen for h is too small, a fragmented home range may be generated (**Figure 2(b)**). Alternatively, if h is too large, an overly smoothed single peak may be generated (**Figure 2(d)**). The choice of h is important because home range size estimates derived from fragmented KDEs may underestimate the home range area, while estimates from overly smoothed KDEs may overestimate the home range size and habitat use. The two major categorizations of smoothing parameter are fixed and adaptive. Fixed kernels utilize the same smoothing parameter across the entire area, while adaptive kernels use a different bandwidth for each location. KDEs are currently the standard methodology in home range estimation, and have relatively few disadvantages. The primary weakness of the method is the large influence that the choice of smoothing parameter can have on home range estimates.

Mechanistic home ranges

Both the MCP and KDE home range estimators are primarily descriptors of space use. Unfortunately, such statistical home range estimates have little or no predictive value. In contrast, the recently developed mechanistic home range analysis methods reviewed by Moorcroft and Lewis allow researchers to use correlated random walk movement models to integrate empirical field studies and predictive theory about animal space use. For example, researchers may make predictions about how space use will change over time, given changes in predator presence or food availability. However, this approach is still very new, and implementing mechanistic home range methods is considerably more involved than utilizing existing statistical home range methods.

Social behavior

In addition to determining the size of an area that individuals use, home range analyses may also be used to investigate social behavior. Researchers often use home range overlap to quantify the strength of association between two animals. For example, in **Figure 3**, a female prairie vole's (animal A) home range is overlapped by the home ranges of two males (animals B and C). The relative strength of A's association with B versus C may be quantified by comparing the percentage of A's home range that is overlapped by the home range of animal B with the percentage of A's home range that is overlapped by animal C's home range. Many home range analysis programs will output pair-wise home range overlap for any two animals in a data set. It is important to consider the directionality of home range overlap measures – because any two home ranges are likely to contain different amounts of area, the percent overlap of home range A on home range B will often be different from the overlap of B on A (**Figure 3**).

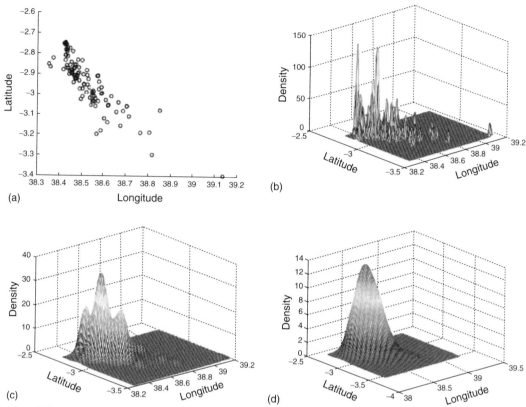

Figure 2 Quantifying the home range of an African elephant using a KDE. (a) The spatial locations of the animal in two dimensions. (b) KDE with $h = 0.008$. (c) KDE with $h = 0.03$. (d) KDE with $h = 0.08$. The figure was created using Matlab. Details of the functions used can be found in Martinez W and Martinez A (2002) *Computational Statistics Handbook with MATLAB*. Boca Raton, FL: Chapman & Hall/CRC.

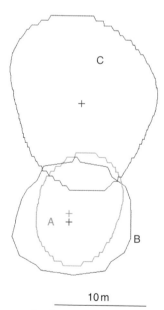

Figure 3 An example of two male prairie vole home ranges (B and C) overlapping a single female's range (A). Male B's home range overlaps 93.87% of female A's, while male C's home range overlaps just 24.80% of female A's. Note that the overlap value depends upon which animal is the focal: overlap of B on A = 93.87%, but the overlap of A on B is just 65.64% (because B's home range is larger than A's).

It is also relatively simple to determine the number of conspecifics with which an animal shares space, by counting the number of home ranges (or core areas) that coincide. If the objective of a study is to calculate a mean value of home range overlap (e.g., mean percent overlap of male home ranges on female home ranges) for an entire population, researchers must be careful to simultaneously track all animals of interest, because untracked individuals will bias the results.

Habitat use

Home range analysis is also useful for describing habitat use. By overlaying a home range on a habitat map of the study area, researchers can determine whether an animal uses particular habitat types more or less than would be expected based on the availability of those habitat types. Such an analysis of habitat use requires some type of geographic information system (GIS) habitat database. Digitized habitat maps may already be available for a study area; if not, researchers may digitize existing maps, or carry out habitat surveys. Alternatively, many sources now provide satellite images that can be used as the basis for creating digital maps. Two examples are Landsat and Satellite Probatoire d'Observation de la Terre (SPOT),

comparing movement paths measured on the same spatial scale. FRACTAL is a freely available computer program that calculates D, along with several other measures of path complexity.

Dispersal range analysis

Home range methods offer a relatively standardized way to quantify area use by resident animals. But how should researchers quantify the area used (or perhaps more appropriately, searched) by nonresident animals? Some researchers have applied home range methods to analyze data on the movements of dispersing animals, but several problems with this approach have been pointed out. For example, many home range analyses begin by eliminating outliers (i.e., 95% MCPs); however, those 'outliers' are often the most important and interesting data in dispersal studies. Erik and Veronica Doerr developed a suite of variables that can be used to quantify search behavior from tracking data. Here we present two of these variables: search area and search thoroughness. The calculation of search area is based upon the assumption that as animals move through an area, they are able to assess the surrounding habitat within a certain distance, dictated by their perceptual range (the assessment radius (AR)). If both the movement path and the width of the AR are known, these data can be combined to generate an assessment corridor (AC) of width $= 2(AR)$ surrounding the movement path. The search area can then be estimated as the area contained within the AC (**Figure 4(a)** and **4(c)**). Search thoroughness may be quantified as fractal D, or as the ratio of the AC area to the area of the MCP containing all of an animal's locations (AC/MCP; **Figure 4(b)** and **4(d)**). Low AC/MCP values indicate low thoroughness, and ratios approaching 1 indicate very thorough search. DRAP, or Dispersal Range Analysis Program, is a freely available program that calculates these and other measures of dispersal behavior.

Theoretical Models of Individual Movement Behavior

Much effort has been devoted to the theoretical modeling of animal movements, primarily from an ecological (rather than a behavioral) perspective. The modeling literature is vast, and much of it concentrates on questions of population redistribution, rather than on the mechanisms influencing individual movement behavior. An awareness of recent advances in the field of modeling animal movements is beneficial but somewhat beyond the scope of the current article. For an excellent introduction to models of movement behavior, we direct the reader to Turchin's book.

See also: Remote-Sensing of Behavior.

Further Reading

Batschelet E (1981) *Circular Statistics in Biology*. New York, NY: Academic Press.

Buskirk SW and Millspaugh JJ (2006) Metrics for studies of resource selection. *Journal of Wildlife Management* 70: 358–366.

Doerr ED and Doerr VAJ (2005) Dispersal range analysis: Quantifying individual variation in dispersal behaviour. *Oecologia* 142: 1–10.

Laver PN and Kelly MJ (2008) A critical review of home range studies. *Journal of Wildlife Management* 72: 290–298.

Manly B, McDonald L, Thomas D, McDonald T, and Erikson W (2002) *Resource Selection by Animals: Statistical Design and Analysis for Field Studies*. Dordrecht: Kluwer Academic.

Millspaugh JJ and Marzluff JM (2001) *Radio Tracking and Animal Populations*. New York, NY: Academic Press.

Moorcroft PR and Lewis MA (2006) *Mechanistic Home Range Analysis*. Princeton, NJ: Princeton University Press.

Thomas DL and Taylor EJ (2006) Study designs and test for comparing resource use and availability II. *Journal of Wildlife Management* 70: 324–336.

Turchin P (1998) *Quantitative Analysis of Movement: Measuring and Modeling Population Redistribution in Animals and Plants*. Sunderland: Sinauer Associates, Inc.

Relevant Websites

http://anatrack.com/ – Anatrack Ltd. Analysis and Tracking Software.

http://www.absc.usgs.gov/glba/gistools/animal_mvmt.htm – Animal Movement ArcView Extension.

http://www.esri.com/ – ESRI (GIS software).

http://nsac.ca/envsci/staff/vnams/Fractal.htm – FRACTAL (free software to analyze movement paths).

http://grass.ibiblio.org/index.php – GRASS GIS (free GIS software).

http://landsat.gsfc.nasa.gov/ – NASA Landsat Program.

http://www.noaa.gov/ – National Oceanic and Atmospheric Administration (NOAA).

http://www.movebank.org/ – MOVEBANK (online repository of movement data).

http://kovcomp.co.uk/oriana/index.html – ORIANA (circular statistics software).

http://www.r-project.org/ – The R Project for Statistical Computing.

http://www.spatialecology.com/index.php – Spatial Ecology software.

http://www.spot.com/ – SPOT IMAGE.

Specialization

D. J. Funk, Vanderbilt University, Nashville, TN, USA

Introduction

It is interesting to contemplate a tangled bank, clothed with many plants of many kinds, with birds singing on the bushes, with various insects flitting about, and with worms crawling through the damp earth, and to reflect that these elaborately constructed forms, so different from each other, and dependent upon each other in so complex a manner, have all been produced by laws acting around us ... There is grandeur in this view of life, with its several powers, having been originally breathed into a few forms or into one; and that, whilst this planet has gone cycling on according to the fixed law of gravity, from so simple a beginning endless forms most beautiful and most wonderful have been, and are being, evolved (Charles Darwin).

In this famous passage, Darwin expresses his admiration for the evolution of biodiversity via natural selection. One might further note, however, that the existence and coexistence of this wondrous biodiversity owes, in good part, to the evolution of ecological specialization – the use of a specific subset of locally available resources. Ecological specialization commonly allows otherwise similar species to coexist by partitioning available resources (e.g., food and habitat), thus preventing interspecific competition for them. Complementarily, the competitive exclusion principle predicts that two species using the same resources (or niches) cannot stably coexist, as competition will eventually yield the extinction of one or the other.

Classic examples of ecological specialization and resource partitioning include Darwin's own Galapagos finch species, which behaviorally specialize in their foods (e.g., different-sized seeds) and exhibit divergent morphological specialization in the form of beak structures allowing the efficient consumption of species-specific food items. A more unusual example is provided by the cichlids of the African rift lakes. These fishes have rapidly evolved species of specialist feeders on foods as diverse (and bizarre) as algae, plants, zooplankton, hard-shelled mollusks, insects, fish, the babies of mouthbrooding cichlids (which they force the mother to disgorge), fish parasites, fish scales, and fish eyes! As with the finches, these behavioral specializations are often accompanied by anatomical ones. Other cases are provided by groups of closely related herbivorous insect species that coexist on the very same host plant species. This is accomplished by partitioning the plant itself, with different species specializing on different plant parts (e.g., flowers vs. seeds vs. stems, etc.) and even on different plant sexes.

The opposite of the ecological specialist is the ecological generalist, a species that uses a considerable variety of resources. Generalists are not exposed to strong and specific directional selection pressures to the degree that specialists are. Rather, they are simultaneously subject to disparate sources of selection associated with the use of each of their resources. Thus, they less frequently exhibit the 'extreme' and diverse traits that often characterize the intimate ecological relationships of specialists.

The above issues are illustrated by the effects of the end-Cretaceous mass extinction (of all animals >50 lbs.) on mammalian evolution. True mammals had existed for ~100 My preceding this extinction event. Yet their evolutionary diversification had been unimpressive, with their modest taxonomic diversity primarily restricted to small, morphologically generalized (more or less rat-like) insectivores. This paucity of early mammalian variety has commonly been attributed to their competitive exclusion from most niches by ecologically dominant reptile taxa, such as dinosaurs. This hypothesis receives support from the spectacular phylogenetic and phenotypic 'mammalian radiation' that occurred during the ~10 My immediately following the mass extinction. Most major mammalian orders first appear in the fossil record at this time, as do many of the most specialized mammalian forms such as bats, whales, primates, and enormous herbivores. This belated mammalian profusion was seemingly promoted by access to suddenly unexploited resources/niches and their subsequent partitioning.

A different, coevolutionary, scenario of adaptive radiation was offered by Ehrlich and Raven to explain the evolutionary proliferation of two ecologically associated groups of organisms: plant-feeding insects (hereafter referred to simply as 'herbivores') and flowering plants. The scenario begins with an herbivore lineage overcoming the defenses of an unexploited group of plants (e.g., by evolving digestive enzymes that neutralize plant toxins). The herbivore lineage exploits and partitions this new resource via the evolution of new herbivore species, each specialized on a different plant. Subsequently, a lineage of these plants evolves a newly impenetrable defense, allowing it to analogously diversify by filling the niches it had been excluded from by the presence of its herbivore attackers. Iteratively repeating this coevolutionary cycle could thus explain much about specialization and species richness in these two groups. This model helped establish

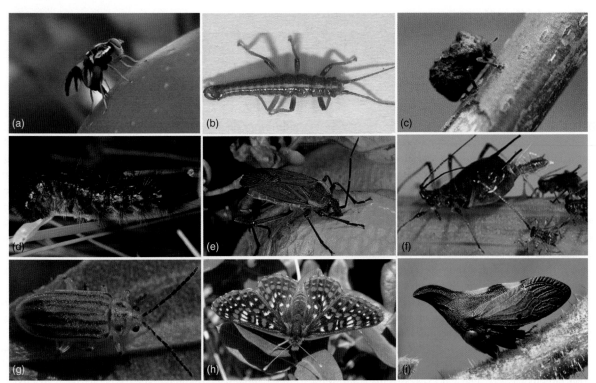

Figure 1 Herbivorous insect species that have provided insights on the evolution of ecological specialization. Most are invoked in the article. (a) *Rhagoletis pomonella* fruit flies, whose host race formation via specialization on apple versus hawthorn trees has informed our study of speciation through work by Jeffrey Feder and colleagues. Courtesy of J. Feder. (b) *Timema cristinae* stick insects, whose two ecotypes exhibit specialized morphologies (body shape and stripe presence/absence) that render each ecotype best camouflaged from predators when on its native host plant. Studied especially by Chris Sandoval and Patrik Nosil. Courtesy of P. Nosil.
(c) *Neochlamisus bebbianae* leaf beetles, whose populations each specialize on one of several unrelated tree species. These populations prefer, and survive best on, their native host plant and appear to be speciating as a consequence of this divergent host adaptation. Developed and employed as a study system by Daniel Funk. Courtesy of C. Brown. (d) *Grammia crotalaria* caterpillars, generalists that are predators on various small herbaceous plants, perhaps due to the nutritional value of such a mixed diet. Studied by Michael S. Singer, who provided the photo. (e) *Jadera haematomola* soapberry bugs, whose beak length has evolved markedly within a matter of decades in response to the introduction of plant species whose seeds (the bug food source) deviate in their depth from the fruit surface as compared to seed depth in the insect's normal host plant. Research and photo provided by Scott Carroll. (f) *Uroleucon ambrosiae* aphids, whose evolution of generalization from specialized ancestors was the focus of work by Daniel Funk and colleagues, and is detailed in the last section of this article. (g) *Ophraella notulata* leaf beetle, one of the species studied by Douglas Futuyma to evaluate the role of genetic constraints on the evolutionary history of host shifts and specialization across this beetle genus.
(h) *Euphydryas editha* butterflies, whose populations exhibit complex patterns of geographic variation in host plant specialization, including host preference changes that have evolved over a matter of years. Devotedly studied by Michael C. Singer, who provided the photo. (i) *Enchenopa binotata* treehoppers. The eggs of *Enchenopa* populations associated with different host plant species begin development at different times that specifically correspond to variation in initial spring sap flow among these hosts. Thus, treehoppers on different hosts mature at different times, reducing interpopulation mating opportunities and illustrating environmentally based temporal isolation. Investigated by the late Tom Wood.

the ecological specialization of insect herbivores as a major topic in evolutionary ecology, hence the focus on this phenomenon by the present article.

Herbivorous Insect Exemplars

Herbivory has independently evolved dozens of times across nine insect orders. Compared to other taxa, insect herbivores frequently exhibit strong ecological specialization (**Figure 1**). The degree of specialization of an herbivore species is commonly evaluated in terms of the variety of taxa used as host plants. Whether the criterion for specialist status is having hosts restricted to a single plant family, a single genus, or a single species, a sizeable proportion of herbivores qualify as ecologically specialized. At the same time, more 'generalized' herbivores that use a considerable diversity of host plants also exist. Thus, although most herbivores are specialists, considerable variation in the degree of specificity/generalism exists among herbivore taxa. For example, while Orthoptera (grasshoppers and kin) and certain Lepidoptera (butterflies and moths) families are dominated by highly polyphagous generalists that use many plant families as hosts, leaf beetles, and aphids exemplify groups dominated by monophagous specialists that use but a single host plant species. In between, are oliphagous species that use a modest number of host taxa. Other informative terms for characterizing herbivore specialization include 'preference hierarchy,' a ranking of the relative degree to which an herbivore prefers each of its multiple hosts, and 'host range,' which refers to the suite of particular plant species an herbivore uses as hosts.

In thinking about herbivore specialization, it is important to recognize that for many herbivores, the host plant is effectively its entire habitat. There, all life activities occur, from oviposition through larval development and adult feeding and mating. For such insects, host specialization thus imposes extremely strong selection on diverse aspects of biology. This may influence the evolution of sensory structures and behaviors governing the location and acceptance of the host as a site of oviposition, feeding, and mating; of anatomical structures allowing adherence to the host substrate and feeding on host tissues; of the physiological capacity to tolerate or avoid host toxins and digest host tissues; of behavioral strategies for avoiding desiccation and natural enemies; and so on.

With this in mind, imagine that a population specializing on host A becomes geographically isolated on novel plant B in an area devoid of host A. Since herbivore host-associations are often determined by behavioral preferences that belie a physiological capacity to develop on nonhost plants, our hypothetical population may plausibly survive on novel plant B. If indeed it manages to establish a population on B, strong selection would thereafter promote adaptation to it, resulting in its adaptive divergence from the ancestral population on host A in diverse phenotypic traits relevant to ecological specialization. Recent molecular studies reveal wide-ranging genomic effects of this divergent host adaptation, supporting the enormous influence of host-associated selection on herbivore evolution. For example, sampling hundreds of genetic loci from *Neochlamisus bebbianae* leaf beetle population specializing on maple versus willow trees indicated that 10–15% of their genomes may be evolving under the influence of this 'host-related selection.' And such selection can yield rapid results. For instance, within the last 100 years, soapberry bugs have evolved considerably longer beaks to penetrate the broader fruits of an introduced plant species, allowing them to feed on the seeds inside.

Because of their great diversity, discrete hosts/habitats, and tendency to exhibit phenotypic divergence, as well as their small size, fast generation times, and ready propagation, many herbivores provide great opportunities for controlled and informative studies of ecological specialization. Such studies may even have practical benefits, given the great economic costs imposed by herbivorous pests of crops, most of which are not native hosts of these pests. A notable example is the sowing of an herbivore's native host plants amidst crop plants. This provides the herbivore with a preferred alternative to the crop plant, thus ameliorating the otherwise intense selection pressures on such herbivores to adapt to crops grown in large monocultures. This strategy reduces the likelihood that a native herbivore will evolve into a (greater) pest.

Evaluating Host Preference: A Principle Determinant of Host Use

A herbivore's capacity to feed and develop on a plant is influenced by various anatomical and physiological factors, such as having appropriate mouthparts and digestive enzymes. However, host preference is governed by critical behavioral components. This section introduces some of these components and various means of studying them.

Since most herbivores must locate new host individuals during their life cycle (e.g., for oviposition), a herbivore's initial manifestation of host preference is often based on long-distance cues. These are generally olfactory and detected by receptors (e.g., on the antennae) sensitive to very small concentrations of air-borne plant volatiles. Locating the source of those volatiles eliciting interest brings the herbivore within range of potential visual cues. A plant acceptable to this point may be landed upon, such that touch (via mechanoreceptors) and taste (via chemoreceptors) come into play as the herbivore assesses external plant anatomy (e.g., surface waxes, plant hairs, etc.) and chemistry. This may be followed by the sampling of internal constituents via feeding and the assessment of postingestive feedback. Plants produce a diverse array of secondary metabolites that often play critical roles in plant acceptance versus rejection as a host. These include nonvolatile chemicals that contribute to the stimulation versus deterrence of herbivore feeding and oviposition.

Herbivore researchers have evaluated the stages and factors involved in host identification, acceptance, and preference using a heterogeneous array of approaches: Wind tunnels are used to evaluate long to medium distance host orientation and preference. A herbivore flying upwind from one end of the tunnel toward odor sources blown from the other end allows the assessment of its orientation behavior in the odor plume and of odor attractiveness. Another common technique uses 'Y-tube olfactometers,' whereby an insect is placed in the tube at the base of the 'Y' while alternative odors are wafted through the two distal tubes, providing the herbivore with a choice when it reaches the tube's fork. Visual cues may be evaluated by investigating herbivore responses to variables such as alternative leaf shapes and colors. Behavioral experiments are often supplemented by fine-scale, electrophysiological assays. The electroantennogram technique involves removing a herbivore antenna, inserting silver wires in each end and measuring the antenna's electrical output when exposed to an odor, thus quantifying relative

Male Behaviors Associated with Sperm Competition

Behaviors and morphological adaptations that reduce the risk of future sperm competition once a male has mated with the female or increase his fertilization success when the female mates again are termed '*defense*' mechanisms. There are three main ways in which males can reduce female remating. Males can transfer material during copulation that forms a mating plug. For example, in some spiders males mutilate themselves using their copulatory organs as a mating plug, effectively sterilizing themselves in the process. Secondly, males can guard the female, most commonly after copulation thereby preventing her from mating again before fertilizing her eggs. When females have a short window of receptivity or mate only once in their lifetime, males may guard the female before mating. Precopulatory mate guarding is also expected where the first male sires the majority of offspring, as mating with nonvirgin females do result in fertilization. Finally, males of multiply mating species often transfer compounds in the ejaculate that suppress female receptivity or render her unattractive to rival males thereby reducing the risk of sperm competition.

Adaptations to sperm competition when sperm are competing with other males' sperm from previous matings, when mating with already inseminated females, are termed '*offence*' mechanisms, which can occur in a variety of ways.

Mechanical Sperm Removal

Males of several species such as rodents, snakes, insects, and spiders possess intromittent organs deigned to physically remove sperm from previous matings. Alternatively, they may function to stimulate the female and induce ovulation, as copulation-induced ovulation is documented in mammals and marsupials, which may be beneficial to both sexes. Males increase the chances of siring offspring and females receive a reliable cue that successful insemination has occurred.

Sperm Stratification

The last male to mate with the female can also increase paternity by depositing its sperm in the most favourable place inside the female. Males can push previous ejaculates to the back of the females' sperm storage and placing his ejaculate closer to the site of fertilization.

Sperm Number

One of the most important determinants of sperm competition is the relative number of sperm that compete with each other. Not all sperm that males inseminate will end up in the female tract and hence be available for fertilization. The timing between the two inseminations can directly affect the number of sperm in competition. If there is a long time interval between copulations, the majority of sperm from the first mating may have been lost, or have died because of old age, or have been used by the female in fertilization.

The commonly observed last male sperm precedence indicates that there may be stronger selection on males to evolve effective sperm offence mechanisms compared to investing in defense mechanisms when in sperm competition. On the other hand, positive genetic correlations between males' sperm defense and offense abilities occur in some insects, suggesting that the same or linked genes are involved. Also, there may be stronger selection on males to reduce the risk of sperm competition altogether, which would preclude the need to invest in sperm offence mechanisms.

Male Strategic Ejaculation

Sperm production is associated with costs and males should therefore be careful in how they spend their limited sperm supply to maximize their reproductive success. There are numerous examples showing that males are sensitive to the expected fertilization returns and tailor the number of sperm provided accordingly. If males are able to discriminate the mating status of females, they should provide virgin females with fewer sperm since the risk of sperm competition is reduced. In contrast, when mating with a previously inseminated female, the male 'knows' that his sperm will face competition and hence should increase the number of sperm ejaculated. There is plenty of empirical evidence that males are sensitive to cues revealing female mating status and vary the number of sperm ejaculated. For example, male sticklebacks (*Gasterosteus aculeatus*) ejaculate more sperm when viewing videos of rival males courting females than of a male tending eggs. Similarly, human males viewing images depicting sperm competition ejaculate higher quality semen. Males can also vary the number of sperm provided in relation to female quality. For example, female size often covaries with fecundity, and males often provide larger, more fecund females with more sperm resulting in higher fertilization returns. However, bigger females often mate more frequently and hence may also represent an increased risk of sperm competition. Female age is another aspect directly affecting male fertilization returns. Older females are often less fecund as they have fewer eggs left to lay, and are more likely to have already mated therefore representing a higher risk or even higher intensity of sperm competition.

Male ejaculate tailoring can represent a cost to females in terms of reduced fertility. For example, females are often not able to discriminate between males in terms of their mating status. This can result in extended copulations as males may require a considerable recuperation time before being able to produce a new ejaculate, during which time he remains in copula before transferring his sperm, or in reduced fertility if mating with a recently mated male. Females mating with particularly popular males may also suffer sperm limitation, as these males experience dramatically higher copulation rates than less attractive males and can run out of sperm. In these cases, female must tradeoff reduced fertility against the benefit of mating with popular males, which may result in sons with higher mating success. For example, dominant male Soay sheep (*Ovis aries*) run out of sperm at the end of the rutting season leading to subordinate males having a chance of gaining some paternity in sperm competition. Similarly, in blue-headed wrasse fish (*Thalassoma bifasciatum*), males with the highest mating success confer the lowest fertility to females. There is evidence that females can compete with each other for access to males' sperm supplies in such circumstances. For example, in the lekking great snipe (*Gallinago media*), females compete for access to preferred males. Popular males can suffer sperm limitation because of repeated copulations and discriminate against females with which they have already mated, which may result in sexual conflict over males' sperm supply.

Female Behavior in Relation to Sperm Competition

Females mate multiply for a number of reasons, the most obvious being to ensure a steady sperm supply to fertilize her eggs, and may remate to counter inseminations from potentially infertile males. However, females also remate for several other reasons related to both direct and genetic benefits. Females may also remate if they are prevented from freely exercising mate choice to increase the chances that her offspring are fathered by 'preferred' males.

There are several examples from a range of species in which females directly interfere with sperm transfer and thus reduce the number of sperm stored by unfavored males. For example in many cricket and bush cricket species, males attach their sperm packet externally to the females' genitalia, which means that females can directly interfere with sperm transfer by removing the sperm packet before sperm transfer is complete (**Figure 1**). Following mating, females can eject sperm which may be a consequence of insemination by nonpreferred males, as seems to be the case in fowl (*G. gallus*) in which females are more likely to eject sperm after mating with subordinate males compared to a dominant male. Once sperm have been deposited inside the female, there is limited scope for males to manipulate females' sperm use. Females, however, have several mechanisms to increase the chances that sperm from preferred males are used in fertilization. She could preferentially store sperm from

favored males and shunt sperm from other males to areas where they are less likely to be utilized. Female yellow dungflies (*Scatophaga stercoraria*), for example, have multiple sperm storage organs and appear to differentially use sperm from different sperm stores (**Figure 2**). In crickets (*Gryllus bimaculatus*), there is evidence that females bias sperm use against related males by favoring unrelated males' sperm, as this increases egg hatching success. However, the precise mechanism whereby this is achieved is not known. Females may also ingest or kill off sperm after insemination. In hermaphrodites, individuals frequently ingest sperm and even have specialized organs with the sole function of sperm digestion. For example, in *Helix aspersa* snails only ~0.1% of all transferred sperm escape! Similarly, in mammals females commonly kill off a large proportion of the sperm by phagocytosis before they reach the ova. It is possible that females could selectively kill off sperm as a way to bias paternity against certain males. In general, it is often very difficult to separate the impact of female-generated from male-generated mechanisms responsible for differential male fertilization success. It is likely that both male and female mechanisms affect sperm use.

There are also several behaviors adopted by females that indirectly affect the paternity of her offspring. In many species, females risk being inseminated by males adopting a sneaker strategy. In both chickens (*G. gallus*) and elephant seals (*Mirounga angustirostris*) for example, females omit a stress call when a subordinate males manages to sneak a copulation, which alerts the attention of the dominant male, which will chase off the sneaker male and re-inseminate the female, reducing the risk of sneaker males fathering the young.

Ejaculate Evolution

An alternative way to reduce the need for producing costly sperm is to evolve other aspects of the ejaculate

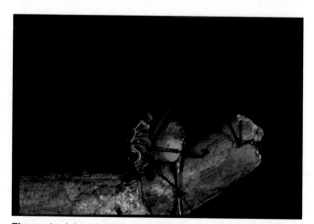

Figure 1   A female bush cricket (*Austrodectes monticolus*) busy feeding on a large externally attached spermatophore during sperm transfer. Courtesy of D. Rentz.

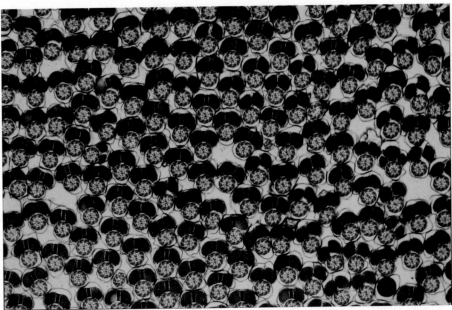

Figure 2   The sperm storage organ of the female yellow dungfly (*Scatophaga stercoraria*) is full of sperm (seen in cross section) following mating. Courtesy of D.J. Hosken.

that either reduce the risk of sperm competition, or increase the chances of fertilization in sperm competition. Reducing the risk of sperm competition can be achieved either by preventing the female from remating, or suppressing her receptivity and stimulating oviposition. Males have evolved various factors transferred in the ejaculate that function to achieve this aim; as mentioned, mate guarding and mating plugs are common in several animals. Another way in which males can reduce female receptivity is rendering them unattractive to rival males following mating. Male butterflies and moths can transfer chemical compounds, or antiaphrodisiacs, to females during mating that reduce their attractiveness. Males of the majority of polyandrous species often transfer compounds in the ejaculate that suppress female receptivity and stimulate oviposition and egg maturation rate.

Increased ejaculate efficiency may, instead of producing larger sperm, be achieved by grouping sperm together. In the firebrat (*Thermobia domestica*) for example, sperm are paired and motile only when coiled around each other, which is also found in possums, millipedes, beetles, and molluscs. A fascinating example of sperm cooperation is present in the wood mouse (*Apodemus sylvaticus*) where hundreds or thousands of sperm link together as a train using hooked structures on their heads, which enables them to swim at almost twice the speed of a single sperm. These trains break up prior to fertilization as only one sperm can fertilize an egg, requiring the majority of sperm to sacrifice themselves. Such hook-like structures appear to be common on rodent sperm and can vary dramatically in both size and shape (**Figure 3**).

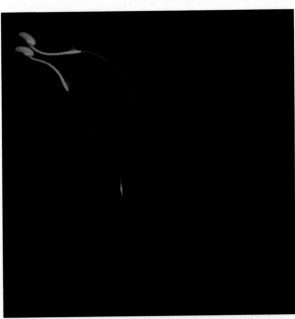

Figure 3   Sperm of the rodent *Hydromys chrysogaster* have hook-like structures (stained green) on their heads (stained blue) that can link up to form large 'sperm trains.' Courtesy of S. Immler.

Instead of clumping together more than one sperm type into units that may be more efficient in transporting the sperm, males can also produce different sperm types. Most commonly, only one of the sperm types in sperm heteromorphic species are able to fertilize the egg. There are several hypotheses for the function of nonfertile

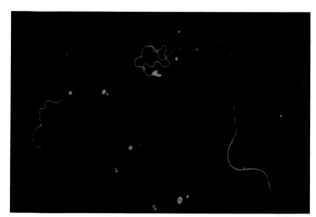

Figure 4 The ejaculate of the green-veined white butterfly (*Pieris napi*) contains both fertile (longer) and nonfertile (small and squiggly) sperm. Courtesy of N. Wedell.

sperm. They are suggested to aid the transfer of the fertile sperm or may represent a nutrient source for the female, the developing zygote or the fertile sperm while in storage, although there is no real support for this idea at least in insects. Nonfertile sperm can also play a role in sperm competition by reducing female receptivity or by interfering with rival males' sperm from previous mating. In butterflies, males produce a large number of nonfertile sperm, and in the green-veined white (*Pieris napi*), these sperm appear to fill the female's sperm storage and thereby reduce female receptivity (**Figure 4**). Nonfertile sperm are also suggested to actively incapacitate rival males sperm. In marine sculpin fish (*Hemilepidotus gilberti*), nonfertile sperm appear to enhance the delivery of sperm to the site of fertilization, but may also hamper subsequent rival males' sperm from reaching the egg. Both sperm types are shed in water where the nonfertile sperm form clumps that 'swallow up' fertile sperm.

Male Female Coevolution

Whenever females mate multiply and hence sperm competition occurs, there is a conflict between the sexes over female remating, the number of eggs laid before remating, and sperm use, among other things. Males have evolved several means aimed at manipulating females into making a larger investment into his offspring, even if this is costly to females. Male insects often transfer compounds in the ejaculate that manipulate female reproductive physiology by increasing immediate egg output, suppressing receptivity, or preventing female remating. This is beneficial for the male as it results in higher paternity. However, this often comes at a cost to the female. For example, in poecilid fish with internal fertilization, males' intromittent organs frequently damage the female genitalia, which may serve to reduce female remating due to cost of mating to females. Male ejaculate tailoring can also represent a cost to females

in terms of reduced fertility, and the production of nonfertile sperm may also potentially pose a cost compromising female fertility. For example, transfer and storage of large number of nonfertile sperm may interfere with successful storage and utilization of fertile sperm.

The cost to females of manipulative ejaculates favor female traits and behaviors that minimize such costs. However, this will impose further selection on males to come up with new ways to increase their paternity when in sperm competition, which may result in continuous cycles of adaptation and counter adaptation between males and females, potentially leading to elaboration of traits involved in this sexual conflict and coevolution of such traits. For example, in some butterflies, males try and suppress female receptivity by providing many nonfertile sperm, as they are cheaper to produce than fertile sperm. However, some females avoid this manipulation and regain control over their receptivity, imposing selection on males to further increase the number of nonfertile sperm delivered to suppress female receptivity. It is possible that sexual conflict over female receptivity, using nonfertile sperm, is responsible for the ejaculate in many butterflies consisting predominantly of nonfertile sperm. In addition, comparative studies of moths suggest nonfertile sperm function as 'cheap filler.' There is a positive correlation between the number of nonfertile sperm and female sperm storage organ size, whereas no such relationship was found for fertile sperm, indicating that female sperm storage volume and nonfertile sperm number have coevolved. In some flies, nonfertile sperm appear to function to protect the fertile sperm from female spermicide. The higher the proportion of nonfertile sperm transferred by *Drosophila pseudoobscura* males, the greater the likelihood his fertile sperm will survive in the females' reproductive tract and hence have a higher change of fertilizing the eggs. This may also be the outcome of coevolution between males and females, whereby males are able to withstand female spermicide by producing a less vulnerable sperm morph, which today make up about half the ejaculate in these flies.

In flies, there is also evidence that coevolution between the sexes affect sperm length, as *D. melanogaster* females with longer sperm-storage ducts favor longer sperm. This suggests that females have evolved longer sperm ducts to have some control over which males' sperm is used in fertilization, which favors males with longer sperm. Finally there is also evidence of rapid coevolution between males' seminal component and female reproductive behavior in several species of insects and mammals, including man.

Concluding Remarks

Sperm competition is ubiquitous throughout the animal kingdom and is responsible for the tremendous diversity seen in sperm shape and function. The challenge now is to

extremely diverse array of habitats in Africa, including savanna, desert, swamps, woodland and montane forest up to 4000 m of elevation. Although spotted hyenas need water for drinking, they can make do with very little water, and seldom require access to it. Even lactating females can survive without water for over 1 week. Plasticity in all of these domains permits spotted hyenas to be unusually resilient to environmental perturbations. Thus, because their behavioral plasticity far exceeds that of other large African carnivores and because spotted hyenas are relatively easy to monitor, their responses to environmental change should represent convenient and conservative indicators of ecosystem health.

Global expansion of human populations has caused alarming declines in, or extirpation of, many carnivore populations. Despite their remarkable plasticity and the

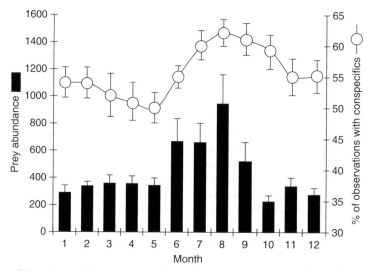

Figure 20 Monthly mean ± SE numbers of local prey animals counted each month during biweekly ungulate censuses within one hyena territory in the Massai Mara Reserve (left vertical axis and histogram bars) and percentage of observation sessions in which spotted hyenas were found in subgroups containing more than one individual (right vertical axis and open circles). Reproduced from Smith JE, Kolowski JM, Graham KE, Dawes SE, and Holekamp KE (2008) Social and ecological determinants of fission-fusion dynamics in the spotted hyaena. *Animal Behaviour* 76: 619–636.

Figure 21 Like many large carnivores, spotted hyenas are in conflict with humans over resources, and they sometimes kill livestock such as the cow shown here. Spotted hyenas are commonly killed in response to livestock depredation, even within protected areas (Photo by Joseph M. Kolowski).

fact that they are more abundant than any other large carnivore in Africa, populations of spotted hyenas living outside of protected areas are also declining. Disease is an important mortality source for hyenas in some areas, but humans and lions cause most adult mortality. Spotted hyenas are commonly killed by humans in response to, or in fear of, livestock depredation (**Figure 21**). Humans also have nonlethal effects on spotted hyenas that occur before hyena population size declines. Recent work suggests that anthropogenic activity, particularly that occurring in the form of pastoralists grazing livestock within the boundaries of protected areas, alters the behavior of spotted hyenas and forces them to make energetic compromises not observed where pastoralists are absent. To avoid potentially lethal encounters with humans, hyenas in areas used by pastoralists alter their activity and movement patterns, and they also spend more time being vigilant than do hyenas living in undisturbed areas. Furthermore, presence of pastoralists affects the stress physiology of spotted hyenas, as indicated by higher concentrations of glucocorticoid hormones. Interestingly, these negative effects do not occur in response to visitation by tourists, perhaps because tourists do not represent a threat to spotted hyenas.

The disappearance of this extremely resilient species from an African ecosystem indicates that the habitat has become very severely degraded, perhaps irreversibly so. In areas where spotted hyenas still occur, their behavior, stress physiology, and demography can be monitored to reveal warning indications of deleterious trends. If such trends can be identified and quantified, then they can potentially be halted or reversed. This is particularly important in sub-Saharan Africa, where loss of large carnivores would remove an important incentive for tourists to visit from abroad, and thus eliminate a key source of foreign exchange for many developing nations. As human population density continues to increase, and habitats continue to be modified by human activity, spotted hyenas will continue to serve as visible and conservative indicators of how these and more specialized large carnivores in the same ecosystems are responding to environmental change.

See also: Aggression and Territoriality; Body Size and Sexual Dimorphism; Communication Networks; Conflict Resolution; Conservation Behavior and Endocrinology; Cooperation and Sociality; Cryptic Female Choice; Decision-Making: Foraging; Dominance Relationships, Dominance Hierarchies and Rankings; Ethology in Europe; Group Living; Hormones and Behavior: Basic Concepts; Kin Selection and Relatedness; Maternal Effects on Behavior; Multimodal Signaling; Predator's Perspective on Predator–Prey Interactions; Sex Change in Reef Fishes: Behavior and Physiology; Social Cognition and Theory of Mind; Social Selection, Sexual Selection, and Sexual Conflict; Wintering Strategies: Moult and Behavior.

Further Reading

Dloniak SM, French JA, and Holekamp KE (2006) Rank-related maternal effects of androgens on behaviour in wild spotted hyaenas. *Nature* 440: 1190–1193.

Drea CM, Weldele ML, Forger NG, et al. (1998) Androgens and masculinization of genitalia in the spotted hyaena (*Crocuta crocuta*). 2. Effects of prenatal anti-androgens. *Journal of Reproduction and Fertility* 113: 117–127.

East ML and Hofer H (1991) Loud calling in a female-dominated mammalian society: I and II. *Animal Behaviour* 42: 637–669.

East ML, Hofer H, and Wickler W (1993) The erect 'penis' as a flag of submission in a female-dominated society: Greetings in Serengeti spotted hyenas. *Behavioral Ecology and Sociobiology* 33: 355–370.

Engh AL, Funk SM, Van Horn RC, et al. (2002) Reproductive skew among males in a female-dominated society. *Behavioral Ecology* 13: 193–200.

Frank LG (1986) Social organization of the spotted hyaena (*Crocuta crocuta*): I and II. *Animal Behaviour* 34: 1500–1527.

Hofer H and East ML (1993) The commuting system of Serengeti spotted hyaenas: How a predator copes with migratory prey. I, II, and III. *Animal Behaviour* 46: 547–589.

Holekamp KE (in press) *Evolutionary Ecology of the Spotted Hyena*. Princeton, NJ: Princeton University Press.

Holekamp KE, Sakai ST, and Lundrigan BL (2007) Social intelligence in the spotted hyena (*Crocuta crocuta*). *Philosophical Transactions of the Royal Society B-Biological Sciences* 362: 523–538.

Holekamp KE and Smale L (2000) Feisty females and meek males: Reproductive strategies in the spotted hyena. In: Wallen K and Schneider J (eds.) *Reproduction in Context*, pp. 257–285. Cambridge, MA: MIT Press.

Höner OP, Wachter B, East ML, et al. (2007) Female mate-choice drives the evolution of male-biased dispersal in a social mammal. *Nature* 448: 798–801.

Kruuk H (1972) *The Spotted Hyena: A Study of the Predation and Social behavior*. Chicago: University of Chicago Press.

Mills MGL (1990) *Kalahari Hyaenas: The Behavioral Ecology of Two Species*. London: Unwin Hyman.

Smale L, Frank LG, and Holekamp KE (1993) Ontogeny of dominance in free-living spotted hyaenas: Juvenile rank relations with adult females and immigrant males. *Animal Behaviour* 46: 467–477.

Smith JE, Kolowski JM, Graham KE, Dawes SE, and Holekamp KE (2008) Social and ecological determinants of fission–fusion dynamics in the spotted hyaena. *Animal Behaviour* 76: 619–636.

Tinbergen N (1963) On aims and methods in ethology. *Zeitschrift für Tierpsychologie* 20: 410–433.

Van Meter PE, French JA, Dloniak SM, Watts HE, Kolowski JM, and Holekamp KE (2008) Fecal glucocorticoids reflect socio-ecological and anthropogenic stressors in the lives of wild spotted hyenas. *Hormones and Behavior* 55: 329–337.

Wahaj SA, Place NJ, Weldele ML, Glickman SE, and Holekamp KE (2007) Siblicide in the spotted hyena: Analysis with ultrasonic examination of wild and captive I individuals. *Behavioral Ecology* 18: 974–984.

Relevant Websites

http://www.hyaenidae.org/ – IUCN Hyaena Specialist Group.
http://hyenas.zoology.msu.edu/ – Kay E. Holekamp Laboratory.
http://msuhyenas.blogspot.com/ – Michigan State University Hyena Research Blog.

Stress, Health and Social Behavior

R. M. Sapolsky, Stanford University, Stanford, CA, USA

Introduction

The ancients, replete with their togas, understood the mechanisms underlying pathophysiology – they were the whims of the gods. A smile upon your countenance by one of them you're good to go for a long, healthy life. Have some Olympian frown come your way and it is tapeworms and rickets from then on. As documented by the gerontologist Caleb Finch, somewhere between the times of Homer and Herodotus, a shift occurred, and the explanation for a long life was taken from the gods and given, in part, to people, and *how* they lived their lives. This ushered in a concept that is now one of the pillars of medicine, namely that behavior can influence health.

This can be obvious and not all that interesting. If we eat to excess, our fat cells are more likely to become insulin-resistant. If we consume large amounts of alcohol, our liver will pay. And if we regularly bungee jump off bridges with cords that are tattered, we will shorten our life expectancy.

But this relationship between behavior and health is both deeper and broader. Broader in that it includes not only behavior, especially social behavior, but also its underpinnings, such as thoughts. emotions, memories, personality, and temperament. Deeper, as it helps one understand how life's adversities can influence something as reductive as the length of telomeres, the stabilizing ends of chromosomes.

This is not pertinent to all domains of medicine. For example, our internal emotional life has little to do with the health outcome following drinking typhoid-riddled water from our medieval town well, being grievously injured, or starving. But it is extremely important for understanding the diseases of Westernized life, in which we live well enough and long enough to have our bodies slowly accumulate damage over time. This relevance has answered all sorts of unlikely questions pertinent to our health: What is our psychological makeup?, What is our social status?, How do people with our social status get treated in our society?, Or even a question such as, When we feel unloved, do we eat lots more carbohydrates? And, one might ask a non-human primate, "Are we the sort of individuals who behave in a way that gets us groomed frequently?"

An encyclopedia such as this reflects not only the intrinsic fascination of animal behavior but also the continuity between our own species and others, and the insight into humans that can be derived from the study of animals. This certainly extends to the physiology and pathophysiology that we share with other species, and to the ability of behavior to impact on health. Thus, the goal of this article is to consider how behavior can adversely affect health in a range of species, and the central role of stress in this relationship. (As a gross simplification but as an expository convenience, 'behavior' will also subsume thoughts, emotions, memories, personality, and temperament throughout the article. Moreover, despite the phylogenetic myopia, the article is heavily biased toward considering mammals, particularly primates.)

Stress, Homeostasis, and Allostasis

A concept that has dominated issues of stress and disease since the word 'stress' entered the medical literature (approximately a century ago) is that of 'homeostasis.' Classically, this is the notion that there are optima of physiological function – an ideal body temperature, level of glucose in the bloodstream, and so on. Framed this way, a 'stressor' is anything in the environment that disrupts the homeostatic balance, and the 'stress-response' is the array of physiological responses that reestablish homeostasis.

The homeostasis concept has been expanded in recent years into the newer concept of 'allostasis,' which differs from the older term in two important ways. The first is that ideal homeostatic set points can differ over time. For example, optimal heart rate differs dramatically, whether one is sleeping or sprinting, and optimal circulating estrogen levels differ dramatically, whether one is in the follicular or luteal phase of the ovulatory cycle. The second difference between the homeostasis and allostasis concepts is that the latter incorporates the fact that physiological balance can be reattained by multiple adaptations across wide ranges of physiological systems, including behavior. This can be illustrated with a nonmedical example: imagine that the state of California is suffering from the maladaptive situation of chronic water shortages. A local 'homeostatic' response might be to require all new homes to have low-flow showerheads. In contrast, a more extensive 'allostatic' response would be to do that, but to also negotiate for the state to get a larger share of water out of the Colorado River, cut back water subsidies to farmers growing tropical crops such as rice in a semi-arid state, promote the idea that cactus gardens look better than lawns, and so on. Similarly, in the case of an individual being dehydrated, homeostatic thinking would focus on the changes in renal filtration rates, while allostatic thinking would additionally incorporate far-flung responses including changes in sweat-gland function, energy utilization in muscle, and the neurobiology of thirst. As is relevant throughout this article, framing problems of stress in the context of allostasis helps appreciate not only the role of behavior in solving problems of allostatic imbalance, but also the potential for behavior to *generate* problems of allostatic imbalance, often thanks to creating a problem in the process of behaviorally solving another (e.g., solving a problem with social anxiety by drinking heavily).

The Physiology and Pathophysiology of the Stress-Response

Thus, the stress-response is the set of adaptations that are mobilized throughout the body to correct a state of allostatic imbalance. This involves a fairly stereotyped set of neural and endocrine changes. A critical one is the secretion of 'catecholamines' – epinephrine (aka adrenaline) and norepinephrine from the nerve endings of the sympathetic nervous system projecting throughout the body. Another is the secretion by the adrenal glands of a class of steroid hormones called glucocorticoids (GCs) (the primate version of GCs being cortisol). While there is an array of additional changes in levels of various hormones during stress (e.g., generally an increase in circulating levels of glucagon, prolactin, and beta-endorphin, decreases in insulin and reproductive hormones), the secretion of GCs and the activation of the sympathetic nervous system constitute the workhorses of the stress-response (see **Figure 1** for an overview of the stress-response).

This stress-response is adaptive when considering acute physical challenges to allostasis, such as running for our life or running after a meal. Energy is mobilized from storage sites throughout the body and delivered to the circulatory system in forms most readily used by exercising muscle. Blood pressure, heart rate, and breathing rate increase to deliver that energy, along with oxygen, to the exercising muscle as rapidly as possible. Various unessential processes are inhibited until more auspicious times. For example, the gastrointestinal tract is silenced; the energy is mobilized within seconds from fat cells and liver, whereas digestion is a slow, energetically costly process that can be deferred. As such, secretion of saliva is inhibited (i.e., our mouth gets dry), stomach contractions and small intestinal peristalsis cease, blood flow is diverted from the gut, and appetite is suppressed. Following the same logic, growth and tissue repair are inhibited. Reproductive physiology is damped in both genders for obvious reasons – the thickening of uterine walls or the production of sperm can wait until after you have successfully sprinted for your life. The possibility of an injury demonstrates the adaptive logic of other features of the stress-response – immunity is enhanced as a defense

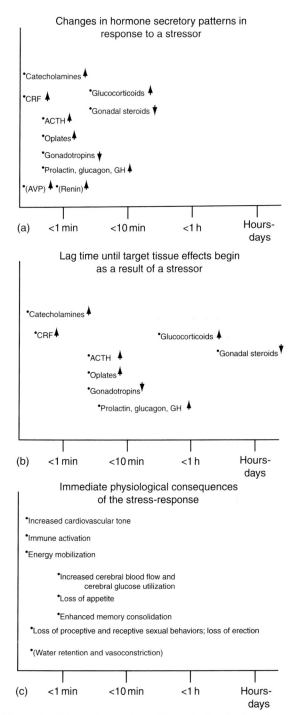

Figure 1 Schematic overview of the typical endocrine stress-response. (a) The time course of changes in hormone-secretory patterns in response to a stressor. (b) The lag time until target tissue begin as a result of a stressor. (c) Immediate physiological consequences of the stress-response. Asterisks approximate where on the time line the particular hormone is first having its effect (a and b) or when on the time line the physiological consequence is initiated (c). There is no formal *y* axis – hormones or consequences are simply spaced vertically to facilitate reading. Reproduced from Sapolsky R, Romero M, and Munck A (2000) How do glucocorticoids influence stress responses? Integrating permissive, suppressive, stimulatory, and preparative actions. *Endocrine Reviews* 21: 55–89.

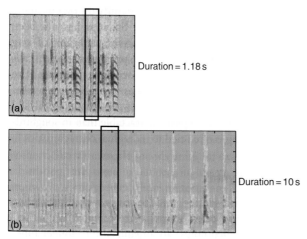

Duration = 1.18 s

Duration = 10 s

Figure 2 Spectrograms of (a) zebra finch (*Taeniopygia guttata*) song (duration = 1.18 s and frequency range = 0–22 kHz) and (b) excerpt of European starling (*Sturnus vulgaris*) song (duration = 10 s and frequency range = 0–22 kHz). See audio examples for sound files. Zebra finch song is characterized by short, highly stereotyped bouts of song. Starling song, in contrast, consists of flexible sequences of motifs, which are often repeated. In both figures, time is represented on the horizontal axis and frequency is represented on the vertical axis. A single motif, which is repeated three times, is marked in the starling song, and a single note is marked in the zebra finch song.

others are rarely part of lengthy repetitive sequences. Thus, at least in starlings, there are multiple scales of temporal organization in song.

Markov sequence models have been applied to song production in several different species of North American thrushes, cardinals, *Cardinalis cardinalis*, rose-breasted grosbeaks, *Pheucticus ludovicianus*, and American redstarts, *Setophaga ruticella*. In these species, as in European starlings, most song sequences are best fit by Markov chain models that take into account transition probabilities between ordered pairs of events. Laboratory studies demonstrate that at least in the European starling, sequential transition probabilities between ordered pairs and ordered triplets of motifs aid significantly in the perception and recognition of familiar songs. Sensitivity to song element ordering has also been observed in the field, where for example, swamp sparrows, *Melospiza georgiana*, can recognize differences in syntactic structures of songs from different geographical regions. These studies of song perception are consistent with the idea that perceptual sensitivity of the receiver covaries with the syntactic information content of the signal. Sender–receiver matching is a common property of communication systems.

The phonological syntax of song appears to be culturally transmitted in some species. White-crowned sparrows, *Zonotrichia leucophrys*, normally produce songs composed of three to six 'phrases,' and can learn to sing species-typical songs from tape recordings. When male white-crowned

sparrows are tape tutored with single-phrase song models, the birds learn the phrases and assemble them into species-typical phrase sequences, but not into whole songs. Tutoring white-crown sparrows with phrase-pairs is sufficient for full, species-typical song syntax to emerge. Moreover, birds tutored with reverse-ordered phrase pairs sing songs with reversed phrase order. Thus, phrase sequencing information must be part of the song model experienced during early development. Thus at least in this species, the phonological syntax of song appears to be learned.

Syntax has been extensively studied in the chick-a-dee call of chickadees, *Poecile* sp., which has considerable flexibility in the sequential ordering of its notes. Again, the variability in ordering is not random. Rather, each type of call note, denoted 'A,' 'B,' 'C,' and 'D,' may be repeated a variable number of times or omitted, but the overall sequence of note types is strictly maintained. Calls that violate the note type order are extremely rare, occurring less than 0.5% of the time. This rigid ordering structure in chickadee calls provides a concrete example of a formally computable syntactic rule for call production, and is observed in a number of chickadee species (black-capped Chickadee, *P. atricapillus*, Carolina chickadee, *P. carolinensis*, mountain chickadee, *P. gambeli*, and the Mexican chickadee, *P. sclateri*) and the taxonomically related titmice (e.g., tufted titmouse, *Baeolophus bicolor*). Moreover, violations in the typical call syntax appear to be perceptually salient to receivers, as they often elicit substantially different or no response compared to calls that follow the syntactic rule.

The temporal patterning in the chickadee call system provides a clear example of phonological syntax, as different, well-formed, call sequences can elicit different behaviors. Whether the system meets the criterion for lexical syntax is not clear. Reports from at least two chickadee species suggest that information about the sex and geographic origin of the singer is carried by single notes, but it is not clear how such information is related to the overall meaning conveyed by call syntax. It is possible that such information is coded in the acoustic characteristics of the vocalizations imparted by individually specific morphological variation in the vocal apparatus, and thus cannot be varied by the singer. However, some acoustic properties are probably dynamically regulated, as indicated by changes in some spectral properties of chick-a-dee notes when chickadees form flocks. In either case, spectral information represents a special semantic case. In humans, voice characteristics do not appear to overlap with linguistic components of speech, and they are not altered by syntax under normal conditions.

Primate Vocal Syntax

In contrast to the primarily nonreferential function of songbird vocalizations (i.e., mate attraction, aggression, territoriality, etc.), there is ample evidence that some primate vocal communication signals contain specific referential information, apart from individual identity. Acoustically distinct vocal calls serve as alarms for different predator types, and nearby conspecifics use these alarms to initiate defensive action appropriate to the predator's mode of attack. Acoustically distinct alarm calls can also refer to environmental dangers such as falling branches or trees.

Several species appear to use combinations of alarm (or other) calls in different behavioral contexts. Black-and-white colobus monkeys, *Colobus guereza*, sometimes use a two-call combination made up of a 'roar' introduced with a brief 'snort' to maintain spacing between nearby groups. By itself, the snort serves as an alarm call. Likewise, the 'pant-hoot' given by chimpanzees, *Pan troglodytes*, has components that acoustically resemble a mild alarm call. Male and female titi monkeys, *Callicebus moloch*, incorporate several call types into sequences that differ between behavioral contexts.

Using a variety of behavioral assays, several studies have demonstrated that primates are sensitive to the temporal ordering of sound units. Playback studies indicate that wedge-capped capuchin monkeys, *Cebus olivaceus*, are sensitive to the ordering of sound units in their calls. When calls are arranged so that sound units are ordered naturally, listeners produce fewer moans than when they are arranged in unnatural combinations. Gibbons, *Hylobates* sp., which unlike other non-human apes produce acoustically elaborate songs – typically as duets between mated pairs – appear to show similar characteristics (hear audio **Figure 3**). White-handed gibbons, *H. lar*, produce complex song sequences in response to terrestrial predators and during other normal daily routines. Although composed of the same call note repertoires, the predator-induced songs are assembled differently than other songs. These differences, and potentially their referential meaning, are salient to receivers. The ordering of song in other gibbons, *H. agilis*, is thought to follow a rudimentary set of structural rules, and so further study of vocal syntax in this genus will be very important.

Recent studies provide evidence consistent with lexical syntax in wild primate species. The behavioral response of the wild Diana monkey, *Ceropithecus Diana*, to the alarm call of the Campbell's monkey, *Cercopithecus campbelli*, can be modulated in urgency by introducing boom sound units before the alarm call. These alarm calls are initiated by Campbell's monkeys in response to approaching predators and cause other monkeys to take evasive action appropriate to the predator. Likewise, Putty-nosed monkeys, *C. nictitans*, are sensitive to particular ordered sequences of two loud alarm calls, 'pyows' and 'hacks.' By themselves pyows are a common response to leopards, and hacks (or hacks followed by pyows) are give to eagles. Males sometimes give a series of one to four pyows followed by one to four hacks, either alone or at the start of other call sequences, and these pyow–hack sequences reliably predict group movement. Lexical syntax requires not only that meaningful sound units combine into larger meaningful sequences, but also that the meaning of the sequence is dependent on the ordering of units. Thus, the fact that pyow–hack sequences appear to have a different meaning than hack–pyow sequences and that pyows and hacks are meaningful by themselves is consistent with a rudimentary form of lexical syntax.

Vocal Syntax in Other Mammals

Surprisingly, relatively little work has examined the presence of syntactic complexity in the vocal communication systems of mammals other than primates. Playback studies with Richardson's ground squirrels suggest that the ordering of sound units has little effect on behavior. Some syllables with unique acoustic elements enhance behavioral responses, but do so regardless of where in the call they occur. Mexican free-tailed bats, *Tadarida brasiliensis*, have rich vocal repertoires. While some syllables are unique to specific calls, others are shared among multiple calls, and entire calls associated with one behavior can be embedded in more complex vocalizations used in different behavioral contexts. It remains to be seen whether or not different combinations of simpler call components convey different meanings. It is worth noting that many

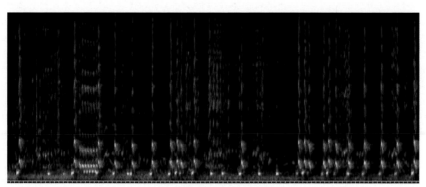

Duration = 10 s

Figure 3 Spectrogram of a gibbon song excerpt (duration = 10 s and frequency range = 0–10 kHz).

other species of bats have rich vocal repertoires, and bats are one of the few mammals in which vocal learning has been documented. Thus, the characterization of syntactic complexity in this order is important.

Conclusion

There is great diversity in the kinds of temporal patterning exhibited in the natural communication systems of non-human animals. Birdsong and the alarm calls of non-human primates have been particularly well studied, and several classes of models have been applied to extract structure in the combination of sound units. Temporal patterning commonly imparts structured combinations of sound units, but only rarely can we clearly identify that variation in this structure carries meaning within the communication system. At present, there is only limited evidence of lexical syntactic structure in animal vocalizations. Interestingly, all such evidence appears to come from a small set of non-human primate species. In contrast, there is abundant evidence for phonological syntactic structures in many species.

Comparative study of how sound units are combined can provide a rich source of evidence for investigating the evolution and neural basis of important cognitive skills involved in language perception and production. And, having emphasized in this article the distinction between phonological and lexical syntax, it may be tempting to view these descriptions as residing along a single continuum of behavioral complexity. We caution against such thinking. It may be that the lexical processes (and associated neural mechanisms) that underlie referential communication are wholly different from those that support temporally sophisticated phonological syntax. As such, lexical syntax may not be the derived form of phonological syntax, but rather may require the rare integration of typically distinct temporal patterning and lexical capacities. Thus, we find one group of very successful animals, namely oscine birds, exploiting phonological syntax without a rich referential lexicon, and another group, non-human primates, using referential signals but in heavily restricted temporal patterns. From this perspective, the human syntactic system might be considered as the rare example of what can happen when complex auditory sequence production and perception co-occur with the use of referential units. The most productive future research is likely to come from work that highlights syntactic and referential characteristics shared among different communication systems, and which is followed by the close study of biological mechanisms as appropriate in each organism.

See also: Acoustic Communication in Insects: Neuroethology; Acoustic Signals; Alarm Calls in Birds and Mammals; Apes: Social Learning; Communication and Hormones; Dance Language; Hearing: Vertebrates; Information Content and Signals; Mammalian Social Learning: Non-Primates; Referential Signaling; Social Recognition; Sound Production: Vertebrates; Vocal Learning.

Further Reading

Arnold K and Zuberbühler K (2006) Language evolution: Semantic combinations in primate calls. *Nature* 441(303).

Gentner TQ, Fenn KM, Margoliash D, and Nusbaum HC (2006) Recursive syntactic pattern learning by songbirds. *Nature* 440: 1204–1207.

Hailman JP and Ficken MS (1986) Combinational animal communication with computable syntax: Chick-a-dee calling qualifies as 'language' by structural linguistics. *Animal Behaviour* 34: 1899–1901.

Lucas JR and Freeberg TM (2007) Information and the chick-a-dee call: Communicating with a complex vocal system. In: Otter KA (ed.) *Ecology and Behaviour of Chickadees and Titmice: An Integrated Approach*, pp. 199–213. Oxford: Oxford University Press.

Marler P (1977) The structure of animal communication sounds. In: Bullock TH (ed.) *Recognition of Complex Acoustic Signals*, pp. 17–35. Berlin: Dahlem Konferenzen.

Suzuki R, Buck JR, and Tyack PL (2006) Information entropy of humpback whale songs. *The Journal of the Acoustical Society of America* 119: 1849–1866.

Yip MJ (2006) The search for phonology in other species. *Trends in Cognitive Sciences* 10: 442–446.

T

Tadpole Behavior and Metamorphosis

R. J. Denver, University of Michigan, Ann Arbor, MI, USA

Amphibian Life History Strategies

Amphibians are geographically widespread, occupy a range of habitats, and show considerable diversity in morphology, behavior, physiology, and life history strategy. There are three living orders of amphibians: the frogs and toads (Order Anura, or Salientia), the salamanders and newts (Order Urodela, or Caudata), and the limbless caecelians (Order Apoda, or Gymnophiona). Most amphibian species have complex life cycles; they begin life as a free-swimming larva that is followed by a metamorphosis to the juvenile adult form. The anurans undergo a dramatic change during metamorphosis, transforming from a fish-like, aquatic larva (tadpole) to a tetrapodal, often terrestrial adult. (**Figure 1**). Metamorphic transitions in salamanders and caecelians are less dramatic than in frogs and toads. The two amphibian life stages are affected by different environmental factors and have different behavioral repertoires. The larvae of most amphibian species are aquatic, and are found in diverse habitats, ranging from water-filled crevices in rocks, logs, or leaves, to larger ponds or streams. Some amphibian species have direct development, in which development proceeds from the embryo to the juvenile adult form without a free-swimming larval period. Some urodeles do not metamorphose, but instead reproduce in the aquatic habitat while retaining the larval morphology (paedomorphosis). No anuran species have a paedomorphic life history.

Amphibian larvae encounter diverse ecological conditions during development, and variations in abiotic factors (e.g., water availability, temperature, photoperiod, etc.) and biotic factors (e.g., intra- and interspecific competition, predation, etc.) can interact in complex ways to influence tadpole morphology, behavior, growth, and development. For example, the presence of predators can lead to increased tail size in tadpoles, and this may enhance the tadpole's ability to escape predator attack. The type of food can have profound effects on tadpole morphology; for example, carnivorous morphs, characterized by dramatic changes in jaw and cranial morphology, may be induced in some species by feeding on invertebrates. Food availability, competition for resources, and predation risk, interact to influence tadpole growth and development, and tadpole growth history and body size determine when a tadpole initiates metamorphosis.

Amphibian species that breed in predictable habitats (i.e., permanent or semipermanent lakes and ponds) tend to have longer larval periods. By contrast, species that breed in unpredictable habitats (i.e., ephemeral pools) generally have much shorter larval periods. Many amphibian species exhibit plasticity in the timing of metamorphosis, which is influenced by growth opportunity and environmental stress in the larval habitat. The proximate mechanisms that govern the timing of metamorphosis involve the production, metabolism, and actions of hormones. Competence to respond to environmental signals depends on the development and activity of endocrine glands that produce the hormones that control metamorphosis. Most research on the endocrinology of metamorphosis and the roles of hormones in larval behavior has been conducted on anuran tadpoles, and, therefore, the following discussion is restricted to anurans.

Tadpole Behavior

Tadpole behavior is directed toward maximizing growth to achieve a minimum body size to initiate metamorphosis. The principal behaviors engaged by tadpoles are feeding and predator avoidance. Tadpoles tend to forage actively, but will reduce their activity if confronted with the threat of predation; there is a tradeoff between the competing needs to grow as quickly as possible and to avoid being eaten by predators. Schooling behavior by some species may be a means to maximize food acquisition and minimize predation.

Feeding Behavior

Tadpoles are specialized herbivores, feeding on planktonic material in the water column (filter feeders), detritus in

Figure 1 Metamorphosis of the South African clawed frog *Xenopus laevis*. Shown are tadpoles in premetamorphosis (top), prometamorphosis (middle), and metamorphic climax (bottom). Photos by David Bay.

pond sediment, or scraping material from submerged substrates such as rocks and reeds. The selection of food type is reflected in a species' buccal anatomy, feeding mode, and microhabitat use. Although tadpoles are predominantly herbivores, most also consume animal tissue, and many species exhibit opportunistic cannibalism. A tadpole's diet can have an important influence on the development of alternate morphologies (i.e., carnivorous morphs in spadefoot toad tadpoles), the rate of growth and development, and consequently, the timing of metamorphosis.

Prior to metamorphosis, tadpoles feed vigorously (in proportion to food availability and predation risk). The ability to sense food and the presence of predators, and to adjust foraging activity accordingly, is critical to survival. During metamorphosis, tadpoles slow, then cease feeding, and this corresponds to a dramatic remodeling of the gut that occurs at metamorphic climax. Similar to other tetrapods, amphibians appear to possess a feeding control center located in the hypothalamus/preoptic area of the brain. Neurons located here form a neural circuit that controls orexigenic (stimulates feeding) and anorexigenic (inhibits feeding) behavioral output, in communication with a major feeding control center located in the hindbrain (the nucleus of the solitary tract). Orexigenic and anorexigenic neuropeptides are produced by neurons located in the hypothalamic feeding control center, and these neurons also respond to blood-borne signals.

In premetamorphic tadpoles, orexigenic pathways that stimulate feeding appear to predominate, but these pathways are poorly understood. More is known about the anorexigenic pathways that inhibit feeding, and that develop/mature during metamorphosis. For example, intracerebroventricular (i.c.v.) injection of the stress neuropeptide corticotropin-releasing factor (CRF) inhibits feeding in prometamorphic tadpoles, but not in premetamorphic animals. By contrast, i.c.v. injection of the CRF receptor antagonist α-helicalCRF$_{(9-41)}$ stimulates feeding, suggesting that endogenous CRF exerts an inhibitory tone on feeding controls in metamorphosing tadpoles. The expression of CRF and its receptors in the tadpole brain increases during metamorphosis. Besides acting as a neurotransmitter in the brain to inhibit feeding at a time when the gut undergoes extensive remodeling, CRF also stimulates the pituitary gland to produce hormones that initiate metamorphosis (described later).

Leptin is a protein hormone secreted by fat cells that regulates food intake. Leptin acts on the brain to signal when the body has sufficient energy stores, thus, inhibiting appetite (i.e., it is an 'adipostat'). Leptin, like CRF, potently inhibits feeding when injected i.c.v. into prometamorphic tadpoles, but not premetamorphic tadpoles. The stage-dependent onset of leptin action on feeding corresponds to an increased expression of leptin receptors in the brain. Leptin may act on CRF neurons to increase their activity, thus, leading to CRF-mediated inhibition of feeding, and CRF-induced secretion of hormones that control metamorphosis. One hypothesis currently being tested is that leptin, produced in proportion to energy stores or body size/condition, signals to the brain to initiate metamorphosis.

Corticosterone, the major stress hormone in tadpoles, stimulates foraging. Corticosterone increases following exposure to different stressors, and while stress leads to rapid inhibition of feeding, mediated by the activation of hypothalamic CRF neurons, the subsequent elevation of corticosterone causes an increase in foraging occurring several hours after exposure to the stressor. This action of corticosterone serves to reiniate feeding after a period of reduced food intake.

Antipredator Behavior

Predator avoidance by amphibian larvae may involve escape behaviors, refuge seeking, and behavioral inhibition (i.e., freezing behavior). When confronted with a predator, tadpoles often display rapid freezing behavior, whereby they settle to the bottom of the pond and remain still. This behavioral response facilitates the avoidance of detection by the predator. While visual cues may play a role in this response, tadpoles also use chemical cues released by conspecifics under predator attack (alarm pheromones), and they may sense chemicals produced by the predator (i.e., kairomones).

Tadpoles release an alarm pheromone from skin cells by a stimulus-secretion coupled pathway upon predator attack. The pheromone comprises at least two components with distinct biophysical properties that must be combined to elicit behavioral inhibition in conspecifics. Tadpoles likely use combinations of chemicals and visual cues in their assessment of predation risk. The antipredator response of tadpoles may be shaped and refined by learned predator recognition, and there is evidence for the cultural transmission of predator recognition among conspecifics. Also, the duration of behavioral inhibition may be influenced by a tadpole's body condition, with tadpoles in good condition capable of maintaining a behaviorally quiescent state for a longer time than tadpoles in poor condition.

Olfactory neurons activated by tadpole alarm pheromone (or other predator cues) lead to the activation of neural pathways that elicit freezing behavior, but the nature of these pathways is not known. Recent findings show that activity of the neuroendocrine stress axis is suppressed by tadpole alarm pheromone, and that this plays a role in the expression of behavioral inhibition. Exposure of tadpoles to alarm pheromone causes a rapid and dose-dependent decrease in corticosterone. The production of corticosterone is controlled by CRF, and the decreased corticosterone may result from the inhibition of CRF neurons. Also, CRF stimulates locomotion, and the suppression of CRF neuronal activity may be linked to the behavioral inhibition. Reversing the decline in corticosterone by adding a low dose of the hormone to the aquarium water blocks the behavioral inhibition caused by the alarm pheromone. Since corticosterone normally induces activity and foraging, sustained suppression of corticosterone may be permissive for the continued quiescent state.

In tadpoles, predation risk extending over days to weeks causes distinct morphological changes, such as a relatively smaller body and larger tail. The tadpole tail is a critical locomotory organ for escape or refuge seeking, and it is maintained until late metamorphic climax owing to its critical role in locomotion. The larger tail may serve as a lure to distract predator strikes from the more vulnerable body, or could allow for enhanced burst swimming for escape.

Transformation of the Central Nervous System Underlies Behavioral Transitions During Metamorphosis

During metamorphosis, the tadpole transitions from an aquatic, fish-like larva that is largely herbivorous, to a tetrapodal, exclusively carnivorous and often terrestrial adult. At this time, the central nervous system undergoes dramatic changes. Certain larval structures degenerate, such as the Mauthner neurons found on either side of the medulla, and the sensory and motoneurons supplying the tail. As larval structures disappear, adult-specific neural

structures are formed. For example, the visual field of tadpoles is panoramic, which allows them to detect movement in the aquatic habitat while actively foraging. By contrast, the frog has a binocular visual field appropriate for a sit-and-wait predator. During metamorphosis, the major portion of the retina develops along with associated visual projections in the di- and mesencephalon to support this transition in behavioral mode. New spinal cord segments develop during metamorphosis to support function of the newly formed limbs. The cerebellum and the mesencephalic nucleus of the trigeminal nerve also develop at this time. The beginning of metamorphosis is marked by a dramatic increase in cell proliferation throughout the brain. Together, the changes in the nervous system prepare the animal for its new behavioral and dietary mode.

Endocrinology of Tadpole Metamorphosis

Hormones orchestrate the diverse morphological and physiological changes that occur during amphibian metamorphosis, and also function as mediators of environmental effects on development. At metamorphic climax, the hormones that promote tissue transformations also influence tadpole behavior, and through their developmental actions on the brain, bring about shifts in behavioral mode between the larva and the juvenile adult.

A striking characteristic of amphibian metamorphosis is that a single signaling molecule, the thyroid hormone, orchestrates the entire suite of molecular, biochemical, and morphological changes. While the thyroid hormone alone is capable of initiating metamorphosis, other hormones such as the corticosteroids function to promote thyroid hormone action. Depending on the tissue, the thyroid hormone may induce cell proliferation, cell death, differentiation, or migration. Thyroid hormone secretion increases markedly during metamorphosis, reaching a peak at metamorphic climax and declining thereafter to reach the juvenile adult level.

Thyroid activity is controlled at multiple levels. The neuroendocrine system, comprising the hypothalamus of the brain and the pituitary gland, serves as an interface between the central nervous system and the peripheral endocrine system. Different sensory inputs are transduced into developmental and physiological responses by neurosecretory neurons in the hypothalamus. The anterior preoptic area of the brain contains neurosecretory neurons that synthesize neurohormones that regulate pituitary hormone secretion. The pituitary gland produces the protein hormone thyrotropin (TSH) that travels in the bloodstream and activates receptors on thyroid follicle cells to increase the secretion of the thyroid hormone. In tadpoles, unlike in mammals where the tripeptide thyrotropin-releasing hormone is the major TSH-releasing factor, TSH secretion is controlled by hypothalamic CRF.

Secretion of pituitary corticotropin (ACTH) is also induced by CRF, and ACTH controls the production of corticosteroids by adrenal cortical cells.

The major hormone secreted by the thyroid gland is thyroxine (T_4), but the hormone with greatest biological activity is 3,5,3′-triiodothyronine (T_3) which is generated by monodeiodinase enzymes within target tissues. The expression of tissue monodeiodinases is an important point of regulation of the thyroid axis. Tadpoles become competent to respond to exogenous thyroid hormone at the time of hatching, and this competence depends on the upregulation of thyroid hormone receptors (TRs). Thyroid hormone receptors are ligand-activated transcription factors that belong to the nuclear hormone receptor superfamily. Hormone binding to the TR induces gene expression in target tissues, and these gene expression programs underlie the tissue-specific changes that occur during metamorphosis.

The thyroid hormone is the primary morphogen controlling amphibian metamorphosis, but corticosteroids produced by adrenal cortical cells also play an important role by enhancing thyroid hormone bioactivity in target tissues. They accomplish this by upregulating TRs and the monodeiodinase that converts T_4 to T_3. Tadpoles often respond to environmental stress (e.g., pond drying, competition for resources, etc.) by accelerating metamorphosis. The secretion of CRF is induced by environmental stress, and, therefore, can mediate the effects of a changing environment on the timing of metamorphosis through its stimulation of TSH and ACTH secretion. The common regulation of the thyroid and the adrenal axes by hypothalamic CRF, and the sensitization of target tissues to low concentrations of the thyroid hormone by corticosteroids, are means for the tadpole to modulate its rate of development in response to environmental change.

See also: Defensive Morphology; Food Intake: Behavioral Endocrinology; Hormones and Behavior: Basic Concepts; Predator Avoidance: Mechanisms.

Further Reading

Benard MF (2004) Predator-induced phenotypic plasticity in organisms with complex life histories. *Annual Review of Ecology Evolution and Systematics* 35: 651–673.

Carr JA (2002) Stress, neuropeptides, and feeding behavior: A comparative perspective. *Integrative and Comparative Biology* 42(3): 582–590.

Crespi EJ and Denver RJ (2005) Roles of stress hormones in food intake regulation in anuran amphibians throughout the life cycle. *Comparative Biochemistry & Physiology A: Comparative Physiology* 141(4): 381–390.

Crespi EJ and Denver RJ (2006) Leptin (*ob* gene) of the South African clawed frog *Xenopus laevis*. *Proceedings of the National Academy of Sciences USA* 103: 10092–10097.

Denver RJ (2009) Endocrinology of complex life cycles: Amphibians. In: Pfaff D, Etgen A, Fahrbach S, Moss R, and Rubin R (eds.) *Hormones, Brain and Behavior*, 2nd edn. San Diego, CA: Academic Press.

Duellman W and Trueb L (1994) *Biology of Amphibians*. New York: McGraw-Hill.

Fraker ME, Cuddapah V, McCollum SA, Relyea RA, Hempel J, and Denver RJ (2009) Characterization of an alarm pheromone secreted by amphibian tadpoles that induces rapid behavioral inhibition and suppression of the neuroendocrine stress axis. *Hormones and Behavior* 55: 520–529.

McDiarmid RW and Altig R (eds.) (1999) *Tadpoles. The Biology of Anuran Larvae*. Chicago, IL: The University of Chicago Press.

Taste: Invertebrates

J. Reinhard, University of Queensland, Brisbane, QLD, Australia

Introduction

Taste, or more formally gustation, is a form of direct chemoreception and is one of the five traditional senses. It refers to the ability to detect the flavor of substances such as food, certain minerals, and poisons. The sense of taste is often closely linked to the sense of smell, or more formally olfaction. In vertebrates like humans, the senses of taste and smell are clearly defined by two separate sensory organs, namely, the tongue and the nose. In invertebrate animals, the distinction is less clear, as the sense of taste is generally not confined to one specific organ. Invertebrate taste receptors can be located on the same organ as olfactory receptors, or they can be distributed on different appendages all over the body, including parts of the mouth, feet, antennae, and ovipositor. Also, many invertebrates live in aqueous environments, which makes a distinction between the senses of taste and smell even more difficult. Generally, invertebrate taste is defined by (1) the morphology of the sensillum housing the sensory neurons; (2) the molecular structure of the receptor proteins that detect the tastants; and (3) the type of chemicals that are detected (e.g., minerals, sugars). This article first gives a brief overview of the anatomy of typical invertebrate taste organs, and the molecular and neural mechanisms underlying the sense of taste in invertebrates. It then describes the types of behavior that are linked to the sense of taste in the invertebrate world.

Invertebrate Taste Organs

The basic unit of the invertebrate taste organ is called 'a sensillum,' which is a hair-like hollow structure that houses gustatory neurons. It looks very similar to the sensilla that house olfactory neurons and are dedicated to the sense of smell; however, the difference lies in the fact that most gustatory sensilla are of the type 'tip-pore *sensillum trichodeum*' (**Figure 1**). That is, they have only one opening to the outside world at the very tip of the sensillum. It is through this pore that the tastant has to enter the sensillum space to make contact with the neurons. For this to happen, the animal needs to get the organs on which the gustatory sensilla are located in direct contact with the substrate or food they are sampling. Therefore, gustatory sensilla are often referred to as 'contact chemoreceptors.' Gustatory sensilla can also come in the shape of stout taste pegs or even without any protruding hair structure at all, merely as an opening in the cuticula.

Even though in invertebrates there is no such thing as one defined taste organ equivalent to the human tongue, the mouthparts of invertebrates are usually considered the main taste organs, as the majority of gustatory sensilla are located on those. For example, in flies gustatory sensilla are found on their labellum, in butterflies on the tip of their tongue, in bees at the base of the tongue, and in many beetles and roaches they are located on the maxillary and labial palps. Furthermore, insects have gustatory receptors inside the oral cavity and on the esophagus lining. Apart from the mouthparts, gustatory receptors are also commonly found on invertebrate feet, legs, or tentacles (e.g., insects, spiders, lobsters, octopus), on their antennae (bees, ants, wasps), wing margins (flies), or on the tip of their abdomen and the ovipositor (locusts, parasitic wasps) (**Figure 2**). Invertebrates thus have an extra sensory dimension that allows them to taste the world with more than just their 'tongue.'

Molecular and Neural Mechanisms

Tastants are chemical molecules including sugars, mineral salts, and bitter or toxic chemicals like alkaloids, or small proteins. When the animal touches its food with its mouthparts, for example, floral nectar, the gustatory sensilla come in contact with the sugar molecules in the nectar, and the sugar molecules enter the sensillum through the tip pore (**Figure 1**). Once inside, they come in contact with the dendrites of the gustatory neurons, and bind to the gustatory receptor proteins (GRs), which are 7-transmembrane proteins located on the dendrite membrane (**Figure 3**).

GRs do not bind any tastant entering the sensillum, but each binds only a specific category of tastant, for example, one specific type of salt or sugar, unlike olfactory receptors, which are often broadly tuned with overlapping specificities. Each gustatory neuron carries one type of GR, and each sensillum houses several different gustatory neurons (**Figure 1**). Each neuron is receptive for a different chemical signal. For example, a typical gustatory sensillum of a fly houses one neuron responsive to sugar, one to water, one to salt, and one responding to bitter tastants. Binding of the tastant to its GR triggers a signal transduction cascade (**Figure 3**), which results in the opening of ion channels on the dendrite membrane and creation of action potentials. These constitute an

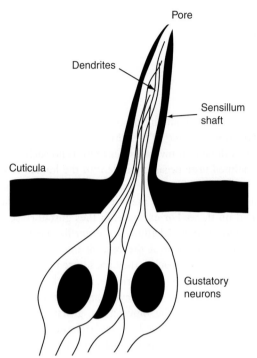

Figure 1   Schematic of typical invertebrate gustatory sensillum (tip-pore *sensillum trichodeum*), containing three gustatory neurons.

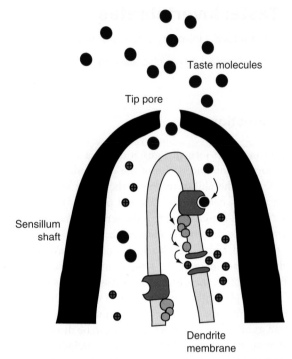

Figure 3   Schematic of the molecular mechanism underlying detection of a taste molecule. Taste molecules enter the sensillum lymph through the tip pore. The gustatory receptor protein (red) binds a taste molecule (blue), which triggers a signal transduction cascade (orange molecules), resulting in opening of ion channels (green) and influx of ions.

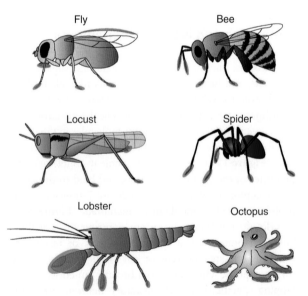

Figure 2   Location of taste receptors (shaded in red) in some invertebrates. Drawings are not to scale.

will extend its legs to grab hold of the food item. All this, from detection of the taste molecule to the motor behavior triggered, happens within a fraction of a second.

Invertebrate Behavior Controlled by Taste

Taste plays a prominent role in the feeding behavior of invertebrates, and to a lesser extent also in oviposition (egg-laying) behavior, courtship behavior, and kin recognition behavior. Many of these are crucial behaviors an animal must accomplish to survive and reproduce. While the sense of smell (olfaction) helps an animal detect a potential food source or oviposition site from a distance and assists in orienting toward it by following the odor plume emanating from the food source, the sense of taste comes into play at close range. This is when the animal makes contact with the potential food source or substrate and has to decide on its quality and whether it is suitable for ingestion or egg laying, respectively. It does so by touching, for example, a possible food source with its gustatory receptors to detect information on its quality, such as sugar content or potential toxins. This leads either to feeding behavior or in case of detected toxins, to food rejection and avoidance behavior. Despite its critical

electric message that is sent to the brain, where the information is integrated. The message is then forwarded via motor neurons to the body parts that control the relevant behavior. For example, when detecting sugar, a bee will extend its tongue to ingest the nectar, or a wasp importance in the everyday life of all animals, the sense of taste has received only limited attention in research, with the focus much more on the sense of smell. Nevertheless, given in the following paragraphs are a number of examples of how the sense of taste can control and modulate behavior in invertebrates in different environments: on land, in the air, and in water.

Proboscis Extension Reflex

One of the best-studied taste-related behaviors of invertebrates is the proboscis extension reflex of flies and bees. The classic work of Vincent Dethier, described eloquently in his book *The Hungry Fly*, showed that stimulation of taste receptors on the feet of the forelegs of flies initiates an extension of the proboscis (**Figure 4**). When the proboscis comes into contact with the same food source, the stimulation of further taste receptors on the labellum initiates drinking. Stimulation of single taste receptors on the labellum is sufficient to evoke activity in the proboscis muscles involved in feeding.

Interestingly, the sugar receptors on a fly's labellum do not respond to all sugars equally, but they respond strongest to fructose, second to sucrose, and least to glucose. This hierarchy of response seems to be linked to the

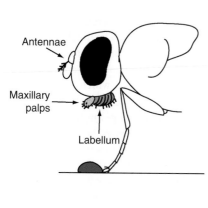

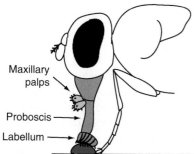

Figure 4 Proboscis extension reflex behavior in a fly showing the withdrawn proboscis (above), and the extended proboscis (below) after stimulation of the taste receptors on the fly's feet with a drop of sugar water. Small hairs on the maxillary palps and labellum are also taste receptors.

perception of 'sweetness' in the same way that the sweet receptor on the human tongue does. In contrast, when bitter receptors on the fly's labellum are activated, that is, gustatory neurons that respond to bitter substances like quinine, the proboscis extension reflex is suppressed and aversive behavior is triggered resulting in the fly rejecting the food. The pathways from the sweet receptors and the bitter receptors to the muscles controlling proboscis extension seem to be hard wired, always eliciting appetitive and aversive behaviors, respectively. Hence, this behavior is called a 'reflex,' crucial for the survival of the animal.

Having gustatory receptors on the feet represents an extremely useful sensory system for a fly, as the feet are likely to be the first body parts to make contact with a potential food source during landing. A fly can thus decide immediately whether the substrate it landed on is suitable as food or not. This is also true for other flies such as the tsetse fly that have chemoreceptors on their feet that respond strongly to amino acids that are found in abundance in the sweat of their primary host, humans. A tsetse fly samples its host for suitability as potential blood source merely by landing on it. If the right amino acids are detected via the gustatory receptor on the fly's feet, the proboscis is extended and pierced through the human skin, and ingestion of blood is initiated.

Bees and honeybees in particular show the same proboscis extension response when their mouthparts or feet are touched with a drop of sugar water. In bees, this behavior is even more pronounced than in flies, probably because of the fact that bees feed exclusively on sugar and thus have more or more sensitive sugar receptors than flies. Bees have sugar receptors not only on their proboscis and feet, but also on their antennae. A bee's antennae are covered in thousands of sensilla, most of them olfactory in nature, that is, for the detection of floral odorants and pheromones. But there are also a number of gustatory sensilla for the detection of sugar. These are mostly concentrated around the tip of the antenna. On close observation of a bee landing on a flower, one can see that she lightly touches the petals with the antennae. If she happens to come across a nectar gland, the bee will detect the sugar in the nectar via the gustatory receptors on her antennae, which triggers proboscis extension and feeding behavior. For a bee, this is a simpler and quicker way for initial sampling of her food, than using her proboscis, which has sugar receptors mostly at the base and not at the tip. The sugar receptors on the proboscis are used in a second stage, during which the bee decides whether she will actually drink and ingest the food.

Honeybees also have receptors for salt and bitter tastants on their antennae. When these are stimulated, the proboscis extension reflex is suppressed and the food is rejected. In case of restrained bees, as commonly used in laboratory experiments, the aversive behavior upon stimulation with salt or quinine is also evident in the bee turning her head away and even actively pushing the salty food source away using her feet.

Leg Avoidance Behavior

The legs are often the first part of the animal to come into contact with a food source, and the taste receptors on the legs play an important role in the assessment of food quality and in food rejection. Plants are a food source for countless invertebrates, many of them insects. They contain chemicals that can be detected on the plant surface, for example, waxes and heavy oils such as eucalyptus oil. Plants also produce noxious chemicals, such as alkaloids, to protect themselves from plant-eating insects. For many insects that feed on plants, such as locusts, what is important appears to be whether the chemical on the plant surface that comes in contact with the leg chemoreceptors is phagostimulatory or aversive. If it is aversive, then the animal takes its leg away (leg avoidance behavior); if it is not or is phagostimulatory, then an animal may go on and sample the food with its mouthparts, for example, the palps, as is the case for locusts. If at this second sampling stage again the chemical stimulant is considered to be aversive, it can be rejected, or if it is a phagostimulant the locust can bite or chew the food. At this third level, decisions can again be made on whether to ingest the food. The chemoreceptors involved in this last stage of sampling are located on the inner surface of the locust's mouth cavity. This chain of rejection/acceptance at different stages suggests a hierarchical organization of behaviors all triggered by gustatory receptors, starting with the leg avoidance behavior, to food chewing and finally food ingestion behavior.

The type of chemicals that trigger leg avoidance behavior when in contact with gustatory receptors on a locust's leg include the secondary plant compound nicotine hydrogen tartrate (NHT), which acts as an antifeedant. Contact with this chemical evokes rapid withdrawal movements especially of the locust's hind legs: the animal remains in the same body position, but moves the hind leg forward on the same plane. Other chemicals that are known to evoke leg withdrawal behavior in locusts are the salt sodium chloride (NaCl), a lysine-glutamate salt, and even sucrose, even though they are all part of a locust's dietary requirements. However, these chemicals trigger leg avoidance behavior, only if they are present in a rather high concentration. This suggest that any chemical, even a nutrient, will be treated as aversive and act to evoke food rejection if present in a high concentration. Such responses are not surprising if we consider our own behavioral responses to a range of chemicals. If we put food in our mouths, some chemicals, such as salt and sucrose, can be considered pleasant at low concentrations but are actively rejected at high concentration.

Food-handling Behavior Under Water

Other invertebrates that have taste receptors on their legs are marine crustaceans, such as prawns and lobsters. They also have olfactory receptors on their antennae, which are used to detect chemicals in the water, that signal the presence of food items, such as fish, molluscs, other crustaceans, worms, and some plant life. Their legs and antennae thus have behavioral functions analogous to the vertebrate senses of taste and smell, respectively, with the gustatory receptors on the legs playing a crucial role in the final stages of feeding behavior.

Feeding behavior in prawns and lobsters starts by antenna flicking to detect any chemicals in the water streams that may signal the presence of food. This behavior is analogous to sniffing behavior in mammals. Once chemicals are detected, the animal orients toward the source of the chemical signals, again using its olfactory receptors. When it reaches the food source, vigorous food-search behavior is carried out with the legs, in particular the chelated first pairs (chelae or claws). These are covered with gustatory receptors that detect L-glutamate, which is an amino acid that signals the presence of available, digestible foodstuffs. The food-search behavior eventually results in the grasp and lifting of the potential food item to the oral region. The lobster or prawn then starts a characteristic cycle of lifting-lowering movements, during which the animal holds the food item close to the mouthparts in an 'up–down' and 'in–out' manner, bringing it in contact with yet a different set of chemoreceptors located in the oral area. The final decision making as to the ingestion or rejection of a food item is dependent on these chemoreceptors. They are sensitive to bitter tastants, such as quinine, which signals that the food item is not suitable for ingestion. If the food item indeed contains quinine, the lifting–lowering movements can be continued for several minutes, before the food is dropped. If the food is suitable, it is ingested more or less immediately. Therefore, not only land-dwelling invertebrates, but also those living in water are equipped with multiple sets of chemoreceptive systems that control and modulate their feeding behavior.

Oviposition Behavior

Feeding is not the only behavior for which the sense of taste is important. Many insects lay their eggs in a substrate that provides the emerging larvae with the nutrients required for growth. These substrates can be fruits, soil, or even other animals, as is the case in parasitic wasps that lay their eggs in the larvae of other insects. It is crucial to pick the right substrate with the right nutrients to lay the eggs on; otherwise, the eggs will not develop and the offspring will not survive. A way to measure a substrate's suitability for egg laying in terms of nutrient content is tasting it. Therefore, behavior connected to oviposition (egg-laying) is also controlled by gustatory receptors. The receptors are located on the insect's ovipositor, which is a needle- or tube-like structure protruding from the end of the abdomen through which the eggs are passed into the substrate.

Locusts are a typical example of insects that use the chemoreceptors on its ovipositor to choose a suitable egg-laying site. They normally lay their eggs in damp soil, and the physical and chemical characteristics of the substrate influence both the selection of an appropriate site in which to lay eggs and the amount of time spent egg laying or digging. For locusts to lay eggs, they must dig down into the sand, using two pairs of hard sclerotized ovipositor valves at the end of the abdomen, to deposit eggs at depths where they are less likely to be exposed to desiccation. During this behavior, the locusts exhibit rhythmic movement of these valves and of the whole abdomen. While digging, the locusts will gradually extend their abdomens by up to twice their resting lengths to dig to depths of 6–8 cm. Contact chemoreception influences the digging movements of the ovipositor valves in such a way that when they come in contact with certain chemicals while moving rhythmically, the valves immediately close and retract. The duration of valve retraction is dependent on the concentration of the chemical. For example, a medium-to-high concentration of the salt sodium chloride (NaCl) will inhibit the digging rhythm and also suppress egg laying altogether. Lower NaCl concentrations are tolerated, and the lower the concentration, the more time a locust will spend digging and laying eggs, and the more eggs it will lay. However, even very low NaCl concentrations in sand are detected by the chemoreceptors on the ovipositor, and the number of eggs laid in sand containing NaCl is always less than the number of eggs laid in clean sand. Salt content is an important factor influencing egg survival, as high salt concentrations will desiccate the substrate and kill the eggs.

Chemoreceptors on the ovipositor are also crucial for successful egg-laying behavior in parasitic wasps. These tiny insects lay their eggs onto or into the larvae of other insects, and the emerging wasp larva feeds on the host larva until the latter dies. Olfactory receptors on their antennae as well as vibratory signals from host larval movement assist parasitic wasps such as the *Lariophagus distinguendus* in finding suitable host larvae from a distance, for example, beetle larvae that live inside grains of wheat. The parasitic wasp then drills into the grain with its needle-like ovipositor and moves it around until it comes into contact with the beetle larva (**Figure 5**). The chemoreceptors on the tip of the ovipositor detect chemicals on the larval surface that inform the wasp whether the host larva is the right species for laying its egg, and whether it is in good enough condition to sustain growth of the parasitic larva.

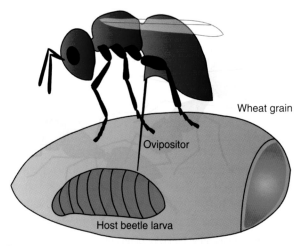

Figure 5 Parasitic wasp laying eggs into a host beetle larva that lives inside a grain. Taste receptors on the tip of the ovipositor give information about the host larva species and condition.

Kin Recognition Behavior

Bees, wasps, and ants are social insects living in large colonies of many thousands of individuals. They all look alike to us, but yet a bee always recognizes whether another bee belongs to her own colony or not, and the same is true for wasps and ants. How do they accomplish this? The hard outer surface of an insect's body (cuticula) is covered with a complex mixture of hydrocarbons, which are long-chain fatty acids, alkanes, alkenes, and esters. The chemical composition of this is influenced by the food a colony collects and eats. As all the members of a colony eat the same food, the hydrocarbon signature on their cuticula is identical. The chemicals composing the hydrocarbon signature are not very volatile, and thus can best be detected via contact chemoreceptors, of which social insects have many on their antennae specific to these hydrocarbons. Every time a bee or wasp or ant meets another bee or wasp or ant, she palpates with her antennae all over the other insect's body (antennation behavior), to detect its hydrocarbon signature, which will tell her whether the other insect belongs to her own colony or is an intruder (**Figure 6**). If the expected hydrocarbon mixture is detected, the other bee or wasp or ant is allowed to enter the colony; if the wrong hydrocarbon mixture is detected, the bee or wasp or ant will be rejected, prevented from entering or even attacked.

Courtship Behavior

The sense of taste even plays a role in sexual behavior in invertebrates as illustrated by the following examples. Discrimination between the sexes is not easy for the fruit fly *Drosophila*, and courtship behavior is believed to be mostly based on chemical signals. Males are thought to

Figure 2 Termite life types and castes. (a) Caste in wood-dwelling termite species, here the drywood termite *Cryptotermes secundus*. Eggs develop via larvae into totipotent false workers, which have the capability to become soldiers, or two forms of reproductives: neotenic reproductives, which breed within the natal nest when the current reproductives die, or winged sexuals (primary reproductives), which disperse and found a new nest. (b) Castes in foraging termite species, here the fungus-growing termite *Macrotermes bellicosus*. Eggs develop into larvae that either become true workers, soldiers, or reproductives. Photos: V. Salewski and J. Korb.

their entire colony life (except for new colony foundation) inside a single piece of wood that serves as both food source and shelter. Thus, there is no need for costly foraging, food is easily accessible to all colony members, and it is a bonanza type food source, whose availability declines predictably. (ii) Foraging termites (Mastotermitidae, most Rhinotermitidae, Serritermitidae, Termitidae; multiple-pieces nesting termites including the intermediate type sensu Abe 1987; **Figures 1** and **2(b)**): These species live in a well-defined nest that is more or less separated from the foraging grounds. To get access to food, individuals sooner or later have to forage outside the nest with the advantage that the colony's longevity is less limited by food availability than in the wood-dwelling species.

This ecological classification of wood-dwelling and foraging species is also reflected in the social organization of the colonies. Wood-dwelling species have a flexible developmental pattern in which 'workers' are totipotent and can develop into any of the caste options. They build the platform from which three permanent castes develop (**Figure 3(a)**): (i) sterile soldiers, (ii) winged sexuals (alates) that leave the nest and found new colonies, or (iii) neotenic reproductives that inherit the natal breeding position when the same-sex reproductive of the colony dies (replacement reproductives) or that breed in addition to other, generally neotenic, reproductives (supplementary reproductives). These 'workers' have also been called pseudergates, helpers, or false workers (the latter in contrast to the true workers of the foraging termites; see Glossary). The flexibility in development in wood-dwelling termites is achieved through a unique combination of progressive, stationary, and regressive molts, reflecting an increase, no change, or a decrease in morphometric size and wing development, depending on ecological conditions and season (see Glossary).

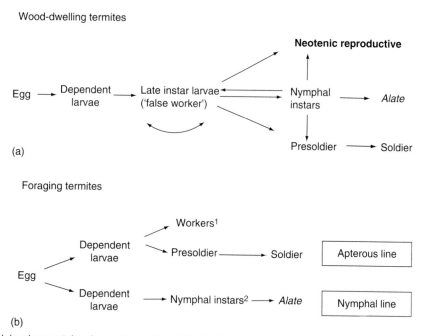

Wood-dwelling termites

(a)

Foraging termites

(b)

Figure 3 Simplified developmental pathways in termites. Italics indicate winged reproductives; bold type indicates wingless neotenic reproductives; →: progressive molt; ←: regressive molt; ↔: stationary molts. [1]Workers develop from different instars depending on the species; they are partly polymorphic: that is major and minor workers; in some species, they still can develop into natal reproductives; [2]in some species, nymphal instars can have regressive molts. Modified from Korb J and Hartfelder K (2008) Life history and development – A framework for understanding the ample developmental plasticity in lower termites. *Biological Reviews* 83: 295–313.

In contrast, the true workers of foraging termite species have more restricted developmental pathways. Their capability for regressive molts is more and more reduced (e.g., within the Rhinotermitidae) or absent (e.g., Termitidae). In these species, there is an early separation into two development pathways (**Figure 3(b)**). In the apterous line, individuals are unable to develop wings and cannot disperse as winged sexuals. They become the workers and soldiers of a colony, although in some species they can still reproduce as neotenic reproductives in the natal nest. In the nymphal line, individuals gradually develop wings through several instars to become alates that found new colonies. The separation of individuals into these two pathways can already be determined in the egg stage and seems to be influenced by season.

Although there have been some debates, generally the wood-dwelling life-type is regarded as the ancestral type in termite evolution. The close phylogenetic relationship of termites with the woodroaches that have a wood-dwelling lifestyle similar to the wood-dwelling termites, suggests that it is ancestral, probably inherited from the common ancestor. Consequently, to understand the factors that might have favored the transition from a cockroach to a social cockroach (in other word, the transition to termites' eusociality), an obvious group to investigate are the wood-dwelling termites.

Social Evolution of Wood-Dwelling Termites: Distinctive Properties of Living in a Single Piece of Wood

The wood-dwelling life-type is associated with some distinctive properties that must have been present when the termites evolved from a cockroach ancestor and which hence most likely shaped the termites' social evolution.

No Selection for Dispersal

A distinctive property of wood-dwelling termites, which they most likely inherited from a woodroach-like ancestor and which largely distinguishes them from foraging species and social Hymenoptera, is their bonanza type food resource. The nest of wood-dwelling termites constitutes a resource that generally largely outlasts the lifetime of the founding reproductives. Hence, offspring can stay with their parents as there is no local resource competition, which normally selects for offspring dispersal. Only when the wood block becomes fully exploited does resource competition occur; in line with this, under limited food conditions, (false) workers of wood-dwelling termites develop into winged sexuals that leave the nest during the annual nuptial flight to found new colonies. This is possible

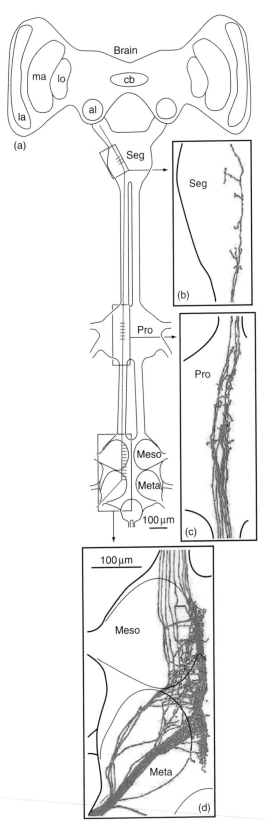

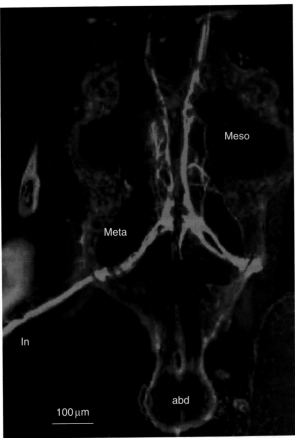

Figure 12 Photomicrograph of IR sensilla afferents in the meso/metathoracic ganglion stained with dextran tracer in *Me. acuminata*. Photomontage of seven individual sections; abdominal ganglion (abd), leg nerve (lg); other abbreviation as in **Figure 11**.

pegs mentioned earlier) on the antenna terminate in the antennal lobe, the first-order olfactory processing neuropil (**Figure 13**).

The antennal lobe is composed of glomeruli (**Figure 13(b)**), spherical neuropil regions that each receive input from a particular class of receptor neurons and process a small subset of odorants. Smoke odors are of extreme importance for the behavior of pyrophilous beetles, similar to the role pheromones play for the orientation of many other insects. In *Melanophila*, one might therefore expect to find glomeruli specialized for smoke compounds among other glomeruli processing more common

Figure 11 Arborization pattern in the central nervous system of afferent neurons originating from the IR sensilla of the beetle *Me. acuminata*. (a) Sketch showing afferents (red) entering the metathoracic ganglion meta and ascending through the mesothoracic meso and prothoracic ganglion pro to the subesophageal ganglion seg and toward the brain; putative presynaptic output in each of the ganglia indicated by short side branches. Insets show graphical reconstructions of actual neurons and their collaterals in (b) the subesophageal ganglion, (c) the prothoracic ganglion, and (d) the fused meso/metathoric ganglia. Brain components indicated in sketch are the optic lobes (lamina (la), medulla (me), and lobula (lo)), the antennal lobes (al), and the central body (cb).

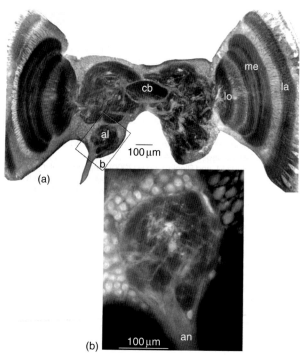

Figure 13 Photomicrographs of (a) the brain of *Me. acuminata*; photomontage of two sections (osmium stained material), the left and right half showing different anterior/posterior depths; antennal lobe enlarged in (b) to show some of the olfactory glomeruli. Abbreviation as in **Figure 11**.

odorants. In analogy to pheromone-processing glomeruli, such 'smoke compound glomeruli' might well be enlarged compared to the other 'ordinary' glomeruli, reflecting the biological significance of those odorants.

Output neurons connecting the antennal lobes to the central brain (projection neurons) show more specific and temporally sharpened responses than sensory neurons. Judging by their anatomical distribution in the central brain, these projection neurons probably do not directly interact with ascending IR information. Rather, one might expect convergence of IR-sensitive neurons and smoke-odor detecting neurons at the level of descending brain neurons, the ultimate integrators of any information processed by the brain. Such neurons would integrate IR and smoke odor information (together with flight-relevant visual and mechanosensory input) and would in turn control the 'flight motor' in the thorax and steer the beetle toward the forest fire. Similar descending neurons have been described to control odor-modulated flight in moths. Future research will have to reveal how exactly the two sensory modalities (IR radiation and smoke compounds) interact to change the arousal status, trigger flight, or modulate the beetles' orientation behavior. The two modalities may also be spatially integrated together with visual information in a consistent fashion. It is known that the tectum of owls calculates a coherent 'map' from visual and auditory information, and the beetles' brain

may perform some comparable, albeit more crude, alignment of different sensory inputs. This would help the beetle to approach the common source of the odor and IR radiation.

Concluding Remarks

Finding a forest fire is a crucial task for the introduced pyrophilous beetles. Therefore, a strong evolutionary pressure has acted upon the sensory systems of the beetles enabling them to detect a fire from distances as large as possible. Forest fires emit light, noise, smoke, and IR radiation. Up to now, there is no evidence that pyrophilous beetles use visible light or sound for the detection and localization of forest fires. A typical forest fire burns with a temperature of about 500–1000 °C. Fires of this temperature, which also emit enormous amounts of smoke, have their maximum emission of electromagnetic radiation in the IR wavelength range of 2.2–4 µm. Since IR radiation of this wavelength is transmitted well through an atmospheric window, IR reception and the smell of burning wood are useful tools for the detection and localization of burning trees. As shown for *Me. acuminata*, certain antennal olfactory receptors have developed into specialized sensors for the detection of characteristic compounds in smoke.

The following hypothetical scenario could explain the evolutionary pathway toward IR organs: insect cuticle absorbs IR radiation very well and this causes heating of the underlying tissue. At a certain temperature, heating may have irritated some sensilla and internal neurones of the body wall. If these 'unphysiological' excitations became combined with reactions, which are meaningful in the behavioral context of a pyrophilous insect, a selective pressure for evolutionary transformation may have become effective. A greater sensitivity seems desirable because it is of advantage to detect an IR source from greater distances. Consequently, the sites on the integument with the best prerequisite of evolving into an IR organ were those in a ventrolateral position facing the ground as well as the lateral environment of the flying beetle. The result was a synorganization of the corresponding part of the cuticle, the epidermis, and the underlying sensory structures (in case of *M. atrata* the multipolar neuron) as well as neuronal connections within the central nervous system.

See also: Thermoreception: Vertebrates.

Further Reading

Apel K-H (1989) Zur Verbreitung von *Melanophila acuminata* DEG. (Col., Buprestidae). *Entomologische Nachrichten und Berichte* 33: 278–280.
Bullock TH and Fox W (1957) The anatomy of the infrared sense organ in the facial pit of pit vipers. *Quarterly Journal of Microscope Science* 98: 219–234.

Time: What Animals Know

J. D. Crystal, University of Georgia, Athens, GA, USA

Introduction

Time is a fundamental dimension of life, and many animals have evolved the ability to time intervals lasting from a fraction of a second to hours. Events unfold in time, and we experience events over time. The ability to time events is ubiquitous, as illustrated by the assortment of events that are affected by temporal processing – speech, music, motor control, foraging, decision-making, sleep–wake cycles, and appetite. Thus, an understanding of learning, cognition, and performance requires an analysis of temporal information processing. A distinction is traditionally made between interval timing and circadian timing. Interval timing is the ability to time shorter intervals, typically in the range of seconds to minutes. Circadian timing is the ability to adjust to the daily cycle, which has a period of 24 h. A circadian clock repeats itself approximately every 24 h, and this period is set by external cues such as daylight or large meals. This article concerns interval timing and the mechanisms that subserve interval timing.

Timing in Natural Settings

Three examples of timing in animal behavior are described. Free-living hummingbirds time the interval between successive visits to flowers that they visited throughout the day. In experiments in which the flowers were replenished after different intervals of time (e.g., 10 vs. 20 min), the revisits to flowers tracked the replenishment rate. Hummingbirds apparently update their memories of when and where food was encountered for each flower and how long ago they last visited each location. Bees also adjust the time of visiting food sources on the basis of the amount of time elapsed since the last visit. Scavenging birds appear to anticipate food availability and arrive at locations before food has reached its peak amount.

Representations of Time

How are intervals timed? The basic idea is that an interval elapses with respect to the occurrence of some event. With a mechanical artifact such as a stopwatch, we can track the elapsing interval. When the event begins, we reset the stopwatch to zero and start the stopwatch. Reading the stopwatch provides an estimate of time to complete the event. A basic question about temporal information processing concerns the mechanism by which time is represented. Two types of temporal representations are described. One mechanism that may be used to time intervals is a pacemaker-accumulator. A pacemaker emits pulses as a function of time; the accumulator counts or integrates the number of pulses emitted. An hourglass is a familiar example of a pacemaker-accumulator – the elapsed duration, since the hourglass was turned over (i.e., reset), is indexed by the amount of sand in the bottom chamber of the hourglass.

A second mechanism that may be used to time intervals is an oscillator mechanism. Circadian timing is the best known biological oscillator. Circadian timing is based on the completion of a periodic process that is approximately a day. Time of day is indexed by the phase of a circadian periodic process. Circadian timing is widespread. Oscillator properties of timing intervals of approximately a day may be extended to intervals in the range of seconds to minutes as discussed in the section 'Oscillator Properties of Interval Timing'.

Formal Properties of Interval and Circadian Timing

In 1997, Gibbon and colleagues provided a classic description of the operating characteristics of interval and circadian timing systems. Gibbon's outline is briefly described in this section. According to the classic description, the interval timing system is based on a pacemaker-accumulator mechanism, and the circadian system is based on an endogenous-oscillator mechanism. Endogenous means that the oscillator does not require continued periodic input to produce ongoing periodic output. For example, when an animal is exposed to daily periodic light cycles such as 12 h of light followed by 12 h of darkness, activity patterns occur at species-typical times of day. After termination of the periodic light cycle (e.g., constant dim illumination), behavior 'free runs' with a period that typically departs slightly from 24 h. Free running behavior after the termination of periodic stimuli provides evidence that the timing system is endogenous because this pattern of data is important to rule out the possibility that observed behaviors are linked to the occurrence of daily environmental changes (e.g., temperature or noise fluctuations). In contrast, Gibbon and colleagues outlined the interval timing system as requiring resetting. The timing system operates on an elapsing interval timed with respect to the occurrence of some stimulus; a single presentation of the stimulus is necessary and sufficient to reset the interval timing system (i.e., one shot reset).

The circadian system operates within a limited range of entrainment. In particular, presentation of a periodic input entrains the endogenous oscillator only if the periodic input is within a limited range of periods near 24 h. By contrast, the interval timing system has a broad training range covering 3–4 orders of magnitude from seconds to hours.

A hallmark feature of the circadian system is its slow adjustment to a phase shift. A phase shift is an abrupt change in the initiation of a periodic process. Jet lag is a familiar example; we experience a phase shift in the unusual wake-up times after flying to a destination across several time zones. It usually requires several days before activities are synchronized to the new time zone. By contrast, the interval timing system undergoes immediate adjustment to a phase shift; the response to a single shift in a cycle is complete adjustment or complete resetting of the timing processes (i.e., one-shot reset).

Temporal performance based on a circadian oscillator is highly precise as measured by cycle-to-cycle variation. Precision is typically measured relative to the interval being timed. In particular, the coefficient of variation (CV) is the standard deviation of time estimates divided by the mean of time estimates. The CV of circadian performance is ~1–5%. By contrast, interval timing performance is characterized by a much low level of precision (CV of 10–35%). Thus, a characteristic of a circadian oscillator is relatively high timing precision. In particular, a consequence of having an endogenous oscillator dedicated to timing select values within a limited range appears to be relatively high sensitivity to timing these target durations. The variance properties of timing have played an important role in understanding interval timing. By contrast, the analysis of variance properties has had less impact in the study of circadian timing.

Oscillator Properties of Interval Timing

The sections that follow describe a series of empirical tests that were designed to evaluate the hypothesis that interval timing is based, at least in part, on oscillatory processes.

Resetting Properties of Short-Interval Timing

Although the phase-shift manipulation is a classic experimental design for the diagnosis of a circadian oscillator, the same manipulation may be used to assess a short-interval timing mechanism. **Figure 1** shows an example of a phase-shift manipulation in short-interval timing. Rats were trained to time 100 s using a fixed-interval procedure. In a fixed-interval procedure, the first response after the fixed interval elapses produces a food pellet.

To produce a phase shift, an early, response-independent pellet was delivered (i.e., free food). Four food cycles were required before adjustment was complete, which is consistent with the hypothesis that short-interval timing of 100 s is based on an oscillator mechanism.

Endogenous Oscillations in Short-Interval Timing

A critical diagnostic property of a timing mechanism may be assessed by discontinuing periodic input (i.e., extinction) and assessing subsequent anticipatory behavior. The defining feature of an oscillator is that periodic output from the oscillator continues after the discontinuation of periodic input. Similarly, a defining feature of a pacemaker-accumulator system is that elapsed time is measured with respect to the presentation of a stimulus, according to the classic description of this system described earlier. Thus, output of a short-interval system is periodic only when driven by periodic input. Moreover, periodic output is expected to cease if periodic input is discontinued. To test pacemaker-accumulator and endogenous-oscillator mechanisms, rats were trained with a variety of short intervals (e.g., 48, 96, and 192 s). The critical manipulation was the suspension of food delivery. **Figure 2** shows data from a representative individual rat, and **Figure 3** shows group data. Periodic delivery of food produced periodic behavior during training (**Figure 3**, left column), as predicted by both mechanisms. Behavior continued to be periodic after termination of periodic input (**Figure 3**, right column), consistent with an endogenous-oscillator, but not a pacemaker-accumulator, mechanism. The periodic behavior in extinction appears to be based on entrainment to the periodic feeding in training given that the period in extinction increased as a function of the period in training. Short-interval timing is, at least in part, based on a self-sustaining, endogenous oscillator.

Timing Long Intervals

A critical hypothesis about an oscillator is that it functions to confer improved sensitivity to time intervals near the period of the oscillator. To test this hypothesis, a series of experiments investigating meal anticipation was undertaken to identify a local peak in sensitivity to time 24 h. To examine anticipation of long intervals, food was restricted to 3-h meals, which rats earned by breaking a photo beam in a food trough. Critically, the rats tend to inspect the food trough before meals start, thereby providing a temporal anticipation function for each intermeal interval condition. **Figure 4** shows anticipation functions for intermeal intervals in the circadian range (22–26 h) and well outside this range (14 and 34 h). Note that response rates increased later into the intermeal interval for

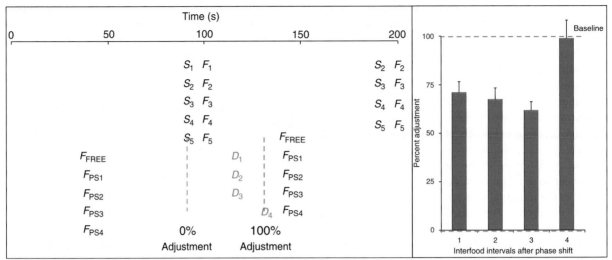

Figure 1 A phase shift produces gradual adjustment in short-interval timing. Left panel: Schematic representation of training, phase-shift manipulation, predictions, and data (double plotted to facilitate inspection of transitions across successive intervals; consecutive 100-s fixed intervals are plotted left to right and top to bottom). Rats ($n = 14$) timed 100-s intervals, and the last five intervals before the phase shift are shown (F = food pellet, S = start time of response burst). A 62-s phase advance (i.e., early pellet) on average was produced by the delivery of a response-independent food (F_{FREE}). All other food-to-food intervals were 100 s (F_{PS} = food post phase shift). Dashed lines indicate predictions if rats are insensitive (0% adjustment, purple) or completely sensitive (100% adjustment, pink) to the most recently delivered food pellet. A pacemaker-accumulator mechanism predicts 100% adjustment on the initial interval after the phase shift on the assumption of complete reset. An oscillator mechanism predicts initial incomplete adjustment. Data (D) indicate incomplete adjustment on the first three trials. Right panel: Start times on the initial three trials were earlier than in preshift baseline. Resetting was achieved on the fourth trial. Each 45-mg food pellet was contingent on a lever press after 100 s in 12-h sessions. The start of a response burst was identified on individual trials by selecting the response that maximized the goodness of fit of individual responses to a model with a low rate followed by a high rate. The same conclusions were reached by measuring the latency to the first response after food. Baseline was the average start time on the five trials before the phase shift. Left panel: Zero on the y-axis (purple dashed line) corresponds to complete failure to adjust to the phase shift; 100% (pink dashed line) corresponds to complete resetting. Error bars represent 1 SEM. Reproduced from Crystal JD (2006b) Time, place, and content. *Comparative Cognition & Behavior Reviews* 1: 53–76, with permission.

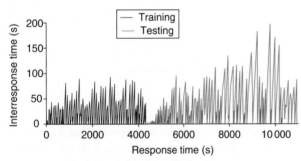

Figure 2 Many small interresponse times in short-interval timing are punctuated by much longer interresponse times, and punctuation by relatively long interresponse times continued after termination of periodic food delivery. Interresponse time (i.e., times of responses $R_{n+1} - R_n$) is plotted as a function of response time for a representative rat. During training, food was delivered on a fixed interval 96-s schedule. During testing, food was not delivered (i.e., extinction). Extinction began at a randomly selected point in the session. The response measure was the time of occurrence of photo beam interruptions in the food trough. Reproduced from Crystal JD and Baramidze GT (2007) Endogenous oscillations in short-interval timing. *Behavioural Processes* 74: 152–158, with permission from Elsevier.

intervals near the circadian range than for intervals outside this range. The response distributions were used to estimate sensitivity to time (i.e., relatively small spreads in the distributions correspond to relatively high sensitivity to time). As shown in **Figure 5**, intermeal intervals in the circadian range produced spreads that were smaller (i.e., lower variability) compared to intervals outside this range. Note that the data in **Figure 5** document a local maximum in sensitivity to time near 24 h, consistent with the hypothesis that a function of a circadian oscillator is improved sensitivity to time.

Endogenous Oscillations in Long-Interval Timing

The examples of timing noncircadian long intervals in **Figures 4** and **5** indicate that rats can time intervals outside the circadian range, but they do not identify the mechanism. In particular, these data could be based on an endogenous oscillator mechanism or a pacemaker-accumulator mechanism reset by meals. By contrast, these examples document endogenous oscillations in

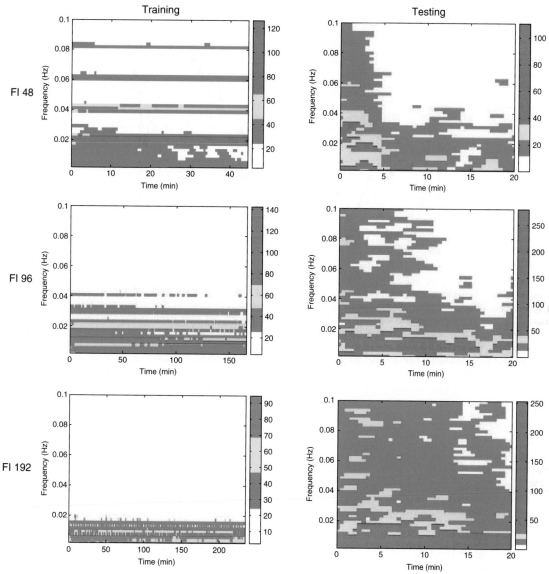

Figure 3 Endogenous oscillations in short-interval timing continue after the termination of periodic input. Short time Fourier transforms are shown for training (left panels) and testing (right panels) conditions using fixed interval 48-, 96-, and 192-s procedures. The three-dimensional images show frequency (period = 1/frequency) on the vertical axis as a function of time within the session along the horizontal axis; the color scheme represents the amount of power from the Fourier analysis. Concentrations of high power occur at a frequency of ~0.02, 0.01, and .0005 which correspond to periods of ~50, 100, and 200 s in top, middle, and bottom panels, respectively. Adapted from Crystal JD and Baramidze GT (2007) Endogenous oscillations in short-interval timing. *Behavioural Processes* 74: 152–158, with permission from Elsevier.

timing short intervals (1–3 min) by demonstrating that behavior continued after the termination of periodic input. The same experimental approach is used in this section to document endogenous oscillations in long-interval timing (16 h).

Rats earned food by interrupting a photo beam in the food trough during 3-h meals using a 16-h intermeal interval. After approximately a month of experience with the intermeal interval, the meals were discontinued. **Figure 6** (top panel) shows that the response rate increased as a function of time prior to the meals, documenting that the

rats timed 16 h, consistent with either oscillator or pacemaker-accumulator mechanisms. When two successive meals were skipped, the rats anticipated the arrival of two successive 16-h intervals (**Figure 6** middle and bottom panels), consistent with the use of an endogenous oscillator. In particular, the response rate was reliably higher during the 3-h omitted meal relative to the earlier 13 h for both first and second nonfood cycles. If timing was based on a pacemaker-accumulator reset by meals, then the rats would be expected to time the first, but not the second, skipped meal. A pacemaker-accumulator does

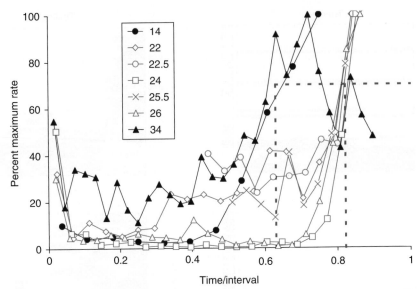

Figure 4 Response rate increased later into the interval for intermeal intervals near the circadian range (unfilled red symbols) relative to intervals outside this range (filled blue symbols); dashed lines indicate width of response rate functions. Anticipatory responses increase immediately prior to the meal for all intermeal intervals except 34 h. Each 45-mg food pellet was contingent on a photo beam break after a variable interval during 3-h meals. Intermeal intervals were tested in separate groups of rats ($n = 3$–5 per group). The end of the meal corresponds to 1 on the x-axis. Testing was conducted in constant darkness. Adapted from Crystal (2001a). Reproduced from Crystal JD (2006b) Time, place, and content. *Comparative Cognition & Behavior Reviews* 1: 53–76, with permission.

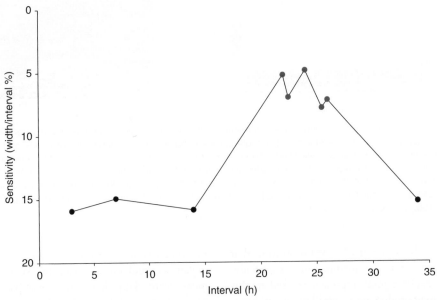

Figure 5 Intervals near the circadian range (red symbols) are characterized by higher sensitivity than intervals outside this range (blue symbols). Variability in anticipating a meal was measured as the width of the response distribution prior to the meal at 70% of the maximum rate, expressed as a percentage of the interval ($N = 29$). The interval is the time between light offset and meal onset in a 12–12 light-dark cycle (leftmost two circles) or the intermeal interval in constant darkness (all other data). The percentage width was smaller in the circadian range than outside this range. The width/interval did not differ within the circadian or noncircadian ranges. The same conclusions were reached when the width was measured as 25%, 50%, and 75% of the maximum rate. The data are plotted on a reversed-order y-axis so that local maxima in the data correspond to high sensitivity, which facilitates comparison with other measures of sensitivity (e.g., **Figure 7**). Mean $SEM = 2.4$. Adapted from Crystal (2001a). Reproduced from Crystal JD (2006b) Time, place, and content. *Comparative Cognition & Behavior Reviews* 1: 53–76, with permission.

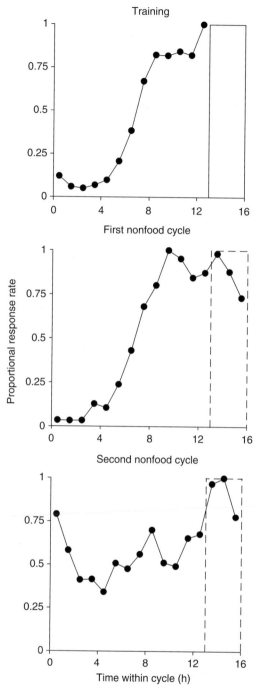

Figure 6 Endogenous oscillations in long-interval timing continue after the termination of periodic input. Response rate increased as a function of time within the 16-h intermeal interval cycle during the first and second nonfood cycle. Response rate (frequency of responses expressed as a proportion of the maximum frequency within the cycle) is plotted as a function of time within the cycle. The cycle included meals (indicated by the solid rectangle) during training (top panel). The meals were omitted (indicated by the dashed rectangles) in the first (middle panel) and second (bottom panel) nonfood cycles. Reproduced from Crystal JD (2006a) Long-interval timing is based on a self sustaining endogenous oscillator. *Behavioural Processes* 72: 149–160, with permission from Elsevier.

not predict an increase in response rate prior to the second skipped meal because elapsed time since the last meal is larger than the intermeal interval during the second non-food cycle (i.e., time since the last meal is unusually long at this point).

The reliability of a periodic trend was assessed and observed periods were estimated with a periodogram analysis; a periodogram analysis involves wrapping a response rate function around different proposed periods to identify the period that best fits the observed data. A reliable periodic trend was observed for each rat, and the mean period in extinction (20.4 ± 0.9 h, mean ± SEM) was reliably different from 16 and 24 h. These data suggest that the natural period of the oscillator that drove behavior was 20.4 h, which is distinct from the circadian oscillator; according to this hypothesis, the two oscillatory systems are dissociated by their different characteristic periods. However, the data are also consistent with the hypothesis that the circadian oscillator's free-running period is modified by the periodic input to which it was previously exposed. According to both of these hypotheses, long-interval timing is based on a self-sustaining, endogenous oscillator; the hypotheses differ in specifying the characteristic period of the oscillators. In either case, long-interval timing is based on a self-sustaining, endogenous oscillator.

Variance Properties in Circadian and Short-Interval Timing

As mentioned earlier, the study of variance properties has historically played a significant role in the development of theories of short-interval, but not circadian, timing. However, the data summarized in **Figure 5** suggest that a function of the well-established circadian oscillator is the relative improvement in sensitivity to time ~24 h. Thus, other putative oscillators may be identified by documenting other local maxima in sensitivity to time. In addition, the observation that short-interval timing in the range of 1–3 min exhibits endogenous, self-sustaining patterns of behavior after the termination of periodic input reinforces the expectation that short-interval timing may be based on an endogenous oscillatory mechanism.

To search for local peaks in sensitivity to time in the short-interval range, a series of experiments were conducted using many, closely spaced target intervals. **Figure 7** shows sensitivity to time plotted as a function of stimulus duration from these experiments. Sensitivity to time short intervals is characterized by multiple local peaks. Each peak in sensitivity to time may identify the period of a short-period oscillator. The procedure involved presenting a short or long stimulus followed by the insertion of two response levers. Left or right lever presses were designated as correct after short or long stimuli. Accuracy was

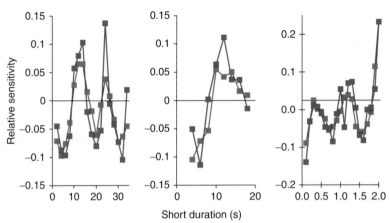

Figure 7   Sensitivity to time is characterized by local maxima at 12 and 24 s (left panel), 12 s (middle panel), and 0.3 and 1.2 s (right panel). Green symbols: average across rats. Red symbols: a running median was performed on each rat's data and the smoothed data were averaged across rats to identify the most representative local maxima in sensitivity. Left panel: Rats discriminated short and long noise durations with the duration adjusted to maintain accuracy at ~75% correct. Short durations were tested in ascending order with a step size of 1 s ($n = 5$) and 2 s ($n = 5$). Sensitivity was similar across step sizes, departed from zero, and was nonrandom. Mean $SEM = 0.03$. Middle panel: Methods are the same as described in left panel, except short durations were tested in random order ($n = 7$) or with each rat receiving a single interval condition ($n = 13$); results from these conditions did not differ. Sensitivity departed from zero and was nonrandom. Mean $SEM = 0.02$. Right panel: Methods are the same as described in left panel, except intervals were defined by gaps between 50-ms noise pulses and short durations were tested in descending order with a step size of 0.1 s ($n = 6$). Sensitivity departed from zero and was nonrandom. Mean $SEM = 0.04$. Sensitivity was measured using d' from signal detection theory. $d' = z[p(\text{short response} \mid \text{short stimulus})] - z[p(\text{short response} \mid \text{long stimulus})]$. Relative sensitivity is $d' - \text{mean } d'$. Adapted from Crystal (1999, 2001b). Reproduced from Crystal JD (2006b) Time, place, and content. *Comparative Cognition & Behavior Reviews* 1: 53–76, with permission.

maintained at ~75% correct by adjusting the duration of the long stimulus after blocks of trials. Sensitivity to time was approximately constant for short durations from 0.1 to 34 s. However, local maxima in sensitivity to time were observed at ~0.3, 1.2, 12, and 24 s.

Figure 8 shows multiple local maxima in sensitivity to time across several orders of magnitude, using data from the experiments described earlier. The data on the right and left sides of **Figure 8** come from **Figures 5** and **7**. **Figure 8** suggests that multiple local peaks in sensitivity to time are observed in timing across several orders of magnitude.

Integration of Interval and Circadian Timing

The summarized data suggest that the psychological representation of time is nonlinearly related to the interval being timed. The existence of a local maximum near a circadian oscillator (**Figure 8**, peak on right side) and local maxima in the short-interval range (**Figure 8**, peaks on left side) are consistent with timing based on multiple oscillators. According to multiple-oscillator proposals, each oscillator is a periodic process that cycles within a characteristic period. Each oscillator can be characterized by its period (i.e., cycle duration) and phase (i.e., current point with the cycle). Thus, each unit within a multiple oscillator system has its own period and phase. Sensitivity to time an interval near an oscillator is expected to be higher than timing an interval farther away

from the oscillator because an oscillator functions to increase temporal sensitivity. Therefore, the multiple local peaks in sensitivity to time shown in **Figure 8** suggest the existence of multiple short-period oscillators.

The data reviewed in this section suggest that interval timing is based on an endogenous-oscillator, rather than a pacemaker-accumulator, mechanism according to the classic distinction discussed at the beginning of this article. The main findings are summarized as follows. The data in **Figure 1** document that short-interval timing exhibits gradual phase adjustment, consistent with an oscillator mechanism. The data in **Figures 2** and **3** suggest that short-interval timing is endogenous and self-sustaining, consistent with an oscillator mechanism. The data in **Figures 4** and **5** document that many long, but noncircadian, intervals can be timed, and the data in **Figure 6** suggest that long-interval timing is endogenous and self sustaining, consistent with an oscillator mechanism. The data in **Figures 5**, **7**, and **8** show that both short-interval and circadian timing are characterized by local peaks in sensitivity to time.

Conclusion

The data suggest continuity of mechanisms in short-interval, long-interval, and circadian-timing systems. The data reviewed in this article may prompt the development of a theory of timing that encompasses the discrimination

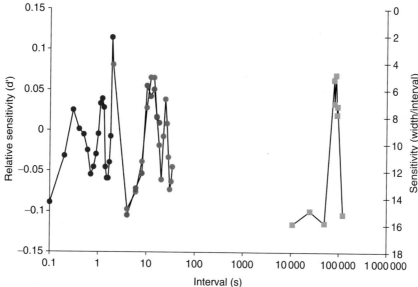

Figure 8 Multiple local maxima in sensitivity to time are observed in the discrimination of time across seven orders of magnitude. The existence of a local maximum near a circadian oscillator (peak on right side; purple squares) and other local maxima in the short-interval range (peaks on left side; blue, red, and green circles) are consistent with the hypothesis that timing is mediated by multiple oscillators. Intervals in the blank region in the center of the figure have not been tested. Left side: Rats discriminated short and long durations, with the long duration adjusted to maintain accuracy at 75% correct. Short durations were tested in sequential order (blue and red circles; $N = 26$) or independent order (green circles; $N = 20$). Circles represent relative sensitivity using d' from signal detection theory and are plotted using the y-axis on the left side of the figure. Right side: Rats received food in 3-h meals with fixed intermeal intervals by breaking a photo beam inside the food trough. The rate of photo beam interruption increased before the meal. Squares represent sensitivity, which was measured as the width of the anticipatory function at 70% of the maximum rate prior to the meal, expressed as a percentage of the interval ($N = 29$). The interval is the time between light offset and meal onset in a 12–12 light-dark cycle (leftmost two squares) or the intermeal interval in constant darkness (all other squares). Squares are plotted with respect to the reversed-order y-axis on the right side of the figure. Y-axes use different scales, and the x-axis uses a log scale. Adapted from Crystal (1999, 2001a, 2001b). Reproduced from Crystal JD (2006b) Time, place, and content. *Comparative Cognition & Behavior Reviews* 1: 53–76, with permission.

of temporal intervals across several orders of magnitude, from milliseconds to days. Such a system is capable of representing when specific events occurred in time, as people describe when events occurred using calendar-date-time systems. This type of a system may underlie the temporal representation of episodic memory (i.e., the memory system that contains memories of unique events from one's past).

See also: Mental Time Travel: Can Animals Recall the Past and Plan for the Future?.

Further Reading

Aschoff J (ed.) (1981) *Handbook of Behavioral Neurobiology, vol. 4: Biological Rhythms.* New York: Plenum Press.

Boisvert MJ and Sherry DF (2006) Interval timing by an invertebrate, the bumble bee *Bombus impatiens. Current Biology* 16: 1636–1640.

Church RM and Broadbent HA (1990) Alternative representations of time, number, and rate. *Cognition* 37: 55–81.

Crystal JD (1999) Systematic nonlinearities in the perception of temporal intervals. *Journal of Experimental Psychology: Animal Behavior Processes* 25: 3–17.

Crystal JD (2001a) Circadian time perception. *Journal of Experimental Psychology: Animal Behavior Processes* 27: 68–78.

Crystal JD (2001b) Nonlinear time perception. *Behavioural Processes* 55: 35–49.

Crystal JD (2006a) Long-interval timing is based on a self sustaining endogenous oscillator. *Behavioural Processes* 72: 149–160.

Crystal JD (2006b) Time, place, and content. *Comparative Cognition & Behavior Reviews* 1: 53–76.

Crystal JD and Baramidze GT (2007) Endogenous oscillations in short-interval timing. *Behavioural Processes* 74: 152–158.

Gallistel CR (1990) *The Organization of Learning.* Cambridge, MA: MIT Press.

Gibbon J, Fairhurst S, and Goldberg B (1997) Cooperation, conflict and compromise between circadian and interval clocks in pigeons. In: Bradshaw CM and Szabadi E (eds.) *Time and Behaviour: Psychological and Neurobehavioural Analyses*, pp. 329–384. New York: Elsevier.

Henderson J, Hurly TA, Bateson M, and Healy SD (2006) Timing in free-living rufous hummingbirds, *Selasphorus rufus. Current Biology* 16: 512–515.

Takahashi JS, Turek FW, and Moore RY (eds.) (2001) *Handbook of Behavioral Neurobiology, vol. 12: Circadian Clocks.* New York: Plenum.

Wilkie DM, Carr JAR, Siegenthaler A, Lenger B, Liu M, and Kwok M (1996) Field observations of time-place behaviour in scavenging birds. *Behavioural Processes* 38: 77–88.

Zhou W and Crystal JD (2009) Evidence for remembering when events occurred in a rodent model of episodic memory. *Proceedings of the National Academy of Sciences USA* 106: 9525–9529.

Niko Tinbergen

R. W. Burkhardt, Jr., University of Illinois at Urbana-Champaign, Urbana, IL, USA

Life and Scientific Career

Nikolaas (Niko) Tinbergen, the third of five children of two Dutch schoolteachers, was born in The Hague on 15 April 1907. Remarkably, two of the family's four sons were eventually awarded Nobel Prizes. Niko's elder brother Jan received the Nobel Prize in Economics in 1969; Niko received the Nobel Prize in Physiology or Medicine in 1973. When reporters asked Niko how it was that his family had produced two Nobel Prize winners, he attributed this not to innate abilities but rather to the supportive conditions in which he and his siblings were raised. The Tinbergen parents allowed their children to follow their own interests. For Niko, this meant outdoor activities – sports, nature rambles, camping out, and eventually a career as a field biologist (**Figure 1**).

Nature studies flourished in Holland in the early twentieth century. Tinbergen joined the Dutch Youth Association for Nature Study (the Nederlandse Jeugdbond voor Natuurstudie or NJN) and took great pleasure in learning about the native flora and fauna, and in particular, watching and photographing animals. However, while living nature inspired him, traditional academic botany and zoology, with their emphasis on taxonomy and anatomy, left him as cold. It was only after a 3-month stay in 1925 at the Rossitten bird station on the Baltic Sea, where he watched the autumn bird migration, that he decided to become a professional biologist.

Tinbergen enrolled as a zoology student at Leiden University in January 1926. Although he found most of the zoological and botanical instruction there to be just as boring as he had feared, there were a few notable exceptions. Jan Verwey, a young field biologist, encouraged Tinbergen to pursue field studies of behavior. Later, Tinbergen's Ph.D. advisor, Hildebrand Boschma, permitted him to do a field study for his thesis. Tinbergen's topic was the orientation behavior of the bee wolf, *Philanthus triangulum*. The experiments he conducted for it were modeled in part on the work of Karl von Frisch. As events transpired, Tinbergen was allowed to submit a dissertation that was exceptionally short – only 31 pages in print – so that he could receive his degree and participate in a Dutch expedition to Greenland. In the spring of 1932, he was awarded his doctorate, he got married (to Elisabeth Rutten), and the young couple set off for East Greenland. Tinbergen had constructed a scientific justification for the trip – a study of the territorial behavior of the snow bunting in spring – but his primary goal was to live in the Arctic among the Inuit and witness the stark, natural beauties of the area.

After 15 months in Greenland, the Tinbergens returned in the fall of 1933 to Holland. Tinbergen took up again an assistantship in the Leiden Department of Zoology (a position to which he had been originally appointed in 1931). In this capacity, he developed a program of research and teaching involving field and laboratory studies of animal behavior. He established in the spring of 1935 a special, 6 week, laboratory 'practical' for third-year undergraduates. In this course, he and his students researched the reproductive behavior of fish, most notably the three-spined stickleback. In summers, he took students to a field camp in the Dutch dunes to study the behavior of insects and birds.

Tinbergen was thus already launched on his career when he met the Austrian zoologist Konrad Lorenz at a small conference on instinct held in Leiden in November 1936. Tinbergen at this time was just 29; Lorenz was 33. The more senior animal psychologists attending the conference were interested in understanding instinct in terms of the animal's subjective experience, whereas Tinbergen and Lorenz were interested in creating a new, objectivistic, science of animal behavior that was biologically rather than psychologically oriented. The two men quickly bonded with each other. Tinbergen was greatly impressed by the way Lorenz was uniting a vast array of disparate information in a single, coherent, theoretical system. Lorenz, for his part, was particularly impressed by Tinbergen's talents as an experimenter. Tinbergen was able to report on the stickleback experiments that he and his student, Joost ter Pelkwijk, had recently conducted. Using 'dummies' to elicit the fish's instinctive reactions, they had investigated the different sign stimuli to which three-spined sticklebacks respond in fighting, courting, spawning, and the like. Lorenz, whose new theoretical system featured releasers, innate releasing mechanisms, and innate motor patterns, regarded Tinbergen's experimental results as just what his new science needed.

The friendship that sprang up between Tinbergen and Lorenz in November 1936 was consolidated the following spring when Tinbergen, with the benefit of a research leave from his department, traveled to Austria to work for 3 1/2 months with Lorenz at Lorenz's home in Altenberg. There the two naturalists conducted their classic study of the egg-rolling behavior of the gray lag goose. There too they experimented on how young birds react to simulated predators. Tinbergen also continued a series of

Figure 1 Niko Tinbergen. Photo by Lary Shaffer. Courtesy of Lary Shaffer.

experiments he had begun in Leiden with his student, D. J. Kuenen, on the gaping response in young blackbirds and thrushes.

Years later, Tinbergen and Lorenz would look back wistfully to their months of being together at Altenberg in the spring of 1937. The Second World War put them on opposite sides politically. After Germany invaded Holland in May 1940, they continued to exchange letters on scientific matters, but this ceased in 1942, when Tinbergen was incarcerated in a prisoner of war camp after resisting the occupiers' attempt to Nazify Leiden University. Lorenz took part in an effort to have Tinbergen released, but Tinbergen would have no part of it. He remained a prisoner for 2 years. After the war, his experiences during the occupation made him unwilling to resume relations with German scientists right away. With regard to Lorenz, Tinbergen's thoughts were especially ambivalent. He was aware that the man whom he had regarded both as a friend and as the pioneer of animal behavior studies had been somewhat 'Nazi-infected.' He was saddened, nonetheless, when the news arrived that Lorenz was missing in action and presumed dead.

It turned out that Lorenz was not dead after all but had been captured by the Russians. However, Lorenz had no prospects of being released soon. Tinbergen concluded that he had to take charge of the postwar reconstruction of

ethology himself. One of his key efforts in this regard was the founding of the journal *Behaviour*, the first issue of which appeared in 1947. He also lectured abroad: in Switzerland, Britain, and the United States. The special series of lectures he gave early in 1947 in New York became the basis of his book, *The Study of Instinct*, which, when it finally appeared in 1951, was ethology's first general text. 1947 was also the year Tinbergen was promoted to a chair of experimental biology that been created especially for him at the University of Leiden. Two years later, when he gave up his post to take up a lecturer's position at Oxford, many of his Dutch colleagues were angry with him. The explanation he offered them was that he felt the need to serve as a missionary for ethology in the English-speaking world.

Tinbergen moved to Oxford in the fall of 1949. There he made his new home with his wife and five children. He would remain at Oxford for the rest of his life, establishing a strong program of behavior studies that attracted many talented PhD and postdoctoral students. He and Lorenz would continue to be the leading figures of the discipline through the 1950s and 1960s, playing conspicuous roles at the international congresses that helped shape ethology's identity. Tinbergen was also a successful popularizer of ethology, most notably with his books, *The Herring Gull's World* (1953), *Curious Naturalists* (1958), and the Time-Life volume, *Animal Behavior* (1965), and with films.

Tinbergen was promoted to a professorship at Oxford in 1966. In addition to the honors he received for his scientific research (which included his election as a Fellow of the Royal Society and his receipt of the Nobel Prize), he was recognized for his achievements as a filmmaker, receiving with Hugh Falkus in 1969 the Italia Prize for their film, *Signals for Survival*. In the 1970s, he joined his wife in a study of child autism. Tinbergen died in Oxford on 21 December 1988.

Practices and Concepts

First at Leiden and then again at Oxford, Tinbergen established a program of researches that featured investigations in both the field and the laboratory. The patient watching of sea gull colonies and field and lab experiments on selected species of insects, fish, and birds characterized his work. In early experiments on the reproductive behavior of sticklebacks and on the gaping behavior of young blackbirds, he used 'dummies' to test the stimuli eliciting the animals' instinctive reactions. His results seemed to correlate nicely with Lorenz's new theorizing about the interrelations of external stimuli, innate releasing mechanisms, and instinctive behavior patterns, and Tinbergen proceeded to employ these concepts in his new work, including the studies he conducted with Lorenz in Austria in the spring of 1937.

In their collaborative experiment on the egg-rolling behavior of the graylag goose, Tinbergen and Lorenz distinguished between two components in the motor sequence by which the goose returns an egg to her nest: an instinctive behavior pattern on the one hand and an orienting response or 'taxis' on the other. In the experiments they conducted on the response of young, hand-reared fowl to simulated predators, their primary interest was the correspondence between external stimuli and innate releasing mechanisms.

Although the latter experiments were never written up in detail, they can serve as a good illustration of the kind of experiments Tinbergen used in his studies of behavioral causation. In this case, he and Lorenz tested the reactions of various species hand-reared fowl to simulated flying predators (cardboard dummies of different shapes, pulled along a rope above the turkeys). The naturalists found that for young turkeys (though not the other barnyard fowl), the shape of the dummies made a difference. Dummies with 'short necks' elicited the young turkeys' alarm calls much more readily than did dummies with 'long necks.' The most striking results involved a single, relatively crude dummy constructed with the 'wings' located off center so as to make one end of the body relatively short and the other end relatively long. Which end appeared to be the neck and which end appeared to be the tail depended simply on the direction in which the dummy was moving. The young turkeys displayed more alarm when the dummy crossed above them with its short end first than with its long end first (**Figure 2**). The appeal of the case was that it appeared to show how well innate releasing mechanisms were tuned to the stimulus situations that triggered them. Though experiments 20 years later would suggest that the young turkeys' behavior was best explained in terms of habituation, Tinbergen and Lorenz interpreted the young turkeys' reaction to the short-end-forward shape as an innate response', forged by natural selection, to an environmental cue signaling 'predator.' The short-necked version, they explained, corresponded to the shape of a hawk, while the long-necked version corresponded to the shape of a goose.

Tinbergen had already found in experiments with male sticklebacks that the optimal stimulus eliciting their fighting response was not just any red fish model but instead a model that was red underneath. Young thrushes likewise directed their gaping responses toward certain models more than others, depending on how the models' 'heads' were presented in relation to the rest of the body. Tinbergen concluded that the behavior of the young turkeys was similar in that they were not responding to the shape of the 'hawk-goose' model per se, but they were responding to the particular configurational stimulus produced when the model moved slowly above them short end first (like a hawk). In the decades that followed, later investigators found it difficult to replicate

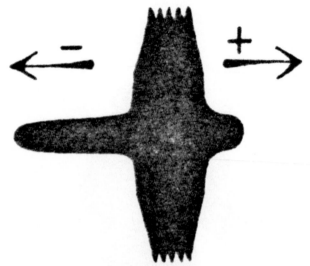

Figure 2 Tinbergen's illustration of a 'card-board dummy that releases escape reactions when sailed to the right ('hawk') but is ineffective when sailed to the left ('goose').' Reproduced from Tinbergen N (1948) Social releasers and the experimental method required for their study. *The Wilson Bulletin* 60: 6–51, 34.

these early experiments on innate releasing mechanisms. Be that as it may, these experiments played an important role, early on, in establishing ethology's credentials as a new science.

Although he endorsed Lorenz's basic system of releasing stimuli, innate releasing mechanisms, and innate motor patterns, Tinbergen's developing thoughts on behavioral causation were not an exact mirror of Lorenz's. In Lorenz's model of instinctive action, instincts were treated as basically independent of each other, at least in the sense that each was fed by its own 'action-specific energy.' Tinbergen, to the contrary, came by the early 1940s to think of instincts as being hierarchically related. Tinbergen' doctoral student Gerard Baerends in this period interpreted the behavior of a digger wasp of the genus *Ammophila* in terms of hierarchically arranged internal states or 'moods.' Hierarchical thinking about instincts also appeared in the thinking of the Dutch animal psychologist, Adriaan Kortlandt. Tinbergen proceeded to describe the instincts of sticklebacks in much the same terms that Baerends used. Stickleback males, by Tinbergen's account, are brought into the reproductive 'mood' through internal, hormonal changes and perceiving an appropriate territory. Once they have reached this general state, they are capable of a number of different activities, notably fighting other males, building a nest, and courting females. Whether a male comes into a fighting, building, or courting 'submood' depends on which stimuli it receives. If the appearance of an opponent brings it into a fighting submood, how the opponent then acts will determine how the male reacts: by fighting, chasing, threatening, biting, etc. (**Figure 3**).

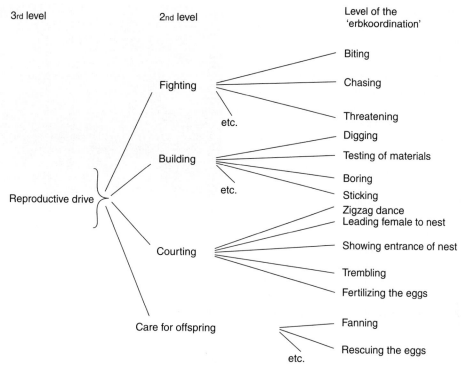

Figure 3   Tinbergen's representation of the hierarchical arrangement of drives. Reproduced from Tinbergen N (1942) An objectivistic study of the innate behaviour of animals. *Bibliotheca Biotheoretica* 1: 39–98, 57.

Tinbergen's work on behavioral causation also involved the identification and explanation of what Tinbergen called 'substitute activities' (later 'displacement activities'). These are stereotypic movements that appear as 'irrelevant' acts, such as birds stopping to feed or preen or collect nest materials while in the midst of fighting or courting. He interpreted these activities as outlets for a strongly activated drive (or two conflicting drives, such as fighting and fleeing) when it is prevented from being normally discharged. He also noted that displacement activities can serve as signals, becoming increasingly distinctive and 'ritualized' in this capacity over the course of evolution.

When Tinbergen moved to Oxford in 1949, he recognized the need to widen the scope of his researches beyond studies of causation. Lorenz had stressed the value of doing comparative studies and had illustrated this with a long, comparative study of the reproductive behavior of ducks. Tinbergen decided to do some comparative work of his own. Having already developed a detailed knowledge of the behavior patterns of the three-spined stickleback and the herring gull, he decided to set his new graduate students to work studying additional species of sticklebacks and gulls. However, he did not want simply to use new species to replicate Lorenz's work. Instead, he wanted to understand how behavior had evolved in particular ecological settings.

The comparative gull studies paid off handsomely for Tinbergen's research team, with the black-headed gull and the kittiwake proving especially instructive. Working on the kittiwake, Esther Cullen, one of Tinbergen's students, identified a whole range of species-specific behaviors that were corollaries of the bird's habit of breeding on the narrow ledges of steep cliffs. These behaviors included particular releasers, specialized fighting movements, distinctive nest-building behavior, and the nonremoval of eggshells from the nest site.

Tinbergen was greatly impressed by Cullen's findings. He was pleased to note how the kittiwake's adaptive characters fit together in a coherent system. From this, he drew two conclusions. One was that a species would often be subject to conflicting selection pressures, from which 'compromises' of various sorts would result. The other was that characters that looked by themselves like the product of random change could prove upon close examination to be the indirect results of selection on the whole adaptive system. To Tinbergen, this constituted a validation of field biology. Field biologists were in a better position than museum taxonomists – or observers of animals in captivity – when it came to observing the effects of selection.

Tinbergen indeed felt that this new research established something that Lorenz had not anticipated. Lorenz had insisted that releasers are especially valuable for taxonomic purposes because they are not closely linked to a species' ecology and thus not subject to convergence. He represented them as arbitrary, historical conventions developed over the course of evolution. Tinbergen concluded to the contrary that the kittiwake's special displays were not arbitrary conventions but were instead intimately linked to the species' cliff-dwelling habit. More generally, he went on to argue that one could expect to find convergence in the threat displays of very different species, because there are only a limited number of postures a bird can adopt in signaling its likelihood of attacking or withdrawing.

In 1957, at the International Ethological Congress in Freiburg, Tinbergen reported that he was feeling with respect to his work like a butterfly emerging from a chrysalis. Although he was not very explicit what he meant by this, the context of his remark seems clear. The previous several years had been a complex time for ethology. A number of its original concepts had come under fire. Within ethology, a new generation of students had identified problems with Lorenz's psycho-hydraulic model and Tinbergen's hierarchical model of instinctive action. From the American comparative psychologist Daniel Lehrman came in 1953 a scathing critique of Lorenzian ethology, claiming, among other things, that Lorenz's notion of 'innate' behavior was hindering the study of behavioral development. Tinbergen saw the value of establishing a dialog with Lehrman, and he indeed came to agree with some of what Lehrman was saying. But after this period of engaging with critics and trying to decide which of ethology's original concepts were still viable and which were not, Tinbergen wanted to get on with new work. His comparative gull studies were leading him to new studies of behavioral function, that is, how behavior patterns contribute to an animal's survival. In the late 1950s and early 1960, as he engaged in experimental field studies of behavioral function, he concluded that this type of work appealed to him more than any other. This was the basis for his sense of metamorphosis.

Studying the function of eggshell removal in the black-headed gull revealed to Tinbergen and his students how complex and how beautifully adapted the bird's behavior was. What was safest for the parents was not necessarily what was safest for the brood. Likewise, what helped protect against one kind of predator, such as carrion crows, did not necessarily work against other kinds of predators, such as foxes. The timing of removing eggshells from the nest area was itself a kind of compromise: while predation by herring gulls and crows constituted a selection pressure favoring the rapid removal of eggshells, leaving newly hatched chicks briefly unprotected exposed them to being eaten by neighboring black-headed gulls. Through his experimental field studies of selection pressures and adaptations, Tinbergen helped stimulate the development of modern behavioral ecology (**Figure 4**).

Fittingly enough, when Tinbergen in the early 1960s took an opportunity to comment on how his research program had developed at Oxford, he described the

Figure 4   Tinbergen in the field (in 1972 on Skomer Island, off the coast of Wales). Photo by Lary Shaffer. Courtesy of Lary Shaffer.

choices he had made in setting up his program in much the same terms that he used when describing adaptive radiation in gulls. His choices in building his program, he said, were influenced not only by general considerations about where ethology should be heading, but also by local conditions. He had been keen to connect with Oxford's strong traditions in ecology and evolutionary biology. At the same time, he knew that he had to be economical with the limited budget at his disposal. By his account, the character of the Oxford program ultimately reflected a compromise between breadth of approach and close attention to specific subjects.

The Four Questions of Ethology

Tinbergen is remembered for having defined and promoted ethology as 'the biological study of behavior.' His classic statement of what he meant by this appeared in his 1963 paper, 'On aims and methods of ethology.' There he offered his famous formulation of the 'four questions of ethology.' To understand behavior biologically, he said, one needs to ask of it: (1) What is its physiological causation? (2) What is its function or survival value? (3) How has it evolved over time? (4) How has it developed in the individual? This, he went on to suggest, not only provided the best way to characterize the nature of ethology, but also held the key to ethology's future. For ethology to continue to flourish, he said, it is needed to address all the four of these questions in a balanced and coordinated fashion.

Tinbergen's definition of ethology as of 1963 was a definition that had taken shape in his thinking over the course of some 30 years. Both he and Lorenz back in the 1930s had emphasized that their new approach was bringing biological perspectives to bear on the questions of animal psychology, but what they had actually stressed at any given time generally depended on the particular audience they were addressing. In 1942, for example, when he was contrasting the objective nature of ethology with the subjectivist approach of his animal psychologist countryman, J. A. Bierens de Haan, Tinbergen maintained that the ethologists' primary interest was behavioral causation, that is, understanding innate behavior in physiological terms. (Lorenz for his part, while equally committed to the study of behavioral causation, was inclined to say that what made ethology most distinctive was its comparative, evolutionary nature.) In 1951, in his book, *The Study of Instinct*, Tinbergen in effect identified ethology's 'four questions,' but without calling them that and certainly without giving them balanced treatment. Tinbergen devoted more than half of the book to the study of behavioral causation. He provided only a scant chapter each on behavioral development, function, and evolution. As it was, when he published his 'Aims and Methods of Ethology' paper in 1963, he did not feel that a balanced approach to the four questions of ethology was anywhere near being achieved. Studies of behavioral causation still far outweighed studies of behavioral function (despite the work that he and his students were doing on the latter topic). Ironically, perhaps, within a decade and a half, the tide would turn dramatically and functional studies would come to enjoy a disproportionate share of the research in ethology.

To view Tinbergen's 1963 paper in historical context is also to consider what it reveals about Tinbergen's ongoing relations with Konrad Lorenz. Tinbergen dedicated the paper to his old friend as part of a *Festschrift* commemorating Lorenz's 60th birthday. Tinbergen's contribution radiated friendship and goodwill, and it gave full credit – indeed in some ways exaggerated credit – to Lorenz's pioneering efforts in the field. Writing the paper required a certain amount of diplomacy on Tinbergen's part, for on the subject of behavioral development in particular, Tinbergen and Lorenz no longer saw entirely eye-to-eye. Lorenz was angry with Tinbergen for having been too accepting of Daniel Lehrman's critique of ethology's early assumptions about 'innate' behavior. Tinbergen acknowledged that he and Lorenz held different opinions with regard to behavior development – Tinbergen had come to believe that applying the word 'innate' to behavioral characters was harmful heuristically – but he did not want to press the issue too hard at that moment. A more subtle critique of Lorenz can be detected in what Tinbergen said in the paper regarding behavioral function. While he lauded Lorenz as one of the first students of behavior to be interested in survival value, he also made it clear that if scientists were to gain a better understanding of how natural selection actually operates, hard fieldwork,

including the experimental demonstration of survival value, needed to be done on the subject. In these ways and others, Tinbergen's 'Aims and Methods' paper testified both to the evolution of his own thinking and to his ongoing relations with Lorenz.

Tinbergen's 'four questions' of ethology were simultaneously a vision for the future development of ethology and an affirmation of his life-long commitment to fieldwork. At the end of his career, he felt that if he were to be remembered for anything, it would not be for any particular discovery as much as for his long-term promotion of a fully biological approach to behavior.

See also: Behavioral Ecology and Sociobiology; Ethology in Europe; Future of Animal Behavior: Predicting Trends; Herring Gulls; Integration of Proximate and Ultimate Causes; Neurobiology, Endocrinology and Behavior.

Further Reading

Burkhardt RW (2005) *Patterns of Behavior: Konrad Lorenz, Niko Tinbergen, and the Founding of Ethology*. Chicago: University of Chicago Press.

Hinde RA (1990) Nikolaas Tinbergen. *Biographical Memoirs of Fellows of the Royal Society* 36: 547–565.

Kruuk H (2003) *Niko's Nature: A Life of Niko Tinbergen and His Science of Animal Behaviour*. Oxford: Oxford University Press.

Lehrman DS (1953) A critique of Konrad Lorenz's theory of instinctive behavior. *The Quarterly Review of Biology* 298: 337–363.

Lorenz KZ (1935) Der Kumpan in der Umwelt des Vogels: der Artgenosse als auslösendes Moment sozialer Verhaltungsweisen. *Journal für Ornithologie* 83: 37–215; 289–413. Translated as Companions as factors in the bird's environment: The conspecific as the eliciting factor for social behaviour patterns (1970). In: Lorenz KZ (ed.) *Studies in Animal and Human Behaviour*, vol. 1, pp. 101–258.

Lorenz KZ and Tinbergen N (1938) Taxis und Instinkthandlung in der Eirollbewegung der Graugans. *Zeitschrift für Tierpsychologie* 2: 1–29; Translated as Taxis and instinctive behaviour pattern in egg-rolling by the Greylag goose (1970). In: Lorenz KZ (ed.) *Studies in Animal and Human Behaviour*, vol. 1, pp. 316–150. Cambridge, MA: Harvard University Press.

Ten Cate C (2009) Niko Tinbergen and the red patch on the herring gull's beak. *Animal Behaviour* 77: 785–794.

Tinbergen N (1942) An objectivistic study of the innate behaviour of animals. *Bibliotheca Biotheoretica* 1: 39–98.

Tinbergen N (1948) Social releasers and the experimental method required for their study. *The Wilson Bulletin* 60: 6–51.

Tinbergen N (1951) *The Study of Instinct*. Oxford: Clarendon Press.

Tinbergen N (1953) *The Herring Gull's World*. London: Collins.

Tinbergen N (1958) *Curious Naturalists*. London: Country Life.

Tinbergen N (1963) On aims and methods of ethology. *Zeitschrift für Tierpsychologie* 20: 410–433.

Tinbergen N (1967) Adaptive features of the black-headed gull *Larus ridibundus* L. In: Snow DW (ed.) *Proceedings of the XIV International Ornithological Congress*, pp. 43–59. Oxford, 24–30 July 1966. Oxford: Blackwell Scientific Publications.

Tinbergen N (1972–1973) *The Animal in Its World: Explorations of an Ethologist, 1932–1972*, 2 vols. London: George Allen and Unwin.

Tinbergen N (1989) Watching and wondering. In: Dewsbury DA (ed.) *Studying Animal Behavior: Autobiographies of the Founders*, pp. 431–463. Chicago: University of Chicago Press. Originally published in 1985 as *Leaders in the Study of Animal Behavior*. London: Associated University Presses.

Trade-Offs in Anti-Predator Behavior

P. A. Bednekoff, Eastern Michigan University, Ypsilanti, MI, USA

Introduction

A trade-off is an inescapable compromise between two conflicting demands. A classic example is the removal of the shells of hatched eggs by black-headed gulls (but not kittiwakes, which nest on cliffs) to make their nests harder for crows and other predators to locate. However, the parents do not remove the eggshells until the time their chicks dry after hatching and cannot easily be swallowed by other gulls. The timing of the removal involves a trade-off between danger from roving predators such as crows and danger from neighboring gulls.

In the lives of organisms, there are inescapable compromises between different components of fitness. The best response maximizes overall fitness and where doing more results in fitness losses through one component of fitness that are equal to fitness gains through another component. Changes in costs and benefits are called 'marginal costs and benefits' and are often represented by the derivative in calculus. The use of the word 'marginal' comes to behavioral ecology via economics, where it was first used to refer to food production.

The way for an animal to minimize predation risk would be for it to never expose itself in the open by never feeding and never reproducing. Obviously, any life history strategy without feeding or reproduction would not maximize fitness. In general, pursuing more food and reproductive opportunities will expose animals to greater predation risk. (Other behaviors including play, grooming, and sleep may also involve trade-offs with predation risk.) The first reason for such trade-offs is that doing more can simply take more time. If animals choose to venture out during the safest times first, any further increase in time spent out must be done during more dangerous times. Thus, predation risk will increase more than proportionally with time spent foraging or pursuing mates. Birds may begin foraging in the dawn twilight and begin singing even earlier. Starting so early increases their risk of being captured by nocturnal owls. Therefore, birds might limit their danger by singing from behind a screen of branches or leaves, even though this may lower the effectiveness of their song.

Where animals go also affects the degree of danger they are exposed to. In lakes, weedy portions provide more places for small fish to hide, while the open water produces more photosynthetic phytoplankton and zooplankton. Small fish can always find more to eat in the open water, but they also are more likely to be eaten themselves. Many fish (and other organisms) limit this compromise by moving into productive open waters at night. Although this movement can be from the shoreline into the open, in big bodies of water much of the movement is from the depths up to the surface. Many spawning events also involve movements to surface waters or other exposed locations. Exposed places make it easier for eggs and larvae to spread, but also expose adults and young to predators.

Regardless of location, just the act of moving may bring more danger, as well as more opportunities to feed and mate. Whether searching for food or for potential mates, an animal can cover a greater area by moving faster, and thus may encounter more predators. Tadpoles move less when in the presence of predators and are less likely to be killed by dragonfly larvae, which usually sit and wait for them. Male prairie dogs move between burrows of potential mates causing them to die at far higher rates than females.

The details of how animals behave in a particular time and place also affect their risk of predation. Some behaviors make it less likely that animals will detect and escape from predators. For example, dugongs are less likely to dig for rhizomes in sea grass beds when tiger sharks are around, as digging stirs up the substrate and prevents scanning for predators. In general, animals often pause during foraging to scan the environment. Such scanning, called 'vigilance,' functions largely to detect potential danger. Some animals can, however, consume food and maintain vigilance simultaneously. Many small birds husk seeds with their heads up; when in more dangerous situations, juncos pick up bits of seed that require husking rather than consume tiny bits that require them to attend to the substrate. Finally, watching to see where others are feeding is more compatible with looking for predators than is searching the substrate for food.

Mating displays and predation risk may attract attention and require focused attention. Guppies court at dawn and dusk even though their displays are somewhat less conspicuous to females in these times. Critically, displays at these times are much less conspicuous to predators and attacks occur at twice the rate in the middle of the day than at dawn and dusk. Furthermore, males in areas with more predators shift away from mating displays and toward sneaky matings. In this way, predators both raise the costs of displaying and lead to greater conflict between the sexes over mating.

Finally, trade-offs between behavior and predation risk can come about because animals deplete resources while diluting predation risk. When animals repeatedly use a relatively safe area, such as the vegetation near a burrow, they create a gradient of habitat with richer vegetation at greater risk farther away. If predators have a central place, danger is generally higher near this place and prey may avoid this area. For example, deer spend more time away from where wolves den and toward the territory boundaries of wolves.

Often, grouping together reduces predation risk, though gathering with others can also alter individual encounter rates with food, rivals, and potential mates. For example, a colony of seabirds may be able to effectively drive off predators, and seabirds may add nests to colonies rather than use suitable substrates elsewhere. But, such a colony may have to fly considerable distances to get food. Similarly, small fish congregate in the shallows because they are safer from predators and deplete the food in the shallows. This depletion accentuates the differences in feeding between the shadows and open water. In both cases, grouping reduces predator success – increasing the inherent safety – and local food depletion reduces feeding success – increasing the feeding difference between habitats. Thus, trade-offs can be both cause and consequence of animal behavior.

Beyond foraging, animals often face trade-offs with predation when competing for or displaying to potential mates. Male songbirds may sing from the tops of vegetation, from where their calls can travel without obstruction to the largest area. However, when predators are nearby, they may sing from within a shield of vegetation, or even stop singing for a spell. Both options are likely to make them less effective in attracting mates and repelling rivals. Similarly, male guppies show off their color patterns during courtship, but the same behaviors that make them obvious to females are likely to catch the attention of predators. Many elements of reproductive behavior are likely to be inherently risky. How much risk they carry, however, depends on where and when they are done. For example, cock-of-the-rock males are richly orange-colored birds that display in spots of sunshine on the floor of South American rain forests. Their displays include bouncing movements that make them seem to glow in the sunshine. These striking displays within the general gloom of the forest are no doubt visible to predators as well as potential mates. However, when out of the sunspots, these orange birds are not very conspicuous because the light penetrating the leaves largely lacks the wavelengths that their feathers reflect brightly. Thus, the costs (and benefits) of the plumage are great only when coupled with the behavior of dancing in sunspots.

Behavioral Ecology and Models of Rate Maximization

Behavioral ecology examines the fitness consequences of behavior, and different environments often change these fitness consequences. Much of behavioral ecology involves how changes in environment predict changes in behavior. Behavioral ecologists think about fitness, but often do not measure it directly, instead often making assumptions about the relationship between behavior and fitness.

Within behavioral ecology, optimal foraging considers how foraging contributes to fitness. Early studies assumed that more foraging resulted in more fitness and tested whether animals maximize their net rate of food acquisition. Their tests showed that animals respond strongly to rate, but rarely maximize it. I take this as a success of the optimality approach. Only by having quantitative predictions could the tests have yielded this result. The results could be interpreted thus: foraging rate is related to fitness, but not in a simple one-to-one manner. Another way of saying this is that animals seem to trade off foraging rate against some other good. Avoiding predators is the other good that we will consider here. Resource acquisition is necessary for fitness, but it is not sufficient. Simply put, food is generally good for the forager, but not if the forager is dead.

With the benefit of hindsight, we can examine how accounting for predator avoidance altered the results of three classic models of rate maximization – the optimal diet model, marginal value theorem, and ideal free distribution.

Optimal Diet Model

In its environment, an animal encounters various things that it could eat. What should its diet include? In the optimal diet model, items differ in their energetic content and handling time. Foragers face a trade-off between handling food and searching for other food. Food should be refused that is slower to process than the average rate of searching for and processing of more profitable food. The classic algorithm for determining the optimal diet involves ranking the possible diet items by the ratio of energy gain per unit of handling time. Obviously, the highest-ranked item should be included. The second ranked should be included if it increases the overall rate of gain compared to just eating the top ranked. The third-ranked item should be included if it increases the overall rate of gain compared to eating the two top-ranked items. One prediction of this classic model is that each item is either in or out of the diet: It should either always be eaten when encountered or it should never be eaten.

The optimal diet model can be modified to maximize the energy gain per unit of mortality cost during a foraging bout. This criterion maximizes the total energy gained over the lifetime of an animal (so long as nothing changes with age or state). The general effect of predation risk on diet selection depends on what is dangerous about foraging. When searching is more dangerous than handling, predation risk broadens diets so that foragers spend less time searching and more time handling. Conversely, when handling is more dangerous than searching, predation risk narrows diets as foragers focus on items with the lowest handling times.

Marginal Value Theorem

In the marginal value model, foragers should leave a patch when the return rate for continued foraging matches the return rate for traveling to and exploiting a new patch. The marginal value theorem also predicts that the optimal time spent in a food patch should increase with travel time between food patches. In support of this classic prediction, animals often stay longer in patches when travel time between patches is greater. Even here, however, animals often stay longer than the time that would maximize their overall rate of gain indicating that some other costs besides energy gain influence patch time. More surprisingly, travel time also affects patch time for nondepleting patches, against the predictions for energy gain. These findings suggest that some other cost is important and that this other cost accelerates over time. For example, parent birds may leave the nest unattended to forage. Where parental visits prevent nest predation, longer times between visits may involve increasing risk of damage to the nest contents. Here, we would have a trade-off between energetic gain and nest defense.

Jim Gilliam has worked out the basic effects of predation risk on foraging in patches and provides an approximation for optimal diet and patch use. If searching is more dangerous than harvesting, foragers should stay longer in the patch than predicted by the marginal value theorem. Conversely, if harvesting is more dangerous than searching, foragers should shorten their harvesting in patches and return to searching more quickly than predicted by the marginal value theorem. The observation that animals often stay longer in patches than would maximize their overall rate of energetic gain suggests that searching is often more dangerous than harvesting.

A related approach proposes that predation risk imposes a cost on foraging in depleting patches and that foragers will leave when the harvest rate equals the sum of metabolic, predation, and missed opportunity costs. Here, the predation cost involves the risk while harvesting in that patch and also other terms that express the long-term consequences of predation risk and foraging gain. This general approach has been very successful in predicting and interpreting how animals forage from trays of seeds mixed with inedible substrate. For example, scientists have quantified how much extra food it takes to entice small animals to venture into the open. Gerbils demand 4–8 times as much food in order to make a patch 1 m into the open equally attractive as a patch under bushes. Similarly, blue tits require a much higher feeding rate in order to move 1.5 m from the edge of the trees than to feed under the trees. Such dramatic differences in habitat use are not predicted by the energy expenditure of going the extra distance to reach food.

Ideal Free Distribution

Another classic model in behavioral ecology is called 'the ideal free distribution.' This concerns how animals should distribute themselves between habitats. Habitats differ in their richness of resources, and the resources in a habitat are divided between individuals in that habitat. When new resources continuously arrive into the environment and animals divide up these resources via scramble competition, the intake for each individual is on average the input rate of new resources divided by the number of individuals in that patch. Thus, the basic outcome of the ideal free distribution is that organisms distribute themselves in proportion to the inputs of resources. At the solution, individual intake rates are equal across the patches and individuals have no incentive to move between patches.

When predators are present, however, the situation may be quite different. For example, Will Cresswell's research group has found that saltmarsh habitat is rich in food yet inherently dangerous to redshanks because it allows attacks at close range by sparrowhawks and peregrine habitat. Redshanks in the study area spend most of their time on the safer and less productive mudflats. When we add predation risk, intake rates may vary across patches and individuals may do well to move between patches, depending especially on how risk is diluted within patches, and how freely predators are to move between patches. When predation risk does not depend on the number of potential prey in that habitat, imposing the same predation risk on each patch does not affect the distribution of foragers. Safety can often be shared, however, and the presence of others can increase safety. A colony of gulls for instance, can keep many predators at bay. Gulls increase the safety of their nests by nesting next to other gulls and may be safest by being in the center of the colony. When one patch is more dangerous, it is used less often. When predation risk is diluted by the number of foragers in a patch, however, a patch is safer when it has more foragers in it. Here foragers may not simply distribute themselves between habitats in a stable way. If predators are also free to move between habitats, potential prey may move to get away from predators, and predators move to find prey. This is ecologically realistic, but moves these models far away from the ideal free predictions of stable distributions in which no individual has incentive to move.

Integrating Foraging and Predation Risk

The μ Over g Approach

The approaches described earlier started with foraging rate maximization and later added the effects of predation. An approach that integrated foraging and predation risk from the outset was originally developed for animals such as frogs that move between habitats and phases in their life history. Here, each habitat has a growth and a mortality rate. If animals are not rushed for time, it is optimal for the animal to be in the habitat with the lower ratio of mortality (μ) to growth (g) rate. Mortality and growth rates may depend on size. This 'μ over g' framework helped explain why animals remained in juvenile habitats when they could grow faster in (more dangerous) adult habitats.

The μ/g rule is a special case of a more general minimization of $\frac{\mu+r\frac{b}{V}}{g}$, where r is the intrinsic rate of growth for the population, b is current reproduction, and V is expected future reproduction. This simplifies to μ/g for juveniles in a stable population: b is zero as juveniles are not yet reproducing and r is also zero since the population is neither growing nor shrinking. Modifying other assumptions, such as time limits, can modify the solution somewhat. Surprisingly, many problems in foraging under predation risk have solutions that are not far different from the μ/g rule, even if the initial assumptions seem very different.

Condition Dependence and Asset Protection

Another general formulation of the trade-off between foraging and predation risk finds that it is optimal to maximize the net intake rate multiplied by the marginal rate of substitution of predation risk for energy gain, minus the predation rate. We can easily accept that foraging rate increases fitness and predation rate decreases it. The unfamiliar term is 'the marginal rate of substitution.' This term converts the gain of foraging to the same scale as the losses due to predation risk and focuses our attention on how food and predation risk affect fitness. Importantly, both foraging gain and predation risk depend on size. The optimal behavior will often depend on some state or condition of the organism. Since predation brings future fitness to zero, foraging under predation risk has an element of inherent state dependence: the cost to the forager is its life and this cost will be higher for foragers with better reproductive prospects. This logic has been labeled 'asset protection.' Foragers engage in antipredator behavior to protect their future prospects, so those with better prospects are expected to engage in more antipredator behavior. In nature, it is likely that foragers also face diminishing fitness gains on foraging success. Both asset protection and diminishing returns predict that antipredator behavior will depend on state.

Condition dependence can affect the predictions for foraging models. For the diet choice and patch use models, the predictions depend on how prey estimate their long-term rate of foraging gain. If this estimate varies with state, diet and patch time will vary. For example, bumblebees increase their use of flowers that vary in reward if their colony honey stores are depleted. One consequence

is that foragers will include items in their diet some of the time, but not in others. Thus, state dependence could explain why partial preferences, despite the predictions of the optimal diet models that each item should either always or never be accepted when encountered. For the ideal free distribution, state dependence allows a new sort of solution: individuals may mostly gather together in safety in a low-yielding patch, but leave to forage elsewhere when their energetic state is low. Thus, the combination of predation risk and condition dependence can lead to qualitatively new predictions for foraging.

Formulating Hypotheses about Trade-Offs

Survival is a full-time job, so antipredator behavior will permeate all activities. Therefore, we may expect that trade-offs with predation risk occur at all times and on all scales. Nonetheless, we learn most by specifying clear alternative hypotheses. This may include specifying different 'currencies' that could be maximized by behavior. Choosing any one currency assumes that that measure is approximately equal to fitness. Fitness always involves surviving and reproducing, with growth also important at times. Therefore, we cannot expect fitness to ever be due only to survival or only to reproducing. At times during an animal's life, however, the approximation can be good. For example, for small birds, surviving the winter can approximate fitness because the breeding season is far away. Nonetheless, some birds do act to protect and favor potential mates even during winter. Behavior can be predicted from maximizing survival, subject to an energetic requirement. We cannot view animals as maximizing energy gain subject to a constraint of predation risk. Predation risk directly relates to survival in lifetime measures of fitness, so it is often better to regard animals as minimizing predation risk, subject to an energy requirement. Also, we should not assume that animals minimize their time spent exposed to predation. For animals that can rest in a very secure refuge, limiting time exposed may maximize overall survival. Nonetheless, overall survival is a far better approximation of fitness, and overall survival will depend on both time and behavior across risky and relatively safe situations.

Constraints in behavioral ecology are assumptions about what organisms can and cannot do behaviorally. They define the problem to be solved. For example, the optimal diet model treats handling time as a constraint, though studies later showed that foragers can control handling time to some extent. (Unless these changes are great, however, they are not likely to obscure our understanding of optimal diet decisions.)

When we investigate animal behavior, we must take various features of the animal's morphology, physiology, and development for granted. This article and encyclopedia concentrates on behavior. Whether the nonbehavioral features are constraints in a larger evolutionary sense is a subject for investigation on a larger scale. Testing the degree of trade-off between two traits is informative in any case, as I hope to illustrate by example. Animals may generally behave to limit the total costs for achieving a combination of good things. That is, they act where and when the trade-off is least severe.

Behavioral trade-offs are important, yet we should remember that they exist within the larger functional biology of organisms. For example, birds must put their heads up or down when feeding from the substrate. Pecking for seeds limits detection of predators and attending to the substrate likely does also. Because their eyes and their bill are both on their head, birds face a fundamental trade-off between pecking and scanning for predators. As we mentioned earlier, birds can limit this trade-off by keeping their heads up while husking seeds and by relying on others to search for food patches. They also reduce this trade-off by having eyes that are somewhat specialized: one for near vision and the other for scanning in the distance. When birds have specialized eyes, they may choose to position themselves with the distant vision eye facing the open horizon and use the close vision eye for finding food. A wonky-eyed jewel squid goes one further by having very different eyes, a huge one of which looks up into open water for prey and much smaller one which look toward the substrate for predators. Finally, some fish have divided eyes so that each can effectively look in two places at once. A recently described species has reflectors to bring images from above on to its retina. These examples remind us that morphology also may be selected to adequately perform several partially incompatible functions, and that behavioral trade-offs are part of a larger set of trade-offs subject to selection. Potentially, asymmetry in eye function might be good for combining vigilance with feeding, for example, but may impair flying ability in birds.

See also: Defensive Avoidance; Group Living; Habitat Selection; Life Histories and Predation Risk; Optimal Foraging Theory: Introduction; Patch Exploitation; Vigilance and Models of Behavior.

Further Reading

Bednekoff, PA (2007) Foraging in the face of danger. In: Stephens, DW, Brown, JS, and Ydenberg, RC (eds.) *Foraging: Behavior and Ecology*, pp. 305–329. Chicago: University of Chicago Press.

Brown, JS and Kotler, BP (2007) Foraging and the ecology of fear. In: Stephens, DW, Brown, JS, and Ydenberg, RC (eds.) *Foraging: Behavior and Ecology*, pp. 437–480. Chicago: University of Chicago Press.

Clark, CW and Mangel, M (1999) *Dynamic State Variable Models in Ecology: Methods and Applications*. Oxford: Oxford University.

Gilliam, JF (1990) Hunting by the hunted: Optimal prey selection by foragers under predation hazard. In: Hughes, RN (ed.) *Behavioral Mechanisms of Food Selection*, pp. 797–818. Berlin: Springer.

Houston, AI and McNamara, JM (1999) *Models of Adaptive Behaviour: An Approach Based on State*. Cambridge: Cambridge University Press.

Stephens, DW and Krebs, JR (1986) *Foraging Theory*. Princeton: Princeton University Press.

Training of Animals

L. I. Haug, Texas Veterinary Behavior Services, Sugar Land, TX, USA
A. Florsheim, Veterinary Behavior Solutions, Dallas, TX, USA

For centuries, animals have been trained for a wide variety of purposes, including agricultural work, transportation, and companionship. Why and how people chose to train animals reflected not only the animals' roles in society but also the community's understanding of animal behavior. Current motivations for training animals have changed to reflect the change in animals' functions in society.

To better understand choices made in modern training practices, one needs an understanding of the modern views of animal behavior. During the nineteenth century, the application of scientific principles to the study of animal behavior dramatically impacted theories of why animals behave the way they do. Several schools of study have emerged, sometimes with diverging viewpoints. The two most prominent are ethology and behaviorism. While ethology focuses on an animal's instinctual or innate behavior – patterns that are not a product of learning – behaviorism focuses on the role learning plays in behavior. We will first take a more in-depth look at these two schools of thought and then discuss how these viewpoints influence modern animal training.

Ethological Perspective

Ethology focuses on the study of animals' behavioral patterns in a natural environment. Ethologists study what is termed proximate and ultimate causes of behavior: the 'how' and 'why' of behavioral responses. These behaviors are generally associated with feeding, reproduction, territorial defense, and social interaction between and among species. Much of the focus is on instinctual or innate behavior patterns (modal or fixed action patterns) that are not a product of learning. Ethologists argue that animals of different species behave differently because they act within a different set of rules. These rules are determined by the animal's physiological makeup: the form determines the function.

Modal action patterns (MAPs) are defined as innate, highly stereotyped motor responses that appear consistently across all individuals of a species. They are physiological or motor sequence responses elicited by well-defined, simple stimuli (stimulus–response relations). Once the pattern has been activated by the appropriate stimulus, the response or behavioral sequence is performed in its entirety. A classic example is the elicitation of begging behavior (the MAP) in gull chicks in response to a red dot (the releaser) on the parent gull's bill. The fox's mouse pounce is another example of a MAP; the pouncing motor pattern is highly stereotyped across all individuals.

From this discipline arises, in part, the trend for using construct labels, such as dominance and submission, to describe behavioral patterns or the animal's personality traits. One animal in a social group may be labeled as dominant or alpha, while another is called submissive or omega. Wolves are described as territorial because they show aggression to unfamiliar wolves that intrude upon their core range. As we shall note, this tendency for labeling has had a dramatic impact on animal training paradigms.

Behaviorism

Behaviorists focus on measurable behavior within the context of the environment in which the animal behaves without inference about the animal's underlying cognitive or emotional state. Behavior is classified in two ways, respondent and operant, which roughly corresponds to involuntary and voluntary behavior. Respondent behaviors are reflex responses and species-typical behaviors – behavioral processes that are most studied by ethologists.

In contrast, operant (instrumental) behaviors depend on consequences – Thorndike's Law of Effect: favorable/desirable consequences predict that the behavior will be more likely to be repeated in the future whereas unpleasant/undesirable consequences predict that the behavior will be less likely to be repeated in the future.

Respondent Learning

Respondent learning centers on automatic stimulus–response (S–R) patterns and stimulus–stimulus (S–S) associations. Respondent learning, also termed classical conditioning, is best illustrated by Ivan Pavlov's famous salivation experiments with dogs. Dogs would begin to salivate (an unconditioned response – UR) at the sight of food (an unconditioned stimulus – US). Pavlov then began ringing a bell (a neutral stimulus – NS) prior to presenting food. After a number of trials of this S–S pairing, the dogs began to salivate (the conditioned response (CR)) at the sound of the bell (the conditioned stimulus (CS)).

Respondent learning is also seen with physiologic (internal) processes such as hormonal changes. For example, teat stimulation (US) by milking machines elicits oxytocin release (UR) in dairy cattle in milking parlors (previously an NS). Over time, oxytocin release (CR) and milk letdown occur as the cattle begin to approach the milking parlor (CS).

Similarly, in a lab setting, J. M. Graham and Claude Desjardins studied the effect on male rats of pairing the scent of wintergreen (NS) with sexually receptive female rats (US). Male rats exposed to female rats that are emitting pheromones (US) associated with sexual receptivity experience a reflexive rise in hormones in the bloodstream (UR), which indicates sexual arousal in the male rat. When the rats were exposed to the wintergreen scent (CS) alone, the rise in the measured hormones of the male rats equaled the rise in hormones of the male rats exposed to the sexually receptive female rats (Chance, 2008, p. 65).

Respondent learning is far more complicated than just a simple pairing of two unrelated stimuli. Various studies since the 1960s have demonstrated that not all stimuli are equally associable and that the differences in associability of the CS–US pairings are key in how well respondent learning occurs.

Operant Learning

In operant learning, the behavior performed is considered voluntary rather than reflexive, and the behavior is controlled by its consequence. Consequences can be either reinforcing or punishing. Reinforcement increases the future probability of the behavior and punishment decreases the future probability of the behavior. The effectiveness of both reinforcers and punishers is dependent on how closely the behavior and the consequence are paired in time and on the salience of the reinforcer or punisher to the animal.

Reinforcement and punishment are further classified as positive or negative. These are mathematical terms and indicate whether something is added (positive) or subtracted (negative) from the environment within which the behavior is performed. With positive reinforcement, a stimulus is added to the system, which results in the behavior being more likely to occur again. For example, if a dog sits, it is given a treat. Negative reinforcement describes a situation where a typically aversive stimulus is removed from the environment when the animal performs the target behavior. A rider applies spurs to a horse's sides until the horse begins to move, at which point the spurring stimulus ceases. When a behavior is positive punished, a stimulus is added to the system: if a rat is shocked every time it presses a lever, the rat is less likely to press the lever in the future. Negative punishment refers to the removal of a stimulus (including the opportunity for reinforcement) making a behavior less likely to occur. If a sea lion exhibits an aggressive response during a training session, the trainer may step away and ignore

the animal for a short period of time. This removes the trainer's attention and also the animal's opportunity to earn further reinforcement.

Reinforcers and punishers are categorized as primary or secondary. A primary reinforcer or punisher reinforces or punishes innately and does not depend on an association with other stimuli. Examples of primary reinforcers include food, water, reproduction, and relief from environmental stress (e.g., uncomfortable temperatures). Examples of primary punishers include pain and fear-inducing stimuli. Secondary, or conditioned, reinforcers and punishers are dependent on their previous association with primary reinforcers or punishers. For example, the reinforcing effect of a light or sound is dependent upon how frequently that light has been paired with a primary reinforcer such as food. The word 'No' acquires punishing properties by pairing it with other unpleasant stimuli – hitting the dog or applying a collar correction. Secondary reinforcers are acquired by respondent conditioning (S–S pairings); therefore their continued strength as reinforcers is dependent on consistent contiguity with a primary reinforcer.

In his experiments with pigeons in the 1960s, Herrnstein evaluated the use of secondary reinforcers by examining the relative rate at which pigeons would peck a disk to obtain a secondary reinforcer. Four pigeons were trained to peck at either of two response keys. Pecking at either key occasionally produced a secondary reinforcer. Then, in the presence of the secondary reinforcer, further pecking occasionally produced the primary reinforcer, food. He determined that the rate at which each pigeon pecked to obtain a secondary reinforcer equaled the relative rate that the secondary reinforcer was paired with a primary reinforcer. As they decreased the number of times the secondary reinforcer was paired with the primary reinforcer, the pigeon worked less to obtain the secondary reinforcer. Secondary reinforcers, or bridges, such as clickers and whistles, have become very popular in modern animal training, as discussed in the following paragraph.

Reinforcement schedules influence the variability and persistence of behavior over time. There are several types of reinforcement schedules and each has a distinctive effect on the target behavior. The simplest schedule is termed continuous reinforcement – the target behavior is reinforced each time it occurs. Continuous reinforcement leads to the most rapid learning of new behavior, and for this reason, it is the most appropriate choice when teaching new behaviors to animals.

Intermittent schedules include duration/interval schedules and ratio schedules. Each of these can be applied on a fixed or variable schedule. Variable schedules produce the most enduring and persistent behavior and are most applicable to animal training outside the laboratory. Duration schedules are most appropriate for long-duration behaviors such as teaching a dog to heel or stay.

The dog is reinforced after a variable period of time for successfully staying in position. Another example is a predator waiting for the appearance of prey; the predator is reinforced at variable intervals by the appearance of potential prey.

Ratio schedules can be fixed or variable and depend on the number of behaviors offered. Reinforcement occurs after a specified number of correct behaviors are emitted. A bear fishing for salmon in a stream might have to dunk his nose in the water multiple times before he is reinforced by successfully catching a salmon. This is an example of a variable ratio – the number of times the bear has to dunk his head before catching a fish varies.

Impact of the Study of Behavior on Training

Perspectives on behavior and learning studied in research institutions have greatly influenced modern training concepts and techniques. The ethological influence on past and current training techniques may be known more for its misapplication than for its legitimacy. For decades, ethologists chose to focus only on wild animals because domestic animals were considered inappropriate subjects to study since they did not live within a natural environment. Domestic animal behavior was interpreted by extrapolating from studies of wild animal ancestors and related species. Ethologists defined behavior terms and constructs, and, lacking other resources, trainers usurped these terms and constructs and applied them to domestic animals. For example, because dogs were considered descendents of wolves, dog behavior was defined by wolf pack structure. Dogs were assumed to live in strict pack structures with a defined linear dominance hierarchy.

A few problems arose with this approach. Because wolves are reclusive in the wild, wolf behavior was often studied in captive or semiwild situations, not the actual natural setting. This meant that some aspects of wolf behavior were misinterpreted and then these misinterpretations were further distorted when the information was 'passed on' to nonethologists working with dogs. Additionally, current research on dogs has shown that dogs actually do *not* live in pack formations, and while some overlap occurs, dogs do not behave identically to wolves.

On the basis of this erroneous approach, dogs showing objectionable behavior, particularly aggression, were then declared 'dominant' because they did not submit to the trainer's will as the alpha member of the relationship. This philosophy was applied indiscriminately across all contexts and has led to the use of disturbingly abusive training practices. Rather than evaluating the actual *behavior* in the context in which it occurred, the *animal* was given a label (i.e., dominant) and then its entire behavioral repertoire was interpreted within that framework.

Behaviorists, on the other hand, attempted to avoid the pitfalls of labels and constructs by focusing on objective, quantifiable measures of behavior. By objectively observing behavior, behaviorists avoided the trap of trying to assume the animal's underlying thoughts, feelings, and motivations. Freed from the distraction of worry about an animal's potential nefarious intentions, a trainer was able to focus on the actual behavior and the consequences driving it, thereby effectively designing a program to manipulate the behavior–consequence contingencies to shape desired behaviors.

Despite these advantages, behaviorism has its limits. Animals are not blank slates. As ethologists have demonstrated, instinctive behavior is part of the animal's repertoire and it can influence operantly learned behaviors. In Breland and Breland's 'The Misbehavior of Organisms' (1961), the Brelands describe unrewarded, inappropriate behavior during training. For example, after successful conditioning, raccoons that had learned to pick up a coin and put it in a piggy bank would start to rub the coin between their paws for seconds to minutes at a time, delaying their reward. 'Misbehavior' of the raccoons was actually an expression of this food manipulation behavior. The Brelands attempted to use hunger as a motivator thinking that the hungrier the raccoons were, the faster they would seek food reward. In fact, the hungrier they were, the more they exhibited the coin rubbing behavior and further postponed food (reinforcement) delivery.

> The general principle seems to be that wherever an animal has strong instinctive behaviors in the area of the conditioned response, after continued running the organism will drift toward the instinctive behavior to the detriment of the conditioned behavior and even to the delay or preclusion of the reinforcement. In a very boiled-down, simplified form, it might be stated as "learned behavior drifts toward instinctive behavior" (Breland and Breland).

While behaviorism has contributed a great deal to our understanding of animal behavior, and had a significant influence on modern training methodologies, it cannot stand alone in practice, as the Brelands discovered. Instinctual patterns matter. The perspective highlighted respectively by ethologists and behaviorists does not, and should not, argue the existence or importance of the other. Each paradigm has strengths and weaknesses in terms of devising training programs.

Applied Animal Behavior and Training

Animal training is, simply, the manipulation of behavior. Behavior is not the tool with which the animal is trained, but rather the measure of the training procedure: if the animal's behavior changes, then training (learning) has occurred.

There exists an argument that species-specific differences necessitate devising unique training approaches for that particular animal, as though the principles of learning differ between species. Similarly, trainers are often informed that there is a need to assess the animal's personality (e.g., horsonality profiles in horses) in order to know which techniques to use on a particular animal. Labels such as friendly, nervous, shy, dominant, spooky, etc. may serve to enhance communication between professionals if these are clearly defined, but these labels themselves do not help us define the actual *behaviors* that the animal is exhibiting or that we wish to train. For example, a bird bites a person's hand when asked to step up on it. Some might see the bird's biting as a sign that the bird is asserting its dominance over the owner. Someone else might believe the bird is afraid of the hand and therefore biting it defensively. Still others might just say it is a 'bad' bird. None of these labels actually helps us devise a plan to change the bird's behavior. In fact, they are likely to inhibit our efforts and perhaps lead to the use of inappropriate and inhumane training methods.

A prime example is the aforementioned and ubiquitous use of dominance constructs during training. This philosophy has been a central approach in training programs for many species including dogs, horses, cattle, elephants, and others. In order to make the animal perform the desired behavior, the handler/trainer must establish physical dominance over the animal, typically by some form of confrontational interaction. Under the guise of training, handlers pin dogs to the ground, chase horses with a rope or whip, and subdue elephants with an ankus (hook).

Viewed under the umbrella of operant and respondent learning, the prudence of these approaches fails to draw merit. The bird bites the hand because biting makes the hand go away. By manipulating the consequences of biting (and of stepping up willingly), the bird's behavior can be altered with equal efficacy whether we think the bird is dominant, fearful, or just 'bad.'

Most animal professionals recognize that species show a repertoire of MAPs. These innate behaviors serve certain functions for the animal and the behaviors are influenced and shaped further by the animal's experiences. As the Brelands noted, these MAPs can at times interfere with training goals. Nevertheless, these innate behaviors also can be used to enhance the success of training goals. Natural horsemen have done this with the manipulation of the horse's instinct to run when threatened. Escape behaviors (i.e., running away) are easy to condition in an animal that is already predisposed to running away. The weakness in the approach is that too few trainers have an understanding of the principles of learning. They rely solely on ethological concepts to explain behavior (i.e., the horse runs because it is a prey species and/or is acting submissive) rather than understanding that avoidance behavior is conditioned via negative reinforcement.

A common example is horses that crowd close to handlers during training. They are often deemed pushy or dominant but is there another explanation? Natural horsemanship training techniques often use round-penning and lunging exercises to encourage compliance and perceived submission from the horse. The horse is compelled to move away from the trainer by 'pressuring' or threatening the horse by waving a rope or whip at the animal. A significant number of horses will kick out at the trainer as they move away, especially during the early stages of the training. For safety, the trainer typically uses the whip or rope only in a threatening manner when the horse is several feet away (outside the range of the horse's kick). So when the horse is close to the trainer, the trainer does not generally threaten the horse, but when the horse is several feet away, the trainer feels safe enough to swing the rope or whip at the horse. In this situation, crowding is not about dominance – it is a simple case of negative reinforcement. The horse learns that staying close to the handler prevents the occurrence of aversive stimuli. Essentially, this training practice can actually *teach* the horse to crowd the handler.

Conditioning New Behaviors

New behaviors can be developed in three general ways: capturing, prompting/luring, and shaping. Each technique has strengths and weaknesses and the best approach will vary with the animal and the specific target behavior.

Capturing is the process of reinforcing the goal behavior when the behavior is spontaneously offered by the animal. The trainer simply observes the animal, and when the animal emits the behavior, reinforcement is provided. Capturing can work well for behaviors that are a frequent part of the animal's normal behavioral repertoire and when the behavior occurs in a relatively invariant form. For instance, during breeding season, Atlantic Harbor Seals frequently slap the water with their flippers. Skilled trainers can quickly capture this behavior and put the behavior on cue as this behavior does not vary frequently in the form it is offered. Capturing is not an optimal technique to use if the behavior occurs infrequently, as the resulting reinforcement rate would be so low that the animal would become frustrated or the training process would take too long.

Prompting and luring are two techniques using various stimuli to trigger the appearance of the target behavior or an approximation of the target behavior. Luring, typically using food or a toy, is a popular method for teaching basic obedience behaviors to pet dogs. A treat is used to manipulate the dog's body into a position, which maximizes the likelihood that the dog will sit. Once the dog sits, it is immediately given the food treat as a reinforcement. Horse owners often use food to try to lure horses into trailers or stalls. Placing honey or molasses on the horse's bit encourages the animal to open its mouth to accept the bit.

Prompts are stimuli used to trigger a desired behavior. For example, a trainer can prompt a sea lion to turn its head to one side by making a noise off in the desired direction and reinforcing the animal when the head turn occurs. A cat owner can prompt the cat to wipe at its face by placing a small piece of tape on the cat's nose. Once the behavior is triggered by the lure or the prompt, the trainer then reinforces the behavior. Over time, the lure or prompt is gradually faded and the behavior is transferred to a different cue (discriminative stimulus) so the behavior can be elicited without the lure or prompt.

A distinct advantage of luring and prompting is that behavior can be generated quickly, often within minutes. This is also a relatively humane and compulsion-free technique; however, one disadvantage of prompting is that the animal may habituate to the prompt stimulus. For example, after a few repetitions, the sea lion may no longer orient to the sound used to prompt a head turn. The major disadvantage of using lures and prompts is the process of fading the stimulus. Lures are generally easier to fade than prompts, but in both cases, the process can take time. Another criticism of luring is that the learning process is trainer driven, rather than animal driven. Animals trained only by luring may offer little spontaneous behavior, or behavioral experimentation, during training sessions. This can become problematic if the trainer wishes to develop a behavior that is outside the animal's natural behavioral repertoire. This problem can be avoided by using the last method, shaping, because behavioral experimentation and variability are essential to this technique.

Shaping, also called differential reinforcement of successive approximations, is a process where a behavior is developed by initially reinforcing any behavior that remotely resembles the target behavior. Over subsequent training sessions, the final target behavior is shaped by progressively reinforcing behaviors that more closely resemble the target behavior. A bird can be trained to elevate its wings for examination by first reinforcing the bird for any tiny lift of the wing away from the body. When this behavior is being offered with relative reliability, reinforcement is withheld for this criteria and the bird is reinforced only if it lifts its wing away from the body and begins to slightly extend the carpal joint. Wing lifting is progressively shaped by only reinforcing greater extensions of the wing until the bird is lifting the wing completely away from the body and extending the wing out to its full span. The smaller the approximations, the more seamless the learning process will be.

When shaping criteria are carefully planned, the animal often offers a new level of behavior before the trainer actually changes reinforcement criteria. Shaping is most effective when done with a bridge, or secondary reinforcer. This is typically a unique sound such as a click or whistle that has been previously paired with a primary reinforcer such as food. The target behavior is 'marked' with the bridge signal and then a primary reinforcement is delivered to the animal. The bridge signal increases the accuracy of the reinforcement process particularly for rapid behaviors or situations where the animal is not close to the trainer (thereby making timely delivery of a primary reinforcer difficult). As seen with Hernstein's pigeons, the effectiveness of a secondary reinforcer relies on its consistent association with a primary reinforcer. If a bridge is used in the absence of a primary reinforcer, it loses its effectiveness as a secondary reinforcer and 'marking' stimulus.

Shaping produces the most learner participation as the process is entirely learner driven: the trainer does not prompt the animal in any way but rather reinforces behaviors that the animal offers spontaneously. Shaping does require the trainer to have excellent observation skills and knowledge of the topography of the behavior being shaped. As with other training methods, shaping can be frustrating to the animal if the reinforcement rate is too low or the trainer's criteria are unclear.

Shaping can be combined successfully with prompts in some circumstances. For example, hoof care is an important part of equine husbandry so it is essential that the horse reliably lift its feet when cued to do so. A shaping procedure combined with a prompt would begin with the trainer gently squeezing the tendons of the horse's leg. When the horse shifts its weight even the slightest degree to the off foreleg, the trainer reinforces the horse by removing the prompt and offering food reinforcement (a combination of negative and positive reinforcement). Over trials, the horse is reinforced only for greater weight shifts, then lifting the foot slightly, etc.

Changing Existing Behaviors

Training is not only used to teach new behaviors but also as a tool used to alter or reduce existing behaviors. Problematic behaviors include normal species typical behaviors that are undesirable for the situation in which the animal lives, as well as behaviors that are considered abnormal. For example, weaving is an abnormal, stereotypical behavior in horses that is generally caused by social isolation and confinement, but, unlike weaving in horses, rooting in pigs is a normal species-typical behavior. While rooting is normal behavior for the pig, it may be considered inappropriate for a pig kept as a house pet because rooting can cause damage to the owner's home. Depending on the type of problem faced, trainers can emphasize the use of antecedent changes, operant, or respondent techniques to alter behavior.

Antecedent changes

Antecedents are learned signals for the behavior-consequence contingency to follow. Opening a food door may become the antecedent for a chimpanzee to lunge because the lunging has been reinforced with the delivery of food. The strength of the antecedent to cue a particular behavior is related to the strength of the reinforcer that follows the behavior. There are three ways to manipulate antecedents to alter behavior patterns: adding/removing a cue, changing the setting events, and strengthening/weakening the motivation for the behavior (termed establishing operations).

Antecedents can be changed or removed to alter the behavior/consequence pairing and prevent the elicitation of the problematic behavior. For example, some dogs will bark and run to the door when they hear the doorbell in anticipation of people entering the home. If the doorbell is turned off or guests enter the home without ringing the doorbell, the dog may be less likely to bark.

The context, conditions, or situational influences that affect behavior are often some of the easiest antecedents to change. Horses that weave when placed in a stall can be moved to a pasture with other horses. Cats that urinate outside of the box only when the box is dirty can have their boxes cleaned more regularly. If a cat bites when it is petted while on someone's lap, the owner can refrain from letting the cat on laps or from petting the cat once it is there.

Motivating or establishing operations are antecedents that temporarily alter the effectiveness of consequences. For example, certain toys may be highly motivating to a dog if it rarely has access to them but less so if the dog has unlimited access to the toys all the time. A horse may more readily go back into its stall and rest quietly once it has had a long workout and time at pasture with social interaction with other horses. Similarly, food reinforcers are more effective if the animal is hungry, so training will be most effective if the training session is schedule just before the animal's next mealtime.

Manipulating consequences to change behavior

While antecedent behavior-change strategies are preventative solutions, consequence changes rely on learning. Manipulation of consequences necessitates that the trainer determine what reinforces a particular behavior. Once the reinforcement is identified, the trainer can withhold or eliminate it. When the association between the behavior and its consequence is severed, the number of times the behavior is offered decreases and eventually disappears or returns to baseline. This process is known as extinction.

Extinction can be problematic as a behavior change strategy. Current studies on extinction indicate that extinction does not destroy the original learning but instead generates new learning that is very dependent on context. A change of context after extinction can cause a return of the initial CR, termed the renewal effect. Some behaviors may be maintained by internal reinforcement processes (e.g., physiologic changes); therefore, the reinforcement cannot be withdrawn easily. Other behaviors may have external reinforcements that are very difficult or even impossible to control. For example, many dogs receive substantial reinforcement for chasing small and birds. Because the dog must be let out into the yard for elimination each day, the dog is likely to continue to receive some level of reinforcement for this behavior even if the dog is kept on leash.

Extinction can be a very slow process and the problematic behavior may recover over time. Cessation of the behavior is often preceded by an *extinction burst*, a sharp increase in the frequency and intensity of the problematic behavior. This can be a very frustrating process for those working with animals and many individuals reinforce the animal during this burst, thereby creating an even stronger response. Extinction can also result in frustration-induced aggression.

Another behavior change strategy is differential reinforcement of alternative behaviors (DRA). The alternative behavior selected should be one that replaces the function of, and generally is incompatible with, the problem behavior. The new behavior should receive more reinforcement than the problem behavior. If a chimpanzee charges the feeding door to receive its food, the chimpanzee would be trained to sit away from the door. The animal would be highly reinforced for this behavior such that he receives more reinforcement for sitting away from the door than he does for charging at the door when it opens. The Matching Law states that organisms will apportion their behavior in accordance with how much reinforcement each behavior receives. If a pig roots out high value reinforcements in a specific area of the yard, the pig is significantly less likely to root in other parts of the yard where it receives less reinforcement.

The alternative behavior should ideally be something the animal already knows how to perform. It will be easier to replace a problem behavior with a behavior that the animal already knows well, rather than trying to teach the animal a brand new response. If a dog is highly proficient with sitting on cue, but is just learning lie down, then sit is the more appropriate choice as an alternative behavior for lunging at people on walks.

Respondent techniques for altering problem behaviors

Respondent techniques are appropriate choices for problematic respondent behaviors. The most frequent situation involves fear behaviors that are significant enough to interfere with an operant training paradigm. While fearful behavior is normal in all species, animals exhibiting fear in captive or domestic settings may pose a variety of problems, including difficulty maintaining health and husbandry, destructive escape behavior, or aggression. Respondent fear can interfere with operant learning. Trainers often choose to work through the respondent behavior problem first to help clear the way for operant learning. There are three well-established procedures to reduce respondent fear: systematic desensitization, counterconditioning, and response blocking (flooding).

Systematic desensitization is a process in which a conditioned emotional response (CER) such as fear is extinguished by exposing the animal in a graduated manner to the fear-eliciting stimuli. When creating a program for systematic desensitization, a stimulus hierarchy is created which ranges from a level that elicits a mild response to a level that elicits an extreme response. The animal is exposed to the first step of the hierarchy (mild to no response) until the mild response is extinguished. Once this occurs, the animal is exposed to the next step on the stimulus hierarchy. This process continues until the animal no longer shows a fear response to the stimulus even at full intensity.

Systematic desensitization's effectiveness is often improved when it is paired with a process called counterconditioning. The animal's initial CER is replaced with an alternative, competing response by pairing it with an eliciting stimulus that will trigger an opposing emotional or physiologic response. For example, a show cat may need to be bathed and dried prior to a show. If the cat is afraid of the sound of the blow dryer, the sound of the dryer can be paired with a favored food (the eliciting stimulus), which elicits pleasure and potentially a reduction in heart rate.

Counterconditioning occurs only if the new eliciting stimulus triggers a response powerful enough to supersede the original CER. If the cat is extremely afraid of the sound of the dryer, it is very likely that it will not eat in the presence of the blow dryer. Hence, it is often advantageous to pair this counterconditioning with systematic desensitization. By minimizing the intensity of the original CS, the new eliciting stimulus is likely to be salient enough to overcome it.

Response blocking (flooding) is an exposure-based technique in which an animal is exposed to a fear-evoking stimulus at full intensity until the animal's fear to that stimulus is extinguished. The animal is prevented from escaping the stimulus, typically by some form of restraint, hence the term response blocking. A dog that is afraid of the sound of fireworks would be exposed constantly to the sound until the fireworks no longer elicited a fearful response. Similarly, a draft horse that is afraid of having a harness on would be placed in the harness until it no longer showed a fearful response.

While a favored technique in the past, the use of response blocking has largely fallen out of favor with educated behavior professionals because of the problems that can arise. If the session is aborted prior to extinction, the process actually can exacerbate the animal's fearful response. Additionally, during the session, there is a very real risk of injury to the animal and any

human participants if the animal shows severe panic behavior or aggression. Response blocking may also lead to a condition known as learned helplessness. The animal learns that its behavior has no effect on its environment; this results in decreased response to the stimulus even when the opportunity to escape the stimulus is returned. Learned helplessness can induce a global suppression on responding such that the animal also fails to respond to other previously learned cues. Learned helplessness has been associated with wider reaching detrimental side effects including physiologic problems such as gastric ulceration.

Summary

Regardless of the purpose behind the training program, animal caregivers should rely on a humane hierarchy when devising behavior programs. These programs should focus on environmental (antecedent) management and positive reinforcement-based interventions. It is inappropriate to base training and behavior change programs on outdated ethological models to the exclusion of the laws of learning. Learning plays the primary role in shaping behavior on a daily basis. Animal educators must have a clear understanding of these principles in order to be effective at their tasks.

See also: Animal Behavior: Antiquity to the Sixteenth Century; Animal Behavior: The Seventeenth to the Twentieth Centuries; Behavioral Ecology and Sociobiology; Comparative Animal Behavior – 1920–1973; Ethology in Europe; Future of Animal Behavior: Predicting Trends; Game Theory; Imitation: Cognitive Implications; Integration of Proximate and Ultimate Causes; Neurobiology, Endocrinology and Behavior; Psychology of Animals; Punishment; Social Cognition and Theory of Mind; Social Learning: Theory.

Further Reading

Bouton ME (2004) Context and behavioral processes in extinction. *Learning & Memory* 11: 485–494.

Breland K and Breland M (1961) The misbehavior of organisms. *American Psychologist* 16: 681–684.

Chance P (2008) *Learning and Behavior – Active Learning Edition*. Belmont, CA: Cengage Learning.

Davison M and Baum WM (2006) Do conditional reinforcers count? *Journal of the Experimental Analysis of Behavior* 86(3): 269–283.

Herrnstein RJ (1964) Secondary reinforcement and rate of primary reinforcement. *Journal of the Experimental Analysis of Behavior* 7(1): 27–36.

Ramirez K (1999) *Animal Training: Successful Animal Management Through Positive Reinforcement*. Chicago, IL: Shedd Aquarium Society.

Rescorla RA (1988) Pavlovian conditioning: It's not what you think. *American Psychologist* 43: 151–160.

Staddon JER and Cerutti DT (2003) Operant conditioning. *Annual Review of Psychology* 54: 115–144.

Tryon WW (1995) Resolving the cognitive behavioral controversy. *The Behavior Therapist* 18: 83–86.

Tribolium

A. Pai, Spelman College, Atlanta, GA, USA

A Familiar Beetle

The name 'tribolium' comes from the Latin verb for threshing, 'tribulo.' This is likely a reference to the fact that the first *Tribolium* beetle to be discovered, *Tribolium castaneum*, is a pest of stored grains and was found near where grain was threshed. Because of their association with stored grain products, these beetles are likely one of the most familiar beetles known to man. The various beetles of the genus *Tribolium* have a long history of being used as a model organism in laboratory studies in the areas of ecology, evolution, genetics, developmental biology, and animal behavior.

Taxonomy and Diversity

Tribolium beetles belong to the group of 'darkling beetles' or Tenebrionidae. As the name suggests, they have a characteristic dark body color (**Figure 1**). Like many other tenebrionids, *Tribolium* beetles are saprophagous, phytophagous, or mycophagous, and may be found in leaf litter, under the bark of trees or near fungi. In addition, like some of their close relatives of the tribe Triboliini, *Tribolium* beetles too may be associated with the nests of other species such as insects and some vertebrates. For example, some beetles such as *T. myrmecophilum* may be ecologically associated with insects' nests, whereas others such as *T. castaneum* are associated with human dwellings and their stored grain. Thus, there is significant diversity in the ecological habitats of the three dozen species described in this group.

Evolutionary History

Tribolium beetles may have diverged from other holometabolous insects as early as the beginning of the Cretaceous period. Since then, these beetles may have evolved in five separate lineages, each of which is associated with a particular region of the world. Each has distinctive morphology. Thus, the *brevicornis* group evolved in South America, the *confusum* group in Africa, the *alcine* group in Madagascar, the *castaneum* group in Indo-Australia, and the *myrmicophilum* group in Malay-East Indies. Of these, the *confusum* group is the most speciose with 14 known members and the *myrmecophilum* group is the smallest with only two known species.

Stored Product Pest

Of the dozens of species in this genus, as few as ten are storage pests associated with human grain stores and come from the *castaneum*, *confusum*, and *brevicornis* groups. Evidence from ancient Egypt indicates that the association between *Tribolium* and humans may be as old as 4000 years. These insects are incapable of attacking whole grain and are therefore secondary pests of grains such as wheat, maize, rice, barley, rye, oats. As their typical habitat is in stores of the flours of these grains, they are commonly called 'flour beetles' or 'bran bugs.' Interestingly, although *Tribolium* are regarded as a major pest of grain products, they may also be found in other types of foods such as beans, peas, nuts, chocolate, and even spices like ginger and red pepper.

Because of their synanthropic nature, many of these pest beetles such as *T. castaneum, T. confusum, T. destructor,* and *T. madens* have a worldwide distribution. Others such as *T. anaphe, T. audax, T. brevicornis, T. freemani, T. parallelus,* and *T. thusa* have a more restricted range. The biology of the cosmopolitan storage pest beetles such as *T. castaneum* and *T. confusum* is the best understood and will be the main focus of this article.

Morphology

Adult *Tribolium* beetles are ~3–5 mm in size. Some such as *T. brevicornis* (~4 mm) and *T. freemani* (~5 mm) are visibly larger than other species such as *T. audax, T. anaphe, T. castaneum,* and *T. confusum* (~3 mm or smaller). Adults are sexually dimorphic. The sexes may be significantly different from each other in body size. In some species, such as *T. brevicornis,* males may be larger than females but in others including *T. confusum,* there may be a tendency for females to be larger. *T. castaneum* adult males are distinguishable from females because of the presence of setiferous glands ('sex patches') on their first pair of legs, which are absent in females. Similarly, in *T. confusum,* only males have these glands but they appear on each pair of appendages.

Life Cycle

Beetle life cycles include four stages: egg, larva, pupa, and adult. The microscopic eggs of tenebrionids are ovoid and

Figure 1 Adult *Tribolium castaneum* in flour. (Photo credit: Mark Lee, DOP images, Atlanta, GA).

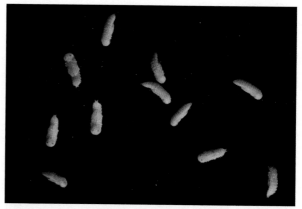

Figure 3 Pupae of *Tribolium castaneum*. (Photo credit: Mark Lee, DOP images, Atlanta, GA).

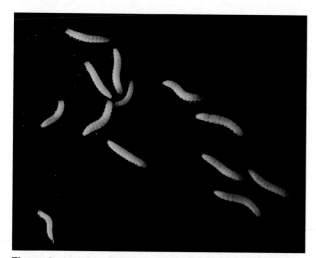

Figure 2 Larvae of *Tribolium castaneum*. (Photo credit: Mark Lee, DOP images, Atlanta, GA).

bright white. In *T. castaneum* and *T. confusum*, the egg stage typically lasts 4–5 days in ideal laboratory conditions. The larvae are worm-like and yellowish (**Figure 2**). In optimal conditions, this stage is typically about 2–3 weeks for *T. castaneum* and *T. confusum*. The pupae are light colored, and either white or yellowish (**Figure 3**). Pupae are not capable of locomotion though they may wriggle (with the movement originating in the abdomen). The sexes are dimorphic and easily discernable as pupae, because the females have larger genital papillae compared to those found on the male pupae. In a favorable environment, the pupal stage may last 5–6 days in *T. castaneum* and *T. confusum*.

The typical life cycle from egg to adult in laboratory strains of *T. castaneum* and *T. confusum* may be as short as 4 weeks in optimal conditions of food, temperature, and humidity. However, the rate of development from egg to adult varies vastly among species and may be as long as

almost 9 weeks for some species of flour beetles. Developmental time varies even among different strains of a species. In addition to genetics, environmental factors such as temperature, humidity, and food also influence development. Adults may live more than 3 years, though a life span of 1–6 months is more typical in laboratory settings. In general, the life span of virgin adults in laboratory conditions for *T. castaneum* and *T. confusum* beetles is 7–11 months.

Tribolium as a Model Organism

Several features about their biology make flour beetles a convenient model system. Their affinity for stored grain products makes it easy to find and collect them from pantries, feed mills, feed stores, silos, etc. At these sites, beetles are easily detectable because they leave tunnel shaped tracks in the flour dust (**Figure 4**). When brought to the laboratory, beetles do not require much space and basically can be stored in glassware such as vials and jars because of their small size. They can be reared rather inexpensively in flour medium (typically a mixture of wheat flour and brewers' yeast). Further, they can be indefinitely kept in dark conditions; this tolerance for dark facilitates their lives as storage pests in grain products.

Tribolium cultures do not require constant maintenance, since all the life stages of these beetles can be grown in the same flour medium. Easy separation of the life stages also facilitates experimentation. Eggs and young larvae are significantly smaller than pupae and adults, and can be easily separated from each other by means of sifters with different mesh sizes. Although pupae and adults are of similar size, they are also easily sorted from each other, by taking advantage of the lack of mobility of the pupae. These insects are hardy and easily withstand being handled for experimental manipulation.

Figure 4 Beetles-infested flour with tunnel shaped tracks. (Photo credit: Mark Lee, DOP images, Atlanta, GA).

Flour beetles, like many pest species, have a high reproductive capacity and remain reproductive throughout the year. Not only do males and females readily mate in a laboratory setting, but also, because adults have a long life span, a single female may produce hundreds of eggs in her long reproductive life, which enables high number of experimental trials. Because of their relatively short life cycle, these insects are suitable for multigenerational studies that ecology and evolutionary biology necessitate. All of these listed attributes have made *Tribolium* a well-studied insect on which there is a plethora of information.

History as Laboratory Model Organism

The flour beetle first came to be used as a laboratory model organism in population ecology studies initiated by Chapman in the late 1920s. In the subsequent four decades, from the 1930s to 1970s, Thomas Park and colleagues at the University of Chicago used this insect extensively as a laboratory model system to examine questions in ecology, genetics, behavior, and evolution. Indeed, some classical works illustrating such fundamental concepts in ecology and evolution, as population regulation and interspecific competition have employed flour beetles as model organisms.

Many notable discoveries in genetics have also been made in *Tribolium*. For example, selfish genetic elements

were first discovered in the confused flour beetle in the 1990s. The *Tribolium* system has been used to study many major ideas in evolutionary biology such as mechanisms of reproductive isolation, group selection. In the last three decades, red flour beetles have proved to be invaluable as models for sexual selection studies and evolutionary–developmental biology (evo-devo) studies.

The prolific literature on this genus provides a wealth of background information on the biology of these insects and has led to the development of new tools and methodologies that will further promote experimentation in this organism. Several authors have developed microsatellite markers in the red flour beetle, which can be applied to population genetics and parentage analysis studies. Also, the genome of *T. castaneum* was recently sequenced, which has positioned this species as one of the most significant models for evo-devo, genetics, and genomics studies.

Behavioral Repertoire of *Tribolium*

Aside from being one of the foremost model organisms for population processes and developmental genetics at present, these beetles also make an appealing animal in which to study behavioral ecology and behavioral evolution, given their remarkable repertoire of behaviors. This article discusses only a small subset of some of the more intriguing of these behaviors.

Cannibalism

Tribolium beetles at various life stages may eat their own kind. Because of this taxon's use as a laboratory model of population regulation, cannibalism is one of the best-studied aspects of beetle behavior and serves as an example of interference competition within a species. In general, the inactive stages, egg and pupa, are cannibalized by the active stages, larva and adult. However, it is also possible for adults to feed on very young larvae and for the old larvae to feed on callow adults the exoskeleton of which is not sclerotized. Studies in the confused flour beetles report that adults may devour an egg in a matter of minutes. This behavior is thought to be of great adaptive significance because it may facilitate beetles' colonizing a novel food environment.

Cannibalism is known to be under polygenic control. Thus, the tendency for cannibalism is variable and strongly depends on the species as well as the genetic background of beetle strains. Cannibalism in various flour beetle species does not correlate with phylogenetic relatedness. Apparently, it has been shaped by the evolutionary–ecological history of each individual species.

Similarly, because of differences in selective pressures on different life stages that practice cannibalism, the larvae and the adults too show differences in cannibalism behavior. The larvae in general may feed preferentially on eggs, whereas the adults may feed preferentially on pupae. One explanation for this difference may be that larval nutritional requirements are more completely met by feeding on eggs, whereas adults may possibly eliminate competition in the near future by consuming pupae. Thus, each life stage may be maximizing their chance of survival through preferential cannibalism. Environmental variables such as amounts and quality of food, as well as population density, also influence cannibalism rates. Beetles tend to be more cannibalistic in environments with greater population densities and lower food quantity and quality. Therefore, cannibalism in flour beetles is a classic example of population regulation through density-dependent processes. Cannibalism of eggs and pupae has a significant impact in regulating population sizes. In the confused flour beetle for instance, population sizes would be tenfold larger without cannibalism.

Competition

Different species of flour beetles may occupy a similar niche in an ecosystem. Because they require similar types of abiotic conditions as well as feed on similar foods (stored grain products), they may co-occur within a space and consequently compete for resources in the same environment. Interestingly, the outcome of competitive interactions between the red flour beetle and confused flour beetle depends on abiotic factors such as temperature and humidity, as well as biotic factors such as initial densities of beetles, and presence or absence of parasites. Competition experiments on flour beetles conducted by Park and his colleagues are a classic study in ecology. Their experiments revealed that the confused flour beetle is likely to dominate in environments that are cool and arid, whereas the red flour beetle is likely to prevail at higher temperatures and in humid environments.

Furthermore, parasitism also influenced the outcome of competition between the two flour beetle species. When beetle cultures were infected with a microparasite, *Adelina*, it was found that *T. confusum*, which is better able to withstand *Adelina*, was likely to 'win' in the interspecific competition even in abiotic conditions that typically favored *T. castaneum*. When a different parasite, *Hymenolepis diminuta* was used, the results were the opposite; when beetles were infected with the abovementioned rat tapeworm, the red flour beetle tended to be the superior competitor.

One of the chief ways competing species of flour beetles interact with each other is preying on their competitor's eggs and pupae. Beetles are able to distinguish between eggs and pupae from their own species and those of their competitors. Adult beetles may preferentially feed on the heterospecific eggs and pupae.

Chemical Communication

Beetles produce various chemicals used in communication with conspecifics as well as with other species. Flour beetles produce a host of chemicals that are secreted to their external environment; for example, both *T. castaneum* and *T. confusum* produce over half a dozen such chemicals. However, the function of these compounds is not very well understood. Some compounds such as 4–8 dimethly decanal (4–8 DMD) are produced only by males, others such as Z-2–9-propionate are produced by females, and still others, such as various toluquinones and benzoquinones are produced by both sexes.

Possibly, the best-known compound is 4–8 DMD, which is an aggregation pheromone. It is a common observation that beetles in laboratory cultures and in feed mills are found in aggregations. Typically, beetle behavioral response to 4–8 DMD includes walking toward the source accompanied by a movement of the head and antennae. So strong is this behavioral response that many commercial lures use synthetically produced 4–8 DMD to trap beetles in stores of grain. While this is primarily an aggregation pheromone, it also functions to attract potential mates at closer ranges. Female mate choice in *T. castaneum* may partially depend on male pheromone cue at least in some populations, and male sperm precedence was shown to depend on female's response to this olfactory signal among those populations.

Toluquinones are among the suite of compounds produced by beetles and function as allomones. It is likely that they serve as a defense against the microbes found in the flour and predators in the environment.

Defensive Behaviors

In addition to chemical defense, beetles may resort to thanatosis or feigning death as means of defense against predators. Beetles often feign death by lying still for as few as a fraction of a second to up to a few minutes upon sensing a predator. This behavior is known to vary among different genetic backgrounds of flour beetles. Recent studies show that death feigning behavior is correlated with another type of predator avoidance behavior, fleeing. Beetles that feigned death for longer fled shorter distances than beetles that feigned death for shorter durations. This suggests that beetle strains may be genetically predisposed to one of the two alternative strategies to avoid danger from a predator, either feign death or flee. Death feigning behavior and its evolutionary significance warrants further investigation.

Interactions with Parasites

Flour beetles have been used as a model system to study host–parasite dynamics especially in the recent past.

P. pustulosus is a small (snout-vent length ca. 30 mm), brown, unassuming species of frog – until it calls. The call sounds as if it emanates from a video game. It contains two components: an ever-present whine, which can be

Figure 1   Two calling male túngara frogs, *Physalaemus pustulosus*. Courtesy of Alex T. Baugh.

followed by 0–7 chucks (**Figure 3**). Males gather in choruses from a handful to a few hundred individuals during the rainy season, which in Panama, where most of these studies were conducted, is from May to December. Most males at the breeding site call most of the time, and there appears to be no long-term noncalling or satellite mating strategies. Females move, mostly unimpeded, among the chorusing males and express their mate chioce by making physical contact with a male. At this point, the male clasps the female from the top and the pair remains in this state, known as amplexus, for up to several hours. In a study spanning 152 consecutive nights, which was most of the breeding season, 617 males were marked and 751 matings documented. Each night an average of 27 males and 10 females occupied the breeding site; thus, there was strong competition among males to mates. A male's chance of mating increased with the number of nights he spent at the breeding site. On any given night, females were more likely to choose larger males as mates. The choice of a larger male resulted in a reproductive benefit. Female túngara frogs are larger than males, as is true for most

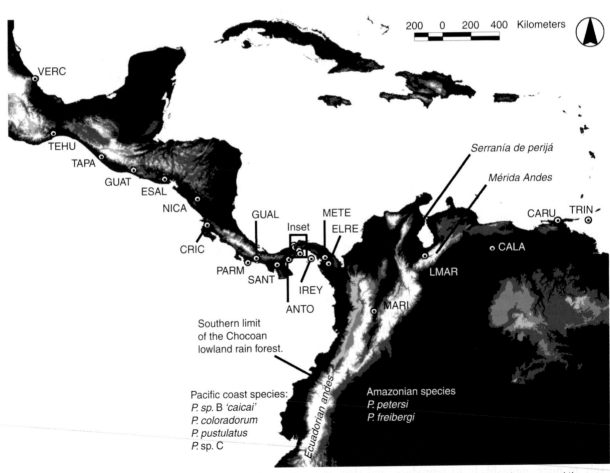

Figure 2   Locations of studies populations of túngara frogs, *Physalaemus pustulosus*, throughout the species' range, and the general location of other species in the same species group. Modified from Weigt LA, Crawford AJ, Rand AS, and Ryan MJ (2005) Biogeography of the túngara frog, *Physalaemus pustulosus*: A molecular perspective. *Molecular Ecology* 14: 3857–3876.

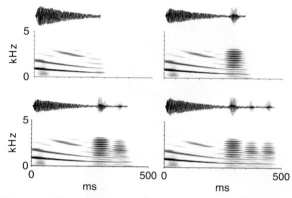

Figure 3 Waveforms (blue, top) and spectrograms (gray, bottom) of a whine followed by 0, 1, 2, and 3 chucks (from top-bottom, left to right).

anurans. As the size difference between a female and her mate decreased so did the number of unfertilized eggs. This benefit seems to derive from a mechanical advantage. When the male and female are similar in size, their cloacas are more likely to be in close juxtaposition during external fertilization so the sperm is released closer to the eggs; similar effects have been reported in other species. Thus, a simple rule of thumb for a female to increase her reproductive success is to choose a large male.

Before the night ends, the mated pair constructs a foam nest, which contains the female's entire clutch of more than 200 eggs. Nest construction typically takes more than an hour and the oviposition site is not necessarily the male's calling site. Nests can be constructed singly, or groups of frog-pairs can produce a larger communal nest. There is no parental care of the nest; eggs hatch out and fall into the water in about 3 days, and in about 3 weeks the tadpoles metamorphose into froglets. In nature, the frogs do not to live for more than 1 year.

The Mating Call

As with most anurans, the long-distance mating call is the primary sexual display. Túngara frogs do not have a short-distance courtship call: a feature found in some other frogs. The mating call of this species is unusual in its varying complexity and its two distinct call components.

The fundamental component of the mating call is the 'whine' (**Figure 3**). The typical whine's fundamental frequency sweeps from 900 to 400 Hz in about 300 ms with a dominant frequency of about 700 Hz. The call has substantial energy in each of the five harmonics of the call, although about half of the call energy is in the fundamental. The whine can be produced alone (the simple call), or it can be followed by up to seven 'chucks' (complex calls). The typical chuck is a short, high-amplitude burst of

sound, about 45 ms in duration with 15 harmonics and a dominant frequency of 2500 Hz (**Figure 3**). A diagnostic feature of the chuck is a fundamental frequency that is one-half of the whine's fundamental, about 200 Hz in many chucks. The whine decreases in amplitude substantially before the production of the chuck, but the whine's fundamental frequency grades into the second harmonic of the chuck. In the wild, the whine transmits over greater distances than the chuck.

Besides the túngara frog, the only other species known to make a similarly complex mating call are found in the túngara frog's sister clade in the same genus, which contains *P. petersi* and *P. freibergi*. Males of these species can add a secondary component to their whine known as a 'squawk.' They never add more than one squawk. In these species, males in some populations are able to make complex calls while males in other populations are restricted to simple calls.

There appears to be an unusual morphology underlying the production of these unusual complex calls. Most frogs vibrate the vocal folds in the larynx to produce sound. Túngara frogs possess a large larynx with a large fibrous mass that hangs from the vocal cords and projects from the larynx into the bronchi that connect the larynx to the lungs (**Figur 4**). Other frogs can have these fibrous masses but they are usually much smaller. In *Physalaemus* males that produce complex calls (*P. pustulosus* and some populations of *P. petersi* and *P. freibergi*), the larynges and fibrous masses are large, while they are small in the species and populations that do not produce complex calls. Thus, the large fibrous mass seems to play some role in the production of the chuck. This correlation between structure and function is supported by ablation experiments. When the fibrous mass is excised from a male, he is unable to produce a chuck. He still attempts to produce a chuck, as he increases the amplitude of the call after the whine. However, the resulting sound has only the frequency harmonics of the whine and not the 'half' harmonics in the chuck. Females do not respond to the calls of the unfortunate males as if they hear complex calls, as discussed in the following paragraphs.

When males call by themselves they usually produce only a simple call, while most males in choruses produce complex calls. In a series of recordings of call bouts of 85 individual males, 53% of the calls had no chucks, 36% had one chuck, and 10% had two chucks. In experiments in which calls are broadcast to males, either in the field or in the lab, males increase their call complexity in response to calls of other males. In addition, the presence of a female causes the male to increase his chuck number. She does this by swimming in front of the male or bumping him and then quickly retreating.

Female preferences for calls can be measured using phonotaxis experiments. In a typical experiment, a female is placed equidistant between two speakers, each of which

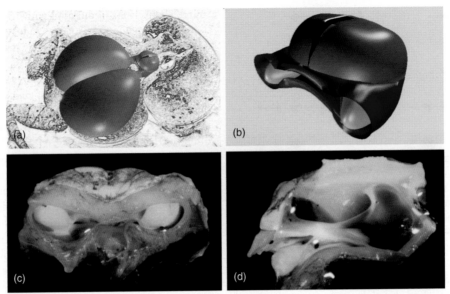

Figure 4   The larynx of the túngara frog, *Physalaemus pustulosus*. (a) An illustration showing the location of the larynx (gold) relative to the lungs (copper) and the bronchi which attach the lungs and larynx (green). (b) An illustration of the larynx showing the protrusion of the fibrous masses from the túngara frog larynx. Photographs of a larynx showing the location of the fibrous mass (c) from the perspectives of the bronchi, and (d) a sagittal section through the larynx. Courtesy of Marcos Gridi-Papp and Cristina O. Gridi-Papp.

broadcasts a test call. The calls are broadcast in sequence, rather than at the same time. A large number of these phonotaxis experiments, more than 4000, have shown that in 85% of the experiments, females preferred a whine with one chuck to a simple whine: a more than fivefold preference.

Females exhibit more subtle preferences than just favoring complex over simple calls. As noted earlier, females are more likely to choose larger males than smaller males. In most animals that vocalize, larger individuals produce sounds of lower frequencies because they have more massive vibrating structures, such as vocal cords, which vibrate at lower frequencies. The same is true in túngara frogs. Larger males produce lower-frequency chucks than do smaller males. In phonotaxis experiments in which females were given identical whines that were followed by a single chuck of lower or higher frequency, females preferred the call with the lower-frequency chuck. Females also preferred lower-frequency whines to higher-frequency ones.

Sensory Biases and Female Preferences

Frogs begin to analyze the mating call in their peripheral auditory system. Unlike most other vertebrates, frogs have two inner-ear organs that are sensitive to airborne sound, the amphibian papilla (AP) and the basilar papilla (BP). The AP is most sensitive to sounds below 1500 Hz and the BP to sounds above 1500 Hz. If a species' mating call has energy within the range of only one of the inner ear

organs, there is generally a good match between the frequencies that have the most energy in the call and the tuning of that inner ear organ. If the call has substantial energy in both low and high frequencies, then usually both the low and the high peaks will match the tuning of the AP and BP, respectively. In túngara frogs, the tuning of the AP is about 700 Hz and matches the dominant frequency of the whine. The BP is tuned, on average, to about 2200 Hz and is a bit below the average chuck's dominant frequency of 2500 Hz (**Figure 5**).

Auditory processing does not stop in the inner ear, of course. In one large auditory nucleus in the midbrain, the torus semicircularis, studies using gene expression as a measure of neural activity show that there is enough information for females to differentiate between the conspecific call and a heterospecific call and between the whine and a whine-chuck (**Figure 6**). Such studies also show that hearing conspecific calls increases correlated neural activity between anatomically distant brain divisions that are involved in social decision making and in the behavioral-motor output directed by such decisions.

Studies of the auditory system provide insights both into the types of call preferences exhibited by females and into the evolution of these preferences. For example, there is a mismatch between the tuning of the BP and the average dominant frequency of the chuck in the population; this also means that on average the BP is more sensitive to chucks with dominant frequencies lower than the population average. In nature, females choose larger males, which have lower-frequency chucks, and phonotaxis experiments confirm that females prefer

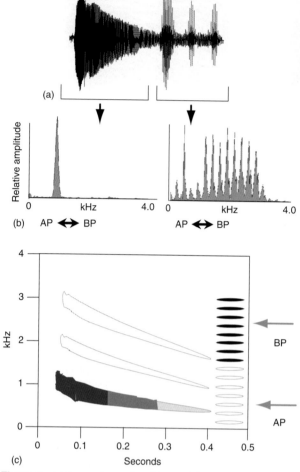

Figure 5 (a) A waveform of a whine and three chucks. (b) Power spectra showing the relative amount of energy in frequencies of the whine (left) and the chuck (right) and on each abscissa the range to which the AP and the BP of most frog species are sensitive. (c) The frequencies to which the AP and the BP are most sensitive (indicated by red arrow). The sonogram indicates the portions of the whine that are necessary to elicit phonotaxis from females (black), portions that increase the probability of phonotaxis if added to the necessary portions (gray), and portions that do not influence female phonotaxis (white). The sonogram also shows for the chuck the efficacy of the upper-half and lower-half of all the harmonics in making the call more attractive than a simple whine. The upper-half of the harmonics are necessary (black) to make the call more attractive than a whine only, while the lower-half harmonics have no such effect (white).

whether in túngara frogs the frequency characteristics of the chuck coevolved with the BP tuning. Besides *P. petersi* and *P. freibergi*, the other species in the *P. pustulosus* species group do not produce complex calls. These other species all produce whine-like simple calls with dominant frequencies in the range of their AP sensitivity. Interestingly, the BP tunings of eight species of *Physalaemus*, five of which are in the *P. pustulosus* species group, are statistically indistinguishable, with the exception of one poorly studied frog, *P. pustulatus*. This comparison shows quite clearly that the tuning involved in the detection of the chuck evolved long before the chuck.

Evolutionary matching of male traits with preexisting sensory biases is known as sensory exploitation. There are several reasons why the tuning of the BP is similar in species with and without complex calls. First, phylogenetic inertia could cause the BP trait that was useful in a distant ancestor to be maintained with no current function in species with only simple calls. Second, the BP is used in detecting other sounds, such as predators. Third, parts of the whines of some of the other species sometimes encompass frequencies to which the BP is sensitive.

Knowledge of the frog's auditory system can also guide us in determining the salient aspects of mating calls. The concept of the 'sign stimulus' cautions that just because we can accurately measure and quantify a signal it does not mean that all aspects of the signal are meaningful to the receiver. The whine has five harmonics. The fundamental frequency, which has about half of the whine's total energy best matches the sensitivity of the AP (**Figure 5**). A synthesized call containing only the fundamental frequency sweep is as attractive as a synthetic version of the entire call and is more attractive than a synthetic version of the upper four harmonics. When only the fundamental frequency is compared with natural calls, it is just as attractive. Females respond similarly to the chuck; as long as a synthetic version of the chuck stimulates the most sensitive frequencies of the BP, the females respond to it as a chuck.

Cognitive Aspects of Mating Call Recognition

The whine is necessary and sufficient to elicit female reproductive behavior. Although a chuck makes the whine more attractive to females, females are not attracted to a chuck by itself. The female also prefers the conspecific call to the call of other species they live with and to calls of their closely related species.

If females were not able to discriminate between conspecific and heterospecific calls, they might then choose heterospecific mates and most likely, mate but not produce viable offspring. Thus, both the sender and the receiver appear to be under strong selection to avoid the

these lower-frequency chucks to higher-frequency ones. Thus, the three-way relationship between BP tuning, the dominant frequency of the chuck, and male size provides a mechanistic explanation for why females prefer larger males and lower-frequency calls.

Comparative studies can be used to ask about how calls and preferences evolve by examining the relationship between BP tuning and the presence of complex calls among closely related species. Specifically, we can ask

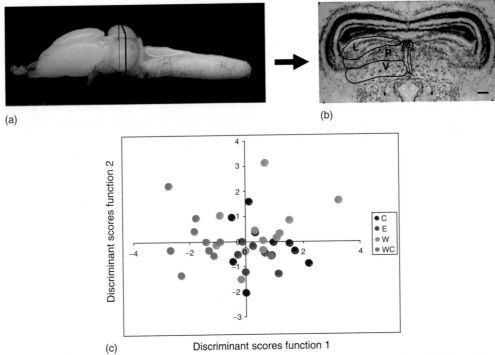

Figure 6 (a) A brain of the túngara frog (*Physalaemus pustulosus*) showing the location of the section which in (b) illustrates the various nuclei of the main auditory nucleus the torus semicircularis. (c) The results of a discriminant function analysis. The analysis compares a proxy for neural excitation, the amount of expression of an immediate early gene, across the torus in groups of females that are exposed to different calls: WC, a whine plus one chuck; W, a whine; E, the mating call of a different species, *Physalaemus enesefae*; and C, a chuck alone. The results show there is sufficient information in the torus alone to allow females to discriminate among these call types ($P < 0.01$).

costs of heterospecific matings: the male's whine evolves to transmit the conspecific status and the female's auditory system to decode this information. The limited degree to which the whine can be manipulated without disrupting this very basic function is not surprising. As noted earlier, the whine needs to contain the fundamental frequency, and the fundamental is also necessary to elicit female phonotaxis. But even within the fundamental frequency sweep, not all portions are perceived as equally important to females. Within the fundamental, stimulation in a high-frequency region between 900 and 560 Hz is necessary, followed by stimulation in a partially overlapping low-frequency region between 640 and 500 Hz. No single frequency or constant-frequency band suffices.

Species-specificity of the chuck, however, is not critical to increase call attractiveness. Although the chuck occurs at the end of the whine, its precise placement can vary and it will still make the whine more attractive. Ninety per-cent of the chuck's energy is in the upper-half of its harmonics. This part of the chuck by itself makes a call more attractive while the lower-half harmonics by them-selves, given their natural energy content, do not influ-ence female preferences. If all the energy is shifted to the lower harmonics, this part of the chuck alone makes the whine more attractive. The chuck can also be replaced with the squawk, the secondary call component of its close

relative *P. petersi*, while noise, or even bells and whistles, and the addition still enhance the attractiveness of the whine. It appears that the chuck might do little more than add sensory stimulation to the female's auditory system once the female has recognized the call as being conspe-cific, rather than the chuck being a message with a partic-ular meaning to the female túngara frog.

The chuck must be heard with the whine for the chuck to be perceived as part of the mating call. That might not seem an issue, since both call components are produced in a specific order from the same source. Frogs, however, congregate in choruses to advertise for females. The cacophony of mating calls is somewhat akin to a human cocktail party. But the túngara frogs do not have quite the same abilities described in human as the 'cocktail party effect,' in which we can sort out, in one auditory stream, the words from a particular voice. A chuck by itself is not recognized as a mating call by a female, so if a whine and a chuck are displaced spatially from one another and the female approaches the chuck, this is evidence of percep-tual linkage, or binding, of the whine and the chuck despite the fact they emanate from different sources. In túngara frogs, perceptual binding of these two call com-ponents takes place over considerable spatial separation, up to 135° (**Figure 7**). This is true, although to a lesser degree, even if the temporal position of the whine and chuck are varied. By altering the spectrum of the chuck to stimulate primarily the AP or BP, the results from phonotaxis experiments suggest that auditory grouping over large spatial separation results from processes in the brain rather than in the peripheral auditory system.

Another issue in perception is how females perceive signal variation. For example, do females perceive calls as being more or less similar to some ideal of a conspecific call, or alternatively, are calls perceived as either conspecific or heterospecific? To explore these questions, females were tested with a series of synthetic calls that were intermediate between the conspecific call and one of several heterospecific calls. In most cases, the female's response to the calls changed gradually; the less similar the calls are to the conspecific, the less likely females responded to them.

There were instances, however, in which females exhibited a response pattern similar to that common in humans: categorical perception. This occurred when gradual variation among stimuli was perceived categorically. There were two components of categorical perception. One is that continuous variation is labeled into categories, and the other is that discrimination between stimuli within a category is weaker than discrimination of stimuli between categories, even though in both the cases the stimuli are as physically different from one another. In a series of intermediate calls between *P. pustulosus* and *P. coloradorum*, there is a category of calls that were all recognized as conspecific and another category of calls that are not recognized as conspecific (which, operationally, is akin to being recognized as heterospecific). There was little discrimination between calls within the same category but strong discrimination between stimuli in different categories even though the acoustic differences between all pairs of stimuli were the same. It is not known

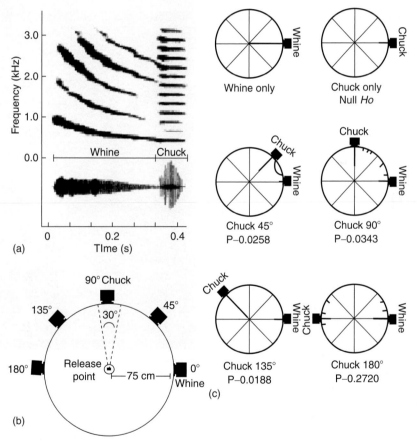

Figure 7 (a) Spectrogram and waveform of the whine plus one chuck stimulus. (b) Diagram of the phonotaxis arena and example of one stimulus condition. Five speakers at 45° separation were configured along the perimeter of a 75 cm radius circle on the floor of a sound chamber. Stimuli consisted of a whine and/or a chuck presented together or alone. After release in the center of the arena, female position and exit angle were recorded using an infrared camera and video recorder. For the categorical analysis, because frogs exiting the arena within 13 cm of the center of a speaker could still make contact with a speaker-case, all responses 10° of the center of a particular speaker were scored as a positive response to that speaker. (c) Each point represents the exit angle (re: whine position) for one female *P. pustulosus* presented with a whine or a chuck alone or in combination with varying spatial separation. Probability values are shown for a Fisher's exact test comparing chuck attractiveness when presented with the whine to that when presented alone. Chuck amplitude was 6 dB re. whine amplitude (90 dB SPL). Adapted from Farris HF, Rand AS, and Ryan MJ (2002) The effects of spatially separated call components on phonotaxis in túngara frogs: Evidence for auditory grouping. *Brain, Behavior and Evolution* 60: 181–188.

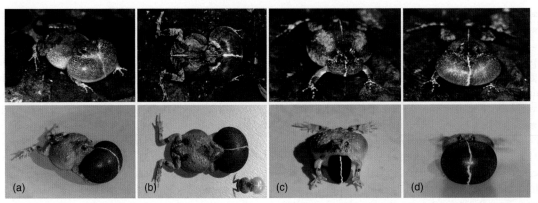

Figure 8 Photographs comparing real (top) and robotic (bottom) túngara frogs. Views: (a) lateral, (b) dorsal, (c) anterior with deflated vocal sac, and (d) anterior with fully inflated vocal sac. Vocal sacs on all robotic frogs were part of a catheter except for the one inset (b), which was a latex balloon. Courtesy of B.A. Klein, J. Stein, R.C. Taylor.

how common is categorical perception of mating signals, but if it is common it could have important consequences for the tempo and mode of sexual selection and species recognition.

Visual Communication

One of the better known features of frogs is the extendable vocal sac that inflates when a male calls. The vocal sac probably evolved as a means of shuttling air back into the lungs during calling. Thus, the air can be reused for multiple calls, and a frog does not have to pump air into its lungs for each call. But the vocal sac also makes males visually more conspicuous when they call, and in some species, the sac serves as a visual cue for females. When given a choice between two calls, one call associated with a video of a stationary male and the same call associated with a video of a male with his vocal sac inflating and deflating with each call, females prefer the latter. Also, physical models of frogs, 'robo-frogs,' with inflating vocal sacs make a call more attractive compared to a call with no associated visual cue (**Figure 8**). The vocal sac inflation, however, must be synchronized with the call; otherwise, not only do females not perceive it as part of the male's courtship display, but also they avoid it.

While vocal sacs serve as visual cues in diurnal frogs, most frogs breed at night. Behavioral measures of the visual sensitivity of túngara frogs show that they are able to see under the low-light levels that characterize their breeding sites.

Eavesdroppers

Communication signals evolve because they influence the behavior of a receiver. This is the 'intended receiver.' But there might also be 'unintended receivers' or 'eavesdroppers'

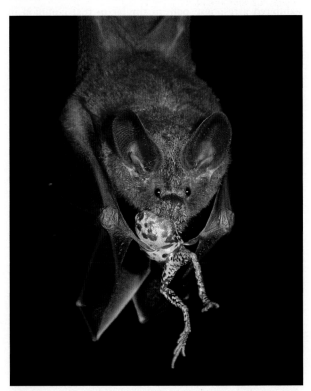

Figure 9 A frog-eating bat, *Trachops cirrhosus*, with a túngara frog, *Physalaemus pustulosus*, in its mouth. Courtesy of Alex T. Baugh.

that can detect these signals. Eavesdroppers can have an important influence on the evolution of communication, and nowhere is this more apparent than in túngara frogs. Bats, flies, opossums, turtles, crabs, and other frogs all eat túngara frogs and can use information from eavesdropping to locate their prey.

The frog-eating bat, *Trachops cirrhosus*, is unusual in that frogs are an important part of its diet (**Figure 9**). At one site in Panama, *Trachops* captured and ate 30 calling males in less than 3 h. Even more unusual is this bat's mode of hunting.

Figure 10 A calling male túngara frog, *Physalaemus pustulosus*, surrounded by a swarm of blood-sucking flies, *Corethrella* sp. Courtesy of Ximena Bernal.

In addition to using their echolocation system to navigate through the forest, the bats rely on the frog's call to locate its prey. Thus, when male túngara frogs call to advertise their presence to females, they inadvertently advertise their presence to frog-eating bats. When male túngara frogs escalate from simple to complex calls to increase their attractiveness to females, they also increase their predation risk to *Trachops* since the bats also prefer complex to simple calls. One reason for the bats' call preference is that they are better able to localize complex calls, although there is no evidence that the same is true for the female frogs.

The bats also use the calls of frogs to determine which frogs are edible and which are unpalatable. *Trachops* readily approach the calls of túngara frogs and other edible species but they do not fly toward the calls of unpalatable toads. The bats are able to learn this association between the frog's call and its palatability. Within a single night, bats from the wild can be conditioned to respond to toad calls and avoid túngara frog calls. They can also pass this information about prey cue and prey quality to other bats. The flexibility of foraging behavior of *Trachops* should allow them to capitalize quickly on encounters with new species of frogs.

Male túngara frogs are also tormented by blood-sucking flies of the genus *Corethrella* (**Figure 10**). These flies are close relatives of mosquitoes and buzz around many species of frogs in the tropics. These flies typically land on a calling male túngara frog, walk on his back until they reach his nares, and then take a blood meal. Like the frog-eating bats and female frogs, the blood-sucking flies are attracted to the male's call, and they are preferentially attracted to complex calls over simple calls. It is not clear what costs to the frog, other than loss of a small amount of blood, are incurred from the flies, but the flies might transmit parasites to the frogs. It is also not clear how the flies hear the call. Mosquitoes have receptors on their antennae that are sensitive to low-frequency sounds but this type of receptor would probably not serve the *Corethrella* flies well if they needed to locate túngara frogs from a substantial distance.

Conclusion

This review illustrates why túngara frogs are a useful system for studying sexual selection and communication. The main advantage derives from the integrative nature of studies that have merged knowledge of the frog's brain and behavior and evolution toward an understanding of how the communication system is influenced by sexual selection.

See also: Acoustic Signals; Agonistic Signals; Mate Choice in Males and Females; Mating Signals; Social Selection, Sexual Selection, and Sexual Conflict; Sound Production: Vertebrates.

Further Reading

Bernal XE, Page RA, Rand AS, and Ryan MJ (2007) Cues for eavesdroppers: Do frog calls indicate prey density and quality? *The American Naturalist* 169: 412–415.

Bradbury JW and Vehrencamp SL (1998) *Principles of Animal Communication.* Sunderland, Massachusetts: Sinauer Associates, Inc.

Darwin C (1871) *The Descent of Man and Selection in Relation to Sex.* London: Murray.

Page RA and Ryan MJ (2005) Flexibility in assessment of prey cues: Frog-eating bats and frog calls. *Proceedings of the Royal Society, London Series B* 272: 841–847.

Ryan MJ (1985) *The Túngara Frog, A Study in Sexual Selection and Communication.* Chicago, IL: University of Chicago Press.

Ryan MJ and Rand AS (1995) Female responses to ancestral advertisement calls in Tungara frogs. *Science (Washington DC)* 269: 390–392.

Turtles: Freshwater

R. M. Bowden, Illinois State University, Normal, IL, USA

Introduction

Turtles are an ancient vertebrate lineage, dating back nearly 220 My; throughout their evolutionary history, they have maintained their characteristic shell, making even extinct species readily identifiable. Today, turtles are among the most imperiled groups of vertebrates, with 62% of species currently listed as vulnerable, endangered, or critically endangered by the International Union for Conservation of Nature and Natural Resources. The most commonly cited reasons for their decline include habitat degradation or loss, their status as a commodity in the pet trade, and their widespread use as a food item by humans.

Turtles are generally secretive animals. Freshwater turtles mostly inhabit murky backwaters, swamps, and marshes, making it challenging to study their behavior in the wild. Nonetheless, researchers have been investigating turtle behavior for well over a century, using a combination of laboratory and field observations to amass a diverse body of literature on these long-lived animals. The vast majority of research on freshwater turtles focuses on a limited number of common and widely distributed North American species including painted, common snapping, and red-eared slider turtles, but there is a growing body of research on a more diverse array of species from around the globe.

In the mid-twentieth century, studies by Cagle, Ernst, and Gibbons provided a wealth of information on the natural history of several North American species including home range size, homing, overwintering, and reproductive behaviors. In recent years, research on turtles has built upon these foundations, and has expanded to explore the interface of behavior with physiology, evolutionary biology, and conservation. Researchers are currently addressing questions such as: How are some turtles able to survive in subzero temperatures while others cannot? How do turtles tolerate hypoxic conditions? Why is sex determined by temperature in many species of turtles, and how will climate change affect population sex ratios? Despite years of study, it is clear that we have much to learn about these fascinating animals.

Mating Behavior

Freshwater turtles have a variety of stereotypical mating behaviors that are almost exclusively carried out in aquatic environments. Mating is typically initiated by the male, with the initial approach oriented towards the hind end of the female. In some species, the male may give chase to the female for some time prior to being able to approach her, but once the approach is made, it is often linked to an investigation of the cloacal region. The cloaca contains glands that secrete chemical compounds that may be used by the male to determine the sex of the individual; this may be particularly useful in species that are size monomorphic. If a male happens to approach another male, they rarely attempt to court, and occasionally, the approached male may respond aggressively towards the pursuing male. If the male approaches a female, he may immediately attempt to mount her, but in many species, the male courts the female prior to mounting.

Courting behaviors appear to be species-specific, and may be used as species recognition in areas where distribution ranges overlap for similar species, such as the map turtles that inhabit the upper Mississippi River. Courting in the common map turtle includes direct contact between the snouts along with head-bobbing, while in the false map turtle, the male vibrates his forelimbs near the ocular region of the female. In painted and slider turtles, males stroke the head of the female using the backside of their forelimbs and claws. In musk turtles, males nuzzle the area where the carapace and plastron meet (bridge), a region that possesses additional glands in these species. Females of many species are passive participants in the courting process, unless they are unreceptive or uninterested in mating and attempt to leave. If a male is successful at courting, he then proceeds to mount the female, grasping her carapace firmly with his foreclaws. He might also bite the female's head or neck region, or place his chin over her snout to prevent her from extending her neck. Additionally, males may hold the female underwater during much of copulation. Females sometimes struggle to swim to the surface with the male latched onto them.

For most North American species, mating occurs throughout the active season, with the majority of mating occurring shortly after emergence in the spring and again in the fall. Multiple males may congregate around and attempt to mate with a single female, with larger males typically displacing the smaller males. In the red-eared slider, older, larger males are darker (melanized), and these melanized males have greater reproductive success. Males and females mate multiply, and multiple paternity has been documented in the painted, wood, Blanding's, and common snapping turtle as well as the European pond turtle (**Figure 1**). In addition to mating multiply

Figure 1 Common snapping turtle (*Chelydra serpentina*) spotted walking across a field (Rock Island Co., IL). Leeches, seen here attached to the skin and shell, are commonly found on North American freshwater turtles. Courtesy of MCN Holgersson.

within a season, females of many species store sperm for extended periods of time. Although females can store sperm, why they do so is less clear. Sperm storage capability would be useful if males and females are widely distributed in space, limiting their rate of encounters. Alternatively, sperm storage may allow females some measure of control over reproduction if they are able to selectively use sperm from individual males while not using (or limiting the use of) sperm from others. Further work is needed to determine the extent to which sperm competition may occur in turtles, and if females are able to exert postcopulatory female choice through selective sperm utilization.

Sex Determination and Nesting Behavior

Sex determination, or the generation of males and females, is not typically thought of as a behavior, but an individual's sex certainly influences many of their behaviors. In turtles, there are two primary methods for determining sex – genetic and environmental pathways. Although some turtles clearly possess sex chromosomes, most turtles with genetic sex determination lack identifiable sex chromosomes. In these instances, it is presumed that there are sex specific differences in genes, but to date, these genes have not been identified.

Many turtles employ temperature-dependent sex determination (TSD) in which the incubation temperature of the embryo determines sex. In turtles, there are two patterns of TSD, the most prevalent being pattern I, with males are formed at cool incubation temperatures and females at warm temperatures, while pattern II (males formed at intermediate temperatures and females formed at either cool or warm temperatures) is less common.

For species with TSD, females have the potential to have a large degree of control over the sex ratio of their offspring through the selection of a nesting location. For example, if a female places her nest in an exposed area with little or no overhanging vegetation, her nest will experience warmer conditions during development. Conversely, nesting in areas with increased cover produces cooler incubation temperatures during development. This variation in nest temperature is sufficient to produce sex ratios that range from 100% male to 100% female. Although females may adaptively choose nesting locations to manipulate offspring sex, a more proximal factor that likely drives nest site choice is offspring survival. In this model, females appear to select nesting locations that are most suitable for the successful development of their eggs. A wide range of conditions are conducive to successful embryonic development, and depending upon geographic location, females alter their nesting behavior to provide their eggs with suitable conditions. In North American species with a wide latitudinal distribution such as painted and snapping turtles, nests are generally located in open, sunny areas in northerly populations and in more shaded areas in southerly populations. This is necessary to ensure that nests developing in the north receive adequately warm conditions to complete development prior to the onset of winter, while those in the south do not reach excessively high temperatures that exceed the thermal maximum for development. Such a pattern could result in highly skewed sex ratios by latitude for species with TSD, however, the range of temperatures that produce males versus females shifts with latitude, allowing for the reliable production of both sexes across the geographic range.

Females should also select nest sites that minimize the risk of predation to both herself and her offspring. Nests placed closer to the water appear to be more prone to predation than those further inland, but trekking further onto land may increase the likelihood of predation on the female as she cannot easily return to the relative safety of the water. Although there are few non-human predators of adult turtles, there are reports of mammals, including fox and raccoons, preying upon nesting females. Much more commonly, the eggs are the target of predation, which frequently occurs within minutes or hours of a nest being laid. There is a long list of known egg predators; some of the more common animals include raccoons, skunks, fox, coyote, and various snakes, but presumably any terrestrial vertebrate capable of unearthing the nest could take advantage of this resource. Should the eggs in a nest survive to hatch, hatchling turtles are vulnerable to terrestrial predators including birds and mammals. Despite the perils of dispersal, hatchling turtles and can traverse large distances from their nest to the nearest source of water.

Nest construction is similar among all freshwater species. Females leave the water to search for a suitable nesting location. Nest searching behavior varies greatly among females. Some spend a few minutes searching, while others may spend hours. Once the female has decided on a location, she begins the process of nest excavation. The female excavates the nest using her hindlimbs and claws to loosen and remove the nesting substrate until she has formed a flask-shaped nest with a narrow opening and a rounded cavity to hold the eggs. Upon completion of digging, eggs are laid into the nest and then the female replaces the nesting substrate, firmly packing the nest using her hindlimbs, and in some species, the plastron. Once finished, the female walks away from the nest, giving no subsequent maternal care. There is some variation in the choice of nesting substrate among species, but most nests are placed in areas that are well drained, consisting of either loamy soil or sand.

In many north temperate freshwater turtles, nesting occurs over a relatively constrained period between May and July. In some species, females produce only a single clutch of eggs in a season. In females of multiclutching species, latitude appears to play a role in how many clutches are produced within a nesting season, with northern populations laying one or two clutches and southern populations producing as many as five clutches. Also, females do not necessarily lay eggs every year. The number of clutches produced by a female within a year appears to depend upon several factors including her energetic reserves and environmental conditions.

Overwintering Behavior

Because freshwater turtles build their nests on land, hatchling turtles either leave the nest and disperse to water prior to the onset of winter or spend the winter on land inside their subterranean nest. The pattern of overwintering behavior seems to be consistent within a species across its geographic distribution. For example, hatchling painted and red-eared slider turtles almost exclusively overwinter terrestrially while hatchling snapping and Blanding's turtles leave their nests in the fall in search of a suitable aquatic environment.

Regardless of their overwintering habits, all hatchling turtles must complete embryonic development and subsist using nutrition gained from their yolk for some period of time. The duration of subsistence varies dramatically depending upon the overwintering strategy. Hatchlings that move to water in the fall rely on their yolk reserves for a much shorter posthatching period as they have foraging opportunities available to them once they leave the nest. On the other hand, hatchlings that remain in the nest are forced to subsist off of yolk reserves for many months. Fortunately for these hatchlings, their activity is severely limited (both by being confined within the nest cavity and by the cooler winter temperatures), which should allow them to maximize their survival time based on their limited energetic stores. Also, these hatchlings may be able to strategically utilize their yolk reserves so that they save energy-rich lipids for use during periods of high activity such as movement to water in the spring.

Overwintering behavior also has physiological ramifications for hatchlings, especially for those in more northerly populations. Unfortunately, very little is known about the behavior or physiology of hatchlings that overwinter aquatically, but it would seem that a primary issue for them is lack of access to oxygen, and the consequent hypoxia that can occur in ice-covered aquatic environments. It appears that young turtles faced with hypoxic conditions have some ability to exchange gasses across their softer epithelial tissues, such as that on their limbs, but the shell imposes a strong barrier to gas exchange across most of the body.

Far more is known about hatchlings that overwinter terrestrially. In particular, these hatchlings risk exposure to subzero temperatures. Research by Ken Storey, Gary Packard, and Jon Costanzo and colleagues has demonstrated how these small, ectothermic animals survive harsh winter conditions. In the painted turtle, hatchlings can withstand freezing conditions within the nest provided that temperatures do not fall far below zero for extended periods of time and that the nest cavity is free of any debris that may serve to initiate ice crystal formation. Blanding's turtle hatchlings are also reported to be freeze tolerant, but they do not necessarily overwinter within the nest. In the red-eared slider, freezing is not an option and animals exposed to subzero temperatures succumb to chill injury rapidly. Although few aquatically overwintering species have been studied, it appears that they are not generally freeze tolerant. For all terrestrially overwintering species, depending upon local weather conditions, nests also may face desiccation or inundation with water. These conditions provide additional challenges to hatchlings.

Many adult freshwater turtles overwinter in aquatic environments. For those residing in warmer climates, they may experience periods of reduced activity during the winter months, and they may or may not continue to forage throughout this time depending upon climatic conditions. Animals in northern climates are exposed to the same general environmental conditions as aquatic hatchlings, and thus must cope with long periods of cold temperatures and submersion under ice. The result is that animals can be forced to spend extended periods of time below the water where they have limited access to oxygen, but are protected from freezing temperatures. Adult turtles are well known to have a high tolerance for hypoxic conditions, with a suite of behavioral and physiological adaptations that can be employed under low oxygen conditions. These include decreased movements, heart rate, and respiratory rate. Further, many adult turtles capture oxygen using nontraditional respiratory surfaces. Of particular note is the ability to perform gas exchange across their cloacal surface. This so-called 'cloacal breathing' does not appear to be as efficient as respiration across the lungs, but it does supply a modest amount of oxygen to the animal. Because turtles are ectothermic, during these overwintering periods essentially all of their activities are decreased. Animals do not appear to feed, and in preparation for this period, they may also clear their gut to prevent rotting of food in their digestive tract as digestion all but ceases at cold temperatures.

Thermoregulation

Turtles have many patterns of activity that are highly dependent upon temperature. One of the most iconic behaviors of turtles is basking, a behavior that can be performed when ambient air temperatures exceed the temperature of the water (aerial basking) (**Figure 2**), or when the upper layers of water are warmer than the lower layers (aquatic basking) (**Figure 3**). Turtles of all ages and both sexes take

Figure 2 Aerial basking behavior in the painted turtle (*Chrysemys picta*, Rock Island Co., IL). Courtesy of MCN Holgersson.

Figure 3 Aquatic basking behavior in a juvenile red-eared slider turtle (*Trachemys scripta elegans*, Tazwell Co., IL). Courtesy of MCN Holgersson.

advantage of basking as a means of thermoregulation, with seasonal variation in this behavior being associated with physiological requirements. For example, female turtles tend to bask more extensively during the spring and fall when they are maturing follicles in preparation for egg laying, and it is thought that elevating body temperature may help with the mobilization of energetic reserves. Males also engage in frequent basking in the spring, suggesting that they too have high energetic demands during this time, possibly related to mating behavior.

A comparison of two populations of painted turtles found that body temperature decreased with increasing latitude. Whether this variation can be attributed to thermal preferences of the animals, or a constraint on available thermal niches is not known. Evidence for a latitude-related preference for body temperature was found in the terrestrial ornate box turtle; turtles from a Wisconsin population selected lower body temperatures than a population from Kansas. Interestingly, it appears that the lower preferred body temperatures in the Wisconsin population actually increases the range of temperatures over which the turtles can remain active, and consequently extends their active season.

Through selective microhabitat use, it is possible for an individual to maintain a fairly consistent body temperature throughout the day. To regulate body temperature, turtles can shuttle in and out of the water to elevate or lower body temperature. On warm spring days, it is common to find multiple turtles lined up on any available substrate that allows access to the sun. Softshell turtles of the genus *Apalone* appear to prefer basking along the banks of aquatic habitats, while most other species that actively bask seek out logs or other partially submerged substrates as basking platforms. The key to selecting a basking site appears to be ready access to water. If approached, basking turtles rapidly return to water, presumably to avoid potential predators. Alternatively, turtles can also bask within the water by maintaining a position at or just below the surface to take advantage of radiant heating. Both aerial and aquatic basking are effective in maintaining body temperature in the painted turtle, but there is variation among individuals in how they utilize the environment for thermoregulation. Some animals spent more time in open water, while others preferred shallower habitats that were less conducive to thermoregulation, but with increased foraging opportunities.

Thermoregulatory behavior is well developed in juvenile turtles of several species. Given a choice of water temperatures within an ecologically relevant range, juvenile wood turtles actively selected the warmest available temperatures. Wood turtle hatchlings emerge in the fall, and thus they must begin foraging prior to the onset of colder winter temperatures. Similar thermal preferences exist in other fall emerging species, including spiny and smooth softshell and snapping turtles, and in the red-eared slider, which overwinters terrestrially. As is true for adult turtles, many physiological processes should be enhanced at warmer temperatures, and thus by actively searching for warmer habitats, juvenile turtles may increase their ability to forage and process food in preparation for winter hibernation which should enhance their probability of survival.

Turtles can also modify their thermoregulatory behaviors to create a fever when ill. To elicit a fever, the turtle seeks out a warmer habitat to elevate body temperature and spends more time with an elevated body temperature than would be expected for a healthy individual. Behavioral fever has been documented in several freshwater species and appears to serve the same function as fever in endothermic animals.

Chemical Communication

Like many animals, turtles rely upon a variety of modes of communication, including chemical signaling. Given that aquatic turtles often live in murky habitats with low visibility, chemical communication may be one of the most reliable means of conveying and receiving information. Turtles have glands in their integument, and many possess structures known as Rathke's glands at the anterior and posterior margins of the shell bridge (the region where the upper carapace and lower plastron meet). Rathke's glands produce a complex secretion that contains lipids and glycoproteins; this provides the characteristic musky smell of mud, musk turtles, and snapping turtles. Other sources of chemical secretions include the mental glands that release their secretions near the chin area and glands within the cloaca. Turtles also have well-developed sensory systems including the vomeronasal and olfactory systems, suggesting that they can detect and process a variety of chemical signals. While in water, turtles appear to acquire chemical cues primarily through the process of gular pumping, or throat expansion, which moves cue-laden water across chemically sensitive surfaces.

Chemical secretions have been implicated in mediating a number of behaviors. When musk turtles were provided with the choice between dechlorinated water only, or dechlorinated water spiked with tank water that had held either the predatory alligator snapping turtle or a nonpredatory species, they avoided the water with the predatory cues. Moreover, they did not appear to avoid the water with nonpredatory turtle cues, suggesting that the musk turtles distinguish between chemical cues of different turtle species. Similarly, three-toed box turtles show antipredator behavior when exposed to urine from a natural predator of box turtles, the coyote, but show no such behavior when exposed to deer urine or to water. There is some evidence that interspecies competition may also be mediated by chemical cues in turtles. In a series of studies, Polo-Cavia and colleagues tested how the native Spanish terrapin responds to invasive red-eared sliders. Their studies revealed that the native terrapins prefer water with conspecific chemical cues and avoid water that contains red-eared slider chemical cues, while the sliders did not avoid water with terrapin cues present. Further, when presented with a simulated predation event, Spanish terrapins engaged in escape behaviors more quickly than did red-eared sliders, which remained withdrawn into their shells rather than actively moving to deeper water. Because the red-eared slider does not appear to change its behavior in the presence of the native species, and is bolder, available habitat for the native species is in decline. Red-eared sliders have been introduced throughout Europe, Asia, and Australia as part of the pet trade, and may be having similarly negative impacts on a variety of native species.

Mating behaviors can also be driven by chemical cues. In many species of turtles, mating is often preceded by investigation of the female's cloacal region by the male. If a male approaches and investigates another male, they rarely exhibit any further attempt at mating, and this has been interpreted as evidence that turtles produce sex-specific chemical signals that can be used for mate recognition. In the European pond turtle, males appear capable of distinguishing between cues from female and male conspecifics, with males showing a strong preference for water with female cues present. Males also appear to prefer cues from larger females over those from smaller females and show avoidance behavior when presented with cues from larger males. These findings suggest that turtles can make mate choices based upon fecundity. Larger females are more fecund; therefore, they are preferred mates and males compete for access to larger females. Data on mate choice and competition for mates are limited and further study is needed to firmly establish the role of chemical cues in mate choice and mate competition in freshwater turtles.

Waterborne cues can also aid in homing behavior for freshwater turtles. Presently, we do not know how important chemical cues are for navigation, but mark-recapture studies indicate that turtles can return to their point of release when displaced. Certainly, chemical cues are only one of the potential suites of cues available for navigation in freshwater turtles. Other possible sources of information include visual, phototaxic, and geotaxic cues, but how these cues are ultimately integrated to produce homing behavior is not well understood.

Territoriality and Aggression

Home range sizes vary considerably within and among species of freshwater turtles. In some species, females occupy larger areas than males, but the reverse is also true. Home range size is influenced by the available aquatic habitat (one large body of water or several smaller, nearly contiguous ponds), the density of conspecifics and heterospecifics, and resource availability. Female turtles show philopatry to nest sites, and as a consequence, they must have the ability to navigate their aquatic environment to arrive at their preferred nesting location and then back to their home territory once nesting is completed. As in females, males should benefit from the ability to successfully navigate through complex aquatic environments in search of foraging and mating opportunities.

Despite abundant evidence that freshwater turtles occupy home ranges, there is little evidence to suggest that they actively defend these spaces. For most aquatic species, the home ranges of males may overlap the home ranges of multiple females, and possibly many males as well. When males go out in search of mating opportunities, they may encounter other males also searching for females, which can result in agonistic behavior. Alternatively, aggressive interactions may be avoided through chemical signaling in some cases. The generally murky habitats of freshwater turtles make it difficult to directly monitor many behavioral interactions in nature.

When basking, turtles are either found solitary, spaced well apart from one another, or oriented in a manner that avoids direct eye contact. These have been interpreted as strategies to avoid aggressive interactions. There is evidence of aggressive interactions when turtles are kept in close proximity, including scratching and biting, as well as pushing other individuals off of basking platforms. Turtles may develop dominance hierarchies when held under artificial conditions, and there is some evidence for dominance hierarchies occurring in natural populations, particularly with regards to mating behaviors.

Foraging Behavior

Gut content analysis of a variety of North American species indicates that most freshwater turtles have extremely varied diets, and can be best described as omnivorous. Essentially, all freshwater turtles consume some combination of plants and animals, with the latter being dominated by invertebrate prey. Many species are opportunistic scavengers, and there are numerous reports of cannibalism – primarily of hatchlings and juveniles. Although most foraging takes place within the aquatic habitat, there are reports of terrestrial foraging in red-eared sliders and map turtles. Terrestrial foraging appears to be rare, however, and if an aquatic species forages on land, they return to water to consume their forage. Despite their reputation for being slow, turtles can move rapidly within the water. Softshell turtles can catch swimming fish, and snapping turtles can prey upon waterfowl resting at the surface of the water as well as small mammals. In many freshwater species, there is a tendency for diet to shift with age, with hatchlings and juveniles tending to be more carnivorous than adults.

The alligator snapping turtle can employ a sit-and-wait strategy by using a lingual lure. Alligator snappers are the only turtle known to possess such a structure, which is an effective means of attracting fish as prey. The lure extends from the tongue and is anchored by complex musculature that allows for rapid movements that are thought to mimic the movements of small worms or insect larvae. Most of the available information on this luring behavior comes from laboratory studies conducted by Hugh Drummond and Elizabeth Gordon in which juveniles were tested with visual, chemical, and tactile stimuli. Animals appear to orient toward prey items using very slow movements, and to strike at prey when the prey approaches the lure. Alligator snapping turtles do not rely solely upon 'fishing' as their only mean of obtaining food, as gut content analysis reveals a wide variety of prey that could not be captured by luring including mussels and clams, and a variety of plant material. Alligator snappers also prey upon numerous turtle species, which has apparently caused prey species to actively avoid these predatory turtles though chemosensory stimuli.

Future Directions

Despite the large amount of information that has already been collected on freshwater turtles, it is clear that there are still substantial gaps in our knowledge. Studies aimed at investigating behaviors in the wild would be particularly useful for understanding how turtles interact with one another and utilize their ever shrinking habitats, and advances in technology are making such studies tractable. Researchers are also benefitting from increasing study of a wider variety of species from around the world. In species with TSD, we can only predict how these species may respond to global climate change. Given that so many turtle species are imperiled, largely due to human impacts, efforts to gather as much information as possible on these species is necessary for successful conservation of these interesting animals.

See also: Behavioral Endocrinology of Migration; Body Size and Sexual Dimorphism; Olfactory Signals; Sea Turtles: Navigation and Orientation; Smell: Vertebrates; Sperm Competition.

Further Reading

Costanzo JP, Lee RE, Jr, and Ultsch GR (2008) Physiological ecology of overwintering in hatchling turtles. *Journal of Experimental Zoology* 309A: 297–379.

Ernst CH, Lovich JE, and Barbour RW (eds.) (1994) *Turtles of the United States and Canada.* Washington, DC: Smithsonian Institution Press.

Ewert MA and Nelson CE (1991) Sex determination in turtles: Diverse patterns and some possible adaptive values. *Copeia* 1991: 50–69.

Gibbons JW (ed.) (1990) *Life History and Ecology of the Slider Turtle.* Washington, DC: Smithsonian Institution Press.

Jackson JF (1990) Evidence for chemosensor-mediated predator avoidance in musk turtles. *Copeia* 1990: 557–560.

Janzen FJ (1994) Climate change and temperature-dependent sex determination in reptiles. *Proceedings of the National Academy of Sciences of the United States of America* 91: 7487–7490.

Quinn VS and Graves BM (1998) Home pond discrimination using chemical cues in *Chrysemys picta. Journal of Herpetology* 32: 457–461.

Rowe JW and Dalgarn SF (2009) Effects of sex and microhabitat use on diel body temperature variation in midland painted turtles (*Chrysemys picta marginata*). *Copeia* 2009: 85–92.

Shine R (1999) Why is sex determined by nest temperature in many reptiles? *Trends in Ecology & Evolution* 14: 186–189.

Wyneken J, Godfrey MH, and Bels V (eds.) (2007) *Biology of Turtles.* Boca Raton, FL: CRC Press.

Unicolonial Ants: Loss of Colony Identity

K. Tsuji, University of the Ryukyus, Okinawa, Japan

Introduction

The behavior of ants is considered typical of kin selection theory, because a striking reproductive division of labor exists among genetically highly related colony members, which is maintained by the colony-membership discrimination. However, this premise collapses in some species described as unicolonial, in which the colony borders are obscure and there are many nests containing many fertile queens in a supercolony covering a wide area. Unicoloniality is a challenge to kin selection theory, because the estimated nestmate relatedness often falls to a low value that is statistically indistinguishable from zero. Therefore, the maintenance of eusociality – characterized by the presence of reproductive altruism among workers — seems to be difficult to explain by kin selection.

Unicoloniality, Multicoloniality, and Supercolonies

Unicoloniality and its opposite, multicoloniality, are population characteristics. As mentioned in the following paragraph, however, in real ants there are many exceptions and intermediate cases between unicoloniality and multicoloniality. Therefore, the following descriptions should be regarded as two extreme stereotypes.

Let me first describe multicoloniality. Colonies of stereotypical multicolonial ants are characterized by monogyny (the presence of a single queen in each colony) and monodomy (single nest in each colony; for definition of an ant nest, see the reference, Debout et al., 2007). The workers defend their nest and other resources against alien conspecifics. Therefore, a multicolonial ant population consists of many mutually hostile colonies. A new colony is founded by a winged queen after the nuptial (mating) flight and dispersal on the wing. Due to this nuptial flight, mating occurs usually between nonnestmates (i.e., is out-bred).

At the other extreme, unicolonial ants are characterized by polygyny (more than one queen in a colony) and polydomy (more than one nest in a colony). There is no hostility among conspecific individuals and therefore exchange of members (workers, queens, brood, and males) among nests is frequent. New queens do not perform a nuptial flight but mate in the natal nest, usually with a nestmate male. Nests are founded by budding, in which one or more queens move to a new nest site on foot accompanied by some workers, rather than involving flight.

There seems to be no general consensus on the definitions of unicoloniality, however. One reason for this is that real ants do not always have the set of traits described earlier. For example, army ants of the genera *Eciton* and *Dorylus* found a colony by budding, but their typically monogynous colonies are hostile to conspecific ones, and therefore exchanges of individuals do not take place. Furthermore, army ant queens are outbred, copulating with multiple nonnestmate males. The deepest disagreement on the definition of unicoloniality may particularly involve the difference between unicoloniality and polygyny–polydomy, and how large supercolonies should be before they are 'unicolonial.' Here, I define unicoloniality as a phenomenon in which a population – an assemblage of conspecific individuals living within a special range in which they can naturally interact through ecological processes such as competition, cooperation, and mating – becomes a single polygynous and polydomous colony. I also define multicoloniality as the situation where a population consists of at least two mutually hostile colonies. I propose to use the term polygyny–polydomy when the population contains multiqueened and multinested colonies that are mutually hostile (i.e., muliticolonial). Finally, I suggest to use the term 'supercolony' flexibly, so as to include the situations of both polygyny–polydomy and true unicoloniality. Myrmecologists tend to describe a polygynous–polydomous colony as a supercolony, when it stretches spatially over an extremely large scale, to the degree that direct interactions of individuals from distant

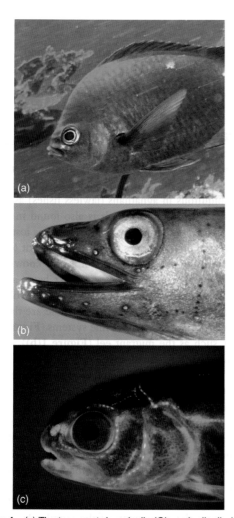

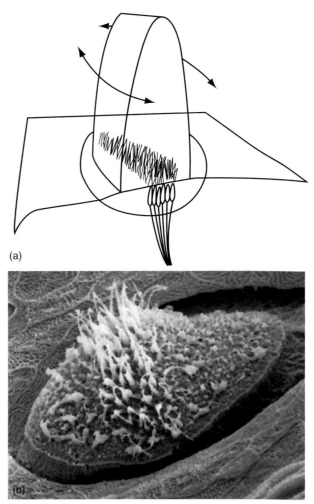

Figure 1 (a) The two-spot demoiselle (*Chromis dispilus*) sitting in the current over a reef. This picture serves to illustrate the fact that in many fish the lateral line can be scarcely seen. The modified scales of the trunk lateral line can be just made out as a curved line between the eye and the dorsal white spot. The picture also serves to illustrate rheotaxis and to introduce particle streak photography. The current flowing over the reef from left to right can be visualized by the movements of the bright particles in the water, and the fish is shown holding station against the current. In experimental situations, the procedure is formalized by illuminating the particles in a discrete plane so that the particle movements can be reliably measured. (b) Superficial neuromasts of the eel (*Anguilla diefenbachii*) are located as an array of receptors on the surface of the head. The rows of pigmented dots below and behind the eye are superficial neuromasts. The white openings of the lateral line canals are visible above and below the mouth. (c) Steelhead trout treated with DASPEI and viewed with a fluorescent microscope. The lateral line is visible, composed of neuromasts (green dots). Courtesy of Cech, Joseph, Jr., and Timothy Mussen (2006) *Determining How Fish Detect Fish Screens and Testing Potential Fish Screen Enhancements.* California Energy Commission, PIER Energy-Related Environmental Research Program, CEC-500-2006-117.

Figure 2 Anatomy of a superficial neuromast. (a) Diagrammatic view of a superficial neuromast. Water flows (arrows) across a gelatinous structure called 'a cupula.' Movement of the cupula is sensed by the cilia of the hair cells. (b) Scanning electron micrograph of a superficial neuromast showing the exposed cilia of the hair cells. The cupula is removed in the preparation process for electron microscopy.

neuromasts do not respond to direct flows but are sensitive to the acceleration of oscillating flows and respond best to vibrations in the range of 10 s to a few 100 s of Hz.

The essential idea is that canal neuromasts are less influenced by low-frequency, large-scale flows (noise) such as those generated by currents in the water surrounding the animal, or movements of the animal itself. This enables canal neuromasts to respond more specifically to the higher-frequency signals generated by other animals such as prey. The division of labor between superficial and canal neuromasts is beautifully illustrated by the work of Engelmann et al. showing the effects of a background flow rate on the responses of these two submodalities to a small vibrating sphere. This study can also serve as an illustration of the standard way in which lateral line function is typically recorded.

For simple technical reasons, almost all the electrophysiology done with the lateral line is done on restrained animals. Following ethically suitable procedures, the fish is held and a portion of the lateral line nerve exposed. Single unit recordings are then made with microelectrodes, and the neural recordings related back to a 'controlled' stimulus. The most common stimulus employed has been a vibrating bead of known size, at a known location, moving backward and forward at a known distance and frequency. The term 'controlled' is put in inverted commas since although we can know with some precision the movements of the bead, a bead moving in this fashion is a dipole stimulus and produces quite complex 3D water movements. This means that the actual stimulus at the neuromast depends on the precise geometry of the relationship between the stimulus and the neuromast, and is also influenced by the presence of the fish. In the Engelmann study, the response of a single nerve fiber is recorded to the vibrating stimulus, while the background flow rate in the tank is systematically varied. At zero background flow, all afferent fibers respond with phase-locked responses to the vibrating source; however, with increasing background flow rate, the response of one class of afferent is progressively masked, whereas another class continues to encode the

high-frequency stimulus from the vibrating source. It is reasonable to equate superficial neuromast input with the class of afferent masked by the flow 'noise' and canal neuromasts as the unaffected class.

Lateral-Line-Mediated Behavior

At the most basic level, biological behavior is all about mating, feeding, and moving about safely (and cheaply) in between times. Hydrodynamic information encoded by the lateral line has been shown to play a critical role in each of these behaviors. For example, lateral-line-based communication has been shown to be important in the mating behavior of a few species of salmonids. However, most of the work on lateral-line-mediated behavior has concentrated on feeding and movement which will be covered in separate sections.

Prey Detection

In many species, and in particular those that inhabit turbid or low light environments (i.e., caves, the deep sea, Antarctica), the lateral line is extensively used in prey detection and capture (**Figure 4**). Understanding the anatomical and functional dichotomy between superficial and canal neuromasts is the base from which to explore the behavioral role of the lateral line and its

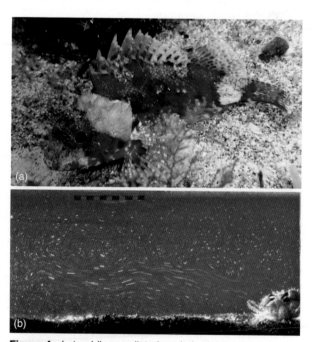

Figure 3 Anatomy of a canal neuromast. (a) Diagrammatic illustration of a canal neuromast. The hair cells and their cilia are shown located within a canal, which opens to the exterior through the pore at the top of the diagram. (b) Scanning electron micrograph looking down on a canal neuromast. The canal has been transacted longitudinally to reveal the neuromast. The oval area of the sensory strip can be seen with an inner oval of cilia. As in **Figure 2**, preparation for electron microscopy has removed the cilia.

Figure 4 Lateral-line-mediated predation in the dwarf scorpionfish (*Scorpaena papillosa*). (a) Picture taken in daylight of the dwarf scorpionfish. Scorpionfish feed at night using their lateral line to detect prey. (b) Particle streak photography of the respiratory current of a crab. This species is one of the common prey items of the scorpionfish.

submodalities in such habitats. For example, in low light conditions, the lateral line canal system has been shown to be principally responsible for prey detection and capture, where the prey is a small vibrating source such as a *Daphnia*. Much of this work has been done on the mottled sculpin which can be readily trained to orient toward, and strike at, a small vibrating bead, which mimics its prey. Using this behavior, Coombs and her associates have been able to show that this orienting behavior is mediated by the lateral line canals on the head and trunk of the fish. They have also shown that the job of locating and tracking targets can be related to the high spatial organization of the lateral line system. Their studies on the activation of separate lateral line units by discrete sources clearly show the potential for the encoding of location parameters such as azimuth, elevation, and distance. The response patterns of single afferent neurons provide a coarse indication of target location. One way of describing this is that individual afferent neurons have relatively broad receptive fields centered on their position on the body surface. However, it is likely that the CNS can utilize the pattern of activity across the sensory array to effectively sharpen target location both in terms of which part of the body surface is closest to the target, and the distance of the target away from the surface. The information to perform this task is clearly available in the pattern of activity generated by the sensory array, but detailed psychophysical measurement of behavioral capability in regard to spatial acuity, and the location and mechanisms of receptive field refinement in the CNS, is largely unknown.

It is interesting to note that target location using surface waves is a special case which provides us with some of our best evidence for the ability of the lateral line to determine target direction and distance. A number of fish and amphibians have been shown to have behavioral and anatomical specializations to detect insects and other prey that fall onto the water surface. Surface waves radiate out from the source of the disturbance, and the resulting wave fronts and wave dispersal characteristics provide enough information for the fish to turn and approach the source. In an elegant series of behavioral experiments, Bleckmann and colleagues were able to show the precise basis of distance discrimination by producing stimuli with the appropriate characteristics to 'fool' the fish into overshooting the target.

One of the new exciting findings in lateral line research is that some fish are able to follow turbulent trails and can use this hydrodynamic trail following to track down prey. Nocturnal catfish have been shown to adopt this 'hydrodynamic trail following' tactic to hunt prey in darkness. The characteristic element of the behavior is that once the catfish encounters the trail, it then follows the trail to the prey. This clearly results in a circuitous route that follows the wake left behind by the prey, rather than a path that would result from detecting and orienting to the prey directly.

Rheotaxis, Orientation, and Hydrodynamic Imaging

Superficial neuromasts respond to water flow over the surface of the body, so it should not be surprising that they contribute to rheotaxis or orientation to water flow. However, it was not until relatively recently that a role for the lateral line in rheotaxis was demonstrated. Behavioral experiments conducted on a range of different species showed that fish with an intact lateral line will orient to flow at quite low flow velocities, but that following pharmacological blockade of the entire lateral line or physical ablation of superficial neuromasts, the threshold for rheotaxis is elevated. For example, in the Antarctic fish *Pagothenia borchgrevinki* (**Figure 5**), the rheotaxic threshold with the lateral line intact was between 1 and 2 cm s^{-1}, whereas

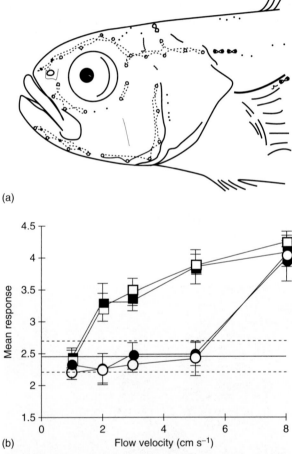

(a)

(b)

Figure 5 Rheotaxis behavior of the Antarctic notothenioid fish *Pagothenia borchgrevinki*. (a) The arrangement of the superficial (dots) and canal organs is shown on the head of the fish. (b) Orientation into the current in a flume is shown as a function of flow velocity in the flume. With the lateral line intact (filled squares), or canal neuromasts pharmacologically blocked (open squares), the rheotactic threshold is between 1 and 2 cm s^{-1}. With the entire lateral line block (closed circles) or the superficial neuromasts ablated (open circles), the rheotactic threshold is raised to somewhere between 5 and 8 cm s^{-1}.

with the superficial neuromasts ablated this threshold rose to between 5 and 8 cm s^{-1}. Pharmacological blockade of the canal lateral line submodality had no effect on rheotactic thresholds. Thus, these findings support the idea that superficial neuromasts respond best to slow and uniform currents and play a role in orientation to these currents.

However, it is worth noting when it comes to more complex orientation behaviors, information is probably integrated from both superficial and canal neuromasts. The canal system has also been shown to be important in orientation to objects by the blind cave fish (*Astyanax fasciatus*). In this case, the self-generated flow fields form the basis of an active 'imaging' system, where the fish senses distortions in the flow field created by nearby obstacles. Blind cave fish show a characteristic tail beat and glide swimming behavior. Our studies show that cave fish are able to avoid head-on collision with a wall and react to the presence of the wall at the relatively short distance of about 0.1 body lengths (**Figure 6** + video). Both flow visualization techniques and computational fluid dynamic modeling show that the stimulus to the lateral line begins to climb steeply at this point. Kinematic studies, of the head-on approach to the wall, show that collisions become much more prevalent if the approach to the wall occurs during a tail beat rather than during the glide (**Figure 7** + video). It is apparent that the added complexity of flow field during the tail beat, and/or the direct effects of the movement on the lateral line seriously impair lateral-line-mediated hydrodynamic imaging.

With respect to more complex rheotaxis in turbulent flows, information from both superficial and canal systems has been shown to be necessary. This is the case when there is an obstacle in the flow; and the fish maintain station behind the object, either by occupying the low-pressure zone behind and to one side of the object, or by exhibiting a distinctive 'von Karmen' swimming gait that enables them to swim more efficiently by extracting energy from the regular vortices generated by the object.

Lateral Line Diversity and S/N

Demonstration of the anatomical and functional dichotomy between canal and superficial neuromasts still begs the question as to the extent to which lateral line endorgans are adapted to the particular hydrodynamic environment in which the fish lives, or to particular prey of interest. There is certainly a huge diversity of fish species and a corresponding diversity of the lateral line structure. Some of the more extreme lateral line morphologies are found in fishes inhabiting low light environments such as the deep sea. Unfortunately, it is very difficult to conduct physiological studies on these animals, so we still lack a functional understanding of the widened membranous canal systems, or the stalked superficial neuromasts, found in deep sea fishes (**Figures 8** and **9**).

In thinking about the lateral line system, one of the key issues is signal and noise. Hydrodynamic noise may come from the presence of the animal in a flow, either steady flow or turbulence, but it may also come from the animal's own movements. Self-generated noise can be recognized as an issue for many sensory systems, but given the

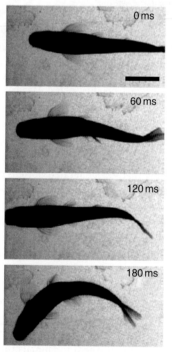

Figure 7 Series of images with a fish approaching the wall and colliding with the wall. The fish starts a tail beat as it approaches the wall and shows no sign of detecting the wall before the collision.

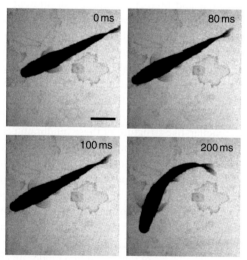

Figure 6 Series of images with a fish approaching the wall and avoiding collision. The fish glides toward the wall, then at 100 ms extends the pectoral fins away from the body and turns to follow the wall without making any contact.

Figure 8 Photograph of the skinned head of *Poromitra crassiceps* showing the extensive array of canal neuromasts. In life, these have a membrane covering forming the so-call widened membranous canals system seen in a number of deep sea species. Photograph courtesy of N.J. Marshall.

Figure 9 Photograph of a living specimen of the deep sea anglerfish *Phrynichthys wedli* with lines of papillate superficial neuromasts. Photograph courtesy of N.J. Marshall.

sensitivity of lateral line receptors and the observation that 'fish are seldom still,' self-generated noise must be a particularly significant problem for the lateral line system. It is worth noting that self-generated noise is often under-recognized since almost all functional studies are made on immobilized animals. For the purposes of improving the signal–noise ratio, the one advantage that self-generated noise has is that, to a degree, it is under the control of the animal. The simplest strategy for noise reduction is to cease movement. Motionlessness can be seen in sit-and-wait predators, but can also be a key component of the search strategy in lateral line-based predation.

The search strategies of visual predators have received extensive study. There is a recognized continuum from sit-and-wait predation through to animals that 'move and pause' (saltatory search) to cruise searchers, which exhibit continuous movement. Each strategy has particular costs and benefits associated with prey and environmental characteristics (prey density, prey size, encounter rates, environmental complexity, prey conspicuousness, etc.). Recent research on the dwarf scorpionfish is the first systematic study on the search strategy of a lateral line predator. The fish were observed searching for randomly located crustacean prey in darkness using infrared video. Under these conditions, the fish adopts a saltatory search behavior. The characteristics of this behavior include an average move distance of 18 cm, with fish pausing every 1.25 s for ~6.7 s. Prey are located mostly during the pause period, and movement will be terminated early upon the detection of a target. Prey are also located throughout the search space which can be defined in terms of the reactive distance (48 cm) and the reactive angle (96°). After the pause period, the fish moves to a new location to gain new search space. The characteristics of saltatory search are similar between lateral line and visual predators, but occur for somewhat different reasons. In visual predators, remaining stationary improves the visibility of the minute movements of cryptic prey against a complex but still visual background. In the case of lateral line predators, the pause period minimizes self-generated noise, making hydrodynamic signals more detectable. Consistent with this interpretation, we have shown that under the same circumstances, another nocturnal predator, the southern bastard cod, using a cruise search mode (video), has a shorter lateral line reactive distance. The cod uses a mixed lateral line/chemosensory search strategy, which has a different set of costs and benefits when compared with the dwarf scorpionfish. One of the costs is that the constant motion of the cruise reduces the distance at which lateral line stimuli can be detected, and in the cod, the lateral line reactive distance was only a third of that exhibited by the dwarf scorpionfish.

Some patterns of movement (e.g., breathing) will provide a regular and predictable pattern of afferent input. Since the movements are generated by the animal itself, it has, in effect, *a priori* knowledge of movement and the potential to predict and cancel the associated afferent inputs. Studies in both lateral line and electrosense, particularly electrosense, show that the hindbrain processing centers for both these senses form an adaptive filter which learns to cancel predictable input. The basis of this ability is the cerebellar-like structure of these hindbrain centers. The *crista cerebellaris*, which overlies these structures, comprises what is called 'a molecular layer of parallel nerve axons' which carry information about ongoing movements. This information comes as efference copy from motor centers, proprioceptive information about movement, and a number of other sources. In effect, the molecular layer contains a rich matrix of information about movement. The principal cell type of the lateral line hindbrain center is called 'a crest cell.' Crest cells have dorsal spiny molecular layer dendrites that receive parallel fiber information, but also receive direct lateral line afferent input on their ventral dendrites. A rather simple synaptic plasticity learning rule allows the input from the parallel fibers to generate a 'negative image' of the reafferent noise arriving at the ventral dendrites. In this way, the reafferent noise is cancelled, yet the crest cells remain sensitive to external, biologically important, signals. Ventilation is but one example of a movement that produces unwanted sensory reafference. Recordings from lateral line afferents, particularly in the area of the gills, show strong ventilation-mediated responses. By comparison, the crest cells show greatly reduced responses to ventilation movement.

Reptilian, Avian, and Mammalian Analogs

The mechanosensory lateral line is found in all fishes and in aquatic amphibians but was lost in the transition of vertebrates onto land. However, there are mechanosensory systems in reptiles, birds, and mammals that, though not evolutionarily related, do mediate some remarkably similar behaviors to those described earlier.

For example, alligators hunt at night and show a well-refined ability to detect and orient to surface waves. This behavior is mediated by small 'dome-shaped' pressure receptors on their snouts that are innervated by the trigeminal nerve. Dome pressure receptors are also evident in fossil reptiles dating back to the Jurassic, but only in groups thought to have had a semiaquatic lifestyle.

Birds have specialized mechanosensory feathers called 'filoplumes.' Though not well studied or understood, they have been implicated in sensing contour feather deformation and hence, airflow and separation over the wing. In this sense, they may be analogous to the use of the lateral line in rheotaxis and orientation in turbulent flows. Recent studies also show the use of head filoplumes in obstacle avoidance.

Mammalian vibrissae (or whiskers) are another understudied vertebrate sense. They, too, provide a sense of touch at a distance. The observation that in some mammals very significant proportions of the sensory cortex are devoted to vibrissae input gives some indication of their potential importance. Their use has been studied in prey capture in the dark, in some marsupial species where in the auditory localization of a prey rustling in grass produced a pounce to locate the array of whiskers over the prey, allowing precise orientation for prey capture. Rats have been used as a model species for many years, but they are now being recognized as nocturnal animals that rely on their sense of touch, and particularly their vibrissae, to explore the world. Active 'wiskering' where rats sweep their vibrissae back and forth in the air and against objects has similarities with the hydrodynamic imaging employed by cavefish. Both behaviors allow the extraction of information about surrounding objects. The analogy between vibrissae and lateral line function is further strengthened by the interesting recent discovery that seals, like fish, can perform hydrodynamic trail following, and that this ability is dependent on their vibrissae.

Clearly, the use of vibration detectors is vitally important in many behaviors across a wide variety of vertebrate taxa. The study of lateral lines and other vibration detection systems has enjoyed a resurgence in recent years. Firstly, because these studies provide insights into ways of perceiving the world that are widespread across the animal kingdom, but alien to us as humans. Secondly, these systems also provide inspiration for a new range of technological devices. Biomimetic sensors, based on our understanding of how animals sense their world, will become an increasingly important component of robotics technology.

See also: Empirical Studies of Predator and Prey Behavior; Hearing: Insects; Hearing: Vertebrates; Sound Localization: Neuroethology.

Further Reading

Bassett D, Carton AG, and Montgomery JC (2007) Saltatory search in a lateral line predator. *Journal of Fish Biology* 70: 1148–1160.

Bleckmann H (2008) Peripheral and central processing of lateral line information. *Journal of Comparative Physiology A* 194: 145–158.

Coombs S, Braun CB, and Donovan B (2001) The orienting response of lake Michigan mottled sculpin is mediated by canal neuromasts. *Journal of Experimental Biology* 204: 337–348.

Coombs S, Hastings M, and Finneran J (1996) Modeling and measuring lateral line excitation patterns to changing dipole source locations. *Journal of Comparative Physiology A* 178: 359–371.

Dehnhardt G, Mauck B, Hanke W, and Bleckmann H (2001) Hydrodynamic trail-following in harbor seals (*Phoca vitulina*). *Science* 293: 102–104.

Engelmann J, Hanke W, Mogdans J, and Bleckmann H (2000) Hydrodynamic stimuli and the fish lateral line. *Nature* 408: 51–52.

Liao JC (2007) A review of fish swimming mechanics and behaviour in altered flows. *Proceedings of the Royal Society of London, Series B: Biological Sciences* 362: 1973–1993.

Marshall NJ (1996) Vision and sensory physiology – the lateral line systems of three deep-sea fish. *Journal of Fish Biology* 49 (supplement A): 239–258.

Montgomery JC, Baker CF, and Carton AG (1997) The lateral line can mediate rheotaxis in fish. *Nature* 389: 960–963.

Montgomery JC and Bodznick D (1994) An adaptive filter cancels self-induced noise in the electrosensory and lateral line mechanosensory systems of fish. *Neuroscience Letters* 174: 145–148.

Montgomery JC, Mcdonald F, Baker CF, Carton AG, and Ling N (2003) Sensory integration in the hydrodynamic world of rainbow trout. *Proceedings of the Royal Society of London, Series B: Biological Sciences* 270(supplement 2): 195–197.

Pohlmann K, Atema J, and Breithaupt T (2004) The importance of the lateral line in nocturnal predation of piscivorous catfish. *Journal of Experimental Biology* 207: 2971–2978.

Soares D (2002) An ancient sensory organ in crocodilians. *Nature* 417: 241–242.

Solomon JH and Hartmann MJ (2006) Sensing features with robotic whiskers. *Nature* 443: 525.

Windsor SP, Tan D, and Montgomery JC (2008) Swimming kinematics and hydrodynamic imaging in the blind Mexican cave fish (*Astyanax fasciatus*). *Journal of Experimental Biology* 211: 2950–2959.

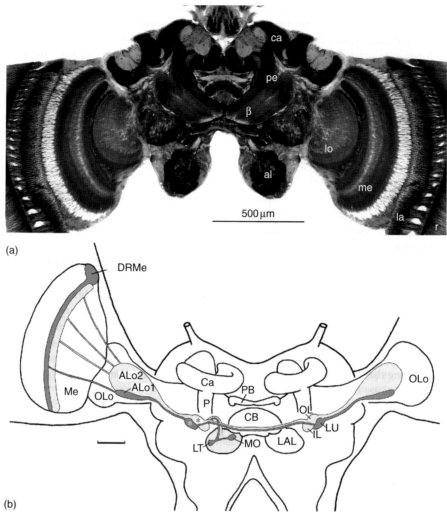

(a)

(b)

Figure 4 Visual neuropils and pathways in the insect brain. (a) A frontal section of the entire bee brain. The retina r and various neuropils of the optic lobe (lamina, la; medulla, me; and lobula, lo) are shown with neuropils of the brain: the mushroom body (comprising the calyx, ca; the peduncle, pe; and the β-lobe of the peduncle, β), and the antennal lobe al. Section courtesy of Wulfila Gronenberg. Reproduced from Ehmer B and Gronenburg W (2002) Segregation of visual input to the mushroom bodies in the honeybee (*Apis mellifera*). *The Journal of Comparative Neurology* 451: 362–373. (b) A schematic drawing of the optic lobes and brain of the locust *Schistocerca gregaria*, with the polarization pathways shown in dark gray. Me: medulla; DRMe: dorsal rim area in the medulla; ALo1–2: layers 1 and 2 of the anterior lobe of the lobula; OLo: outer lobe of the lobula; Ca and P: calyces and penduncles of the mushroom bodies; CB: central body; LAL: lateral accessory lobe; PB: protocerebral bridge; OL and IL: outer and inner lobes of the upper unit of the anterior optic tubercle; LU: lower unit of the anterior optic tubercle; LT and MO: lateral triangle and median olive of the lateral accessory lobe. Scale bar = 200 μm. Diagram courtesy of Uwe Homberg. Reproduced from Homberg U, Hofer S, Pfeiffer K, and Gebhardt S (2003) Organization and neural connections of the anterior optic tubercle in the brain of the locust, *Schistocerca gregaria*. *The Journal of Comparative Neurology* 462: 415–430.

This is a useful and easy metric for comparing the light-gathering capacities of different eyes, with a lower F-number indicating a brighter image. The huge camera eyes of the nocturnal net-casting spider *Dinopis subrufus* have an F-number of < 0.6, a much lower value than in the anterior-median (AM) eyes of the diurnal jumping spider *Phidippus johnsoni* (F-number = 2.0). The dark-adapted human eye has an F-number of around 2.1. Thus, the eyes of *Dinopis* are clearly constructed for high sensitivity. Among compound eyes, the superposition eyes of the nocturnal hawkmoth *Deilephila elpenor* have an F-number of around 0.7.

This high sensitivity to light has permitted many invertebrates to have remarkably good vision in dim light. Nocturnal hawkmoths and bees can see the colors of flowers and negotiate dimly illuminated obstacles during flight. Nocturnal bees can also home using learned terrestrial landmarks, while moths and dung beetles can navigate using constellations of stars or the dim pattern of polarized light formed around the moon.

Temporal Vision

The ability of animals to see things that move at different speeds is a reflection of their temporal vision, that is, how 'fast' they are able to see. A human observer is unable to discriminate a bullet fired from a rifle simply because it moves too rapidly to be seen. The fastest objects that can be seen by an animal depend on many factors, including the physiological properties of the photoreceptors and the ambient light intensity. Moreover, the speed of vision varies from species to species, with some species having very fast vision (like the fast-flying aerobatic diurnal insects) and others having very slow vision (like sedentary nocturnal toads). This indicates that the speed of vision, like every other aspect of vision, is matched to the ecologies of animals: those that move rapidly, or need to detect fast-moving objects (e.g., mates or prey in full flight), tend to have fast vision, while those that are sedentary or slowly moving tend to have slow vision.

The speed of vision varies widely between animals because the dynamics of phototransduction, and the identities and proportions of ion channels present in the photoreceptors (and the membrane kinetics they thus establish), differ substantially from species to species, and these differences have evolved in response to the various ecological needs of animals. In fact, the cell membrane, via its electrical properties, acts as a 'matched filter' that is able to match the response properties of the eye to the lifestyle that the animal possesses, such as its locomotion speed or its preferred light intensity niche. In addition to these ecological influences, the filtering properties of the photoreceptor membrane and the transduction cascade are influenced by the state of adaptation and the temperature.

The fastest visual systems found in the animal kingdom are possessed by invertebrates, and the fastest are likely to be those of calliphorid flies, which in a light-adapted state are able distinguish light stimuli that flicker at rates of up to around 300 Hz (for humans, in comparison, the limit is around 50 Hz). These flies engage in high-speed pursuit and interception of mates, a behavior that involves rapid and unpredictable changes in flight trajectory and which requires a rapid visual response. For all species, vision slows down as light levels fall (due to adaptive changes in the physiological properties of the photoreceptors), and nocturnal and deep-sea species tend to have intrinsically slower vision than species active in bright sunshine. Slower vision in dim light significantly improves the reliability of vision because it filters out faster visual details that tend to be inherently noisy and degrade visual performance.

Spatial Vision

Most visual scenes viewed by animals on land or in the upper depths of the ocean are extended in nature, meaning that light reaches the eye from many different directions at once. The spatial details of such a scene – defined by local contrast differences between areas of light and dark – are imaged by the eye onto the underlying retina. The finest spatial details that can be seen by an eye are determined by two main factors: (1) the quality of the optical system that images the scene (i.e., the cornea and lens(es)) and (2) the density and visual fields of the photoreceptors that receive the image. Both these factors are in turn subservient to the amount of light that the eye can collect from the scene. As observed earlier, as light levels fall, visual reliability declines because of a decreasing signal-to-noise ratio. This is particularly true for the smaller contrast differences typical of finer spatial details, which tend to be lost in the noise as intensity falls, a limitation that is equally problematic for all eyes, irrespective of their optical quality or photoreceptor density. Thus, in general, spatial resolution declines with light intensity.

The optical quality of an imaging system depends on the extent it suffers from various aberrations, particularly spherical and chromatic aberrations. Moreover, in very small lenses such as those found in compound eyes, it also depends on diffraction of light waves entering the aperture. The effect of all these optical imperfections is to blur the image formed on the retina, that is, to reduce its contrast. Even though there are many invertebrate species whose eyes lack an imaging system altogether (e.g., the cephalopod *Nautilus*), those that do possess one very frequently have surprisingly crisp optics and good optical image resolution. For example, among the single lenses of camera eyes, those of tiny cubozoan jellyfish eyes are remarkably sharp, as are those of most gastropod and cephalopod molluscs and most arachnids. What is typical for most of these lenses is the presence of a powerful radial (and often parabolic) gradient of refractive index from the center to the edge of the lens, which tends to almost exactly cancel spherical aberration. The lenses of jumping spiders are an excellent example, which together with a densely packed retina, allow jumping spider eyes to have among the best spatial vision for their size in the animal kingdom. In compound eyes, the small ommatidial lenses are typically only 20–100 times wider than the wavelength of visible light, a fact that invariably leads to image degradation due to diffraction. In superposition eyes, optical image quality is additionally limited by the accuracy of superposition of light rays on the target rhabdom, and in species where the rhabdoms are not shielded by screening pigments or a tapetal sheath, by the spread of more steeply incident rays through neighboring rhabdoms (which blurs the image neurally).

Once the image is focused on the retina, the underlying matrix of photoreceptors must reconstruct it. Like the pixels of a digital camera, the density of photoreceptors in an eye sets the finest spatial detail that can be reconstructed: more densely packed photoreceptors can

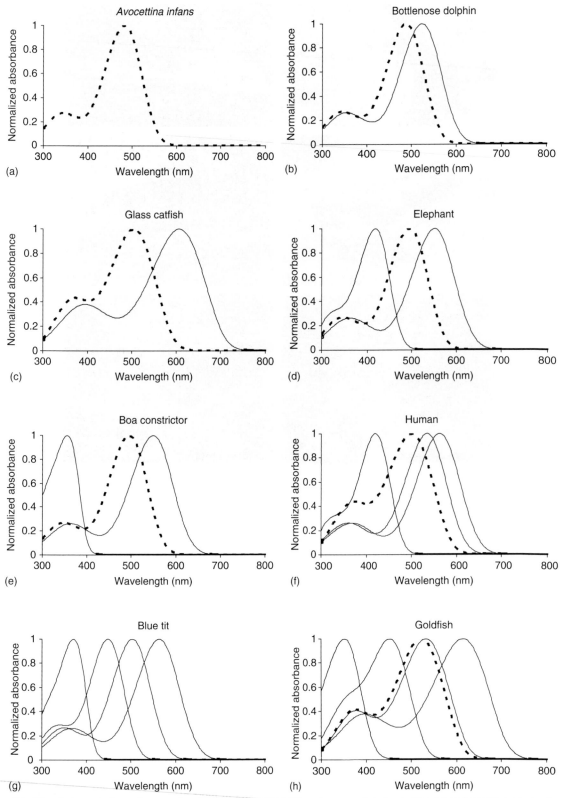

Figure 21 Visual pigment absorption spectra of various animals. (a) The deep-sea fish *Avocettina infans* has only a single class of rods (λ_{max}: 482 nm). Such animals will see no color as, in this example, a dim 480-nm light would bleach as much pigment as a brighter one of 600 nm. (b) The bottlenose dolphin (*Tursiops truncatuse*), like many marine mammals, possesses rods (λ_{max}: 488 nm) and a single LWS cone type (λ_{max}: 524 nm). Intuitively, one might expect the cones of mammals that inhabit the 'blue' open ocean, to possess short-wave-sensitive cones. The presence of LWS cones might reflect the 'coastal evolution' of these species, where long wavelengths are more prevalent. (c) Some freshwater catfish, such as glass catfish (*Kryptopterus bicirrhis*), also have only rods

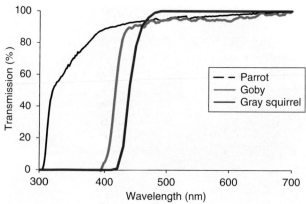

Figure 22 Spectral transmission of the lenses of a parrot (*Platycercus elegans*), a goby (*Gobiusculus flavescens*), and a gray squirrel (*Sciurus carolinensis*). The lens of the parrot is as transparent as a biological structure can be and transmits significant amounts of UV light, potentially allowing UV sensitivity. The colorless lens of the goby contains specific UV-absorbing pigments and prevents these wavelengths reaching the retina, and the lens of the squirrel cuts out enough blue light to have a yellow tinge.

fish and amphibia. However, many diurnal teleost fish, reptiles, and birds express visual pigments based on all four cone opsin genes and potentially have tetrachromatic color vision (**Figure 21(g)** and **21(h)**).

Intraocular Filters

Shortwave-absorbing filters in the cornea, lens, or retina serve primarily to enhance acuity and protect the retina from the most damaging light, but they will also decrease the perception of UV and blue (**Figure 22**). The filters of most significance to color vision, however, are oil droplets within the inner segments of photoreceptors (**Figure 23**). Although they are present in a few fish such as lungfish, some nonplacental mammals, and amphibia, they are particularly widespread in diurnal birds and reptiles in which they sharpen the spectral sensitivity of the photoreceptors, thereby improving wavelength discrimination.

Ultraviolet Sensitivity

Biologists have known for over 100 years that animals are able to perceive ultraviolet light (300–400 nm). However,

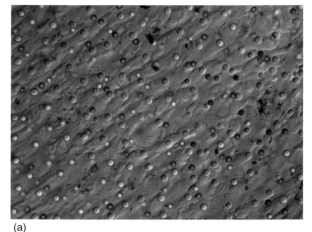

(a)

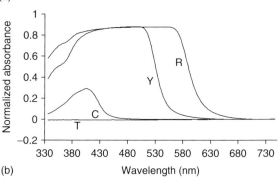

(b)

Figure 23 Bird oil droplets. (a) Wholemount of a kookaburra (*Dacelo novaeguineae*) retina showing colored oil droplets in photoreceptor inner segments. (b) Absorbance spectra of transparent (T), colorless (C), yellow (Y), and red (R) oil droplets in the blackbird (*Turdus merula*) retina, showing that they act as cut-off filters. Photo and data courtesy of Nathan Hart.

within the last 10–15 years, interest in such shortwave sensitivity has been rekindled following its rediscovery by behavioral scientists.

Many birds and fish, as well as some reptiles, amphibia, and mammals have visual pigments maximally sensitive in the UV (**Figure 21(e)**, **21(g)**, and **21(h)**), and its perception is often regarded as a 'secret' form of communication. However, in truth, so many animals see UV that it is only secret from a few.

Nonetheless, this perhaps misplaced enthusiasm for UV sensitivity has resulted in some useful demonstrations of its significance, which are lacking for other parts of the spectrum. Thus, controlled experiments have shown UV

(λ_{max}: 504 nm) and a single LWS cone type (λ_{max}: 607 nm) in their retina. The dissolved material in freshwater usually absorbs short wavelengths more readily than longer wavelength light, accounting for the brown appearance of much freshwater and explaining the presence of only LWS cone pigments in these species. (d) The elephant, *Elephas maximus*, has one spectral class of rods (λ_{max}: 496 nm) and is a cone dichromat (λ_{max}: 419 & 552 nm). It will therefore probably see the world in a similar way to a red/green color blind human. (e) The Boa constrictor is also a cone dichromat (λ_{max}: 357 & 549 nm) with high sensitivity to UV light. (f) Humans have one class of rod (λ_{max}: 496 nm) and trichromatic cone vision (λ_{max}: 420, 535, and 563 nm). (g) The blue tit, *Parus caeruleus*, has a single class of rods (λ_{max}: 503 nm) and four spectral types of cone (λ_{max}: 371, 448, 503, and 563 nm) (the absorption spectrum of the rod pigment is not visible as it is identical to that of one of the cones). (h) The goldfish, *Carassius auratus*, is also a cone tetrachromat (λ_{max}: 355, 452, 532, and 614 nm). All the visual pigments shown are rhodopsins, except those depicted in (c) and (h) which are porphyropsins.

either the PRR or F_0. As described earlier, the CNS of fish has direct control over the contraction rate of the vocal muscles and hence, PRR and F_0. This is readily shown by monitoring the activation pattern of the paired VPGs on either side of the CNS by placing electrodes on the surface of the two vocal nerves that excite the paired vocal muscles. Such neurophysiological recordings show that the two VPGs are activated at the same time (synchronously) in some species, but in alternation (asynchronously) in others. Thus, a PRR/F_0 of, for example, 100 Hz could be achieved by activating the two VPGs on each side of the midline of the CNS (all vocal nuclei are bilaterally symmetrical) simultaneously at a rate of 100 Hz. However, the VPGs could be activated out of phase with each other at 50 Hz, with their combined firing pattern producing a combined vocal muscle contraction rate of 100 Hz and hence a PRR/F_0 of 100 Hz. One predicted tradeoff in mechanisms would be in call amplitude. Synchronous activation of the paired vocal muscles should summate into a louder sound. However, by having an asynchronous system, an individual should be able to produce a more variable PRR/F_0 that might increase the degrees of freedom for this one parameter in terms of individual identification regarding, for example, age, sex, and body size.

The Hearing/Auditory Brain

As with vocal pathways in the CNS, comparative studies of all the major groups of vertebrates have delineated auditory nuclei in the brain that encode the temporal and the spectral properties of sounds in the firing pattern of action potentials. Many of these brain nuclei are common across all vertebrates, providing strong support for the hypothesis that both vocal and auditory systems of vertebrates are highly conserved.

The peripheral hearing organs used by fishes to detect sounds are homologous with those of tetrapods, and are appropriately referred to as 'ears.' Fish and tetrapod ears share a similar structure, both containing three semicircular canals lined with an epithelium of hair cells and three otolithic regions (see later) with dense beds of hair cells arranged along an epithelium in the saccule, the utricle and the lagena (in tetrapods, the lagena forms the cochlea). A fish's body is roughly the same density of water so that propagating pressure waves from a sound source will travel through the aquatic medium and then pass through the fish undetected. However, all actinopterygian fishes have three calcified bones (otoliths) associated with the saccule, utricle, and lagena that are of higher density than water and stimulate hair cells by a shearing motion in response to propagating sound waves. While fishes have a diverse array of otolith shapes and sizes, it is unclear what effect this variety may have on their hearing abilities.

Paralleling their diversity of vocal organs, fishes have evolved a variety of structures that have different densities than water to provide mechanical transduction mechanisms to translate the pressure wave into a detectable signal, thus broadening the perceivable range of acoustic pressure waves. In many cases, fish swimbladders serve as a pressure transducer; this low-density, air-filled structure vibrates in response to sound waves and increases the fish's frequency sensitivity. Some fishes have independently specialized their swimbladders with rostral projections (often referred to as 'bullae') that position the anterior portion of the swimbladder up against the skull and ears, and further increase the fish's hearing ability. Other fishes, the Ostariophysi (includes catfish and goldfish), employ specializations of the vertebral column (the Weberian apparatus) to mechanically couple the ear to the swimbladder. Fishes exhibiting these peripheral auditory adaptations are often referred to as 'hearing specialists,' and have dramatically higher hearing sensitivity at a wider range of frequencies, especially above 1 kHz, than fishes lacking such structures.

Hormones, Vocal Behavior, and Audio-Vocal Coupling

Many species of fish reproduce on a seasonal basis. The plainfin midshipman (*Porichthys notatus*) is one extensively studied species that migrates each spring from deep offshore sites along the northwest coast of the United States and Canada into the rocky, intertidal zone of bays and estuaries. Males excavate nests under rocky shelters that they defend from other males and from which they produce multiharmonic advertisement hums that often last for several minutes to more than one hour to attract females (**Figure 5(a)**). Males also produce briefer agonistic 'grunts' (**Figure 5(b)**) and 'growls' (**Figure 3**) in association with nest defense that differ dramatically from advertisement calls in their spectral and temporal properties. All the evidence to date indicates that females enter a nest for one night, deposit their eggs in a monolayer on the hard, interior surfaces of the cave-like nest, and then depart to return to offshore sites. The nesting male remains to guard newly fertilized eggs and sings on successive nights to attract more females.

As males and females begin to migrate into shallow waters for the breeding season, they exhibit increases in the blood levels of steroid hormones. Fish have the same major classes of steroid hormones as all other vertebrates, including androgens and estrogens. Neuroanatomical and neurophysiological studies show both long-term and short-term influences of steroids on the vocal and auditory systems. Long-term effects occur over a period of days and weeks and typically include structural-related changes such as increases in the size of neurons and the muscle fibers they excite (such as vocal motor neurons and vocal muscles). Short-term effects occur on the order

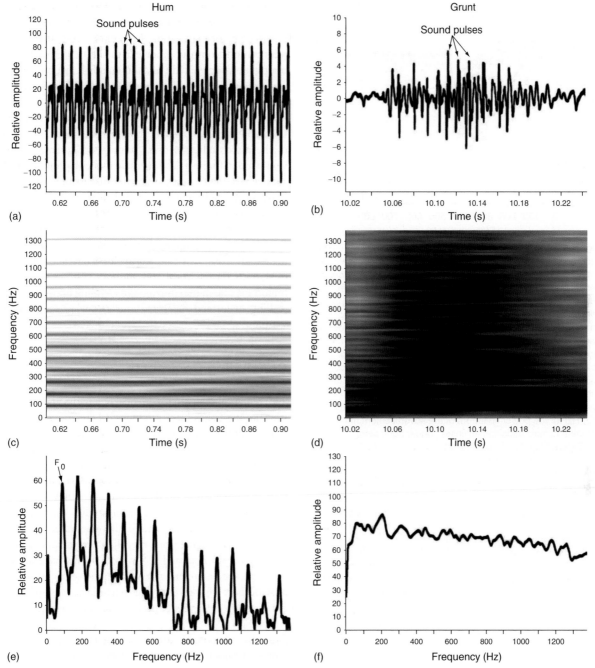

Figure 5 Different vocalizations produced by the plainfin midshipman, *Porichthys notatus*. Grunts are primarily used in agonistic encounters, while hums are used in courtship. Waveform of (a) hum and (b) grunt showing the different patterns of sound pulses. Spectrogram of (c) hum and (d) grunt. Power spectrum of hum (e; F_0 indicates the harmonic peak corresponding to the fundamental frequency) and grunt (f).

of minutes, are functionally referred to as neuromodulation, and include androgen- and estrogen-induced increases in the output of the VPG that directly translates into an increased duration in natural calls (see earlier section). The behavioral evidence in support of this hypothesis is as follows. Studies of toadfish, a close relative of the midshipman fish (same family), in their natural habitat

show that when nesting males are played a call from an underwater speaker that mimics the call of his neighbors, the male quickly (within 5 min) increases his call rate and call duration. The changes in vocal behavior are accompanied by similarly rapid increases in blood levels of the same androgenic steroid that increases the output of the VPG. Other behavioral studies show that the same

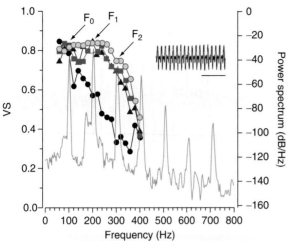

Figure 6 Vocal-auditory coupling in midshipman fish. Female midshipman fish show seasonal and steroid hormone-dependent changes in the encoding of pure tones by individual auditory nerve fibers (afferents) that innervate the saccule, the main auditory division of the inner ear of midshipman and many other teleost fish. Phase locking, one measure for characterizing the pattern of action potential firing by a neuron, best illustrates the frequency-encoding properties of saccular afferents. For the plots shown here, the Y-axis to the left indicates the vector strength of synchronization (VS), a measure of phase locking showing the responses of individual saccular afferents to tone stimuli (a VS value of 1.0 would indicate perfect synchronization). The Y-axis to the right shows a relative amplitude scale (decibels (dB)/Hz) for the power spectrum of the advertisement hum of a humming male midshipman (insert, upper right; also see **Figure 5(a)**). The X-axis indicates frequency (Hz) for both the afferent recordings and the hum's power spectrum (also see **Figure 5(e)**). Median VS values are plotted here for a population of saccular afferents recorded from nonreproductive females that are untreated (black circles), or treated with either an androgen (testosterone, blue triangles) or an estrogen (17β-estradiol, red squares). The highest VS values for nonreproductive females are close to the hum's fundamental frequency (F_0), while steroid-treated females also show robust encoding of the higher harmonics of the hum (F_1, F_2 indicate the second and the third harmonics, respectively). Steroids induce a VS profile like that found for wild-caught females in reproductive condition (yellow circles). Reproduced from Sisneros J, Forlano P, Deitcher D, and Bass AH (2004) Steroid-dependent auditory plasticity leads to adaptive coupling of sender and receiver. *Science* 305: 404–407.

androgens can also induce rapid changes in vocal behavior when nesting males are fed food containing the androgen. Together, these studies establish a causal relationship between hormonal modulation of the vocal CNS and vocal behavior.

Long-term steroid effects are well illustrated by the auditory system. Recall that the inner ear of fish has all the same divisions as that of humans and other mammals except that they lack a cochlea. Fish have adapted the saccule as the main auditory division of the inner ear (terrestrial vertebrates also have a saccule that is sensitive

to sound, but is most important for the sense of balance as is also likely the case for some species of fish). Neurophysiological recordings from the eighth nerve that carries information from the inner ear to the CNS in all vertebrates show that female midshipman fish exhibit seasonal changes in hearing (**Figure 6**). Compared to females in nonreproductive condition collected during the nonbreeding winter season, females in reproductive condition show an enhanced capacity to detect the upper harmonics of the male's advertisement hum (see **Figure 5(a)**) that would aid in finding a calling male in the shallow, intertidal zone where males nest (see earlier discussion about cutoff frequencies). Subsequent studies then showed that treating nonreproductive females for 3–4 weeks with implants that contain either testosterone or estrogen can reinduce the sensitivity to the upper harmonics that characterize a male's advertisement call (**Figure 6**). Together, the effects of steroids on the vocal and auditory systems show their important role in maximizing the coupling between sender (calling male) and receiver (female) during the breeding season.

Concluding Comments

The remarkable diversity of vocal behaviors and mechanisms among fishes briefly reviewed here provides but an initial glimpse of the many avenues for new opportunities for research at the interface of behavior and its underlying anatomical and physiological mechanisms. Given the conserved pattern of the organization of the vocal and auditory systems between fishes and all other groups of vertebrates, the discoveries from studies of fishes will reveal mechanisms relevant to all vertebrates, including humans. At the same time, revealing how diverse adaptations in vocal organ and CNS mechanisms have allowed different species to inhabit new audio-vocal niches will provide new insights into the dynamic process of evolution at multiple levels of biological organization.

See also: Acoustic Signals; Communication and Hormones; Hearing: Vertebrates; Mating Signals.

Further-Reading

Bass AH and Clark CW (2003) The physical acoustics of underwater sound communication. In: Simmons AM, Fay RR, and Popper AN (eds.) *Acoustic Communication*, pp. 15–64. New York: Springer.

Bass AH, Gilland E, and Baker R (2008) Evolutionary origins for social vocalization in a vertebrate hindbrain–spinal compartment. *Science* 321: 417–421.

Bass AH and Ladich F (2008) Vocal-acoustic communication: From behavior to neurons. In: Popper A, Fay R, and Webb J (eds.) *Fish Bioacoustics – Springer Handbook of Auditory Research*, pp. 253–278. New York: Springer.

Bass AH and McKibben JR (2003) Neural mechanisms and behaviors for acoustic communication in teleost fish. *Progress in Neurobiology* 69: 1–26.

Bass AH and Remage-Healey LH (2008) Central pattern generators for social vocalization: Androgen-dependent neurophysiological mechanisms. *Hormones and Behavior* 53: 659–672.

Bass AH, Rose GJ, and Pritz MB (2005) Auditory midbrain of fish, amphibians, and reptiles: Model systems for understanding auditory function. In: Winer JA and Schreiner CE (eds.) *The Inferior Colliculus*, pp. 459–492. New York: Springer.

Bradbury JW and Vehrencamp SL (1998) *Principles of Animal Communication*. Sunderland, MA: Sinauer Associates.

Fay RR and Popper AN (1998) *Comparative Hearing: Fish and Amphibians. Springer Handbook of Auditory Research*, vol. 11. New York: Springer.

Ladich F, Collin S, Moller P, and Kapoor BG (2006) *Communication in Fishes*, vol. 1. Enfield, NH: Science Publishers.

Nelson JS (2006) *Fishes of the World*, 4th edn. Hoboken, NJ: Wiley.

Remage-Healey LH, Nowacek DP, and Bass AH (2006) Dolphin foraging sounds suppress calling and elevate stress hormone levels in a prey species, the Gulf toadfish. *Journal of Experimental Biology* 209: 4444–4451.

Rice AN and Bass AH (2009) Novel vocal repertoire and paired swimbladders of the three-spined toadfish (*Batrachomoeus trispinosus*): Insights into the diversity of the Batrachoididae. *Journal of Experimental Biology* 212: 1377–1391.

Sisneros J, Forlano P, Deitcher D, and Bass AH (2004) Steroid-dependent auditory plasticity leads to adaptive coupling of sender and receiver. *Science* 305: 404–407.

Water and Salt Intake in Vertebrates: Endocrine and Behavioral Regulation

D. Daniels, University at Buffalo, State University of New York, Buffalo, NY, USA
J. Schulkin, Georgetown University, Washington, DC, USA; National Institute of Mental Health, Bethesda, MD, USA

Introduction

The single largest body component of vertebrates is water. For humans, the average adult male contains 42 l of water, representing approximately two-thirds of body mass. This varies slightly between individuals and more so among vertebrate species. Amphibians, for example, have relatively high levels of water (some measures indicating levels as high as 83%), but for the majority of vertebrates (e.g., birds, reptiles, and mammals), the level of body water approximates two-thirds of total mass. The water in the body is kept in separate intracellular and extracellular components, but this separation is not absolute and allows movement of water between the two compartments. In fact, the distribution of water between the intracellular and extracellular space is largely a function of osmosis, driven particularly by the concentration of solutes in either compartment. Irrespective of its distribution, at a very basic level, the amount of water in the body is driven by two factors: the amount excreted and the amount consumed. Regulation of these functions, conservation or excretion of water and water intake, is under the control of a diverse array of systems that coordinate to notify the body of the concentration and volume of the extracellular space. These systems include the heart, liver, kidney, and brain, each of which plays an important role in the coordination and maintenance of body fluid homeostasis. Coordinating these systems relies on both neural and endocrine responses.

Given the importance of solutes for maintaining proper distribution of water within the body, it is not surprising that numerous physiological processes coordinate to maintain proper solute levels, especially sodium. The cell membranes that keep the intracellular and extracellular fluid compartments separate allow for the movement of water from one compartment to the other, but these membranes are mostly impermeable to sodium and other solutes. The resultant chemical gradient generates osmotic pressure that determines the balance of water between the two compartments. Accordingly, maintaining body fluid homeostasis relies on proper levels of sodium. This regulation is achieved by the release of hormones that promote the excretion of sodium, hormones that promote the conservation of sodium, and, perhaps most importantly, by behavior. Indeed, when water and sodium levels are low, physiological conservation can help limit the loss, but restoration of proper levels of sodium and water can only occur through behavioral measures that include drinking and salt intake. The physiological processes that regulate water conservation and blood volume in the face of dehydration have been reviewed extensively elsewhere (see Daniels and Fluharty, 2009). This article focuses on the regulation of the critical behavioral aspects that underlie body fluid and sodium homeostasis.

Thirst and Sodium Appetite Are the Behavioral Regulators of Extracellular Fluid Volume

The response to perturbations of either the intracellular or extracellular compartment involves increased water intake. Perturbations of the intracellular compartment are generally the function of the concentration of the extracellular space. When the concentration of sodium in the extracellular space rises, as it might after eating a salty meal, the resultant change in osmotic pressure draws water from the intracellular to the extracellular space. This movement of water causes a reduction in the intracellular volume that is detected by specific cells called *osmoreceptors*. Although all cells will be affected by the movement of water in the same way, these osmoreceptors are specialized in their ability to respond to the changes, likely through specific membrane proteins that alter ion

Behavior as an Indicator of Welfare

It is generally argued that an animal's behavior is an indicator of welfare. It has also been suggested that if an animal cannot show all of its natural behaviors, it does not have good welfare. Not all animals show all their potential behaviors, and what about normal behaviors shown out of context or to an excess? Bar biting in sows is an example of a behavior that is argued to be abnormal and also a variation of normal. Pigs that are confined to small areas have little to do and some of them will start chewing on the metal bars of their crates or pens. Because pigs in pastures do not chew on metal objects, it has been argued that this is a stress induced behavior and, therefore, an indicator of poor welfare. The other side of the argument suggests that it is a variation of normal chewing behaviors. If a pig in pasture made 2000 chewing motions a day while eating, and a confined pig made 2000 chewing motions, some of which were directed toward metal because there was less volume, but higher quality food, the welfare of the pig is not suffering.

In other situations, the expression of a behavior has an obvious correlation with general welfare. Buller steers in feed lots not only injure other steers that they mount, but the behavior has been associated with the inability to adapt to the crowded, intense social aggregation of cattle. When given a little more space and put in pens with fewer cattle, the behavior is generally self-limiting. Elephants in circuses or working in the forests of Indonesia that suddenly go on a rampage are showing an extreme behavior not seen in the wild. Not only is this behavior dangerous for nearby humans, it is usually deadly for the animal. Stereotypic behaviors are described as expressions of emotional distress. There are many things that can lead to these behaviors, as will be discussed in a later article, but studies have shown that their expression is associated with the production of brain endorphins. These, in turn, are associated with pleasurable feelings. This implies that the development of a stereotypy helps reduce mental stress levels. It also suggests that the simple physical blocking of the behavior, without addressing the causes of the stress, would not be good for the animal's well being.

Behavior Is a Poor Indicator of Welfare

Studies have been designed to teach an animal how to turn on lights, or to change the volume of music, or to make food choices to find what the animal considers to be ideal. These are preference tests and preferences may or may not be related to welfare. Does a horse that has access to crimped oats but not to sweet feed have poor welfare, when testing indicates a strong preference for the latter? Most would agree that welfare needs are being met with a quality diet, even though the top preferences are not included. Some would also argue that sweet feed might

lead to medical problems in some horses that would be bad for overall welfare. Even in the wild, preferences do not always equate with welfare. A bear may be willing to risk multiple bee stings just so that it can raid a honey stash.

While fear of situations is a normal response that results in a situational reaction, that reaction can be excessive and become life threatening in itself. Natural selection has favored horses that shy and run from objects not immediately deemed as safe; such caution is beneficial in the wild. For the suburban horse, shying away from a waving piece of plastic could result in the horse and rider being hit by a car or the horse running into a barbed wire fence resulting in severe cuts.

What about chronic, low stress levels, such as isolation from social peers or poor quality forage? While behaviors may not change significantly, the long-term effects of raised cortisol levels could negatively impact health in several ways. For asocial and antisocial species, the expression of certain behaviors can make individuals more vulnerable to predators. Some animals do not show changes in behavior until they are very ill, or they hide so that behavior changes are not apparent. Again, in nature, exhibiting weakness is not usually favored by natural selection.

Euthanasia

Death is a terminal point for the animal, but how death occurs does impact the animal's welfare if we equate humane life and humane death with good welfare. In the wild, death is often violent, as when a predator catches and eats its prey, but it can be slow and painful, as when an animal is trapped in quicksand. For species that interact with humans, another type of death is by euthanasia. By definition, euthanasia comes from Greek words that mean 'good death.' The concepts here involve techniques that result in the rapid loss of consciousness, which is followed by cessation of brain, heart, and lung function. Anxiety and distress are kept to an absolute minimum. Species variability requires consideration of what agents/techniques are appropriate. Situational and environmental differences also need to be considered. As an example, the use of an overdose of a barbiturate anesthesia would be the ideal way to euthanize a deer that had broken its legs after being hit by a car, but doing so would endanger the person administering the drug, add additional panic to the animal, and render the carcass dangerous to animals that eat carrion. A gunshot to the head would be faster, generally safer, and not damage the meat; however, from the public's perspective, it might be esthetically less acceptable.

Other articles in this section discuss the interconnections of behavior and welfare in animals of various species in various settings. The authors are experts in their respective areas who point out the positives and negatives of relying on behavior to assess the welfare of animals.

Because welfare can be measured scientifically, there are ongoing studies around the world to better understand the complex interactions of behavior and welfare. Knowledge is continuously being generated. This field remains fluid and those who are interested in animal welfare are encouraged to continue to evaluate new findings for a comprehensive understanding.

See also: Avoidance of Parasites.

Further Reading

Beaver BV (2003) *Feline Behavior: A Guide for Veterinarians,* 2nd edn. St. Louis, MO: Saunders.

Beaver BV (2009) *Canine Behavior: Insights and Answers,* 2nd edn. St. Louis, MO: Saunders.

Hart BL (1988) Biological basis of the behavior of sick animals. *Neuroscience & Biobehavioral Reviews* 12: 123–137.

Hart BL, Hart LA, and Bain MJ (2006) *Canine and Feline Behavior Therapy,* 2nd edn. Ames, IA: Blackwell Publishing.

Figure 4   Red squirrel, *Tamiasciurus hudsonicus*, with midden of cones cached as a food supply for the winter.www.washington.edu/burkemuseum/collections/mammalogy/mamwash/rodentia.php#. www.nps.gov/features/yell/slidefile/mammals/redsquirrel/page.htm.

Figure 5   Yellow subcutaneous fat stored in the furculum of a willow warbler, *Phylloscopus trochilus*. Photo by B. SIlverin.

shrubs, and dense herbaceous vegetation. These microhabitats are havens for several ground-feeding species.

Winter at Subtropical and Tropical Latitudes

Subtropical and tropical animals do not experience cold periods or snow cover but food shortages may occur especially during drought periods (**Figure 8**). This may be exacerbated by concentrations of animals in smaller and smaller areas (e.g., with shrinking lakes and rivers) and increased exposure to predators, greater transmission of diseases, and competition for dwindling food resources. Whether animals experience a cold winter or a drought, the critical factor is food shortage, and thus competition.

Figure 6   Five striped palm squirrels, *Funambulus pennantii*, huddle for warmth at times of day when ambient temperature is low. Ranthambore National Park, India, January 2000. Photo by J.C. Wingfield.

Hormonal Regulation of Wintering (Nonbreeding) Strategies

Almost all of the endocrine studies of hormone–behavior interactions in the winter or nonbreeding life history stages have focused on strategies 1 and 2. Field investigations of strategies 3 and 4 are intractable owing to the large distances covered by individuals but as tracking devices become miniaturized, it may be possible to conduct experimental studies in the future. Hormonal

Figure 7 Qamaniq; a microhabitat important for small ground feeding mammals and birds in snow covered regions (Pruit, 1970). Snowfall tends to accumulate in open fields and open deciduous forest (a and b) but along the edges of shrubs (a), under dead vegetation (c) and under coniferous trees (d) snow depth is greatly reduced or even absent allowing access to food. In (e), shallow snow in qamaniq allowed a sparrow to dig up some seeds (indicated by an arrow). All photos by J.C. Wingfield.

regulation of strategies 5 and 6 are covered in the articles on spatial memory and hibernation.

In winter, animals of many species gather in groups where they compete for resources such as food and future mating partners. These groups may hold winter territories or home ranges within which there is often a strict social hierarchy, and social challenges will affect the activity of different endocrine systems that affect the individual's chances of surviving the winter to breed the coming spring. Many other factors such as predator density may also affect these groupings that in turn influence hormonal control of physiological and behavioral responses.

Predators

Groups of elk, *Cervus elaphus*, alter their grouping patterns, vigilance, foraging behavior, habitat selection, and diet when predators such as wolves, *Canis lupus*, are present. This effect occurs throughout the year, breeding and nonbreeding. In the long term, the presence of predators results in reduced progesterone levels, breeding, and ultimately population size. However, glucocorticoid levels

(cortisol or corticosterone – adrenal cortical hormones, also called stress hormones, involved in mobilization of energy stores and many survival behaviors) were not related to predator density, suggesting that effects of wolves on population size, breeding, and nonbreeding, were mediated through changed foraging patterns and resulting nutritional costs.

In showshoe hares, *Lepus americanus*, of the boreal forest of Canada, predator density (e.g., of lynx, *Felis pardus*) results in changed foraging patterns and chronic stress, as indicated by elevated cortisol levels and decreased corticosteroid-binding globulin, which in the long-term results in reduced reproductive function the following spring. Population decline then occurs as a result of predator stress and not decreased food availability per se.

Social Hierarchies

To increase winter survival and to gain priority to breeding resources in spring, it is important to maintain, or improve, position in a social hierarchy in a winter group. This is, however, associated with costs because of

Figure 1   Arctic wolf on Ellesmere Island.

popular literature persist, as remnants of earlier times. The subsequent review subsection is aimed at encouraging critical thinking about what evidence is needed to test hypotheses about wolf behavior and how difficult it has been to obtain.

Review: History of Canid Behavioral Research

Visual signals used by dogs in communication had been noted in the behavioral literature long before wolves were studied either in captivity or the field. Charles Darwin illustrated his hypothesis about the principle of antithesis by contrasting images of the upright posture of an alarmed dog in response to a person approaching in the distance, compared to the crouching posture that the dog switched to as soon as it recognized the person as its 'master.' In terms used in the nineteenth century, this expression of emotion signaled an unambiguous change in motivation, from dominance to submission. In modern terms, the change in visual signal conveyed the information that the dog was unlikely to escalate attack in response to a familiar care-giving companion. Unfortunately, the misperception that some individuals are always dominant and others are always subordinate has persisted despite the original context of the drawings that Darwin used; his drawings illustrate how one individual can rapidly change signals as it gathers more information about a stimulus (e.g., cues about familiarity).

In the popular literature, the anthropomorphic myth persists that a dominant male is needed to enforce order

so that all wolves in a pack know their roles in a dominance hierarchy. For example, Douglas Pimlott quoted Niko Tinbergen's interpretation of a strict hierarchy in the sled dogs he observed in Greenland. At the time, it seemed reasonable to infer that those dogs that look more like wolves would behave like wolves. No published evidence about wolf behavior was available prior to the 1940s, so the hypothesis remained untested for decades. The popular notion of born losers and winners was reinforced by the insightful anecdotes about personal experiences with dogs and wolves published by Konrad Lorenz. However, Lorenz also noted both the persistent personality traits that varied across breeds of dogs, and the extreme changes in one individual deprived of his primary social companion.

Two seminal publications introduced a different interpretation of social interactions in wolf packs, both emphasizing the family structure. In Alaska, Adolph Murie observed a wolf family caring for pups near a den, now in Denali National Park. In a Swiss zoo, Rudolph Schenkel described in more detail how the food begging behaviors of pups developed into solicitous appeasement signals in juveniles interacting with both parents. He interpreted these interactions as the social glue that holds the wolf family together. Both studies emphasized the influence of age on the dynamics of dominance interactions, as both parents and older siblings cared for pups.

Is wolf pack structure more like a pecking order or like a caring family? Understanding the ancestral roots of dog behavior was a compelling justification for the multiple postwar studies that emerged in the 1950s and continued into the 1980s. An American team led by John P. Scott investigated the development and heritability of behavior in dogs, in the context of comparative studies of wolves. In Europe, Erik Zimen examined ontogeny of behavior in wolf/dog hybrids, inquiring in what ways arrested development in dogs might illustrate the principle of neoteny, the persistence of juvenile characteristics into adulthood. Benson Ginsburg's research group examined questions associated with the hypothesis that evolution of social cognition in wolves would have been accelerated if they had been isolated in groups that benefited from helpers caring for young. Among others, Mike Fox and Mark Bekoff studied behavioral development in litters raised without parents, comparing species considered to represent a continuum of solitary foxes, semisocial coyotes, and social wolves.

As evidence from more packs emerged, the variation in group structure became clear (**Figure 2**). Behavior in some groups fit the model of a pecking order and others were more like a caring family, as we will examine in more detail in a later section. Separate research teams examined ontogeny of behavior in long-term studies of hand-reared wolves assembled to form reproductive groups. These lines of inquiry were led by John Fentress, John Rabb, Erik Klinghammer,

Figure 2 The direct stare of the father is enough to interrupt courtship by his adult son in a captive family group.

Erik Zimen, R. Derix, Dave Mech, and Ulysses Seal. Evidence was clear that not all adult-sized wolves reproduce in packs. However, integrated studies of behavior and physiology led to rejection of the hypothesis that nonbreeders were always physiologically stressed because of the behavior of dominants. This left open the question why packs contain adult-sized wolves that do not breed.

Do wolves need to cooperate to kill large prey? On a parallel track, the landmark study of wolves on Isle Royale by Dave Mech opened several lines of research about wolves as predators, including interaction with the dynamics of foodwebs. In the early 1960s, the ecosystem on Isle Royale was relatively stable, and the evidence supported the hypothesis that individual wolves needed to put aside their own interests in reproduction for the sake of group hunting and to avoid wiping out their food supply. The notion that wolves had to cooperate to be successful at hunting large dangerous prey like moose fit nicely with this model, but only for a few years. Over the subsequent decades, Rolf Peterson has monitored dynamic peaks and lows in wolf, moose, beaver, and other carnivore populations on Isle Royale. There have been years when the moose population crashed because of overbrowsing and years when wolves supplemented their diet with beaver and snowshoe hares, then crashed when availability of all prey species was very low. Winter severity and fire ecology have added to the complexity of these ecosystem dynamics.

As we will examine, evidence from additional field studies led to rejection of the hypothesis that wolves are obligated to cooperate in hunting. Under some foodweb conditions, wolves coordinate hunting and pup-rearing activities, but in other conditions they do not. Hunting large prey permits large group size, but does not require it.

Such diverse and dynamic conditions fit evolutionary models that predict that behavioral plasticity would be at a genetic premium for this large-bodied social carnivore that is widely distributed across all biomes of the northern

hemisphere. In addition, the genetic diversity of wolves is illustrated by multiple subspecies isolated in fragmented habitats; one subspecies even lives in ambivalent symbiosis with humans, that is, domestic dogs.

Modern Synthesis: Nested Hierarchical Systems

The social structure of wolves has been analyzed in terms of three levels of selection, each nested within the other: family groups, subpopulations, and ecosystems. Viewed from a systems perspective, at the first level, individual decisions affect survival and reproduction of other wolves within the same family group (wolf pack). At the second level of analysis, the social environment, each family group is influenced by the actions of other family groups within the neighborhood. The groups of individuals that interact most frequently within a neighborhood are technically called a deme, or subpopulation, within a fragment of habitat relatively separated from other fragments. At the third level of analysis, the physical ecosystem, wolf populations are influenced by biotic (e.g., prey, competing species, diseases) and abiotic factors (e.g., winter severity, fire cycles, drought cycles). From a theoretical perspective, each of these systems is viewed as nested, because individuals fit within groups, subpopulations, and ecosystems.

This theoretical framework of nested biological systems becomes important when scientists apply sociobiological models to test hypotheses about wolf social behavior. For example, several general hypotheses about evolution of cooperative breeding have been proposed: (1) *eusocial*: obligate reproductive suppression under extremely harsh conditions in ecosystems where individuals do not survive outside a breeding colony; (2) *conditional suppression*: reproductive behavior is reversibly turned on and off in adults, depending on the social environment (groups and neighborhoods); and (3) *deferred reproduction*: the average onset of first reproduction is delayed by the interaction of social and ecosystem factors (e.g., body size, nutritional condition, olfactory signals, competition for mates). In the following sections, specific evidence from wolf behavior will be synthesized in a manner needed to test each of these three general models.

Individuals Within Family Groups

Food provisioning within a family group is key to understanding the social environment of canids. Although individual wolves may leave their natal group and spend varying amounts of time alone during the transition to another group, all wolves are born into and develop within a family group. Wolf litter size may vary from 1 to 10, usually 5–6 pups. Group size rarely exceeds 15 wolves, depending on whether a breeding pair is disrupted, how many females breed, and how long offspring of varying ages remain in the family group. Although the highest reported count was 42, the usual group size is 6–8 wolves, consisting of a breeding pair with 4–6 offspring (**Figure 3**). However, this varies with the history of each family group as well as environmental changes in the wolf population and ecosystem, for example, prey type.

Unlike most polygynous mammals, pup care by more than the mother matters in monogamous wolves. Pups are unlikely to survive to puberty without the care of parents and/or older siblings. In a study of 148 pooled cases of territorial breeding wolves, at least one pup survived the loss of a parent in 84% of the cases, presumably because pups were cared for by other members of the family group. Pups were more likely to survive the loss of a parent in large groups with auxiliary nonbreeders than in small groups. Older pups were more likely to survive the loss of a parent than younger pups.

Born in an earthen den or shallow scrape on the surface, altricial wolf pups are cared for exclusively by the mother for 2–3 weeks. Their eyes are closed until 12–14 days, and their first reflexive topo-taxic responses are to touch, warmth, smell, and taste. Hearing matures more slowly. Urination by pups in response to the mother licking the urogenital region is an example of a parent–offspring signal. As pups develop coordination to stand and walk, early learning begins to expand from the social context of littermates and soliciting care from the mother to include interactions with physical objects.

Between 3 and 5 weeks, pups explore the entrance of the den, retreating from unconditioned stimuli that elicit an alarm bark and approaching the soft squeaking vocalizations and multimodal stimuli they have learned to associate with the nursing female. In rare circumstances, more than one nursing female may share a den. However, currently there is no evidence that pseudopregnant female wolves

Figure 3   Sibling wolf pups in a family group on Ellesmere Island.

initiate nursing without having previously given birth to a litter, despite speculation in the popular literature.

The social context of learning expands during weeks 5–8, as pups encounter family members that deliver food by regurgitation and carrying pieces of carcass. For example, in a pack on Ellesmere Island, regurgitations were directed to the pups (81%), the nursing female (14%), and other auxiliaries (6%). All adult wolves regurgitated food, including the breeding pair, yearlings, and a post-reproductive female. The breeding female and pups received most regurgitations from the breeding male. Regurgitations by the breeding female were directed exclusively to pups. Wolves respond by regurgitation to muzzle licking by another familiar wolf, a multimodal signal that changes meaning with age, social context, and the presence/absence of food.

During the transition stage from dependency on milk to solid food (6–10 weeks), not only do pups learn to recognize familiar kin, but they are also rewarded by food when they approach or follow adults. Detailed sequence analysis of interactions, in both pups and adults, have illustrated that individuals learn the physical and social consequences of their actions. At this transition stage, bouts of chase and wrestling play are typically 1–3 h between naps and feedings.

At 7–8 weeks, bite strength is sufficient for pups to feed from opened small carcasses, such as arctic hares in the Ellesmere Pack. In the weaning process for one litter, frequency of suckling bouts that occurred outside the den decreased gradually, as the nurser initiated bouts at longer intervals interrupted more bouts and pups persisted less when interrupted. Parent–offspring conflict was not obligate, although it may be conditional on food delivery and food storage by caching. On the average, by 11 weeks, pups no longer suckled and began to follow adults on foraging trips. Activity centers focused on dens and rendezvous sites may change several times, as a litter is carried to a new location by the breeding female or moves in response to disturbance.

Sound analysis indicates the vocal repertoire increases from four to nine call types as pups mature. Barks and howls are examples of vocal signals that have been studied in wolves. Pups bark in response to alarming stimuli and howl when separated from the group or in response to other howls. One hypothesis of the adaptive function of these signals is safety in numbers. Pups are vulnerable to predation by bears and unfamiliar wolves from neighboring packs.

The first agonistic signals used by wolf pups occur in the context of food. When conflict escalates over a large food item, such as a rabbit carcass, pups learn the consequences of uninhibited bites from a sibling. They learn the subtlety of signs (e.g., hard stare, snarl, ear posture, partial lunge) that predict likely escalation to uninhibited biting. Subtle signs of de-escalation include: look-away, lie-down, ears-back, lip-licking, crouch, roll-over, pawing, tuck-tail, and tail-wagging. To the extent that individuals vary in temperament at birth, each learns coping styles influenced by the contingencies of their interactions with siblings and the context of resources, although this complex process is not completely understood because it is so variable.

Current behavioral evidence does not support the hypothesis that the dominance hierarchy within a litter determines which individuals breed later in life. Behavioral profiles of individuals vary on several dimensions in addition to a shyness/boldness continuum. Multivariate quantitative studies determined that affiliate and play behaviors explain more variation in wolf behavior than agonistic actions within intact family groups. Disrupted families that have lost one or more parents are more variable and conflict is more likely to escalate as described in the next section.

Born in late spring, juvenile wolves are not quite adult-sized by their first winter. The synchronized birth season fits the functional hypothesis that the young are born at a time when food is readily available. Those that were born later would have been unlikely to survive the rigors of their first winter. Neonates born in winter would have risked exposure and malnutrition in times of scarce food. This genetic basis for seasonal reproduction has been modified in domestic dogs, which breed year-round. On average, birth dates occur weeks earlier in wolf populations at lower compared to higher latitudes, although the mechanism is still not entirely understood.

Group size expands seasonally, as pups are born, and declines as family members disperse or die. For example, on average in Denali, only half the pups of the year remained with the family through the first winter. Of those that remained, only half were still with the family through the second winter. Only a few wolves remained with their natal group past the third winter.

Despite the popular notion that young wolves are driven out of the pack by conflict with parents and siblings, the data suggest that dispersal mechanisms are a complex interaction of individual maturation, relationships within the group, food availability, and scent marks in the neighborhood. During their lifetimes, individuals may switch among the following categories of tactics: (1) 'biding' auxiliaries are nonreproductive members of a territorial group; (2) 'dispersing' floaters leave the group and wander alone or in transient groups that pass through or between group territories; and (3) 'breeding' parents defend the territory where they forage and reproduce, attacking outsiders of the same sex. Both sexes switch among these tactics.

This evidence of developmental plasticity has led to rejection of the hypothesis that the wolf social system fits the model of eusociality. Dispersing wolves do not cooperate in parental care and do successfully catch prey without the help of others. Transitions among behavioral tactics will be discussed in more detail in the next sections, because they are influenced by factors at both the population and foodweb levels of ecosystems.

Comparing canid species, large body size is correlated with later age of first reproduction. For example, on average, large-bodied wolves reach puberty in their second winter, 1 year later than smaller bodied coyotes. Puberty in wolves may be accelerated or delayed by a couple of years because of interactions of nutritional and social factors.

There is no evidence to support the hypothesis that social stress in wolves turns off the physiological readiness to breed after puberty. For example, fecal cortisol was higher for breeders than nonbreeders in samples from free-ranging wolves in Iberia. In one captive study, nonbreeding adults cycled normally; females ovulated and males produced sperm. In another captive pack, the positive correlation between stress hormones and aggression was skewed by one individual with abnormal adrenal hypertrophy. An early-winter peak in testosterone has been correlated with rates of scent marking and escalated conflict among males.

Deferred reproduction best explains the variation in reproductive tactics of wolves in family groups. In nuclear families, food provisioning shapes asymmetric relations between parents and offspring. During breeding season, parents are more attracted to mating signals from each other than from offspring (**Figure 4**). Adolescent wolves are less attracted to mating signals from siblings than parents, an attraction likely not reciprocated. Older wolves, both parents and siblings, are likely to interrupt sexual activity by younger wolves of the same sex. However, the subtle signs of asymmetric mate choice and same-sex rivalry are only a matter of probability and may shift within weeks when one or both breeders are removed.

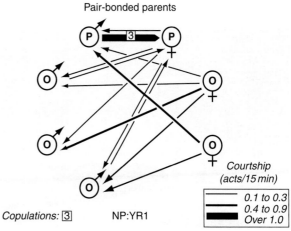

Figure 4 Relations among parents and their offspring during the breeding season in a captive pack.

Wolf mating signals are multimodal, including phero-mones and tactile, auditory, and visual cues. Sexual dimorphism is minor, on average males are 20% heavier; however, no visual signals distinguish the sexes. Howling is an auditory signal with the function of advertising presence of wolves; loners are more likely to howl when traveling in sparsely than densely populated regions. Pheromones deposited in urine and feces are the primary mode for advertising sexual identity in dispersers. Dispersers join up at all times of the year; males are more likely to find single females during proestrus vaginal bleeding, usually 2–8 weeks before ovulation. Plasma estrogen is elevated in females during proestrus and estrus, which occurs during January and February. Synchronized estrus usually peaks mid-February, although the exact timing varies with body condition, the social environment, and latitude.

Proximate mechanisms of reproduction have been well studied in both wolves and dogs. During 1–7 days of estrous, each breeding pair of wolves remains in close proximity (**Figure 5**), exchanging mutually stimulating olfactory, visual, and tactile signs of copulatory readiness, for example, sniffing, sequential-urinating, chinning, darting, ears-together, head-flick, and paws-to-shoulders. Within days of spontaneous ovulation, females signal readiness to stand by averting the tail to the side of the vulva, a reflex that is a fixed action. Ovulation is associated with a peak in plasma luteinizing hormone (LH), which coincides with a drop in elevated estrogen (**Figure 5**).

With experience, male wolves quickly learn to orient mounting to the rear. The ejaculatory reflex follows penetration and penile thrusting. Subsequently, tissues swell in the penile bulb, a reflex keeping the pair locked in a postcopulatory tie that usually lasts 20 min; the range in duration (3–30 min) is conditional on interruptions by familiar rivals. Hypotheses about the function of the postcopulatory tie include (1) oxytocin release that stimulates smooth muscle contraction increasing internal fertilization rate due to sperm and egg movement into the uterine horns and (2) reduced probability of extra-pair copulation during the male postejaculation refractory period.

The seasonal canid reproductive cycle is unusual among mammals because (1) there is only one estrus whether or not a female becomes pregnant, (2) the post-ovulatory growth of the corpus luteum is roughly the same duration for both pregnant and nonpregnant females, and (3) because of elevated prolactin in both males and females during the spring pup-rearing season. The cascade of hormonal changes following ovulation may stimulate growth of nipples, hair loss from the belly, abdominal swelling, denning behavior, and milk production, although these symptoms are highly variable among individuals and change with age. Seasonal peaks in prolactin are associated with den-digging and food-provisioning by both sexes (breeders and nonbreeders). Thus, it is difficult to diagnose pregnancy on the basis of external indicators; more accurate internal indicators may be obtained by sonography and measuring the hormone relaxin.

Although social monogamy is typical of smaller nuclear families of wolves, extended and disrupted families may include polygynous and polyandrous relationships. Since postpubertal wolves retain the physiological readiness to breed, loss of one or more breeding parents may destabilize dominance relationships. For example following deaths of the fathers in two Yellowstone packs, immigration of an unrelated breeding male was followed by multiple litters. A low frequency of plural breeding has been recorded in several field studies. Congenial relations among multiple breeding females are usually unstable and persist no more than a few years.

Agonistic interactions vary with both the immediate presence of resources and the social environment within each group of wolves. Resources include food, mates, and pups. Factors likely associated with escalated conflict are complex, including the quality of the resource, proximity to the resource, motivation (e.g., satiation, reproductive cycle, adrenal activity), personality (e.g., inherited temperament and learned coping styles), and relationships (e.g., learned contingencies of interactions among specific individuals). All these factors influence the complexity of dominance hierarchies (e.g., linear, triadic, age-graded, branched sex-specific, multinodal), which may change within a group over time as well as varying among groups depending on age/sex composition.

The variation in age/sex composition of wolf groups depends not only on internal factors, but also on the interactions among groups within populations and dynamic patterns of food availability within ecosystems. Interactions of internal and external factors are elaborated in the following section.

Figure 5 A male stands near his resting mate, guarding her from rivals while she is in estrous.

Family Groups Within Fragmented Subpopulations

The social structure of wolf populations includes territorial reproductive groups and floaters that move between and through resident territories. In colonizing populations with low territory density, floaters are more likely to join up and start new breeding units in the gaps between territories. In established populations, the number of breeding groups is relatively constant despite turnover as groups break up and new groups are formed. Wolves that are slow to disperse from their natal family are 'biders' (nonbreeding auxiliaries) waiting for a chance to breed when the opportunity arises.

Within wolf populations ranging from those that are recolonizing an area (Yellowstone) to well-established (Denali, northern Minnesota), genetic relatedness is lower between breeders within a group than it is on the average between breeding groups. Wolf groups are semiclosed, usually accepting immigrants only upon loss of a breeder. Mate turnover ranges from 1 to 6 years, varying among populations. As elaborated in the following section, hypothesized mechanisms to explain this pattern of nonrandom genetic dispersal include the following ones: (1) individuals choose mates that are distantly related over those that are close relatives, (2) breeders defend their mates from same-sex rivals in neighboring groups, and (3) dispersal between groups is influenced by mate choice and same-sex rivalry. Given a choice of mates, wolves of both sexes are predicted to be more attracted to unrelated than to related individuals. Although it is unlikely that wolves have a mechanism for directly detecting genetic relatedness, familiarity is highly correlated. Given a choice, unfamiliar individuals are more attractive mates than family members. Contrary to prediction, inbreeding has occurred when a parent had no other choice than offspring and when siblings copulated in the absence of parents.

Same-sex combat may explain the intense, uninhibited conflict between wolf groups, resulting in documented death of breeders, biders, and floaters. Fights between groups have escalated during extra-territorial intrusions. Although breeders may be more likely to escalate conflict with same-sex rivals, in the excitement of a fight, all group members may mob a victim that displays defensive signals, for example, tucked-tail, ears-flat-back, and arched-back. While the sample size is not definitive, small groups are less likely to escalate than large groups. Social monogamy is reinforced and extra-pair copulations are reduced by same-sex combat between groups.

Several categories of dispersal between groups have been documented: (1) biders immigrate from an unrelated neighboring group to one that has lost a same-sex breeder, (2) one group divides into neighboring groups, (3) dispersers may travel distances as long as 1000 km, (4) dispersers meet up and establish a new breeding group within 100 km of their natal group, (5) dispersers return to a familiar group (siblings or offspring) after turnover in the breeding pair, and (6) dispersers immigrate into groups that have lost a same-sex breeder. Overall, genetic variation is likely to be lower between groups than between breeders within each group.

Auditory and olfactory communication influence distance between groups; in contrast to the momentary and ambiguous information conveyed in howls, scent marks may last for days. In response to playbacks of strange-group howls, groups that reply are more likely to remain in place compared to groups that do not reply and retreat. Response rate is positively correlated with group size, breeding condition, and presence of a resource. Single wolves are more likely to approach silently when the playback is a solo howl. The prevalence of scent marks, both urine and feces, on trails near junctions and at the edge of territories has been described as an 'olfactory bowl.' Breeders urine-mark on conspicuous objects at a higher rate than nonbreeders, and the urination rate is highest in newly formed pairs. Pairs deposit urine marks sequentially in the same location, a double-marking behavior that may function in intimidating rivals and stimulating mates.

Group howls occur when resting wolves arise and gather together prior to traveling, as well as when they come together after separation. During a group howl, individuals rub bodies, touch noses, and circle with wagging tails. Individuals that hold the tail high are more likely to respond with an over-the-muzzle bite to nose-licking by wolves with a lower tail posture. Similar to pups soliciting food provisioning, adults that receive an over-the-muzzle bite do not retreat from the group. Whichever individual departs with a confident gait is likely to be followed by those that are more solicitous. However, if a key food provider does not join the departing group, group cohesion may deteriorate.

Group decisions on movements vary between wolf packs, as well as seasonally within each group. Alone, adult wolves can easily travel 40 km in half a day. In general, movements revolve around pups in the spring/summer and the breeding female in the winter. In other seasons, the individual leading a traveling line of wolves is likely to be a breeder. Evidence supporting the hypothesis that the male is more likely to be a leader is ambiguous and depends on the definition of leading behavior. An alternative hypothesis is that variation in leading behavior may relate to which individual is most consistently associated with food acquisition, which likely changes with the age, experience, and personality of group members.

Within a given latitude, group home range size is positively correlated with group size in colonizing wolf populations but only marginally so under saturated conditions. For example, in northwestern Minnesota and Yellowstone, recolonizing groups initially were spaced far enough apart that there were no shared boundaries. As the open areas filled in, groups defended adjacent boundaries and potential for elastic expansion of home range was limited, presumably by encounters with sign left by neighboring packs.

Genetic variability between wolf populations is likely related to the connectedness of habitat patches. In some homogeneous stands of forest, wolves may travel hundreds of kilometers without encountering areas of low wolf density. In other landscapes, they travel hundreds of kilometers through areas without wolves before encountering the sign of wolf presence. From the same litter, some individuals have dispersed short distances and others long distances. Overall, both sexes are equally likely to disperse, although dispersal in some populations has been biased toward males and others were biased toward females.

In summary, an overall model includes neighborhoods of relatively low genetic variation which are nested within habitat fragments that vary in degree of connectivity. Sexual competition limits the openness of family groups to immigration of unrelated individuals; however, disruption of monogamous relationships facilitates movement of individuals between groups. Mate choice tends to favor outbreeding, although inbreeding occurs when choices are limited. Gene flow occurs via dispersal between fragmented habitats.

Populations Within Fluctuating Ecosystems

Sociobiological theory predicts that species adapted to fluctuations of their social and ecological environments will evolve behavioral traits with a high degree of plasticity. Variation in the distribution of resources is most likely to influence the distribution of reproductive females, including group size. In turn, the distribution of females likely influences male tactics for defending females and offspring from the risks of encounters with rival males. Secondarily, predation separately influences evolution of behavioral traits in males and females.

Ecosystems inhabited by wolves range from Arctic tundra (80 °N latitude) to desert mountains (less than 40 °N latitude). Foodwebs within these diverse ecosystems vary from simple to complex. Examples of simple food-webs include (1) blackbuck (India), (2) arctic hares and musk oxen (Ellesmere Island), (3) migratory caribou supplemented by small mammals during denning (Alaskan Brooks Range), (4) white-tailed deer and snowshoe hare (Minnesota), (5) moose and white-tailed deer (eastern Canada), and (6) red deer and wild boar (Spain and Poland). More complex food webs include (1) moose, caribou, and Dall sheep (Denali National Park); (2) moose, snowshoe hare, and beaver (Isle Royale); (3) elk, mule deer, bison, mountain sheep, caribou, mountain goat, and small mammals (western Canada); and (4) red deer, wild boar, roe deer, fallow deer, and mouflon (Appenine mountains of Italy).

Where wild ungulate populations have died out in parts of Israel and Italy, wolves scavenge at garbage dumps in addition to hunting whatever small animals and livestock are vulnerable. Domestic animals (e.g., goats, sheep, pigs, cattle, and dogs) are primary prey in northwestern Spain, and the eastern Caucasus of Russia, wherever wild ungulates are scarce and livestock graze in or near forests. In northern Finland, wolves hunt semidomestic reindeer.

In seasonal environments, wolves may opportunistically feed lower on the food chain when fruits become available in the summer. Seeds of raspberries and blueberries have been found in scats (defecations), as have cultivated fruits (e.g., grapes, cherries, apples, pears, figs, plums, and melon). The frequency of grass in wolf scats ranges from 14% to 43%, based on studies from both continents.

The influences of foraging on wolf populations are evident in the variation of wolf territory size. The correlation between latitude and mean territory size is highly significant. The mean estimated territory size ranges from $137 \, \text{km}^2$ in Wisconsin to over $2600 \, \text{km}^2$ on Ellesmere Island. This variation is also correlated with (1) lower prey biomass at higher latitudes, (2) smaller ungulate body size at higher latitudes, and (3) lower productivity of the plants upon which herbivores feed at higher latitudes.

In a meta-analysis of 38 studies, about one-third of the variation in wolf territory size is positively correlated with prey biomass. Other factors contributing to the variation included (1) wolf density, (2) interaction between wolf density and rate of wolf population increase, and (3) interaction between the mean territory size and the rate of wolf population increase. For example, mean territory size in regions where wolves hunt deer ($199 \, \text{km}^2$) is one-quarter the size in moose regions ($817 \, \text{km}^2$), possibly because moose are harder to catch and wolves travel further between kills.

Do individual wolves benefit from cooperative hunting tactics? In contrast to lions, food acquisition per wolf decreases with hunting group size. Single adult wolves can kill a moose, bison, or musk-ox; however, calves and sick adults are more vulnerable to single wolves. Most hunting sequences described for wolves have been simple and straightforward. Field biologists differ in opinions about whether hunting tactics of wolves show evidence of cooperation, defined in terms of ambushing prey and relay running.

When group size increases, it is most likely due to recruitment of young inexperienced wolves, adding little advantage to capture success by the group. One hypothesis is that young wolves may benefit from group hunting in that they learn the consequences of their own interactions with prey as well as observing the consequences of actions by more experienced group members. However, in the Mexican wolf reintroduction, even inexperienced captive-reared wolves learned to kill elk within 3 weeks after release. Evidence to test this 'trade school' hypothesis about the function of group size would be very difficult to obtain because of welfare issues and limited visibility of wolf hunts in forested ecosystems and rugged terrain.

Alternatively, large groups wolves may have a competitive advantage in interactions with other predators (e.g., black bears) and scavengers at carcasses (e.g., ravens, eagles, foxes, coyotes, wolverines, and bobcats). In one study, the percentage of carcasses consumed by other scavengers was inversely correlated with wolf group size. Groups of wolves were more likely than singles to attack denning black bear or chase them off a carcass. However, grizzly bears usually displace wolves at carcasses independent of the number of wolves present. Single coyotes and foxes are vulnerable to being killed by wolves. Wolves rarely consume carcasses of competitors. Few interactions have been recorded between wolves and felids (e.g., bobcat, lynx, mountain lion, and Siberian tiger).

Do breeding wolves benefit from helpers at the den? One hypothesis is that auxiliaries may contribute more to provisioning under conditions of food abundance than when food is scarce. However, in good times, breeders are likely to be more successful at prey delivery, and auxiliaries are more likely to disperse. More auxiliaries in a group do not always increase the probability that pups will be attended around the clock. Some evidence points to auxiliaries returning to intercept provisioning at times when breeders are likely to return to pups. Older offspring may compete with younger siblings, under scarce food conditions. Further studies are needed to fully answer this question.

Do nonbreeding auxiliaries benefit by inheriting a territory when breeders are displaced? The current working hypothesis suggests that the answer depends again on the interaction of wolf density and prey availability. Under conditions of low wolf density and high food availability, dispersing floaters are more likely to start a new breeding unit than biders are likely to inherit a territory. However, when wolf density is high and prey density is low, mortality is higher in dispersers than biders. Extra-territorial forays and encounters with neighboring groups result in deaths of breeders under these conditions. Under conditions where breeders die, biders are more likely to inherit a territory.

This conditional model of switching tactics and variable pay-offs for wolves foraging in groups has emerged from studies of ecosystem fluctuations. On Isle Royale, the body condition of moose is correlated with browse forage quality. The forage for moose has changed over decades because of plant succession in patches disturbed by forest fire, as well the direct impacts of moose and other herbivores on the plants. When moose are unhealthy, they are more vulnerable to wolves, and wolf predation has more of an impact than when moose are relatively invulnerable. Vulnerability of prey to wolves is also increased by snow conditions and harsh winters.

Since Isle Royale is a closed system on an island, the fluctuations in plants, herbivores, and carnivores are more accentuated. However, similar dynamics exist in other fragments of forested habitat that are more open systems. The linkages among components of each system are harder to measure in regions where wolves and their prey disperse over larger distances. Large expanses of forest are not homogeneous; local conditions function as sources and sinks in terms of the dynamics of wolf populations on a broader scale of analysis.

Theoretical questions about the stability of predator and prey populations due to wolf foraging ecology are still actively debated. However, researchers agree on three generalizations: (1) wolf impact is highest on the juvenile age class of prey; (2) where wolf populations are increasing, the impact of predation is higher; and (3) the combined impact of predation by wolves and bears is more likely to tip prey populations into a declining trend.

Disease outbreaks also contribute to the instability of wolf populations. Over a 30-year study of canine parvovirus in northeastern Minnesota, pup mortality increased 70% in one region and varied from 40 to 60% over a larger scale. The rate of growth for the infected wolf population was 4% as against 16–58% in other wolf populations. Changes in dispersal potentially related to spread of disease included (1) fewer dispersing juveniles, (2) mortality of entire groups, and (3) a higher probability of adults dispersing following disruptions due to death of breeders.

In summary, variation in the canid genome has been shaped over geologic time scales by glacial cycles that repeatedly displaced northern populations and blocked or opened dispersal routes between continents. The behavioral plasticity that permitted wolves to invade ecosystems as diverse as deserts, forests, mountains, and tundra also permits individuals to adapt within lifetimes to ecological changes in prey availability resulting from shorter cycles (e.g., fire, precipitation, plant succession). Interactions of factors within dynamic ecosystems make it very difficult to test behavioral models on the basis of costs and benefits in terms of ultimate fitness. Given the behavioral plasticity in social carnivores, it is all too tempting to infer the adaptive significance of cooperative foraging despite the paucity of definitive evidence.

Some Current Questions

Recent expansion of research in social cognition and the canid genome have opened promising perspectives for

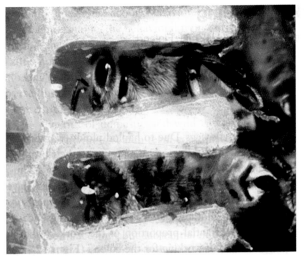

Figure 1 Worker policing in the honeybee. Cells have been constructed against a piece of glass to allow observation of the interaction of a worker with a worker-laid egg that has been experimentally transferred into the cell with forceps. Top. A worker enters the cell and is about to contact the egg; Below. The egg is eaten by the worker that has recognized it as worker-laid. Photos by Francis Ratnieks.

characterized by straight alkanes lacking in nonreproductive workers. Experimental application of synthesized alkanes caused aggression toward treated workers. It is possible that these compounds are unavoidably linked to reproductive physiology, so that egg-laying workers are unable to suppress making them, even if they would benefit from disguising their fertile status. Such compounds are a useful source of information for policing workers.

Policing of Caste Fate

Female larvae in most eusocial Hymenoptera are totipotent with the ability to develop into either a queen or a worker. In a colony rearing both young queens and workers, there is an incentive to develop into a queen rather than a worker because each female is more related to her own offspring than to a sister (**Figure 2**). This incentive, which increases with lower relatedness, can result in the rearing of excess queens. However, in most species, larvae have little power over their caste fate as their feeding is controlled by the adult workers, preventing excess queens

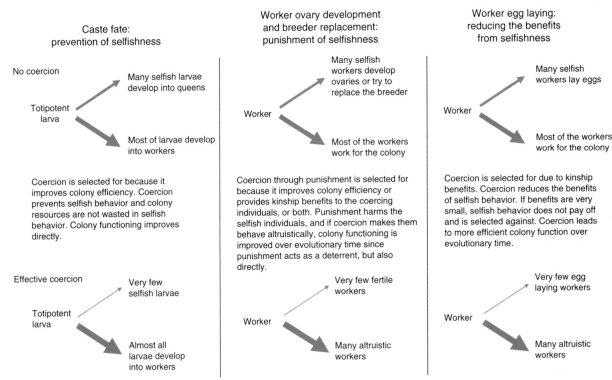

Figure 2 Effects of policing on selfish behavior and colony functioning in three different contexts. The upper row shows the situation in a colony with no coercion. Levels of selfishness (red arrow) are predicted to be high. Level of altruism (green arrow) is predicted to be low. In the absence of coercion only relatedness limits selfishness and selfishness is more common when relatedness is low. On the bottom row, with coercion, selfishness is predicted to be rare, altruism common, and colony functioning efficient. This is when the conflict is resolved so that colony level costs are low. Both coercion and relatedness limit selfishness, and selfishness is predicted to be less common when coercion is effective.

Figure 3 Observations of worker egg laying and policing in natural colony of *Dolichovespula saxonica*. Left, a worker lays an egg into an empty cell. Right, another worker eats the egg soon after it has been laid. Photos by Kevin Foster.

Figure 4 Coercion of caste fate in the honeybee *Apis mellifera*. Uncapped worker cells, each with an egg, a larva, nectar or empty, and capped worker cells each with a fully fed larva or pupa. Queens are reared in special larger queen cells and fed a diet of royal jelly. A larva in a worker cells is unable to develop into a queen because there is not enough space and because it is not fed royal jelly. Each larva is fed progressively in an open cell which gives the adult workers control over caste fate. The developing larva lacks control over its caste fate and acquiesces to its situation, developing into a worker. Photo by Francis Ratnieks.

Figure 5 An opened colony of the stingless bee *Melipona beechei*. On the right, brood cells. On the left, egg-shaped food storage cells. In *Melipona* bees males, queens, and workers are all reared in the same cells. These cells are mass provisioned and sealed after egg laying so the workers do not influence the caste fate of developing larvae. The larvae have a lot of control over their own development, since each cells is large enough, and has enough provisions for a larva to develop into a queen. Photo by Francis Ratnieks.

being reared by withholding the additional food needed to allow a larva to develop into a queen (**Figure 4**). Thus, the caste fate of females is also subject to coercion and policing.

Coercion of female larvae into developing as workers is possible because in most species of social insects, the queens are larger than workers and require more food for their development and sometimes special food or rearing conditions. For example, honeybee queens are reared in special cells on a special diet of royal jelly and queens of the Florida harvester ant *Pogonomyrmex badius*

are reared on a more nitrogen-rich diet than workers. Because larvae are largely immobile, they depend on the adult workers for the additional food needed to develop into a queen. The importance of food control by workers is demonstrated through an exceptional case in which the adult workers cannot coerce the larvae. In *Melipona* stingless bees, workers and queens are similar in size and are reared in identical cells on a provision mass that is placed in the cell before the queen lays an egg and the cell is sealed (**Figure 5**). Each larva has control over her own development as she is sealed away from any interference by the adult workers and has a food mass that is sufficient to allow her to develop into either a queen or a worker. A large excess develop into excess queens, as predicted

Figure 6   Uncapped cells of *Melipona beechei*. Because larvae have control over their own development, several individuals (circled in green) have developed into excess queens. This means that fewer workers develop. Photo by Francis Ratnieks.

Figure 8   Policing of breeder replacement in the dinosaur ant *Dinoponera quadriceps*. The selfish, high-ranking individual in the middle (14) has developed ovaries and has attempted breeder replacement. Low-ranking workers recognize her status and punish her by attacking and immobilizing her. Immobilization may result in the death of the selfish individual, or a lowering of her rank in the hierarchy of workers. Photo by Thibaud Monnin.

Figure 7   Queen execution in *Melipona quadrifasciata*. A newly emerged excess queen is killed by workers, and does not benefit from the selfish behaviour. Photo by Francis Ratnieks.

(**Figure 6**). Because *Melipona* colonies are founded by fission (swarming), very few queens are actually needed, either to head a new colony or supersede a failing mother queen. The excess queens are executed soon after leaving their cells (**Figure 7**).

Policing of Breeder Replacement

The inclusive fitness interests of workers are in conflict in species where the workers and queens are not morphologically distinct so that a worker can take over the queen's role. In some ponerine ants (see Glossary), the queen caste has been lost and the colony is headed by a mated worker, who is the queen (called a 'gamergate') and heads a colony of unmated daughter workers. Even though all workers

are capable of mating, a single gamergate monopolizes reproduction in species like *Dinoponera quadriceps* (**Figure 8**). The high-ranking workers lay some unfertilized eggs but most of these are killed by the gamergate. In addition, it can be in the selfish interest of a daughter worker to overthrow the gamergate, rather than allow one of her sisters to do so. However, most of the workers never challenge the current breeder, and the few workers that challenge only do so once the fertility of the current breeder decreases. Again, the discrepancy between the potential incentive to selfishness and observed occurrence of selfishness can be explained by policing (**Figure 2**).

In the conflict over breeder replacement, workers police by aggression to prevent selfish workers from replacing the queen. In *D. quadriceps*, high-ranking workers are attacked and immobilized by low-ranking females in order to prevent them from challenging the breeder (**Figure 8**). Again, cuticular hydrocarbons, especially compounds that correlate with fertility, likely play a role in recognition of selfish individuals, since the status of the potentially fertile challenger individuals is reflected in their cuticular profile. The queen sometimes helps the workers in recognition. In *D. quadriceps*, the queen uses her sting to mark with glandular secretions the individuals challenging her. These cause nonreproductive workers to punish the challenger. Both the queen, which maintains its status, and the nonreproductive workers benefit by not allowing premature replacement of the gamergate.

What Factors Select for Policing in Social Insects?

Policing suppresses individual selfishness. This benefits the colony as a whole by increasing the number of workers versus excess queens, or working workers versus egg-laying workers (**Figure 2**). But policing is not necessarily selected for to increase group efficiency reasons although this can be a factor. In evolutionary terms, policing needs to be understood in terms of the benefits that policing individuals gain in terms of their inclusive fitness, whether this is indirect, as in worker policing, or direct, as in queen policing and selfish worker policing. In other words, the policing individual must benefit either indirectly through reproduction of relatives, or directly through its own reproduction. Policing is affected by several selection pressures that may be difficult to disentangle, because they can operate simultaneously.

Selection for Policing in Male Production

The theory of worker policing was originally developed to investigate the effects of queen mating frequency on worker–male relatedness in eusocial Hymenoptera. Due to haplodiploid sex determination in the Hymenoptera, the number of males that the queen has mated with affects the optimal source of males for the workers as a whole in a colony. If the mother queen is mated to a single male, the workers are more closely related to the sons of other workers (0.375) than to the sons of the queen (0.25). Thus, on the basis of relatedness alone, workers are predicted to favor male production by other workers. There is also worker–worker conflict over male production, since each worker is more related to its sons (0.5) than to other workers' sons (0.375). If the queen has mated with two or more males the situation turns around. Now workers are more closely related to the sons of the queen than to the sons of other workers' sons. Workers are consequently predicted to favor male production by the queen. But each worker is still most closely related to its own sons (0.5) and so, she still benefits from egg laying. In other words, there is selection on workers both to lay eggs and to police the reproduction of other workers. Following this logic, worker policing was first looked for, and found, in honeybees. Honeybee queens mate with approximately ten males so that workers are more closely related to the sons of the queen (0.25) than to the sons of the other workers (0.15). As predicted, worker policing occurs in honeybees and is in fact very effective as almost all worker-laid eggs are killed. Workers are also more closely related to the sons of the queens than to the sons of workers and thus selected to police worker-laid eggs in species where colonies have multiple related queens.

However, policing also occurs in species where workers are more closely related to the sons of other workers than to the sons of the queen. For example, workers destroy eggs laid by other workers in *C. floridanus*, in which the queen is singly mated, and in the hornet *Vespa crabro*, in which queen mating frequency is only slightly >1 (a small proportion of queens are multiple

mated but most are single mated). There are several inclusive fitness factors that select for policing in addition to kinship differences to males. First, policing is selected for if it enhances the productivity of the colony, as long as the relatedness between colony members is positive. This explanation is quite often invoked when policing has been found in species with single-mated queens. However, direct evidence is lacking and it is not necessarily clear how colony efficiency might improve from policing in this situation. For example, if policing occurs through reducing the benefits from selfishness, such as egg eating, policing does not necessarily increase colony productivity unless worker-laid eggs have low viability. Although this has been suggested for honeybees, the evidence on which this idea was based has been refuted. Unless policing increases colony productivity in the current generation, it is not selected for. If an egg has already been laid, destroying this egg will not cause the worker that laid it to stop laying more eggs and more generally, to work instead of laying. In contrast, if policing occurs through punishment (**Figure 2**), such as aggression toward egg-laying workers, it may cause their ovaries to become less active for them to resume working. In this case, it is more likely that policing improves group efficiency and can be selected for due to immediate benefits in a colony with policing workers.

Other reproductive conflicts within colonies may interact with the conflict over male production and favor worker policing. For example, in a colony with a single-mated queen, the workers benefit from a female-biased sex allocation ratio. If workers kill worker-laid eggs, which are male, then this may benefit them by causing more female-biased sex allocation in their colony. Workers would not benefit by killing queen-laid eggs because most of these are female. Overall, the killing of worker-laid eggs can benefit policing workers, more in terms of biasing female allocation than what they lose by causing the colony to rear additional brother males instead of full nephews. Killing worker-laid eggs could be a low-cost way of biasing sex allocation because it would be an accurate way of recognizing males (assuming that the workers can easily discriminate queen-laid eggs from worker-laid eggs) before much time or energy is invested on rearing them.

Policing can also be carried out for selfish reasons. That is, the policing individuals are themselves laying eggs and increase their direct reproduction through policing. Such 'corrupt policing' by reproductive workers has been found in *Dolichovespula sylvestris* wasps, in which workers that remove eggs laid by other workers themselves lay eggs in the vacated cells after removal. In contrast, workers that specialize in aggressing fertile workers and eating their eggs in the ant *Pachycondyla inversa* are not themselves fertile. The queen may also police workers in order to increase her own reproductive channels in the membranes of the various neurons in the circuit, the synaptic properties of the different connections in the network, and the exact pattern of wiring of the network. Not surprisingly, therefore, mutagenesis studies of zebrafish have revealed genes that disrupt organization at each of these different levels with profound behavioral consequences.

A few examples, all isolated in the large Tübingen screen, serve to illustrate some of the genetic effects that can alter more simple behaviors. This screen looked for larval fish in which the response to a touch was altered. These fell into many categories including those that did not respond at all and a whole series of animals that had aberrant movements subsequent to the touch. Unresponsive animals included those with major deficits in sensory neurons relaying information about the touch as well as in the motoneurons and muscle that are critical for producing the movement. In mutants such as *macho*, the excitability in sensory neurons was altered in a way that blocked their ability to produce action potentials and hence the touch could not trigger a movement. On the motor output side, a whole series of mutants with problems in the muscle or in the connection between nerve and muscle led to an inability to move in response to the touch. These included mutants such as *relaxed* which was a deficit in the dihydropyridine receptor that disrupted the calcium release in the muscle necessary for contraction. Others, such as *sofa potato*, were mutants of the acetylcholine receptors on the muscle, which rendered the muscle unresponsive to neurotransmitter released by the motoneurons.

Many of the more interesting mutants led to altered movement patterns rather than to no movement at all. Disruption of the protein rapsyn that anchors acetylchoine receptors at the neuromuscular junction led to fish that could response to the stimulus, but could not sustain the resulting swimming movements. This mimics human disorders of muscle weakness, one of which involves a mutation similar to that identified in zebrafish. A whole class of mutants, called the accordion class, disrupts the ability of the fish to swim with the normal alternating bends from one side to the other; instead, muscle contractions overlap on the two sides in the mutants leading to an accordion like shortening of the body of the fish. Some of these involve slowed muscle contractions that lead to overlap in activity on the two sides of the body; they also include mutants with alterations in the central nervous system that reveal contributions of molecules to proper circuit function. One of these is the mutant *bandoneon*, which is a mutation in a beta subunit of a glycine receptor. This disrupts inhibitory synapses in the nervous system, including those that usually block activity in motoneurons on one side when the other side is active. The result is bilateral muscle activity.

Mutants can also lead to a rewiring of the neural circuits by altered axonal projections of the neurons or alternatively, by changes in the numbers of particular cell types.

A mutation called *space cadet* produces a whole series of repeated escape-like bends to one side after a single touch. The evidence suggests that this is a consequence of missing pathways in the brain that cross from one side to the other and regulate the activity of neurons responsible for producing normal escapes. A striking mutant called *deadly seven* may offer some insight into the evolution of networks because it is a mutant of a gene notch that is important for controlling cell number in the nervous system. The mutant animals have extra Mauthner neurons, the cells that are a critical component of escape or startle behaviors. Interestingly, these extra neurons are wired into the escape network and the fish can produce robust escape movements. This suggests that extra neurons added by genetic alterations can be incorporated into circuits via normal developmental mechanisms without a dramatic effect on the behavior. This makes it easier to understand how an escape or startle behavior produced by few neurons in early aquatic vertebrates might have evolved into a startle response produced by much larger numbers of neurons in tetrapods.

These are just a few examples of the studied mutants, but many other mutant lines of zebrafish have been isolated, but not yet studied at the genetic or functional levels. The evidence so far documents how even the simplest motor behaviors can be disrupted in many ways by mutants that affect all levels of neuronal and network function. The work serves to highlight the critical interplay between levels of functional organization in the nervous system that is essential to generate a proper behavioral response.

More Complex Behaviors

Complex behaviors are much harder to tackle at the genetic and neuronal level, but are of considerable biological interest. Studies of learning, sleep, aggression, more complex visual-motor behaviors, reward systems in the brain, and even zebrafish models of neurological disorders such as schizophrenia are opening the lines of attack on more complex behaviors. Most of these investigations are still in early stages, although some mutants with deficits in these behaviors have been isolated. The challenge here is to examine how the mutations affect the nervous system to alter the behavior. Analyzing this at the neuronal level depends to a great extent on understanding how the complex behavior is produced normally and this is something that is very difficult when the behaviors involve interactions of networks in many parts of the brain and spinal cord.

The Future

The power of the zebrafish model will assure that it remains at the forefront of genetic and neuronal studies of behavior into the future. There is no other vertebrate model with comparable genetic, optical, electrophysiological, and behavioral accessibility. The most modern optical tools for using light to activate and inactivate neurons play to the zebrafish strength and are only in their infancy. When they are combined with transgenic approaches to target neuronal classes in normal and mutant lines of fish, they promise to forge very compelling ties between the genetic, neuronal, and behavioral level.

Although much of the focus has been on simple sensory or motor behaviors, there is increasing attention to more complex behaviors and this is likely to grow, given the tools in zebrafish may offer the best attack on these behaviors. The links between behavior studied in the laboratory and behavior in the wild are still weak, but foundational work in the natural biology of the animal will hopefully seed other studies of issues such as social hierarchies, territory defense, and mate selection in this model, in which they may eventually be attacked at the genetic and neuronal levels.

See also: Development, Evolution and Behavior; Fish Migration; Fish Social Learning; Genes and Genomic Searches; Nervous System: Evolution in Relation to Behavior; Neurobiology, Endocrinology and Behavior; Threespine Stickleback; Vocal–Acoustic Communication in Fishes: Neuroethology.

Further Reading

Brockerhoff SE, Hurley JB, Janssen-Bienhold U, Neuhauss SC, Driever W, and Dowling JE (1995) A behavioral screen for isolating zebrafish mutants with visual system defects. *Proceedings of the National Academy of Sciences of the United States of America* 92: 10545–10549.

Cui WW, Low SE, Hirata H, et al. (2005) The zebrafish shocked gene encodes a glycine transporter and is essential for the function of early neural circuits in the CNS. *The Journal of Neuroscience* 25: 6610–6620.

Fetcho JR, Higashijima S, and McLean DL (2008) Zebrafish and motor control over the last decade. *Brain Research Reviews* 57: 86–93.

Fetcho JR and Liu KS (1998) Zebrafish as a model system for studying neuronal circuits and behavior. *Annals of the New York Academy of Sciences* 860: 333–345.

Granato M, van Eeden FJM, Schach U, et al. (1996) Genes controlling and mediating locomotion behavior in the zebrafish embryo and larva. *Development* 123: 399–413.

Higashijima S (2008) Transgenic zebrafish expressing fluorescent proteins in central nervous system neurons. *Development, Growth & Differentiation* 50: 407–413.

Liu KS and Fetcho JR (1999) Laser ablations reveal functional relationships of segmental hindbrain neurons in zebrafish. *Neuron* 23: 325–335.

Neuhauss SC (2003) Behavioral genetic approaches to visual system development and function in zebrafish. *Journal of Neurobiology* 54: 148–160.

Nicolson T (2005) The genetics of hearing and balance in zebrafish. *Annual Review of Genetics* 39: 9–22.

O'Malley DM, Kao YH, and Fetcho JR (1996) Imaging the functional organization of zebrafish hindbrain segments during escape behaviors. *Neuron* 17: 1145–1155.

Ono F, Shcherbatko A, Higashijima S, Mandel G, and Brehm P (2002) The zebrafish motility mutant twitch once reveals new roles for rapsyn in synaptic function. *Journal of Neuroscience* 22: 6491–6498.

Orger MB, Gahtan E, Muto A, Page-McCaw P, Smear MC, and Baier H (2004) Behavioral screening assays in zebrafish. *Methods in Cell Biology* 77: 53–68.

Spence R, Gerlach G, Lawrence C, and Smith C (2008) The behaviour and ecology of the zebrafish, *Danio rerio. Biological Reviews of the Cambridge Philosophical Society* 83: 13–34.

Streisinger G, Walker C, Dower N, Knauber D, and Singer F (1981) Production of clones of homozygous diploid zebrafish (*Brachidanio rerio*). *Nature* 291: 293–296.

Westerfield M (1995) *The Zebrafish Book,* 3rd edn. Eugene, OR: University of Oregon Press.

Forced copulation   Contrasts with copulation that individuals seek or freely accept. Most investigators infer that copulation is forced when it is preceded by aggression or the threat of aggression, including 'violent restraint.'

Forward genetics   A phenotype-driven mutant screen.

Forward masking   Reduction of perceptual sensitivity over a given time interval following the perception of a specific stimulus.

Foundress/cofoundress   Foundresses are females that are initiating a nest, or living, on a newly established nest before the emergence of the first offspring. If more than one foundress is present in a nest, they are called cofoundresses.

Fourier analysis   A type of time series analysis that involves fitting a series of sine waves to data. The analysis identifies the amount of strength or power associated with a set of periods.

Fovea   Specifically, a depression in the center of the retina of many vertebrates, providing high-resolution vision. More generally, areas of high visual acuity in vertebrate retinas are called 'area centralis' or 'visual streak.'

Framework   A simplified conceptual structure used to solve complex problems.

Frass   The waste product from an animal's digestive tract expelled during defecation (also known as fecal material, or feces).

Free choice profiling   An experimental methodology in which observers have complete freedom to choose their own descriptive terms and apply them to the observed behavior of animal subjects.

Free-running rhythm   Free-running rhythm refers to fluctuations in physiological or behavioral responses, with a period of about 24h, that recur in the absence of environmental cues.

Freeze tolerance   The ability of an animal to survive freezing of tissues.

Freezing   Remaining motionless upon detection of a predator in hopes of avoiding detection by the predator either through cryptic morphology or habitat cover.

Frequency-dependent selection   Selection that varies depending on trait frequency in the population.

Frequency of sound   The number of cycles of vibration per second of a sound-producing object, expressed in Hz (Hertz, or cycles per second). A good set of human ears can detect frequencies of 20 Hz–20 kHz (a kHz is a kilohertz, or 1000 cycles per second). This physical property of sound is the primary determinant of our psychological experience of sound pitch.

Frequency modulation   Cyclic changes in the frequency composition of a sound over time. The process of modulation produces extra frequencies in the sound, called sidebands.

Frontal cortex   A brain region that (among other functions) plays a key role in long-term planning, executive decision-making, and impulse control.

Functional activity mapping   An analysis of the patterns of neural activity, or its correlates, during the performance of a behavior or in response to a stimulus.

Functional class   A class defined by a common (inherent) function of its members.

Fundamental frequency (f_0)   The lowest frequency component in a harmonic sound.

Future planning   The ability to imagine and preexperience specific personal scenarios that might occur in the future.

GABA–γ   Aminobutyric acid is the chief inhibitory neurotransmitter in the mammalian CNS. The binding of GABA to its receptors causes the opening of ion channels to allow the flow of either negatively charged chloride ions into the cell, or positively charged potassium ions out of the cell, to produce an inhibition of the cell. Receptors to GABA are found in both the central and peripheral nervous systems of several invertebrate phyla. Insect GABA receptors show some similarities with vertebrate GABA receptors.

Gametes   A cell that fuses with another gamete during fertilization.

Game theoretic models   These are mathematical calculations of an individual's success (fitness) in making choices when their choice depends on the choices of others.

Game theory   A mathematical technique for choosing the best strategy given the likely choice of others.

Ganglion   The CNS of insects and other invertebrates comprises a ganglion – a processing center ('brain') – for each body segment connected to the ganglia of adjacent segments by bundles of axons called 'connectives.'

Gap junctions   Specialized intercellular complexes that directly connect the cytoplasm of two cells. Gap junctions allow various molecules and ions to pass freely between cells. Between two neurons, gap junctions form electrical synapses.

Gasterosteidae   Latin name for the family of stickleback fish.

Gating neurons   A type of *command neuron* that must be active during the whole time while a behavior takes place. This term was coined in the study of leech swimming activation to distinguish these neurons from *trigger neurons*, a class of command neurons that is active only for a short time when a behavior begins.

Gene chip   A commercial microarray.

Gene flow   The transfer of alleles of genes from one population to another.

Gene regulation   Relating to the activation (expression) of genes, including both transcription and translation.

Genetically effective population size   The number of reproducing individuals in a randomly mating population; actual population size is usually larger than its genetically effective size owing to the presence of sexually immature or nonbreeding individuals.

Genetic complementarity   The potential for traits on both sides of an ecological interaction to respond evolutionarily to reciprocal selection.

Genetic diversity   The level of biodiversity within a species, in reference to its total existing number of genetic characteristics, which, importantly, provides the raw material for evolution and is critical for long-term sustainability of a population.

Genetic drift   Chance variations in gene frequencies that result from random sampling error.

Genetic monogamy   An exclusive mating relationship between a male and a female resulting in all offspring being genetically directly related to both partners.

Genetic polymorphism   A portion of the genome that is represented by numerous distinct versions in the population. The more polymorphic a given locus is, the greater the number of distinct versions that will exist in the population. Genetic polymorphisms are based on sequence variation at specific loci.

Genetic relatedness   The fraction of genes identical by descent between two individuals. Only the fraction of genes shared above background count. See piece on relatedness.

Genetic structure   The array of alleles and genotype combinations in a population.

Genetic subdivision   Reduced gene flow between populations allows them to differ in the presence and/or frequency of alleles as a result of random genetic drift or natural selection.

Genetic task specialization   Task threshold is genetically influenced. Workers of particular parentage are more likely to engage in particular tasks.

Genic selection   Selection within individual bodies between alleles at a locus.

Genomic imprinting   Form of inheritance in which the expression of a gene depends upon the parent from which the gene is inherited. Because imprinting allows genes to be silenced when inherited from one sex and not the other, it provides a potential mechanism for achieving sex-specific expression. The imprint alters the chemical structure and hence the expression of the gene, but not its nucleotide sequence. Thus, the imprint can be erased and an active gene can be passed down in the next generation.

Genomic library   A collection of fragments of genomic DNA that have been inserted into host cells, typically bacteria or viruses, so that the individual fragments can be replicated in high numbers.

Genotype   The genetic constitution of an organism or one of the loci within that organism.

Geocentric cue   A cue based on information external to the organism.

Geographic mosaic   Ecological interactions vary across space because of the specifics of biotic and abiotic local environments, leading to a spatial mosaic of coevolutionary intensity. Hotspots, where reciprocal selection is strong, and coldspots, where reciprocal selection is weak or absent, characterize the geographic mosaic.

Geolocator   A daylight-level recorder affixed to an animal at capture that can be recovered at recapture up to one year later to estimate the latitude and longitude for each day the device was attached.

Geomagnetic field   Magnetic field associated with the Earth. It is essentially dipolar (it has two poles), the northern and southern magnetic poles on the Earth's surface. Away from the surface, the field becomes distorted.

Geophagy   The ingestion of soil particles which can reduce the potency of ingested toxins.

Geotaxis   Directed movement with respect to Earth's gravitational field. Movement away from Earth is 'negative,' movement toward Earth is 'positive.'

Germinal vesicle breakdown   Dissolution of the nuclear membrane that signals continuation of meiosis.

Ghost experiment   An experiment in which the model who would normally produce some effect in the world is absent, the effect being produced instead by surreptitious ('ghostly') means, such as pulling fine fishing line, allowing a test of how much an observer will learn from this component of the display alone.

Gill operculum   The hard flaps covering the gills of a fish.

Gilliam's rule   The prediction that animals favor using patches that minimize the ratio of predation risk to either expected growth or foraging rates.

Giving-up density and time (GUD and GUT)   Giving-up density is the amount of food or prey items still remaining in the patch, when a forager leaves it. Giving-up time is the length of time a forager will go without encountering a food item before it leaves a patch. Both are important metrics for testing predictions of the marginal value theorem.

Glossopharyngeal nerve   The ninth (IX) of twelve pairs of cranial nerves. It exits the brainstem from the medulla, just rostral (closer to the nose) to the vagus nerve. The glossopharyngeal nerve is mostly sensory and is involved in tasting, swallowing, and salivary secretions.

experienced; it serves as the basis or standard for other members of the same category.

Protozoan   Unicellular microorganisms among eukaryotes. Comprises flagellates, ciliates, sporozoans, amoebas, foraminifers.

Proximal   Closer to a body midline (opposite of distal).

Proximate causation   Explanations of an animal's behavior based on internal and external mediators of behavior including genetic underpinnings, epigenetic forces, maternal effects on physiology, morphology, and development. Questions about proximate causes are sometimes said to be about how animal behavior is expressed or about mechanisms of animal behavior.

Proximate factors   External stimuli (such as specific daylengths) which are used as cues by an animal to trigger preparation for breeding, migration, molt, or other events, or as time keepers to set their endogenous time programs at appropriate times of the year.

Pseudergate   In termites, an alternative technical term that can be found which distinguishes workers with a flexible development and options for direct reproduction from workers with restricted developmental trajectories. Pseudergates are the 'workers' of many lower termites (including wood-dwelling and foraging species) that have broad developmental options, generally including progressive, stationary, and regressive molts. Current use of this term often lacks the precision of its original definition for individuals that develop regressively from nymphal instars to 'worker' instars without wing buds.

Pseudopregnant   Reproductive condition in which a female shows external indicators of pregnancy but is not actually pregnant.

Pseudoreciprocity   The act of increasing another individual's fitness to acquire or enhance the by-product benefits obtained from that individual.

Pseudoreplication   A statistical error in which interrelated observations or measures are treated as though they are statistically independent.

Psychoneuroimmunology   A relatively new field in medicine that explores the ability of the nervous system and psychological states to influence immune defenses, and the ability of the immune system to influence the brain and behavior.

Pterygoid teeth   Small teeth on the roof of the mouth.

Ptilochronology   The study of growth bands in feathers that indicate condition or problems during feather molt in birds.

PTT   A platform transmitter terminal (PTT) sends an ultrahigh frequency (401.650 MHz) signal to satellites.

PTTs are attached to animals in order to track their movements.

Public good   A resource that is costly to produce and provides a benefit to all the individuals in the local group. Public goods systems are often open to exploitation by cheats who benefit, but do not pay the cost.

Public information   Cues produced by animals that can potentially be used by observer animals in making behavioral decisions.

Pulse repetition rate   The rate at which individual sound pulses are produced within a single call.

Punishment   A costly behavior that is negatively reciprocal (decreases harmful behavior in the recipient) (evolutionary biology); any stimulus that reduces the frequency of a behavior (social science); behavior correction and the enforcement of social norms, typically by impartial parties; see also Third-party punishment, Policing (social science).

Pupa   A life stage in some insects that undergo complete metamorphosis that results in the transition between the larval and adult stage.

Purging selection   Mechanisms eliminating deleterious genes from the population.

Pyrophilous insects   Species strongly attracted to burning or newly burned areas, and species that have their main occurrence in burned forests 0–3 years after the fire.

Quality of life   Well-being; a multidimensional, experiential continuum that comprises an array of affective states, broadly classifiably as relating to the states of comfort–discomfort and pleasure; often equated to welfare and well-being.

Quantitative trait   A continuous trait such as body mass that is influenced by many genes and the environment.

Quantitative trait locus (QTL)   A region of DNA that is associated with a particular quantitative trait, containing a gene or genes that influence that trait. Quantitative traits typically have continuous distributions rather than discrete states, and are influenced by several or many loci, each with relatively small or large effects on the expression of the trait.

Quasi-experimental design   An experimental design where a treatment variable may be manipulated but subjects within groups are not equated or randomly assigned.

Quasiparisitism   Occurs when the female that dumps eggs in another female's nest is the resident male's extra-pair partner and her dumping is assisted by that male.

Queen   Reproductive female in a eusocial insect society. She is developmentally and/or behaviorally disposed towards performing all reproductive function for a colony.

Questing   The behavior of ticks, involving an ascent on vegetation that allows for a maximum exposure of sensory receptors on the forelegs to stimuli from approaching hosts.

Quorum decision   A minimum number of individuals required to perform a specific behavior (such as choosing a direction of travel) that results in all of the other members of a group adopting this behavior.

Quorum sensing   A rule under which a social group member's execution of a particular act or behavioral transition is conditioned on the presence of a threshold number of fellow group members.

Radiotracking   The location and tracking of a radiomarked individual from a signal emitted frequently by the radio.

Rape   A legal term and includes other forms of sexual assault as well as forced copulation, including statutory rape, which may appear to be consensual copulation but with a minor; in this case women, not just men, can be rapists.

Rapid-eye-movement sleep   The other basic sleep form in mammals and birds. It is often called 'paradoxical sleep' because the brain activity resembles that of the awake brain. It is characterized by the complete inhibition of muscle tone and suppressed autonomic regulation of most homeostatic functions such as thermoregulation and blood pressure.

Rate of return   The ratio of the amount of food obtained to the time it took to procure the food.

Rationality   A set of consistency principles that decision-makers are expected to follow if they are attempting to maximize some currency such as utility or fitness. Fitness maximization by natural selection is expected to yield rationality, but many instances of irrational choice are known in humans and other animals. Property of individual choice is used both to describe the process of making a choice and to describe the behavioral outcome of choice.

Rayleigh scatter   Light scatter by particles smaller than the wavelength of light.

Reaction norm   A reaction norm describes the production of a range of phenotypes by a single genotype in response to a range of an environmental parameter. Different genotypes may produce different response trajectories in response to a gradient of an environmental parameter. Reaction norms resemble dose-response curves in physiology, for example the effects of a gradient in hormone concentrations. Dose-response relationships are not necessarily monotonic but can include thresholds or show maximal (minimal) effects at low and high doses or medium doses.

Reasoning   A form of logic-based thinking; the cognitive process of looking for reasons for beliefs, conclusions, actions, or feelings.

Receiver psychology   Sensory capabilities of the signal receiver that affect the detectability, discriminability, and/or memorability of signals, and play a role in the evolution of signal design.

Receptivity   Sexual behaviors that are necessary and sufficient for mating.

Reciprocal altruism   Where individual A pays a personal cost to help individual B with the expectation that B will return the favor.

Reciprocal selection   Positive feedback between selection by ecological enemies. Natural selection by predators on prey generates the evolution of increased defense, which in turn causes stronger selection by prey on predators to evolve greater exploitative abilities.

Reciprocity   Delayed exchange of benefits between parties.

Recognition signals   Signals that evolved to make a signaler distinctive.

Recombination   In evolutionary algorithms, a process of crossover that combines elements of existing solutions in order to create at the next generation a new solution, with some of the features of each 'parent solution.' It is analogous to biological crossover.

Reconciliation   Postconflict affiliative reunion between former opponents that restores their social relationship disturbed by the conflict.

Recruitment   Entry of progeny into a population as reproductive adults.

Red queen   Based on the quote from Lewis Carroll's Red Queen, 'It takes all the running you can do, to keep in the same place,' this metaphor describes a coevolutionary dynamic where frequencies of traits or genotypes of ecological enemies cycle through time so that as one type becomes common, it is disfavored and a rare type can spread through the population.

Redirected aggression   Postconflict aggressive interaction directed from the original recipient of aggression to a bystander uninvolved in the conflict.

Redirected behavior   The direction of some behavior, such as an act of aggression, away from the primary target and toward another, inappropriate target.

Redundancy reduction   The reduction in the overlap of information encoded by neurons in the nervous system.

Referent   The on model on which a signal is based.

Reflectance   The ratio of reflected to incident light on a given area (e.g., colored patch in the plumage).

Refraction   Change in direction of light caused by alteration of its velocity on obliquely entering a medium of different refractive index.

Refractive index   A measure of the speed of light in a medium.

Refractive state   The resting refractive state of an animal determines the point at which it is focused without having to expend any accommodative effort.

Regressive molt   A molt that is characterized by a decrease in body size and/or regression of morphological development, generally a reduction of wing bud size in nymphal instars. This type of development is unique to termites.

Regularity   A specific version of independence from irrelevant alternatives. It describes the expectation that the absolute preference for an option should never be increased by the addition of inferior options to the choice set.

Regurgitant   A substance produced in the gut of an insect that is excreted from the mouth as a defensive secretion.

Reinforcement   The evolution of premating isolation after secondary contact as a result of selection against hybrids or hybridization.

Reinforcement/supplementation   Addition of individuals to an existing population of conspecifics.

Reintroduction   An attempt to establish a species in an area which was once part of its historical range, but from which it has been extirpated or become extinct.

Relatedness asymmetries   A group of individuals are more closely related with a certain group of individuals than others within a colony.

Relatedness, _r_   Genetic similarity between individuals, in comparison with randomly chosen individuals in the population, that have a mean relatedness of zero by definition.

Relational class   A class defined by relations between or among its members and going beyond any perceptual similarities or functional interconnections.

Relative risk   An individual's risk of predation given the abundance of its type.

Relaxed selection   This occurs when the sources of natural selection engendering physical or behavioral traits that promote fitness diminish markedly or are no longer present in the environment. In the case of predators, prey species might be separated from their former predators by their isolation on islands. In another context, climate change tolerated by prey might diminish contact with their predators that are intolerant to climate change and eventually disappear.

Reliability   The percentage of signals of a particular type X that are accurately associated with a stimulus (X′).

REMI   Restriction enzyme mediated integration (REMI) is an ingenious method of introducing single gene knockouts in a genome in a way that allows one to identify the actual gene that is knocked out. Used in _Dictyostelium_.

Repeatability   Consistency between different measurements separated in time of a trait of a certain individual, used in population genetics as the upper limit of heritability.

Repertoire expansion   A pattern of temporal polyethism in which workers increase the types of tasks they perform as they age.

Replication   Using more than one observation per observational unit or subject per experimental treatment group.

Reproductive age   The age at which an individual becomes receptive to mating the first time.

Reproductive character displacement   The process of phenotypic evolution in a population caused by cross-species mating and which results in enhanced prezygotic reproductive isolation between sympatric species. Referred to as 'reinforcement' if postzygotic isolation is incomplete.

Reproductive compensation   Refers to any flexible response of constrained individuals that increases the likelihood that their offspring will survive to reproductive age.

Reproductive division of labor   Differentiation of individuals within a eusocial colony into those capable of reproducing, and functionally or physically sterile workers.

Reproductive effort   The proportion of available time, nutrient or energy resources that an adult invests in current reproduction, usually detracting from those available for other functions.

Reproductive groundplan hypothesis   (Originally described as ovarian groundplan hypothesis) Proposes that the evolution of eusociality is based on simple evolutionary modification of conserved reproductive and corresponding behavioral cycles so that during the course of social evolution, reproductive and nonreproductive behavioral and physiological components can be separated and used to build reproductive (queen) and nonreproductive (worker) phenotypes.

Reproductive isolation   Reduced genetic exchange between populations via reduced interbreeding and lower fitness of hybrid offspring; speciation has occurred when reproductive isolation between populations is complete.

Reproductive skew   Asymmetry in the distribution of direct reproduction among individuals within a social group.

Reproductive strategy   An organism's relative investment, behaviorally and physiologically, in offspring, including reproduction and parental care.

Reproductive success (RS)   Refers to the number of offspring an individual produces which survive and go on to reproduce in the next generation. Although 'life-time reproductive success' is the most accurate measure, logistically it is not always possible to obtain this measure.

Consequently, RS may be measured as number of eggs produced, number of young produced, number of young that fledge from the nest (e.g., birds) or survive to weaning (e.g., mammals), or number of young that survive to reproductive age.

Reproductive suppression   A mature individual does not reproduce because of physiological mechanisms that inhibit production of gametes as a direct result of communication with conspecifics.

Reproductive value   The expected reproduction of an individual from its current age onward, given that it has survived to that age. It changes with age, increasing at first and declining until death.

Residual reproductive value   The number of offspring an individual is expected to produce during its remaining lifespan.

Resource competition   A particular form of competition in which members of the same or different species compete for the same resource in an ecosystem (e.g., food, space).

Resource constraint hypothesis (Trivers–Willard effect)   Colonies should invest more in the cheaper sex (i.e., males, which are generally smaller than females in Hymenoptera) when resources are limited.

Resource holding potential   The relative fighting ability of a contestant.

Response blocking   Also called *flooding* – The process of exposing a subject to constant, high levels of a distressing stimulus, while preventing escape from the situation, in an attempt to reduce or extinguish the distress produced by the stimulus.

Retinal disparity   Difference between the images projected on the two retinas when looking at an object that serves as a binocular cue for the perception of depth.

Retinoscopy   A technique used to obtain an objective measurement of the refractive state of the eye, in which a moving light is shone into an animal's eyes and the relative motion of the reflection is observed.

Reverse genetics   A molecule-driven approach to understanding a phenotype.

Rheotaxis   Orientation or response to current flow; moving upstream is positive and downstream is negative rheotaxis.

Rhinophores   Tentacles in some gastropod mollusks that carry the olfactory organ.

Rhodopsin   All visual pigments whose chromophore is retinal, but commonly (although erroneously) used to refer only to rod visual pigments.

$R_{male-male}$ Androgen responsiveness (i.e., the change in testosterone concentrations) during aggressive interactions between territorial males.

R_{season} Seasonal androgen response, reflecting the increase from breeding baseline testosterone concentrations to maximum concentrations during specific parts of the breeding life-cycle stage, that is, during the phase of territory establishment or mate guarding.

Riparian   Interface between terrestrial and aquatic ecosystem. When intact, riparian ecosystems limit soil runoff and are characterized by high biodiversity and thus are an important buffer zone.

Risk effects   Nonconsumptive effects of predators on prey, namely the lost foraging opportunities and lower levels of growth and reproduction experienced by prey investing in antipredator behavior (also known as nonlethal effects). This term avoids the complication that prey that are not directly killed by a predator may in fact be consumed.

Risk history   The frequency, intensity, and duration of predation risk events experienced by prey in the past.

Risk threshold   The level of risk that must be exceeded for the prey to start reducing its antipredator behavior under the risk allocation hypothesis.

Ritualization   Communicative behaviors used in social interactions that evolved from other behaviors with different functions. For example, when attacked an ancestor of the wolf might have flattened the ears, crouched, and tucked the tail to avoid injury; over time these behaviors evolved to communicate submission. Evolutionary modification of a motor pattern used in communication that is thought to improve signal function, often through increased stereotypy and exaggeration.

RNA interference (RNAi)   A technique of molecular biology in which expression of a particular gene is silenced by introducing double-stranded RNA into a eukaryotic organism. RNA interference can provide conclusive proof that a particular gene influences behavior.

Roosting   The act of perching to rest or sleep.

Round-trip migration   A subcategory of migration, with seasonal to-and-fro movements between regular breeding and wintering sites, typical of many birds but rare in insects.

RT-PCR   Reverse transcription PCR, PCR that is performed on DNA that was synthesized from RNA by a reverse transcriptase enzyme.

Rule learning   The ability to infer rule information from a number of different examples connected by a logical operation 'if → then.'

Rules of thumb   Simple measures that animals can use to approximate solutions to optimal foraging problems. An example would be using the number of prey items encountered to leave patches as predicted by the marginal value theorem.

Runaway selection   A theoretical model for the evolution of extravagant traits based on female preference.

The model proposes that female preference for a male trait results in a genetic correlation between preference and trait, such that the trait evolves beyond the level favored by natural selection in a 'runaway' process fuelled by female preference. Also called the Fisher process in reference to Sir Ronald Fisher, who developed the theory.

Saccule   An otolithic subdivision of the inner ear in all vertebrates that has an auditory (hearing) function among many fishes.

Saprophagy   Feeding on dead materials.

Satellite transmitters   These tracking devices are larger than radio transmitters and emit signals that are detected by geosynchronous satellites; these devices carry substantial batteries or are solar powered and continue to transmit for relatively long periods of time (i.e., a year or more); they enable tracking to occur over substantial geographic distances.

Satiation   The feeling of fullness at the end of a meal.

Satiety   The persisting sensation of repletion that results from eating.

Scalar timing   The dominant theory of timing which assumes that the coefficient of variability (i.e., the standard deviation of time estimates divided by the mean of time estimates) is constant across a broad range of temporal estimates (i.e., a specific proposal of the linear timing hypothesis).

Scale-free power-law   A degree distribution described by $p(k) \approx k^{-\gamma}$; demonstrated by a straight line on a log–log plot.

Scan sampling   A type of instantaneous sampling in which a group of individuals is scanned at specified intervals and the behavior of each individual at that instant is recorded.

Scanning   Often synonymous to vigilance.

Scatter hoarding   Hoarding of individual food items in many different locations.

Schistosomiasis (or bilharzias)   A disease caused by a blood fluke of the genus *Schistosoma*, a type of flatworm parasite. The intermediate host is a snail, in which cercariae (larvae) develop and migrate out into water; the cercariae penetrate the skin of hosts which make contact with the water. Symptoms depend on species causing infection, but can include rash, fever, aching, cough, diarrhea, and liver and spleen enlargement.

Schnauzenorgan response   A twitching movement of the elongated chin (Schnauzenorgan) of *Gnathonemus petersii, an electric fish,* evoked by the sudden emergence of a novel object near the animal's head, which is detected through the active or passive electric sense.

Schreckstoffe   Chemical alarm signals released by aquatic injured conspecifics, which is used to warn animals about an imminent danger.

Sclerotized   The hardening of tissue.

Scolopidium   A multicellular sensory structure of arthropods used to detect stretch, vibration, or sound.

Scout   A member of a social group, such as an ant or bee colony, that searches for food sources, nest sites, or other targets of interest. It may exploit its discoveries by itself or recruit other group members to help.

Scramble competition   Organisms use up a common limiting resource but otherwise do not contest or harm each other.

Scrounging   A behavioral strategy that consists of exploiting a resource uncovered by some other individual's efforts.

Seasonal breeder   An organism that breeds only in specific seasons (i.e., not continuously).

Seasonal interaction   When events in one period of the annual cycle, such as timing or condition, of an animal to influence events in subsequent periods.

Seasonality   Changes in hormonal or behavioral status in response to change in seasons.

Secondary defenses   Traits of the prey that influence the action of the predator, subsequent to prey detection, in ways that benefit the prey. Compare with primary defenses that act prior to the predator detecting the prey.

Secondary plant compound   Molecules produced by plants, the presence of which is often characteristic of particular plant taxa and which appear not to be directly involved in primary metabolism.

Secondary polygyny   Polygyny that arises from monogyny, generally through queen adoption.

Secondary predator–prey behaviors   Behaviors concerned with predators capturing prey, or prey escaping from predators, during an attack.

Secondary reproductive   These are produced by many termite species; they are sexually capable individuals who do not have wings, and are capable of superceding sick, injured, or absent parental primary reproductives.

Secondary sexual character   A trait that differs between the sexes and is neither required for reproduction nor related to sex differences in ecology. Most such traits do not develop fully until sexual maturity, are expressed more strongly in males than in females, and are useless or costly for survival. Traits that do not differ between the sexes but share the other two qualities may also be referred to as secondary sexual characters (e.g., ornate plumage in sexually monomorphic birds).

Segregation distortion   Within-individual selection for one or another allele of a diploid body.

Selective attention The cognitive processes of (selectively) concentrating on one aspect of the environment while ignoring others; consciously or unconsciously, the perceiving organism is focused on particular areas of the environment. This is determined by past experience and the skill being performed.

Selective differential The difference in fitness between two or more subsets of a population subjected to different selective pressures with resulting differences in fitness.

Selective sweep Recent and strong positive natural selection on a particular gene which leads to reduced variation in DNA sequence among individuals in a population.

Selective tidal stream transport (STST) Vertical movements of aquatic organisms relative to tides; provides a mechanism for zooplankton and small nekton to move horizontally within and between estuaries and coastal regions.

Self-awareness (self-recognition) Increased self-other distinction, oftentimes indicated by self-recognition in a mirror. Sensitivity to one's own thoughts and feelings; sometimes used to indicate the knowledge that one exists independent of other entities.

Self-control task Experimental situation in which decision-makers must choose between smaller–sooner and larger–later options.

Selfish-herd effect Bunching by foragers to decrease their relative domain of danger when facing predation threats.

Self-medicate The use by animals of secondary plant compounds or other nonnutritional substances in preventing or treating diseases.

Self-organization The idea that the development of complex structures and behaviors in a system can emerge from events taking place primarily within and through the system itself.

Self-propelled particle (SPP) models Models of collective motion in which each group member is treated as a particle that responds to other group members within interaction zones. An individual moves toward or away from other individuals, or aligns itself with them, depending on which zone they occupy.

Semantic memory The ability to acquire general factual knowledge about the world.

Semelparous (semelparity) Reproducing once during a lifetime.

Semiclaustral founding Colony founding procedure in which a founding queen or queens forage outside the brood cell to secure sufficient energy to rear the first generation of workers.

Semi-intact preparation A piece of an animal, along with its nervous system, that produces a behavior or a component of a behavior. Such preparations are normally used primarily to allow access to the nervous system, but can also be used to eliminate sensory input or confounding inputs from other parts of the nervous system.

Semisociality Social groups of same-generation adults and their offspring characterized by cooperative brood care (i.e., alloparental care occurs), and a reproductive division of labor, such that some individuals mainly reproduce while others mainly perform other tasks such as foraging and brood-care.

Senescence The combination of biological processes of deterioration of organismic function in a living organism approaching an advanced age.

Sensillum Hair-like structure that houses sensory neurons.

Sensitive phase A stage of life during which the ability to learn is enhanced. Occurs most commonly early in life.

Sensory drive The hypothesis that sensory systems and sensory conditions in the environment 'drive' evolution in particular directions.

Sensory environment Multiple types of information – signals and cues from other animals and the physical environment – that may be perceived by an animal on the basis of its unique sensory capabilities (i.e., 'umwelt').

Sensory mode The physical characteristics of signal production, on the basis of animal sense organs by which it is perceived (e.g., sound, patterns of light and color, vibration, etc.).

Sensory traps In attempts to induce certain responses in other individuals, the use of stimuli whose effectiveness in inducing these responses evolved in a different context. In a sexual context, the male can produce a stimulus that elicits a particular female response; this female response exists because previous natural selection in another context favored such a response to the same (or a similar) stimulus.

Sentience A general term for the ability to feel or perceive subjectively.

Sentinel An individual in a group that remains vigilant and stands guard while other group members forage or carry out other activities (also called: sentry or guard).

Sentinel cells A newly discovered cell that sweeps through a *Dictyostlium* slug mopping up toxins and bacteria, acting as a kidney, a liver, and an innate immune system.

Sequence divergence Changes in the sequence of DNA bases in different populations or different species. Comparisons of the degree of sequence divergence are used to estimate how long ago the populations or species began to evolve independently.

Sequestering   Accumulation of a chemical in the integument or inner organs of an organism from an outside source (e.g., diet).

Serotonergic basal cells   Round cells at the base of the taste bud, which are immunoreactive to serotonin.

Serotonergic medications   Psychotropic medications that effectively increase the availability of the neurotransmitter serotonin in the brain.

Serotonin (5-HT)   A monoamine neurotransmitter that is derived from tryptophan. It is synthesized in the gut, pineal, and CNS. In the brain, 5-HT influences learning and memory as well as appetite, sleep, and muscle contraction.

Sex allocation   Sometimes used to refer to the process by, or the time at, which a parent bestows gender on offspring (see sex determination and sex allocation sequence, respectively), but more generally used to refer to how resources are apportioned to each gender (also referred to as *investment ratio*). Sex allocation can be thought of as an evolutionarily derived reproductive strategy of the parents and the sex ratio as one of its manifestations.

Sex allocation sequence   The order in which offspring of different gender are produced by a parent. Nonrandom sequences can, but do not always, imply parental control and can influence sex ratio variance.

Sex determination   The genetic basis of an individual's gender. There is an astonishing diversity of sex determination mechanisms among animals, often exerting a profound influence on reproductive behavior.

Sex-limited polymorphism   Occurrence of several discrete forms or morphs within one sex, but not the other sex.

Sex ratio   The proportion of individuals that are male, that is, males/(males + females). Sex ratios are sometimes given as the proportion females (this is not incorrect; there is no strict convention) and sometimes reported as the ratio of males to females, that is, males/females (termed sex ratio *sensu stricto*): this is not a recommended measure as it is not readily amenable to statistical analysis. The sampling unit may be indicated, for example, *population sex ratio*, *clutch sex ratio*, *parental sex ratio* (the sex ratio of offspring produced by a given parent or pair of parents). The developmental stage of offspring may also be indicated: *primary sex ratio* (the sex ratio at offspring production; this may be used to indicate the sex ratios at fertilization or at egg laying), *secondary sex ratio* (the sex ratio at some defined later stage of offspring development, for example, emergence or mating (adulthood)). Developmental mortality can mean that primary and secondary sex ratios are not equivalent.

Sex ratio variance   A measure of the diversity of sexual composition in groups of offspring (e.g., clutches, litters, etc.). Heterogametic sex determination (e.g., the XY system in mammals, the WZ system in birds) leads to the null expectation that distributions of group sex ratios conform to binomial variance. Deviations from the binomial expectation can, but do not necessarily, imply sex ratio control. Under haplodiploid sex determination, there is no particular null expectation of variance, but subbinomial variances have been observed in many haplodiploid species.

Sex-ratio conflict   Conflict between queens and workers over the investment into male versus female reproductives produced by the colony.

Sex role reversal   Occurs when males provide the majority of parental care, resulting in sexual selection on females, who can increase their reproductive success by obtaining additional mates.

Sex-role reversed species   Are those in which females compete for males and males choose among females. Typically, males take care of the young.

Sexual behavior   Behavioral interactions that facilitate the union of eggs and sperm.

Sexual coercion   Occurs when one sex, usually males, use force or the threat of force – forced copulation, harassment, intimidation, restriction of the movement of the other – to increase the probability that mating will occur.

Sexual conflict   Occurs whenever the fitness interests of individuals of different sexes conflict.

Sexual dialectics hypothesis   The idea that whenever the behavior and physiology of one sex decreases the fitness of the other, flexible individuals adaptively modify their behavior or physiology to resist the deleterious effects of interaction(s) with the other sex. Because control and resistance interactions are likely to be dynamic, changing during the lifetime of an individual, the sexual dialectics hypothesis predicts that individuals flexibly adjust resistance behavior in contemporary time.

Sexual dichromatism   A subset of sexual dimorphisms in which males and females of a species differ systematically in coloration or color pattern.

Sexual differentiation   In ontogeny, the anatomical and behavioral differentiation of males and females.

Sexual dimorphism   Refers to differences in morphology, behavior or physiology between males and females. Generally, more intense sexual selection results in greater sexual dimorphism.

Sexually antagonistic selection   A type of selection that is characterized by dynamic interactions – actions and reactions – between individuals of different sexes that can lead to a coevolutionary arms race.

Sexual reproduction   Reproduction involving gamete formation by meiosis and gamete fusion to form new individuals.

Sexual selection Selection for traits that make individuals of one sex better able to compete for individuals of the opposite sex. As a consequence, some individuals have a mating advantage over other individuals of their own sex, such that there is nonrandom differential reproductive success among these individuals.

Sexual signals Advertise the signaler's genetic or phenotypic quality in order to attract mates and deter rivals. Examples include conspicuous traits, such as bright colors and elaborate songs. Signals can be visual, acoustic, olfactory, tactile, or electric.

Sexual size dimorphism (SSD) A subset of sexual dimorphisms in which males and females of a species differ systematically in body size.

Shaping The procedure of reinforcing successive approximations of a desired behavior.

Short day breeder An organism that enters full reproductive capability during short days of winter.

Sibling species Anatomically similar species that are nonetheless reproductively isolated; in herbivorous insects, such species often use different host plants.

Sickness responses The suite of adaptive behavioral and febrile reactions among vertebrate animals associated with the acute phase immune response that includes fever, iron withholding, reduced motivated behaviors such as food and water intake, and lack of sexual, parental, or other social interactions. These responses are critical to survival.

Sign A signal; also anything that gives evidence or trace of something else; also a physical object, usually fixed in space, that is a signal when encountered by a receiver.

Sign stimulus An external stimulus that elicits a specific motor pattern (modal or fixed action pattern).

Signal A character or behavior that has evolved so as to provide information to other organisms.

Signal detection theory A general model of the discrimination of signals from background noise that can be applied to data from psychophysical studies with animals and to situations where an animal must make a discrimination under conditions of uncertainty.

Signal dominance When a multimodal signal generates a response in only one of its component modes in relation to other modes.

Signal enhancement When receiver responses to redundant multimodal signals are increased in their intensity compared to unimodal signals.

Signal equivalence When receiver responses to redundant multimodal signals are the same or equal to unimodal signals in their intensity (equivalence).

Signal independence When the response to a multimodal signal includes the (different) responses to each of its unimodal components.

Signaling mode The physical characteristics of a signal that enables it to be received by a specific type of sensory neuron in a receiver. Signaling modes include chemical, electric, sound, light, and vibration.

Signal parasite An individual that exploits an existing communication system in a way that benefits itself at the expense of a signal giver or a signal receiver.

Signal redundancy When individual components of a multimodal signal presented separately elicit the same response from a receiver and likely contain the same or similar kinds of information about the sender.

Significance level/criterion In statistical analyses it is a criterion of probability below which a statistical test value is said to indicate a significant difference between populations.

Silkie Asiatic breed of chickens characterized by fur-like plumage and dark blue flesh.

Simultaneous hermaphroditism A sexual pattern characterized by individuals possessing both mature ovarian and spermatogenic tissue within the same functional gonad.

Single nucleotide polymorphism (SNP) Variation in a DNA sequence that occurs when a single nucleotide – A, T, C, or G – varies between individuals of the same species.

Sinus gland A neurohemal organ associated with the crustacean X-organ.

Siphon Cylinder created by curling the edges of the mantle in some mollusks. It can be used to forcibly discharge the contents of the mantle cavity.

Sister groups A pair of evolutionary lineages that share their most recent common ancestor and thus are necessarily equal in age.

Site fidelity (see philopatry).

Size constancy The ability to determine the true size of objects despite viewing them at different distances when their images subtend various angles on the retina.

Skylight polarization Due to scattering by particles in the earth's atmosphere, sunlight becomes polarized, with the light wave's electric field oscillating in one direction. The degree of polarization is maximal at $90°$, relative to the direction of incident light.

Sloughing behavior Specific behavior associated with sloughing off skin and associated structures such as hair, feathers, and scales. Often, this involves rhythmic movements to lift off old skin layers (e.g., in snakes), or movements allowing abrasion of skin with substrate (many birds and mammals) to break up and shed skin and its components.

Wintering area   In migratory birds, the area where populations spend the nonbreeding season, usually at lower latitudes.

Wintering dispersal   The distance between the wintering site of an individual in one year and its wintering site in another year.

Winter territory   A home range that an individual occupies and defends its boundaries against others (usually conspecifics but sometimes other species as well). This territory/home range may be held exclusively by the individual or as a pair or as a small group.

Wiring costs   The energetic costs associated with total length neural wiring (axons and dendrites).

Wisdom of crowds   The principle that the collective performance of a group of decision-makers can exceed that of a randomly chosen individual acting alone.

Within-pair offspring (WPC)   Offspring sired by the social father.

Within-sex variance in reproductive success   An operational definition of sexual selection.

Worker   Individual in a eusocial society that primarily performs all nonreproductive tasks in a colony. In primitively eusocial groups, this individual may be physically capable of reproduction; however, in highly eusocial groups, it is effectively sterile.

Xenoestrogens   Chemicals that are produced for agricultural, private, or industrial use that have estrogenic activity in living organisms.

X-organ   A group of neurosecretory neurons in the crustacean eyestalk that synthesize several peptide hormones.

Y-organ   The molting gland of crustaceans that usually secretes ecdysone, the precursor of the active form of the molting hormone, 20-hydroxyecdysone.

Zeitgeber   German word for 'time-giver'; an exogenous cue that entrains an endogenous biological rhythm.

Zoological psychology   A part of animal psychology that lies at the boundary between psychology and zoology. The approach is animal-centered in that the focus is primarily on studying the life of the animal rather than on asking arbitrary questions in a so-called animal model. The emphasis is often upon the natural behavioral repertoire of the animal rather than training the animal to engage in some arbitrary task.

Zoopharmacognosy   The study of how animals use medicinal substances. Interchangeably used by some with the term animal self-medication.

Zooplankton   Small pelagicorganisms in aquatic ecosystems that form central part of the food web. They typically eat algae (phytoplankton) and are consumed by small (planktivorous) fish.

Zugunruhe   Migratory restlessness (hopping or hovering) in caged migratory birds often oriented with respect to seasonal directions of migration (e.g., northward in spring and southward in fall).

Zygote   A newly fertilized egg.

Subject Index

Notes

Abbreviations used in subentries are defined in the main index.

Cross-reference terms in italics are general cross-references, or refer to subentry terms within the main entry (the main entry is not repeated to save space). Readers are also advised to refer to the end of each article for additional cross-references - not all of these cross-references have been included in the index cross-references.

The index is arranged in set-out style with a maximum of three levels of heading. Major discussion of a subject is indicated by bold page numbers. Page numbers suffixed by *t* refer to tables, *f* refer to figures, *b* refer to boxes.